Sustainable Uses and Prospects of Medicinal Plants

Sustainable Uses and Prospects of Medicinal Plants presents information on less known and underexplored medicinal plant species in various regions of the world. The book investigates current advances in medicinal plant science and includes detailed information on the use of green nanotechnology, characterization of plants, conservation, revitalization, propagation, and pharmacological activities of selected plants. A volume in the *Exploring Medicinal Plants* series, it collects information on less-known medicinal plant species in various regions of the world for documentation profiling their ethnobotany, developments in their phytochemistry, and pharmacological activities and provides an indepth look at some specific herbal medicines of importance, threatened and less known species, and addresses sustainable utilization and conservation of medicinal plants to ensure existence and use.

Appropriate for plant and biodiversity conservation organisations, community leaders, academicians, researchers, and pharmaceutical industry personnel, the book comprises innovative works with information of what is expected to address sustainability in the future.

Exploring Medicinal Plants

Series Editor

Azamal Husen

Wolaita Sodo University Ethiopia

Medicinal plants render a rich source of bioactive compounds used in drug formulation and development; they play a key role in traditional or indigenous health systems. As the demand for herbal medicines increases worldwide, supply is declining, as most of the harvest is derived from naturally growing vegetation. Considering global interests and covering several important aspects associated with medicinal plants, the *Exploring Medicinal Plants* series comprises volumes valuable to academia, practitioners, and researchers interested in medicinal plants. Topics provide information on a range of subjects, including diversity, conservation, propagation, cultivation, physiology, molecular biology, growth response under extreme environment, handling, storage, bioactive compounds, secondary metabolites, extraction, therapeutics, mode of action, and healthcare practices.

Led by Azamal Husen, PhD, this series is directed to a broad range of researchers and professionals consisting of topical books exploring information related to medicinal plants. It includes edited volumes, references, and textbooks available for individual print and electronic purchases.

Traditional Herbal Therapy for the Human Immune System
Azamal Husen

Environmental Pollution and Medicinal Plants
Azamal Husen

Herbs, Shrubs, and Trees of Potential Medicinal Benefits
Azamal Husen

Phytopharmaceuticals and Biotechnology of Herbal Plants
Sachidanand Singh, Rahul Datta, Parul Johri, and Mala Trivedi

Omics Studies of Medicinal Plants
Ahmad Altaf

Exploring Poisonous Plants: medicinal Values, Toxicity Responses, and Therapeutic Uses
Azamal Husen

Plants as Medicine and Aromatics: conservation, Ecology, and Pharmacognosy
Mohd Kafeel Ahmad Ansari, Bengu Turkyilmaz Unal, Munir Ozturk, and Gary Owens

Sustainable Uses and Prospects of Medicinal Plants
Learnmore Kambizi and Callistus Bvenura

Sustainable Uses and Prospects of Medicinal Plants

Edited by
Learnmore Kambizi and Callistus Bvenura

CRC Press
Taylor & Francis Group
Boca Raton London New York

CRC Press is an imprint of the
Taylor & Francis Group, an **informa** business

Cover photograph of *Scadoxus multiflorus* (Martyn) Raf. taken by Callistus Bvenura at Ntatshaneni - Masase in a district known as Mberengwat in the Midlands Province of Zimbabwe. November 16, 2021.

First edition published 2023
CRC Press
6000 Broken Sound Parkway NW, Suite 300, Boca Raton, FL 33487-2742

CRC Press
4 Park Square, Milton Park, Abingdon, Oxon, OX14 4RN

CRC Press is an imprint of Taylor & Francis Group, LLC

Library of Congress Cataloging-in-Publication Data
Names: Kambizi, Learnmore, editor. | Bvenura, Callistus, editor.
Title: Sustainable uses and Prospects of medicinal plants/Learnmore Kambizi and Callistus Bvenura
Description: First edition | Boca Raton, FL : CRC Press, 2023. | Includes bibliographical references and index. | Summary: "This book investigates current advances in medicinal plant science, and includes detailed information on the use of green nanotechnology, characterization of plants, conservation, revitalization, propagation, and pharmacological activities of selected plants. It collects information on less-known medicinal plant species in various regions of the world for documentation profiling their ethnobotany, developments in their phytochemistry, and pharmacological activities. The book provides an in-depth look at some specific herbal medicines of importance, threatened and less known species, and addresses sustainable utilization and conservation of medicinal plants to ensure existence and use"— Provided by publisher.
Identifiers: LCCN 2022043153 (print) | LCCN 2022043154 (ebook) | ISBN 9781032071732 (v. 1; hardback) | ISBN 9781032073811 (v. 1; paperback) | ISBN 9781003206620 (v. 1; ebook)
Subjects: LCSH: Medicinal plants. | Materia medica, Vegetable. | Herbs-Therapeutic use. | Botanical drug industry.
Classification: LCC QK99.S95 2023 (print) | LCC QK99 (ebook) |
DDC 581.6/34—dc23/eng/20220916
LC record available at https://lccn.loc.gov/2022043153
LC ebook record available at https://lccn.loc.gov/2022043154

ISBN: 9781032071732 (hbk)
ISBN: 9781032073811 (pbk)
ISBN: 9781003206620 (ebk)

DOI: 10.1201/9781003206620

Typeset in Times
by codeMantra

Access the Support Material: https://www.routledge.com/9781032071732

Contents

Editors

Prof Learnmore Kambizi (Cape Peninsula University of Technology, South Africa) attained his BSc from Cuba in 1998 before enrolling for Masters and PhD at the University of Fort Hare which he completed in 2004. His PhD focussed on "Ethnobotanical Studies of Plants Used for the Treatment of Sexually Transmitted Infections in South Africa". He went on to pursue a postdoctoral fellowship within the fields of Ethnobotany & Phytomedicine. This discipline enabled him to realise his passion for sustainable utilisation and conservation of medicinal plants on a local and international level. He worked as a Lecturer at the University of Zimbabwe in 2005 before moving to the University of the Free State, South Africa in 2007. In the same year, he was appointed by Walter Sisulu University as a Senior lecturer until 2011. He moved to Cape Peninsula University of Technology, South Africa, where he is now a Professor of Botany. In his academic journey to date, he has hosted four post-doctoral fellows and supervised five PhD, 16 Master's and numerous Honours students. In addition, he has published numerous articles in peer-reviewed journals and presented at many national and international conferences. He is a South African National Research Foundation rated researcher. Prof Kambizi has also served as a panel member of the National Research Foundation of South Africa, and has received many research grants to support post graduate students and to assist communities. He also works with Traditional Medical Practitioners in collaboration with the South African Research Council as an advisor. In addition, Prof Kambizi was appointed the first Director of African Centre for Herbal Research based in Nigeria, a position he holds to date. He is a reviewer of many national and international journals.

Dr Callistus Bvenura (Cape Peninsula University of Technology, South Africa) graduated with a BSc in Natural Resources Management and Agriculture in 2006 from the Midlands State University, Zimbabwe and completed his PhD Ethnobotany from the University of Fort Hare, South Africa in 2014. He has served as a postdoctoral fellow at the University of Fort Hare and Tshwane University of Technology, and is currently based at the Cape Peninsula University of Technology. He has made significant contributions in the ethnobotany, conservation, phytochemistry, and physiology of underutilised indigenous plants, including leafy vegetables and medicinal plants. Callistus has published numerous peer-reviewed research articles, book chapters, and review papers in these areas and is constantly presenting his findings at local and international conferences. He also has interests in product development, where he and his fellow colleagues developed Niselo, a probiotic sorghum beverage that has since been commercialised and is being sold in South Africa and the United Kingdom. He is also part of a team that developed broiler chicken feed using indigenous and underutilised grains and insects. In addition, he is a South African National Research Foundation panel expert reviewer and reviews for several local and international journals.

Contributors

Abdullahi Alanamu AbdulRahaman
Department of Plant Biology
Faculty of Life Sciences
University of Ilorin
Ilorin, Kwara State, Nigeria

Sherif Babatunde Adeyemi
C. G. Bhakta Institute of Biotechnology
Uka Tarsadia University
Surat (District), Gujarat State, India

Anthony Jide Afolayan
Medicinal Plants and Economic Development
    (MPED) Research Center
Department of Botany
University of Fort Hare, Alice, South Africa

Omolola Afolayan
Department of Horticultural Sciences
Faculty of Applied Sciences
Cape Peninsula University of Technology
Bellville, South Africa

John Awungnjia Asong
Unit for Environmental Sciences and
    Management
North-West University

Mahikeng Campus
Mmabatho, South Africa

Francis-Alfred Attah
Department of Pharmacognosy and Drug
    Development
Faculty of Pharmaceutical Sciences
University of Ilorin
Ilorin, Kwara State, Nigeria

Rakesh Kumar Bachheti
Department of Industrial Chemistry
College of Applied Science
Addis Ababa Science and Technology
    University
Addis Ababa, Ethiopia

Umar Muhammed Badeggi
Department of Chemistry
Ibrahim Badamasi Babangida University
Lapai, Niger State, Nigeria

Samuel Oloruntoba Bamigboye
Department of Plant Science
Olabisi Onabanjo University
Ago-Iwoye, Ogun State, Nigeria

Callistus Bvenura
Horticultural Sciences
Cape Peninsula University of Technology
Bellville, Cape Town, South Africa

Hlupheka Chabalala
Indigenous Knowledge-Based Technology
    Innovations Directorate
Department of Science and Innovations
Brummeria, Pretoria, South Africa

Farisai Caroline Chibage
National Biotechnology Authority
Newlands, Harare, Zimbabwe

Praise Chirilele
National Biotechnology Authority
Newlands, Harare, Zimbabwe

Barsha Dassarma
University Centre for Research and
    Development
Chandigarh University
Punjab, India
and
Department of Pharmacology
Faculty of Health Sciences
University of the Free State
Bloemfontein, South Africa

Progress Dube
National Biotechnology Authority
Newlands, Harare, Zimbabwe

Ninon Geornest Eudes Ronauld Etsassala
Department of Horticultural Sciences
Faculty of Applied Sciences
Cape Peninsula University of Technology
Bellville, South Africa

Babajide Charles Falemara
Research Coordinating Unit
Forestry Research Institute of Nigeria
Ibadan, Nigeria

Alfred Francis Attah
Department of Pharmacognosy and Drug
    Development
Faculty of Pharmaceutical Sciences
University of Ilorin
Ilorin, Nigeria

Mahboob Adekilekun Jimoh
Department of Plant Biology
Osun State University
Osogbo, Osun State, Nigeria

Muhali Olaide Jimoh
Department of Horticultural Sciences
Cape Peninsula University of Technology
Bellville, Cape Town, South Africa

Nailoke Pauline Kadhila
Zero Emissions Research Initiative (ZERI)
    Division
Multidisciplinary Research Services
Centre for Research Services
University of Namibia
Windhoek, Namibia

Learnmore Kambizi
Cape Peninsula University of Technology
Horticultural Sciences,
Bellville, Cape Town, South Africa

Ishaq N. Khan
Institute of Basic Medical Sciences
Khyber Medical University
Peshawar, Pakistan

Elizabeth Kola
School of Biology and Environmental Sciences
Faculty of Agriculture & Natural Sciences
University of Mpumalanga
Mbombela, Mpumalanga Province, South
    Africa

Ramar Krishnamurthy
C. G. Bhakta Institute of Biotechnology
Uka Tarsadia University
Surat (District), Gujarat State, India

Bongani Petros Kubheka
Plant & Crop Production Research Directorate
Döhne Agricultural Development Institute
Stutterheim, South Africa

Navin Kumar
Department of Biotechnology
Graphic Era University
Clement Town, Dehradun, Uttarakhand, India

Vijay Jyoti Kumar
Department of Pharmaceutical Sciences
Hemvati Nandan Bahuguna Garhwal University
Srinagar Garhwal-246174, Uttarakhand, India

Caroline Peninah Kwenda
National Biotechnology Authority
Newlands, Harare, Zimbabwe

Charles Laubscher
Department of Horticultural Sciences
Faculty of Applied Sciences
Cape Peninsula University of Technology
Bellville, South Africa

Bilqis Abiola Lawal
Department of Pharmacognosy and Drug
    Development
Faculty of Pharmaceutical Sciences
University of Ilorin
Ilorin, Nigeria

Neo Macuphe
Department of Horticultural Sciences
Faculty of Applied Sciences
Cape Peninsula University of Technology
Bellville, Cape Town, South Africa

Cyprian Mahuni
National Biotechnology Authority
Newlands, Harare, Zimbabwe

Innocensia Mangoato
Department of Pharmacology
Faculty of Health Sciences
University of the Free State
Bloemfontein, South Africa

Steven Mapfumo
Biology Department
Chinhoyi University of Technology
Chinhoyi, Zimbabwe

Tafadzwa Phyllis Maranjisi
National Biotechnology Authority
Newlands, Harare, Zimbabwe

Munyaradzi Mativavarira
National Biotechnology Authority
Newlands, Harare, Zimbabwe.

Motlalepula Gilbert Matsabisa
Department of Pharmacology
Faculty of Health Sciences
University of the Free State
Bloemfontein, South Africa

Abhay Prakash Mishra
Adarsh Vijendra Institute of Pharmaceutical
    Sciences
Shobhit University
Gangoh, Saharanpur, UttarPradesh, India

Gugulethu Mathews Miya
Department of Chemical and Physical Sciences
Faculty of Natural Sciences
Walter Sisulu University
Mthatha, South Africa

Moira Amanda Mubani
National Biotechnology Authority
Newlands, Harare, Zimbabwe

Stanley Mukanganyama
Department of Biotechnology and
    Biochemistry
University of Zimbabwe
Mt Pleasant, Harare, Zimbabwe

Dumisane Thomas Muloche
School of Agriculture and Natural Sciences
University of Mpumalanga
Mbombela, South Africa

Davis Ropafadzo Mumbengegwi
Science and Technology Division
Multidisciplinary Research Services
Centre for Research Services
University of Namibia
Windhoek, Namibia

Tatenda Clive Murashiki
National Biotechnology Authority
Newlands, Harare, Zimbabwe

Kola-Mustapha
Department of Pharmaceutics
Faculty of Pharmaceutical Sciences
University of Ilorin
Ilorin, Kwara State, Nigeria

Reward Muzerengwa
National Biotechnology Authority
Newlands, Harare, Zimbabwe

Felix Nchu
Department of Horticultural Sciences
Faculty of Applied Sciences
Cape Peninsula University of Technology
Bellville, South Africa

Prisca Nonceba Ncube
National Biotechnology Authority
Newlands, Harare, Zimbabwe

Peter Tshepiso Ndhlovu
School of Biology and Environmental Sciences
Faculty of Agriculture & Natural Sciences
University of Mpumalanga
Mbombela, Mpumalanga Province, South
    Africa

Joyce Ndlovu
Biology Department
Chinhoyi University of Technology
Chinhoyi, Zimbabwe

Manisha Nigam
Department of Biochemistry, School of Life
  Sciences
Hemvati Nandan Bahuguna Garhwal University
Srinagar, Garhwal, Uttarakhand

Makomborero Nyoni
National Biotechnology Authority
Newlands, Harare, Zimbabwe

Adijat Funke Ogundola
Ladoke Akintola University of Technology
  (LAUTECH)
Ogbomoso, Oyo State, Nigeria

Johnson Olaleye Oladele
Department of Biochemistry
University of Ilorin
Ilorin, Nigeria

Adenike Temidayo Oladiji
Department of Biochemistry
University of Ilorin
Ilorin, Nigeria

Kehinde Stephen Olorunmaiye
Department of Plant Biology
Faculty of Life Sciences
University of Ilorin
Ilorin, Kwara State, Nigeria

Ayodeji Oluwabunmi Oriola
Department of Chemical and Physical Sciences
Faculty of Natural Sciences
Walter Sisulu University
Mthatha, South Africa

Oluwakemi Abosede Osunderu
Forestry Research Institute of Nigeria (FRIN)
Jericho Hills, Ibadan, Nigeria

Wilfred Otang-Mbeng
School of Biology and Environmental Sciences
Faculty of Agriculture & Natural Sciences
University of Mpumalanga
Mbombela, Mpumalanga Province, South
  Africa

Gloria Aderonke Otunola
Medicinal Plants and Economic Development
  (MPED) Research Center
Department of Botany
University of Fort Hare
Alice, South Africa

Adebola Omowunmi Oyedeji
Department of Chemical and Physical Sciences
Faculty of Natural Sciences
Walter Sisulu University
Mthatha, South Africa

Hamza Ahmed Pantami
Department of Chemistry Faculty of Science
Gombe State University
Tudunwada, Gombe, Nigeria

Nikita Patel
C. G. Bhakta Institute of Biotechnology
Uka Tarsadia University
Surat (District), Gujarat State, India

Swetal Patel
C. G. Bhakta Institute of Biotechnology
Uka Tarsadia University
Surat (District), Gujarat State, India

Raffaele Pezzani
Phytotherapy Lab (PhT-Lab), Endocrinology
  Unit
Department of Medicine (DIMED)
University of Padova
Padova, Italy

Iwanette du Preez-Bruwer
Science and Technology Division
Multidisciplinary Research Services
Centre for Research Services
University of Namibia
Windhoek, Namibia

Rohini Mony Radhika
Division of Flowers and Medicinal Crops
ICAR-Indian Institute of Horticultural
  Research
Bengaluru, India

Nishant Rai
Department of Biotechnology
Graphic Era University, Clement Town
Dehradun, Uttarakhand, India

Smitha Gingade Rajachandra Rao
Division of Flowers and Medicinal Crops
ICAR-Indian Institute of Horticultural
    Research
Bengaluru, India

Umer Rashid
Department of Chemistry
COMSATS University Islamabad
Islamabad, Pakistan

Abdur Rauf
Department of Chemistry
University of Swabi
Khyber Pakhtunkhwa, Pakistan

Fanie Rautenbach
Department of Biomedical Sciences
Cape Peninsula University of Technology
Bellville, South Africa

Pramod Rawat
Department of Biotechnology
Graphic Era University
Clement Town, Dehradun, Uttarakhand, India

Ilyaas Rhoda
Department of Horticultural Sciences
Faculty of Applied Sciences
Cape Peninsula University of Technology
Bellville, Cape Town, South Africa

Sefiu Adekilekun Saheed
Department of Botany
Obafemi Awolowo University
Ile-Ife, Nigeria

Dinesh Prasad Saklani
Department of History, Ancient History, and
    Culture including Archaeology
Hemvati Nandan Bahuguna Garhwal
    University
Srinagar Garhwal, Uttarakhand, India.

Sarla Saklani
Department of Pharmaceutical Chemistry
Hemvati Nandan Bahuguna Garhwal
    University
Srinagar Garhwal, Uttarakhand, India

Tanmay Sarkar
Malda Polytechnic
West Bengal Sate Council of Technical
    Education
Government of West Bengal
Malda, West Bengal, India

Deckster Tonny Savadye
National Biotechnology Authority
Newlands, Harare, Zimbabwe.

Syed Uzair Ali Shah
Department of Pharmacy
University of Swabi, Anbar
Anbar, KPK, Pakistan

Chiara Sinisgalli
Department of Science
University of Basilicata
Potenza, Italy

Terence Nkwanwir Suinyuy
School of Biology and Environmental Sciences
Faculty of Agriculture & Natural Sciences
University of Mpumalanga
Mbombela, Mpumalanga Province, South
    Africa

Adeola Tawakalitu Kola-Mustapha
Department of Pharmacognosy and Drug
    Development
Faculty of Pharmaceutical Sciences
University of Ilorin
Ilorin, Kwara State, Nigeria

Satyajit Tripathy
University Centre for Research and
    Development
Chandigarh University
Punjab, India
and
Department of Pharmacology
Faculty of Health Sciences
University of the Free State
Bloemfontein, South Africa

Babatunde Abdulmalik Yusuf
Department of Pharmacognosy and Drug
    Development
Faculty of Pharmaceutical Sciences
University of Ilorin, Ilorin, Nigeria

1 Sustainable Uses and Prospects of Medicinal Plants
A Brief Overview

Callistus Bvenura and Learnmore Kambizi

The best definition of sustainability is perhaps relative to the purpose for which it is intended. As such, different stakeholders and authoritative organisations, such as the United Nations World Commission on Environment and Development (WCED), Food and Agriculture Organization of the United Nations (FAO), United Nations Environment Programme (UNEP), World Wildlife Fund for Nature (WWF), and the 2002 World Summit for Sustainable Development (WSSD), all define sustainability in their own terms. However, there is a consensus among these definitions that points to the exploitation of resources today in a way that ensures that the same resources are available for posterity.

Therefore, sustainable use of medicinal plants in this context is slyly defined as the exploitation of medicinal plants today in a way that ensures their availability for future generations. Concerns over sustainable uses of resources have led to 140 countries to subscribe to what is known as the United Nations 17 sustainable development goals (SDGs). These goals are particularly important in the face of incessant climate change. Reports indicate that global temperatures have risen by 1.1°C between 1901 and 2020, and this has a negative domino effect on both aquatic and terrestrial life. The consequent rise in sea levels had devastating effects on all sea life, including the world's coral reefs. A spike in temperatures has also led to a rise in sea levels, floods elsewhere, and severe droughts, especially in Sub-Saharan Africa. This means that plants that do not adapt well to changing conditions, especially some endemic species, may eventually be forced into extinction.

The world's current known plant species total about 391,000 (Chapman, 2009), with about 28,187 possessing medicinal properties or being used as herbal medicines (Willis, 2017). Furthermore, about 25% of medicines used in the world are derived from medicinal plants (Mohd and Iqbal, 2019). This 100 billion USD herbal industry is a huge business, with an estimated annual growth rate of 15% (Mohd and Iqbal, 2019). It is plausible that this industry grew further during the Covid-19 pandemic due to an increase in the use of herbal interventions the world over. In the advent of this pandemic, there has been a paradigm shift in societies where young people previously shunned the use of medicinal plants for treatment of various ailments. The WHO (2022) estimates that over 80% of the world uses traditional medicines, and most of these medicines are plant based. Although comprehensive data on the regional uses of medicinal plants is scarce and inconclusive, about 90% of Germans are thought to use herbal medicines while this figure was estimated at 72% in South Africa (Williams et al., 2013; Willis, 2017). As much as 70,000 ton of medicinal plant materials are thought to be consumed in South Africa, and this market puts food on the tables of close to 134,000 people and rising (Williams et al., 2013). The continued rise in the use of traditional medicines potentially means that wild populations are being depleted and will continue to do so unless some conservation measures are taken.

Monoculture, human development such as expanding cities, habitat degradation, alien species invasion, among other factors are all contributing to declining species in the wild. The destruction or burning of the Amazon rainforest is particularly worrisome, as some species will be lost forever,

DOI: 10.1201/9781003206620-1

and their potential applications will be lost with them eventually. It is not very clear, yet if efforts to conserve biodiversity, Stop Global Warming, among other initiatives, are exerting a positive effect on biodiversity conservation.

However, in South Africa, for example, two species are now extinct, 82 species are reported as threatened, and 100 species are of concern (Williams et al., 2013). But efforts are being made to ensure the conservation of medicinal plants with records indicating that 1,280 medicinal plants are protected from overexploitation in the world (Willis, 2017). Such efforts could go a long way into the conservation of these important plants.

Perhaps one of the main problems with regard to the efforts facing medicinal plant conservation is the disconnect among the stakeholders along the value chain. Medicinal plant harvesters/traders care less about conservation, because their aim is to put food on their tables, while herbalists and traditional healers and their peers may not readily accept conservation measures such as species protection. These stakeholders view plants as their God-given right and therefore cannot be regulated by any government or organisations of authority. This disconnect can be bridged by engaging these stakeholders with governments, researchers, as well as local traditional authorities. The arguments that are often raised by traditional healers and herbalists, among others, point to the fact that the plant that is in the forest is more potent than the one that has been cultivated. And yet, research has proven that conditions can be manipulated to ensure that the plant that has been cultivated and that in the wild are equally potent.

Therefore, in the face of these medicinal plant conservation issues that the world faces, threats of extinction, due to biotic and abiotic factors, with overharvesting from the wild taking the centre stage, and climate change playing a significant role, it is critical to find ways to mitigate against potential species extinction. Sustainable uses of medicinal plants need to be entrenched in the minds of all stakeholders before more species are added to the list of those that have become extinct. Therefore, this book is dedicated to exploring ways in which medicinal plants can be exploited without driving them towards extinction and robbing the future generations of this green gold.

REFERENCES

Chapman, A.D. (2009). *Numbers of Living Species in Australia and the World*, 2nd edition. A Report for the Australian Biological Resources Study, Australian Biodiversity Information Services, Toowoomba, Australia.

Mohd, S.A.K. and Iqbal, A. (2019). Chapter 1—Herbal medicine: current trends and future prospects. *New Look to Phytomedicine*, Academic Press, pp. 3–13.

WHO (2022). WHO establishes the Global Centre for Traditional Medicine in India: maximizing potential of traditional medicines through modern science and technology, Geneva. (Accessed 18 August 2022). Available at: https://www.who.int/news/item/25-03-2022-who-establishes-the-global-centre-for-traditional-medicine-in-india

Williams, V.L., Victor, J.E., and Crouch, N.R. (2013). Red listed medicinal plants of South Africa: status, trends, and assessment challenges. *South African Journal of Botany*, 86, 23–35.

Willis, K.J. (2017). *State of the World's Plants 2017.* Royal Botanic Gardens Kew, London, UK.

2 Protecting the Endemism of Threatened Cyclotide-Rich Medicinal Plants
The Tropical African Experience

Alfred Francis Attah and Adeola Tawakalitu Kola-Mustapha

CONTENTS

2.1 INTRODUCTION

An estimated 430,000 plant species have been reported to grow on planet earth as human co-tenants (Lughadha et al., 2016; Kougioumoutzis et al., 2021), as sustainable sources of food, and as nature-derived medicines (Butt et al., 2021). The occurrence and spread of these plants appear to be different across continents of the planet (Kougioumoutzis et al., 2021), most probably resulting from the corresponding interactions between ecological and evolutionary processes (Eco-evolutionary processes) that have helped plants to integrate properly to their adaptable habitats (Burak et al., 2018). Eco-evolutionary processes have, therefore, triggered an uneven distribution of plants across the continents, leading to continental geographies with megadiverse plant species and countries within these continents sharing differing levels of overall richness of plant species (Linder and Evolution, 2014).

The most recent report by the World Health Organization (WHO) recognized the increasing role played by the mainly plant-based traditional and complementary medicine achieving the Sustainable Development Goal 3—ensuring healthy lives and promoting well-being for all at all ages—to achieve targeted Universal Health Coverage (UHC) (World Health Organization, 2019; Ganguly and Bakhshi, 2020). A conservative estimation suggests that about 1.2 billion population in Africa depends on plants, especially the therapeutic potentials of plant biodiversity for their primary healthcare needs (World Health Organization, 2019). (The population estimate is based on data from the United Nations, Department of Economic and Social Affairs, Population Division; www.Worldometers.info.) This updated figure, representing about 90% of African population seems to

DOI: 10.1201/9781003206620-2

have been encouraged by the devastating COVID-19 pandemic (Attah et al., 2021); The unavailability of a globally accepted Western medicine has put a greater pressure on plant biodiversity in a continent, whose majority of member states have weak regulations on plant conservation and sustainable utilization. At least 10% of the known global plant species are found growing unevenly across Africa (Senkoro et al., 2020), and about 70% of African plant species are limited to the tropical African region (Sosef et al., 2017), where plant endemism, according to RAINBIO and a recent report, has been estimated to be 50% (Raven et al., 2020). Endemism of plants represents the nativity of plant species to a geographical location, region, country, or continent (Linder and Evolution, 2014). Several of these endemic plant species have found useful applications as food and local medicines to the human populations living around their geographical distribution (Attah et al., 2016b; Burlando et al., 2019; Senkoro et al., 2020; Attah et al., 2021). Scientific investigations of less than 1% of African plants have revealed the presence of several pharmaceutically useful metabolites—alkaloids, glycosides, tannins, terpenes, essential oils, polysaccharides, peptides, and therapeutic proteins (van Wyk and Prinsloo, 2020). Among the therapeutic biopolymers of pharmaceutical interest is the fortuitous discovery of a class of mini protein molecules called cyclotides from African ethnomedicine (Craik et al., 1999) (Figure 2.1). Cyclotides represent a unique class of 2–4 kDa knottin peptides having a circular knot backbone and intercysteine loops defining the peptide structural architecture (Craik et al., 1999). As members of the gene-encoded peptides, cyclotides are ribosomally synthesized and post-translationally modified peptides (RiPPs)

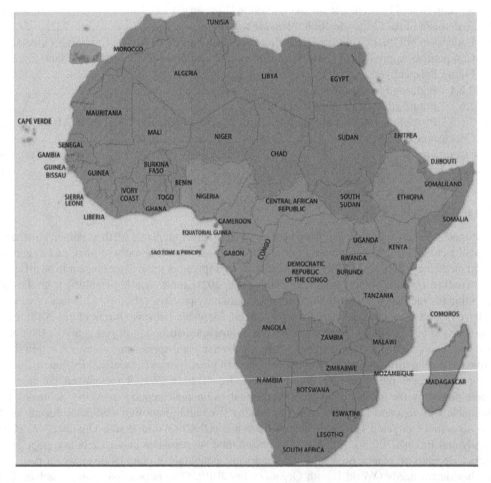

FIGURE 2.1 Map of Africa showing countries where cyclotides have been identified in plants collected; regions of highest endemism of cyclotide-rich flora are in Nigeria, Gabon, Cameroon, Congo, Central African Republic, Democratic Republic of the Congo, Ethiopia, Uganda, Kenya, Rwanda, Burundi, Tanzania, and Madagascar. (Adapted from https://www. freeworldmaps.net/africa.).

whose mature peptides are mostly generated by proteolysis following removal of the leader as well as recognition sequences (Luo and Dong, 2019). Their sequence variability outside the conserved region has encouraged a wide range of bioactivities that extends beyond their suitability for drug development (Camarero and Campbell, 2019). Additionally, cyclotides possess intrinsic phytochemical capability for an ecologically compatible plant protection (Oguis et al., 2019). Meanwhile, host defense-related function has been associated with cyclotide expression in plants (de Veer et al., 2019; Slazak et al., 2021). This means cyclotide expression appears to be common with plants endemic to regions that are more prone to pests, diseases, and climatically extreme environments as found in tropical Africa (Yeshak et al., 2011; Oguis et al., 2019; Gattringer et al., 2021; Slazak et al., 2021). The notable wide variations in conditions associated with rainy and dry seasons common to tropical climates, tropical Africa, in particular, has been proposed to have encouraged more pathogen distribution within the region, as these pathogen carriers (pests) and the disease-causing organisms require water and humid conditions for spread or reproduction (Guernier et al., 2004). Thus, the hypothesis regarding the association between the cyclotide-producing plants and their higher distribution in plants growing around the humid tropics, wetlands, and other extreme habitats remains to be elaborately investigated. For instance, the distribution of a host of plants expressing cyclotides (including *Oldenlandia affinis, Rinorea denata, Rinorea oblongifolia, Rinorea subintegrifolia, Clitoria ternatea, Viola abyssinica, Viola odorata*) are often limited to these extreme conditions that encourage pest/pathogen attack, such as moist and swampy environment (Yeshak et al., 2011; Attah et al., 2016a; Slazak et al., 2021; Gattringer et al., 2021). Plants growing within this region will tend to require additional arsenal of defense molecules to adapt to unstable extreme environments where they grow (Figure 2.1).

Abundant expression of knottin peptides, including cyclotides, within the plant organs in response to the increased level of threat in their immediate environment may provide some explanation to the uneven distribution of cyclotides in plant species and families across Africa—a factor that may closely be associated with their resilience and sustained endemism. For instance, *Oldenlandia affinis* (Roem. & Schult.) DC. Rubiaceae, an endemic herb common to the tropics, especially tropical Africa where it has been linked to the discovery of cyclotides (Figure 2.2), is commonly found growing in proximity to a body of water or during the rainy season in Nigeria. Recently, though this plant has become scarce within this region, just as the species is no longer found growing in locations (such as Sobi Hills, Ilorin) documented by Herbarium samples deposited some decades back at the Forest Herbarium, Ibadan, Nigeria.

Importantly, *O. affinis* and other cyclotide-expressing genera such as Rinorea and Allexis (Violaceae) are used in Traditional Medicine across tropical Africa (Weidmann and Craik, 2016; Attah et al., 2016b; Nworu et al., 2017; Gattringer et al., 2021) where they are endemic; however, no attention has been accorded to their conservation despite myriads of increasing threats from divergent sources endangering

Oldelandia affinis
(Roem. & Schult.) DC
Rubiaceae

Kalata B1

FIGURE 2.2 Cyclotide-rich African *Oldenlandia affinis* and the first cyclotide blueprint, Kalata B1 detected and isolated from the plant. PDB ID: 1JJZ

TABLE 2.1

Conservation Status and IUCN Classification of African Plants Expressing Cyclotides Based on the International Union for Conservation of Nature (IUCN) Red List of Threatened Species

S/N	African Plant	Habit	Endemism Status	Conservation Status	IUCN Classification	Authors' Comment
1	*Clitoria ternatea* L. Fabaceae	Herb	Unassessed	Unassessed	Unassessed	Plant has not received any conservation attention
2	*Oldenlandia affinis* (Roem. & Schult.) DC Rubiaceae	Herb	Unassessed	Unassessed	Unassessed	Threatened; currently very scarce in the wild; no longer found growing in locations of the previous herbarium collection
3	*Rinorea dentata* (P. Beauv.) Kuntze Violaceae	Shrub	Unassessed	Unassessed	Unassessed	Threatened due to logging, habitat loss, and divergent human uses
4	*Rinorea oblongifolia* (C.H.Wright) C.Marquand ex Chipp Violaceae	Understory tree	Unassessed	Unassessed	Unassessed	Threatened due to logging, habitat loss, and divergent human uses
5	*Rinorea ilicifolia* (Welw. ex Oliv.) Kuntze Violaceae	Understory tree	Unassessed	Unassessed	Unassessed	Threatened due to logging, habitat loss, and divergent human uses
6	*Rinorea brachypetala* (Turcz.) Kuntze Violaceae	Shrub/small tree	Not threatened	Unassessed	Unassessed	Vulnerable due to logging, habitat loss, and divergent human uses
7	*Allexis obanensis* (Baker f.) Melch. Violaceae	Understory tree	Vulnerable	IUCN based—Not threatened (Cameroon)	Threatened	Threatened due to logging, habitat loss, and divergent human uses
8	*Allexis cauliflora* (Oliv.) Pierre Violaceae	Understory tree	Vulnerable	Unassessed	Threatened	Threatened due to logging, habitat loss, and divergent human uses
9	*Allexis batangae* (Engl.) Melch. Violaceae	Understory tree	Vulnerable	Unassessed	Unassessed	Threatened due to logging, habitat loss, and divergent human uses
10	*Allexis zygomorpha* Achound. & Onana Violaceae	Understory tree /Shrub	Vulnerable	IUCN based—Vulnerable (Cameroon)	Threatened	Threatened due to logging, habitat loss, and divergent human uses
11	*Viola abyssinica* Steud. ex Oliv. Violaceae	Herb	Vulnerable	Unassessed	Unassessed	Threatened due to climate change and habitat loss
12	*Viola odorata* L. Violaceae	Herb	Vulnerable	Unassessed	Least concern	May soon be threatened due gradual habitat loss

Source: https//www.iucn.org.

the species. The loss of important plant biodiversity endemic to tropical Africa necessitates their conservation to protect their endemism. Emerging evidence suggests that African plant biodiversity faces existential threat (extinction) until the end of the current 21st century of the Gregorian calendar (Stévart et al., 2019). The tropical African flora has been classified under the 'high level of extinction risk' (Stévart et al., 2019), and many cyclotide-rich medicinal plants within the genera Rinorea and Allexis have been listed under the International Union for Conservation of Nature (IUCN) Red list (Table 2.1). Many species within the cyclotide-rich plant taxons have not been assessed, making their status unknown on the IUCN Red list. A typical example of such is *Clitoria ternatea* L. (Fabaceae), native to the humid as well as subhumid tropics, including tropical Africa. *C. ternatea* is a knottin peptide-expressing plant with application in medicine and sustainable agriculture (Oguis et al., 2019). This may indicate that this plant is facing more daring threat of extinction in Africa where relevant regulations are not enforced in many member states. Another instance is the genus *Allexis* whose four known species are restricted to tropical central and Western coast of Africa, yet the three accessed species of this genus have all been categorized as vulnerable on the IUCN Red list, while many more endemic Violaceae and Rubiaceae cyclotide-rich plants have not been evaluated despite their overutilization and overexploitation. To highlight the ever-increasing threats to cyclotide-rich plants endemic to Sub-Saharan Africa, we document for the first time, these endangered and highly valued ethnomedicinal plants whose poor to non-existent conservation status deserves an urgent attention to forestall imminent extinction. In addition, we also present the compelling need to protect the endemism of these important African bioresources whose role in Traditional Medicine, nutrition, and ecological balance buttresses their classification as biofactories of lifesaving metabolites.

2.2   AFRICAN HERBAL ETHNOMEDICINE AND CYCLOTIDE DISCOVERY

The African flora provides a rich source of medicinal plant and plant parts useful in the management of some clinical conditions, and for several decades, the use of herbs, herbal concoctions, and traditional phytomedicines has proved invaluable (Mahomoodally and Medicine, 2013). Evaluating the chronology of primary care and treatment in Africa is incomplete without recognizing the role of herbal ethnomedicine. With a very high biodiversity of plants and cultures, various plant species serve as treatments in different regions and communities. One of such uses led to the early discovery of cyclotides by Lorents Gran—a Norwegian doctor who discovered that African women drank a tea derived from the plant *Oldenlandia affinis* (Rubiaceae) to accelerate the birth of their children (Gran, 1973b; Weidmann and Craik, 2016). The plant was found to contain cyclotides that have uterotonic activity, hence validating its use for such purpose. Complementary and Alternative Medicine (CAM) has been the oldest form of treatment, particularly in Africa. Despite the limited knowledge with regard to the pathophysiology of diseases and mechanism of action of plant extracts, some herbs have been found to be clinically effective in the management of some conditions. Plant biodiversity in Africa provides a rich source of numerous plant species with abundant constituents and metabolites with similar or dissimilar constituents. Some components or peptides may occur across multiple genera and species among endemic plants in the region e.g., cyclotides. Cyclotides are macrocyclic peptides discovered in plants. They are distinguished by their unique cyclic cystine knot (CCK) structural pattern. A head-to-tail cyclic backbone is cross-braced with a cystine knot generated by six conserved Cys residues in the CCK motif (Gran, 1973a; de Veer et al., 2019). Cyclotides are intriguing because of their unusual structure, which confers extraordinary stability, as well as their natural activities as plant defense peptides, and their vast variety of uses as bioactive molecules in agriculture and medicine (de Veer et al., 2019). The presence of cyclotides in some endemic African plants may account for their different uses in ethnomedicine. One example is *Rinorea dentata* (Violaceae family). *Rinorea dentata* (P. Beauv.) Kuntze is a tropical plant that grows in the rainforests of Liberia, Cameroon, Uganda, and Nigeria. Parts of its ethnomedical uses among local settlers in the J4 region of the Omo Forest Reserve in Ogun State, Nigeria, the plant's natural habitat, chew its stem for oral hygiene as a malaria treatment and to shorten labor or make birth easier in farm animals (Attah et al., 2016b). *Allexis cauliflora*—a plant native to Cameroon, Congo, and Gabon belongs to the Violaceae family and has been found to contain cyclotides (Gattringer et al., 2021). Cyclotide-rich extract of the plant has also been found to inhibit human prolyl oligopeptidase (POP) activity (Gattringer et al., 2021). Other members in the same genus that contain cyclotides include *Allexis anensis*, *Allexis batangae*, and *Allexis zygomorpha*. *Rinorea brachypetala* can be tropical Africa, from Guinea East to Kenya, and to Zambia and Angola in the South (Plant Use, Online). Results of Matrix-Assisted Laser Desorption Time of Flight (MALDI-TOF) Mass spectrometry show that it also contains cyclotides (Attah et al., 2016a).

2.3 ENDEMISM OF THE CYCLOTIDE-RICH *OLDENLANDIA AFFINIS*

The Rubiaceae family's genus *Oldenlandia* contains 190 species, several of which are known to be used in traditional medicine. From the Ivory Coast in the north to South Africa Transvaal, *O. affinis* is widely dispersed in Africa's tropical zone (Figure 2.2). It can also be found on Madagascar and in western Asia, with the Malay Peninsula serving as its easternmost outpost. It grows as a weed in the grass at altitudes ranging from sea level to 1,500 m above sea level. It appears to avoid arid areas (Gran et al., 2000). There have been reports and observations on the ethnomedical uses of *O. affinis* in different cultures across the region. An aqueous extract, also known as 'Kalata-Kalata', in Congo is used to facilitate labor and childbirth. It is also used for this same activity in the Central African Republic (Gran, 1973b). *O. affinis* is blended with honey and used to treat learning disabilities in Benin Republic, and the aqueous decoction is used as an anti-colic drink. The decoction of *O. affinis* is utilized as an anti-malarial and anti-febrile medicine in several rural communities in South-Eastern Nigeria (Nworu et al., 2017). These activities are based on the presence of cyclotides peptides, which when compared to other peptides of similar size are extremely resistant to chemical, thermal, and biological degradation (Gould and Camarero, 2017). Protease inhibitory, anti-microbial, insecticidal, cytotoxic, anti-HIV, and hormone-like effects are all biological features of naturally occurring cyclotides (Gould and Camarero, 2017).

As much as cyclotides are important in herbal medicine, they can also serve as useful drug molecules in the synthesis of new drugs, hence, highlighting the importance for preservation of endemic *O. affinis* plants. Since the first isolation of cyclotides from *O. affinis* (Figure 2.1) by Lorent Gran, cyclotides have now been discovered in the plant families Rubiaceae (coffee), Violaceae (violet), Solanaceae (potato), Poaceae (grass), Cucurbitaceae (cucumber), and Fabaceae (pea) (Slazak et al., 2020). The environment has an impact on cyclotide production. Depending on the region or season, different amounts of certain peptides are detected in the same species. Changes in cyclotide gene expression levels appear to be the reason, but breakdown rates may also play a role in forming the final peptide suite (Slazak et al., 2020). The protection of the endemism of threatened cyclotide-rich medicinal plants such as *O. affinis* and other plants is highly imperative. When faced with the growing use of medicinal plants, preserving tropical biodiversity should be prioritized. Most tropical plant species lack extinction risk estimates, restricting the capacity to identify conservation priorities despite their critical importance for terrestrial ecosystems (Stévart et al., 2019).

2.4 ENDEMISM OF OTHER CYCLOTIDE-RICH VIOLACEAE PLANT FAMILY IN AFRICA: EMERGING THREATS

The threat to the endemism of cyclotide-rich plants in tropical African region is not limited to *O. affinis*, the genus *Rinorea* (family Violaceae) appear to be top on the list (The IUCN Red List of Threatened Species: https://www.iucn.org/). *Rinorea* Aubl. is a genus made up of an approximately 250–300 species (World Flora online Consortium—www.worldfloraonline.org), with at least 150 mostly endemic species unevenly distributed across tropical African coast including Angola, Southern Nigeria, Cameroon, Gabon, and Madagascar (Wahlert et al., 2020) (Figure 2.2). This data translates to over 60% of *Rinorea* endemism in the region whose distribution peaks up in Madagascar, recording 80% known species endemism (Wahlert et al., 2020). Following well-informed prediction, it means several or all these endemic cyclotide-producing tropical medicinal plants (Table 2.2) are facing the risk of global extinction before the end of the 21st century.

TABLE 2.2
Continental Distribution of Cyclotide-Rich African Plants. Cyclotides Documented Have Been Retrieved from Cybase—a Database of Cyclic Proteins

S/N	Botanical Source	Traditional Uses	Continental Distribution	Cyclotides Documented in Cybase
1	*Clitoria ternatea* L. Fabaceae		It is used as a source of natural food colorant. It is cultivated as a forage and fodder crop, for nitrogen fixation and improvement of soil nutrient. It is used to aid childbirth	Tropical Africa	Cliotide T1, T2, T4 (Cter P), T6, T8, T9, T10 (Cter 27), T11 (Cter 21), T12, T13, T14, T15, T16, T17, T18, T19a, T19b, T20, T21, T22α, Cliotide T23* -T32*, Cter A, B, C, D, E, F, G, H, I, J, K, L, M, N, O, Q, R
2	*Oldelandia affinis* (Roem. & Schult.) DC Rubiaceae	Facilitation of childbirth.	DR Congo, Central African Republic, Liberia, Sudan, South Africa, Madagascar, Nigeria	Kalata B1–B19; Kalata B20-lin
3	*Rinorea dentata* (P. Beauv.) Kuntze Violaceae	Its wood is used as a chewing stick for oral hygiene. Used as veterinary remedy for aiding parturition in livestock an'd for treatment of neurodegenerative disorders. Its tree trunk can be used as hardwood for tools making, construction, boat building, furniture making, etc.	Tropical rainforests of Liberia, Nigeria, Cameroon and Uganda	R. dentata A (riden A)
4	*Rinorea oblongifolia* (C.H.Wright) C.Marquand ex Chipp Violaceae	In DR Congo the wood is used in construction and for household utensils. In Sierra Leone it is used to make spoons and combs. In the Central African Republic a root extract is drunk as a purgative. It is used to calm the pain in women during childbirth. It is used to treat fever.	Nigeria, Gabon, Central African Republic, Cameroon, DR Congo, Sierra Leone, Uganda.	Unavailable
5	*Rinorea ilicifolia* (Welw. ex Oliv.) Kuntze Violaceae	The hard and heavy wood is used in Uganda to make wooden hammers, tool handles, and walking sticks. In West Africa decoctions of the whole plant are drunk to cure epilepsy. In Kenya the pulped root is steeped in water and the liquid is drunk to treat cough. It is used to calm the pain in women during childbirth and for fever.	Nigeria, Guinea east, Kenya, Angola, Mozambique, KwaZulu-Natal, South Africa, Madagascar, Benin, Burundi, Cameroon, Central African Republic, Comoros, Congo, Ethiopia, Gabon, Ghana, Guinea, Guinea-Bissau, Ivory Coast, Kenya	Unavailable

(Continued)

TABLE 2.2 (Continued)

Continental Distribution of Cyclotide-Rich African Plants. Cyclotides Documented Have Been Retrieved from Cybase—a Database of Cyclic Proteins

S/N	Botanical Source	Traditional Uses	Continental Distribution	Cyclotides Documented in Cybase
6	Rinorea brachypetala (Turcz.) Kuntze Violaceae	Its wood is used in Kenya and Uganda for hut construction, walking sticks, and knobkerries. In DR Congo it is used in construction and to make household utensils. Boiled leaves are eaten as a vegetable and the dried, powdered leaves are sniffed to relieve headache. It is used to calm the pain in women during childbirth and for fever.	Nigeria, Guinea, Kenya, Zambia, Angola, Sierra Leone, Cameroon, Uganda	Unavailable
7	Allexis obanensis (Baker f.) Melch. Violaceae	Used for the treatment of brain disorders	Cameroon, Nigeria (Oban Division and National Park, Cross River)	Alca 1, Alca 2
8	Allexis cauliflora (Oliv.) Pierre Violaceae	Bark is used to treat fever and syphilis	Congo, Cameroon, Ghana, Nigeria	Alca 1 and Alca 2
9	Allexis batangae (Engl.) Melch. Violaceae	Used for the treatment of brain disorders	Congo, Cameroon	Alca 1, Alca 2, Alba 1, 2, 3, 4
10	Allexis zygomorpha Achound. & Onana Violaceae	Used for the treatment of brain disorders	Congo, Cameroon	Alca 1 and Alca 2
11	Viola abyssinica Steud. ex Oliv. Violaceae	In Ethiopia, its leaves are crushed and squeezed to get the juice, which is applied to hard leather to soften it. In Madagascar it is given as an antidote and to stimulate vomiting. Along the eastern border of Congo, it is rubbed with other herbs in butter to make a massage oil.	Fernando Poo (Bioko), South Africa (Transvaal), Madagascar.	Vaby A, B, C, D, E, Varv E (Cycloviolacin O12)
12	Viola odorata L. Violaceae	It is used to treat anxiety, insomnia, skin disorders, and cough. It is used to make violet scones and marshmallows	Morocco, Tunisia and North Africa	Cycloviolacin O1, O2, O3, O4, O5, O6, O7, O8, O9, O10, O11, O12, O13, O14, O15, O16, O17, O18, O19, O20, O21, O22, O23, O24, O25, Varv A, Kalata B1, Vodo N, Vodo M, Violacin A

Source: (https://www.cybase.org.au) (57).

While Africa is unarguably the richest continent in terms of *Rinorea* species diversity, it is the only continent native to the genus *Allexis Pierre* (family Violaceae; Table 2.2), distributed across Tropical West and Central Africa (Olivier et al., 2018; Gattringer et al., 2021). Despite the endemism of a majority of *Rinorea* species and restricted distribution of the four known species of the genus *Allexis* in limited tropical region of Africa, flora diversity within the continent remains the most threatened due to the increasing human populations, combined effect of logging, fuelwood collection, overcollection of wild plants for research, and deforestation for agriculture and mining (van Velzen and Wieringa, 2014). The endemism of Violaceae family to tropical Africa has been demonstrated by the three largest violaceae genera—*Viola, Rinorea,* and *Hybanthus,* which account for around 95% global distribution of Violaceae species (Wahlert et al., 2020; Slazak et al., 2021) and the genus *Allexis* with only four accepted species names (http://www.theplantlist.org/; Table 2.2). While the endemism of the genus *Viola* in Africa and neighboring oceanic island appears to be restricted to extreme geographies, particularly the temperate and mountainous regions (Linder and Evolution, 2014), a few such as *Viola palmensis, Viola cheiranthifolia, Viola anagae* have recently been reported in extreme environments (Slazak et al., 2021) located outside African mainland; however, the next two largest genera based on the recent circumscription, *Rinorea* and *Hybanthus,* are mainly pantropical, occurring more in the humid seasonal tropics found within the most endangered African flora (Wahlert et al., 2020). Exhaustive ethnomedicinal surveys in the recent past no longer document several of these important cyclotide-rich African plants—an observation that further supports the level of threat to their endemism. Deliberate efforts should therefore be made to ensure sustainable utilization of cyclotide-rich African plants that are now becoming extinct in the wild. The sustainability of these threatened plants will mainly be achieved via captivity, cultivation in/or outside native range, based on IUCN Red list description and classification of plants species.

2.5 COMPETITIVE APPROACHES TO SAFEGUARD ENDEMIC CYCLOTIDE-PRODUCING AFRICAN PLANTS

There is an urgent need for the proper documentation, conservation, and utilization of the cyclotide-rich species of African flora. The chemotaxonomic relevance of cyclotides to Violaceae plant family has been emphasized (Burman et al., 2015) as species screened produced the presence of cyclotides, making them the foremost biofactories for cyclotide diversity, abundance, and species distribution (Slazak et al., 2021). The African Violaceae appears to be the least scientifically investigated family despite its apparent endangered status. Meanwhile, African Violaceae has varied uses in the Traditional Mdicine. The potentials of its abundant cyclotides in sustainable agricultural plant protection as biomaterials for drug discovery/development as well as their putative clinical application should form a strong basis for further scientific investigation of the family. Therefore, efforts are needed to implement competitive steps and strategies to sustainably safeguard the endemism of cyclotide-rich African flora. This is particularly urgent in African regions with cyclotide-rich plant diversity, including Tropical West Africa, Ethiopia, Tanzania, and the Democratic Republic of the Congo (Figure 2.1), which have been described to harbor over 40% of potentially threatened endemic plant species (Stévart et al., 2019). Among others, these strategies comprise land/water protection and management, cyclotide-producing species management, education and awareness, livelihood, economic and other incentives, as well as law and policy. The botanic garden approach to plant extinction crises remains one of the easy-to-adopt strategy (Westwood et al., 2021; van Zonneveld et al., 2021). In line with Article 14 of the United Nations Convention on Biological Diversity (www.cbd.int/), adopting environmental impact assessments (EIAs) is germane before implementation of lawful projects that may pose as threats to plant biodiversity. Hence, to minimize risks associated with the environment, the adverse impacts of these projects on plant biodiversity must be identified to circumvent, minimize, or totally prevent negative consequences (Stévart et al., 2019).

2.6 FUTURE PERSPECTIVES

2.6.1 FIGHTING PLANT BIODIVERSITY LOSS IN AFRICA

Africa's fertile ground is host to a wide array of plant species scattered around the continent. An approximate one-quarter of global biodiversity comes from Africa. From tropical regions to temperate areas, and even unto arid regions, different plant varieties are native to these zones, but sadly there is a continuous decline in the number of plant species in the continent, and some species are already going extinct. The importance of preserving this biodiversity hotspot of the world is vital. Many greater numbers of African plants have not received investigative scientific attention despite their inherent potentials for drug discovery. The role of plant biodiversity underlies the pivotal point of almost all ecosystems, and ecosystems will remain adversely affected if the continuous loss of species remains unchecked. The loss of biodiversity has an impact on livelihoods, water availability, food security, and resistance to catastrophic events, especially for the poorest people in rural regions (Bank., 2019).

Tackling plant biodiversity loss is an important issue, and causal factors need to be addressed. Causal factors include overcollection, unsustainable agriculture, and forestry methods, urbanization, pollution, land use changes, the spread of invasive alien species, climate change, a lack of information about the status of most habitats, species, and other vital data (Osawaru et al., 2013). It is evident that there is much room for improvement in biodiversity-related programs and policies in Africa. Plant conservation should be prioritized and not only considered as a subset of wildlife conservation. Governmental institutions and non-governmental agencies need to not only promulgate laws but also to enforce them as well. Land use should be closely monitored, especially in regions with biodiversity of species and endemic plants. There is also a need for the development and update of national plant databases across countries in the region. Plant, animal, forest, and aquatic genetic resources in Africa, like those in other regions, are better documented than associated biodiversity; just a few nations report having national databases or knowledge-management systems dedicated to associated biodiversity (FAO, 2019).

Climate is a strong influencer of the sustainability of plant ecosystems and biomes. Plant variety is shaped by climate on a wide scale, with relatively warm and moist (beneficial, productive) locations sustaining more species (Harrison et al., 2020). Climates are fast evolving toward warmer temperatures, which is predicted to result in less water availability throughout the growing season in approximately half of the Earth's terrestrial area (McLaughlin et al., 2017). Reduction in water availability adversely affects the sustainability and preservation of plant species in the continent. Combatting plant biodiversity loss by responding to climate change requires participation and collaboration across sectors, agencies, and governments (Fernández-González et al., 2005). Climate change is an urgent issue in the 21st century that affects almost every sphere of life, and, as detailed in this chapter, it affects cyclotide-rich African tropical plants as much. While some countries are working consciously to reduce greenhouse emissions, the same level of effort is lacking in under-developed and developing countries. African countries need to put in place structured efforts and plans to reduce carbon footprint and mitigate climate change.

2.6.2 FUTURE OF AFRICAN PLANT BIODIVERSITY

Africa continues to face a reduction in biodiversity and possible extinction of some species. The continent is home to a wide array of plants scattered across towns, villages, communities, and territorial borders. The highlight of this section is to call to mind the need for action to preserve this ecosystem in all its forms and species. The future of African plant biodiversity can be sustained if proper actions are taken. Whether plant biomes will be conserved or not is dependent on the conservation efforts implemented or introduced. It is essential to have a grasp of the full diversity of plant species in a database, characterize the rate of decline, and highlight the indices of contributing factors. According to the mid-term review report from The United Nations Environment Programme World Conservation Monitoring Centre (UNEP-WCMC, 2016), lack of institutional, financial, and technological resources as well as the capacity to implement National Biodiversity Strategies/ Action Plans (NBSAPs) limit efforts in conserving plant biodiversity. A lack of appropriate and harmonized biodiversity indicators to assess conservation needs and NBSAP progress represent contributing factors limiting the conservation of plant biodiversity. Other limiting factors include data and information deficiencies, and national budgetary constraints. (UNEP-WCMC, 2016). On the positive side, some levels of progress have been achieved in Africa towards plant conservation. The Nagoya Protocol on Access and Benefit Sharing has been approved by 30 African nations as of December 2015 (Aichi Target 16). Several other countries in the region are planning to do the same. In addition, at least one NBSAP has been submitted by 44 African Parties (Aichi Target 17) (UNEP-WCMC, 2016).

More major conservation efforts, as well as proactive measures such as adjustments in agricultural techniques, enhanced agricultural commerce and improved land-use planning, and will be required to prevent and decrease threats to world biodiversity (Tilman et al., 2017). There are several biodiversity hotspots in Africa—home to threatened plant species. One example of such spot is the Bale Mountains of Ethiopia. It supports the continent's largest area of afroalpine and subalpine ericaceous plants. Despite it being listed by UNESCO as one of the biodiversity hotspots, it has not received the needed conservation attention. The protection of critical areas like this is essential to maintain the biodiversity of the continent and prevent the extinction of threatened species (Kidane et al., 2019).

2.6.3 Protection and Sustainable Use of Endemic Cyclotide-Rich Species in Africa

Numerous cyclotides have been isolated from the Rubiaceae, Cucurbitaceae, Fabaceae, Solanaceae, Poaceae, and Violaceae families. Many cyclotides are found only in certain plant species, although others are found in multiple species (Slazak et al., 2020, Ravipati et al., 2017). While some families are more abundant than the others and do not face any reasonable threat, some cyclotide-containing species are very rare and may face danger of extinction. Preserving the availability and continued use of these cyclotide-rich species is thus very important. The IUCN Red list provides a comprehensive database on the status of extinction risk of plants, animal, and fungi. It ranks plants on a risk scale starting from least concerned to near threatened, vulnerable, endangered, critically endangered, extinct in the wild and extinct (last on the scale) (IUCN: https://www.iucn.org/resources/conservation-tools/iucn-red-list-threatened-species). According to the IUCN Red list database, *Rinorea tshingandaensis* belongs to the vulnerable group (Ntore, 2021). *Rinorea frisii* and *Rinorea bullata* belong to the endangered group (Alemu, 2018; Ravololomanana, 2020), *Rinorea abbreviate, Rinorea key*, and *Rinorea rauschii* belong to the near threatened group (Bidault, 2020., Faranirina, 2019, Centre., 1998.). All of these species belong to the same genus with *Rinorea dentata* under the Violaceae family, which has been validated to possess cyclotides (Attah et al., 2016b). Despite the poor status of these species, there are little to no conservation efforts put in place. Conservation efforts that can help preserve these species include *ex-situ* conservation, habitat restoration, awareness, legislation, and enforcement among others (IUCN: https://www.iucn.org/resources/conservation-tools/iucn-red-list-threatened-species).

2.7 CONCLUSION AND PROSPECTS

Africa must be deliberate in ensuring the proper utilization of the already endangered cyclotide-rich plants. Considering the threats to the endemism of these cyclotide-producing plants, their intentional cultivation has become necessary to prevent them from getting critically endangered and ultimately going extinct. It is, therefore, pertinent to holistically address the main drivers of biodiversity loss of cyclotide-rich plants in Africa in a timely manner, comprising divergent over-exploitation, increasing reports of invasive species, habitat loss, crude oil pollution, and climate change. Local plant habits and habitats, particularly those persisting in restricted areas must be urgently protected. Besides *in situ* conservation measures, cyclotide-rich plants endemic only to limited and restricted environments such as the genus *Allexis* as well as some endemic *Rinorea* species require additional *ex-situ* conservative strategies such as gene collection from the wild, *in vitro* collections, storage in seed banks, and medicinal plant garden (botanical garden) solutions. In Sub-Saharan Africa, only 13 endemic *Rinorea* species have been accessed and included on the IUCN Red list, representing less than 10% of *Rinorea* species known to the region; yet, over 80% of these accessed species have been classified as critically endangered, endangered, vulnerable, and near threatened (www.iucnredlist.org). One of the main hindrances to the identification and conservation of new *Rinorea* species lies in the close morphological similarity between species of the genus. This has made the identity of over 190 species from Gabon very difficult for taxonomists (Sosef et al., 2017). However, modern methods, such as the application of molecular marker-assisted authentication, are encouraged to resolve the challenges with the classification of endemic *Rinorea* species in a bid to promote their conservation and prevent misutilization (Sonibare, 2021). More importantly, the reported expression of cyclotides in every member of Violaceae family as well as the unique pattern of species cyclotide signature is of chemotaxonomic relevance, as it has the potential to aid in proper species identification, classification, and eventual conservation when morphology alone becomes inadequate (Slazak et al., 2021, Burman et al., 2015). Lastly, cyclotide fingerprint unique to individual species could serve as chemotaxonomic markers for species identification, classification, conservation, and sustainable utilization. A combined effort of molecular marker-assisted authentication and cyclotide-based chemotaxonomic approaches are suggested as deliberate background strategies that could potentially promote the protection of endemic cyclotide-rich African flora diversity.

ACKNOWLEDGEMENT

Authors sincerely appreciate the effort of Akinseinde Ebenezer Akindele (University of Ibadan, Nigeria) in supplying data for a table preparation.

REFERENCES

Alemu, S., Alemu, S., Atnafu, H., Awas, T., Belay, B., Demissew, S., Luke, W.R.Q., Mekbib, E., and Nemomissa, S. (2018). *Rinorea friisii*. The IUCN Red List of Threatened Species. e.T128045953A128045961. https://dx.doi.org/10.2305/IUCN.UK.2018-2.RLTS.T128045953A128045961.en. Downloaded on 28 July 2021.

Attah, A., Sonibare, M., and Moody, J. O (2016a). Chemical detection of cysteine-rich circular peptides in selected tropical Violaceae and Moringaceae families using modified G-250 and mass spectrometry. *Nigerian Journal of Natural Products and Medicine, 20*, 88–95.

Attah, A.F., Fagbemi, A.A., Olubiyi, O., Dada-Adegbola, H., Oluwadotun, A., Elujoba, A., and Babalola, C.P. (2021). Therapeutic potentials of antiviral plants used in traditional African medicine with COVID-19 in focus: a Nigerian perspective. *Frontiers in Pharmacology, 12*, 596855.

Attah, A.F., Hellinger, R., Sonibare, M.A., Moody, J.O., Arrowsmith, S., Wray, S., and Gruber, C.W. (2016b). Ethnobotanical survey of *Rinorea dentata* (Violaceae) used in south-western Nigerian ethnomedicine and detection of cyclotides. *Journal of Ethnopharmacology, 179*, 83–91.

Bank, W. (2019). This Is What It's All About: Protecting Biodiversity in Africa. Available online. https://www.worldbank.org/en/news/feature/2019/02/14/biodiversity. Last Accessed 28 July 2021.

Bidault, E., Paradis, A.H., Texier, N., and Stévart, T. (2020). *Rinorea abbreviata. The IUCN Red List of Threatened Species*. e.T153248970A153255966. https://dx.doi.org/10.2305/IUCN.UK.2020-3.RLTS.T153248970A153255966.en. Downloaded on 28 July 2021.

Burak, M.K., Monk, J.D., and Schmitz, O.J. (2018). Focus: ecology and evolution: eco-evolutionary dynamics: the predator-prey adaptive play and the ecological theater. *The Yale Journey of Biology and Medicine, 91*(4), 481–489.

Burlando, B., Palmero, S., and Cornara, L. (2019). Nutritional and medicinal properties of underexploited legume trees from west Africa. *Critical Review in Food Science and Nutrition, 59*, S178–S188.

Burman, R., Yeshak, M.Y., Larsson, S., Craik, D.J., Rosengren, K.J. and Göransson, U. (2015). Distribution of circular proteins in plants: large-scale mapping of cyclotides in the Violaceae. *Frontiers in Plant Science, 6*, 855.

Butt, M.A., Zafar, M., Ahmed, M., Shaheen, S., and Sultana, S. (2021). *Wetland Plants: A Source of Nutrition and Ethno-medicines*. Springer, New York, USA.

Camarero, J.A. and Campbell, M.J. (2019). The potential of the cyclotide scaffold for drug development. *Biomedicines, 7*(2), 31.

Centre W.C.M. (1998). *Rinorea* keayi. *The IUCN Red List of Threatened Species*.e.T32734A9726300. https://dx.doi.org/10.2305/IUCN.UK.1998.RLTS.T32734A9726300.en. Downloaded on 29 August 2021.

Craik, D.J., Daly, N.L., Bond, T., and Waine, C. (1999). Plant cyclotides: a unique family of cyclic and knotted proteins that defines the cyclic cystine knot structural motif. *Journal of Molecular Biology, 294*, 1327–1336.

De-Veer, S.J., Kan, M.W., and Craik, D.J. (2019). Cyclotides: from structure to function. *Chemical Reviews, 119*, 12375–12421.

FAO. (2019). *Africa Regional Synthesis for the State of the World's Biodiversity for Food and Agriculture*. Food and Agriculture Organisation of the United Nations (FAO), Rome, pp. 68. Licence: CC BY-NC-SA 3.0 IGO.

Faranirina, L.R.M. (2019). *Rinorea urschii. The IUCN Red List of Threatened Species. e.T126344726A126344991*. https://dx.doi.org/10.2305/IUCN.UK.2019-2.RLTS.T126344726A126344991.en. Downloaded on 28 July 2021.

Fernández-González, F., Loidi, J., Moreno, J.C., Del-Arco, M., Férnández-Cancio, A., Galán, C., García-Mozo, H., Muñoz, J., Pérez-Badia, R., and Sardinero, S. (2005). *Impacts on Plant Biodiversity*, Ministerio de Medio Ambiente, Madrid, pp. 183–248.

Ganguly, S. and Bakhshi, S. (2020). Traditional and complementary medicine during COVID-19 pandemic. *Phytotherapy Research, 34*(12), 3083–3084.

Gattringer, J., Ndogo, O.E., Retzl, B., Ebermann, C., Gruber, C.W. and Hellinger, R. (2021). Cyclotides isolated from violet plants of cameroon are inhibitors of human prolyl oligopeptidase. *Frontiers in Pharmacology, 12*, 1737.

Gould, A. and Camarero, J.A. (2017). Cyclotides: overview and biotechnological applications. *Chembiochem: A European Journal of Chemical Biology, 18*(14), 1350–1363.

Gran, L. (1973a). On the effect of a polypeptide isolated from Kalata-Kalata (Oldenlandia affinis DC) on the oestrogen dominated uterus. *Acta Pharmacologcia et Toxicologica, 33*(5), 400–408.

Gran, L. (1973b). Oxytocic principles of Oldenlandia affinis. *Lloydia , 36(2)*, 174–178.

Gran, L., Sandberg, F. and Sletten, K. (2000). Oldenlandia affinis (R&S) DC: a plant containing uteroactive peptides used in African traditional medicine. *Journal of Ethnopharmacology, 70*, 197–203.

Guernier, V., Hochberg, M.E., and Guégan, J.F. (2004). Ecology drives the worldwide distribution of human diseases. *PLoS Biology, 2*(6), e141.

Harrison, S., Spasojevic, M.J. and Li, D. (2020). Climate and plant community diversity in space and time. *Proceedings of the National Academy of Sciences of the United States of America, 117*(9), 4464–4470.

Kidane, Y.O., Steinbauer, M.J., and Beierkuhnlein, C. (2019). Dead end for endemic plant species? A biodiversity hotspot under pressure. *Global Ecology and Conservation*, 19, e00670.

Kougioumoutzis, K., Kokkoris, I.P., Panitsa, M., Kallimanis, A., Strid, A. and Dimopoulos, P. (2021). Plant endemism centres and biodiversity hotspots in Greece. *Biology. 10(2)*, 72.

Linder, H.P. (2014). The evolution of African plant diversity. *Frontiers in Ecology and Evolution, 2*, 38.

Lughadha, E.N., Govaerts, R., Belyaeva, I., Black, N., Lindon, H., Allkin, R., Magill, R.E. and Nicolson, N. (2016). Counting counts: revised estimates of numbers of accepted species of flowering plants, seed plants, vascular plants and land plants with a review of other recent estimates. *Phytoaxa, 272*(1), 82.

Luo, S. and Dong, S.H. (2019). Recent advances in the discovery and biosynthetic study of eukaryotic RiPP natural products. *Molecules: A Journal of Synthetic Chemistry and Natural Product Chemistry, 24*(8), 1541.

Mahomoodally, M.F. (2013). Traditional medicines in Africa: an appraisal of ten potent African medicinal plants. *Evidence-Based Complementary and Alternative Medicine: eCAM, 2013,* 617459.

Mclaughlin, B.C., Ackerly, D.D., Klos, P.Z., Natali, J., Dawson, T.E. and Thompson, S.E. (2017). Hydrologic refugia, plants, and climate change. *Global Change Biology, 23*(8), 2941–2961.

Ntore, S., Simo-Droissart, M., and Tack, W. (2021). *Rinorea* tshingandaensis. *The IUCN Red List of Threatened Species. e.T138014371A138015999.* https://dx.doi.org/10.2305/IUCN.UK.2021-1.RLTS. T138014371A138015999.en. Downloaded on 28 July 2021.

Nworu, C.S., Ejikeme, T.I., Ezike, A.C., Ndu, O., Akunne, T.C., Onyeto, C.A., Okpalanduka, P., and Akah, P.A. (2017). Anti-plasmodial and anti-inflammatory activities of cyclotide-rich extract and fraction of *Oldenlandia affinis* (R. & S.) DC (Rubiaceae). *African Heath Sciences, 17*(3), 827–843.

Oguis, G.K., Gilding, E.K., Jackson, M.A., and Craik, D.J. (2019). Butterfly pea (*Clitoria ternatea*), a cyclotide-bearing plant with applications in agriculture and medicine. *Frontiers in Plant Science, 10*, 645.

Olivier, N.E., Oscar, N.D.Y., Alima, N., François, M., and Barthelemy, N. (2018). Antibacterial properties of the extracts of *Allexis obanensis* and *Allexis batangae* (Violaceae) collected at Kribi (South Cameroon). *The Journal of Phytopharmacology, 7*(3), 275–284.

Osawaru, M., Ogwu, M., and Ahana, C. (2013). Current status of plant diversity and conservation in Nigeria. *Nigerian Journal of Life Science, 3*, 168–178.

Raven, P.H., Gereau, R.E., Phillipson, P.B., Chatelain, C., Jenkins, C.N., and Ulloa, C.U. (2020). The distribution of biodiversity richness in the tropics. *Science Advances, 6(37)*, eabc6228.

Ravipati, A.S., Poth, A.G., Troeira-Henriques, S.N., Bhandari, M., Huang, Y.H., Nino, J., Colgrave, M.L., and Craik, D. (2017). Understanding the diversity and distribution of cyclotides from plants of varied genetic origin. *Journal of Natural Products, 80*(5), 1522–1530.

Ravololomanana, N. (2020). *Rinorea* bullata. *The IUCN Red List of Threatened Species. e.T137832111A137904113.* https://dx.doi.org/10.2305/IUCN.UK.2020-3.RLTS.T137832111A137904113.en. Downloaded on 28 July 2021.

Senkoro, A.M., Talhinhas, P., Simões, F., Batista-Santos, P., Shackleton, C.M., Voeks, R.A., Marques, I., and Ribeiro-Barros, A.I. (2020). The genetic legacy of fragmentation and overexploitation in the threatened medicinal African pepper-bark tree, Warburgia salutaris. *Scientific Report, 10*, 1–13.

Slazak, B., Haugmo, T., Badyra, B., and Göransson, U. (2020). The life cycle of cyclotides: biosynthesis and turnover in plant cells. *Plant Cell Report, 39*(10), 1359–1367.

Slazak, B., Kaltenböck, K., Steffen, K., Rogala, M., Rodríguez-Rodríguez, P., Nilsson, A., Shariatgorji, R., Andrén, P.E., and Göransson, U. (2021). Cyclotide host-defense tailored for species and environments in violets from the Canary Islands. *Scientific Report, 11*(1), 12452.

Sonibare, M.A. (2021). Quest for sustainable healthcare provision: utilisation and misutilization of our co-tenants. *An Inaugual Lecture.* Faculty of Pharmacy, Ibadan University Press, 1–87.

Sosef, M.S., Dauby, G., Blach-Overgaard, A., Van-Der-Burgt, X., Catarino, L., Damen, T., Deblauwe, V., Dessein, S., Dransfield, J., and Droissart, V. (2017). Exploring the floristic diversity of tropical Africa, *Journal of Biology.* 15(1), 15.

Stévart, T., Dauby, G., Lowry, P., Blach-Overgaard, A., Droissart, V., Harris, D., Mackinder, B.A., Schatz, G., Sonké, B., and Sosef, M.S. (2019). A third of the tropical African flora is potentially threatened with extinction. *Science Advances, 5*(11), eaax9444.

Tilman, D., Clark, M., Williams, D.R., Kimmel, K., Polasky, S., and Packer, C. (2017). Future threats to biodiversity and pathways to their prevention. *Nature, 546*(7656), 73–81.

UNEP-WCMC 2016. The State of Biodiversity in Africa: A mid-term review of progress towards the Aichi Biodiversity Targets. *The UN Enivornment Programme World Conservation Monitoring Center,* Cambridge, UK.

Van-Velzen, R. and Wieringa, J.J. (2014). *Rinorea* calcicola (Violaceae), an endangered new species from south-eastern Gabon. *Phytotaxa, 167*(3), 267–275.

Van-Wyk, A.S. and Prinsloo, G. (2020). Health, safety and quality concerns of plant-based traditional medicines and herbal remedies. *South African Journal of Botany, 133*(11), 54–62.

Van-Zonneveld, M., Kindt, R., Solberg, S.Ø., N'Danikou, S., and Dawson, I.K. (2021). Diversity and conservation of traditional african vegetables: priorities for action. *Diversity and Distributions, 27*(1), 1–17.

Wahlert, G.A., Gilland, K.E., and Ballard, H.E. (2020). Taxonomic revision of *Rinorea ilicifolia* (Violaceae) from Africa and Madagascar. *Kew Bulletin, 75*, 1–15.

Weidmann, J. and Craik, D.J. (2016). Discovery, structure, function, and applications of cyclotides: circular proteins from plants. *Journal of Experimental Botany, 67*(7), 4801–4812.

Westwood, M., Cavender, N., Meyer, A., and Smith, P. (2021). Botanic garden solutions to the plant extinction crisis. *Plants People Planet, 3*(2), https://doi.org/10.1002/ppp3.10134.

World Health Organisation (2019). WHO global report on traditional and complementary medicine. *World Health Organization, 11*, 226.

Yeshak, M.Y., Burman, R., Asres, K., and Göransson, U. (2011). Cyclotides from an extreme habitat: characterization of cyclic peptides from *Viola abyssinica* of the Ethiopian highlands. *Journal of Natural Products, 74*, 727–731.

3 Traditional Medicinal Plant Use in Improving the Livelihoods and Well-Being of Namibian Communities

Iwanette du Preez-Bruwer, Nailoke Pauline Kadhila and Davis Ropafadzo Mumbengegwi

CONTENTS

3.1 INTRODUCTION

Much of the world's biodiversity, *i.e.*, the variety of living organisms in a specific area, is located in sub-Saharan Africa (Ngwira et al., 2013; Wilson and Primack, 2020). Biodiversity meets the basic needs of traditional people or communities in resource-poor settings, such as food, shelter, fuel, and medicines (Ministry of Environment and Tourism, 2018). The rich biodiversity in Africa is also a key player in the tourism sector. Therefore, it forms a basis for sustainable development and poverty reduction. A very important development in southern Africa's natural resource governance has been community involvement. This is an attempt to achieve sustainable utilization of natural resources

DOI: 10.1201/9781003206620-3

(Ngwira et al., 2013). Namibia has a rich biodiversity (Ministry of Environment and Tourism, 2018), with over 3,159 plant species (Cheikhyoussef et al., 2011). Plants have been said to form part of a country's biodiversity and culture, and play an important role in traditional medicines (TMs).

According to the World Health Organization, about 80% of the world population uses TM, including herbal medicine as their only system of health care (World Health Organization, 2013). Despite this figure being lower than previously cited (Oyebode et al., 2016), medicinal plants continue to play an important role in the livelihoods and well-being of people in both developed and developing countries (Smith-Hall et al., 2012; Oyebode et al., 2016). They are mainly used to provide treatment of diseases and/or to alleviate symptoms (Mudigonda and Mudigonda, 2010). Especially in African countries, the use of medicinal plants is described in the treatment and or management of diseases such as HIV/AIDS, malaria, cancer, and chronic ailments such as diabetes and hypertension, among others (Mahomoodally, 2013; Ozioma and Chinwe, 2019). Natural therapies may be found in pharmacies and food stores in nutraceuticals or dietary supplements (Ekor, 2014).

In Namibia, herbal preparations are used in rural settings (Ministry of Environment and Tourism, 2018). For some, herbal-based TM is the preferred treatment method; for others, herbs are used as an adjunct therapy to conventional pharmaceuticals. The most common reasons for using herbal TM within communities are that it is more culturally acceptable and fulfills a desire for more personalized health care; it is more affordable; it has fewer adverse effects and higher therapeutic effects; and it allows greater public access to health, for example, communities in remote areas often only have access to TMs but not conventional medicines (Oyebode et al., 2016). There is a need for new medicines for treatment and prevention of diseases, particularly those that are life-threatening such as cancer (Drexler, 2010). Scientific and research interest is drawing their attention toward naturally derived compounds, as they are considered to have less toxic side effects compared to current treatments such as chemotherapy. Some potent plant-based anticancer compounds, such as vinblastine, camptothecin, and taxol, have been isolated and characterized from different plant species, including *Taxus brevifolia*, *Camptotheca acuminate*, and *Catharanthus roseus* (Zlatić and Stankovic, 2015; Lichota and Gwozdzinski, 2018; Changxing et al., 2020). Vinblastine has a market value of US$ 15 million/kg, while camptothecin costs US$ 0.5 million/kg; these are indicators of the possible value of plant-based anticancer agents on the global market. However, an anticancer product resulting from the use of natural resources or any medicinal applications from them will require investment in research and development for any potential market to be realized.

The objective of this review is to shed light on how natural resources are benefiting and improving Namibian communities in terms of health and means of livelihood; and how these resources are being utilized sustainably, both for personal and commercial use using examples of medicinal mushrooms and medicinal plants such as Manketti (*Schinziophyton rautanenii*) and Devil's claw (*Harpagophytum procumbens* and *Harpagophytum zeyheri*). Issues underlying Access and Benefit-Sharing (ABS) are also highlighted.

3.2 MEDICINAL MUSHROOMS

Mushrooms have a long history as a food source (Valverde et al., 2015), and they are important for their healing capacities and properties in traditional medicine (Zhang et al., 2016). According to current estimates, mushrooms constitute at least 12,000 fungi species worldwide; out of those, 2,000 can be eaten, and about 200 wild species are used medicinally (Rathore et al., 2017). Wild and cultivated mushrooms have been documented as being a good source of nutrients and bioactive compounds (Wandati et al., 2013). There are over 700 types of mushrooms that have been identified that can be used as nutraceuticals (Sridhar and Deshmukh, 2021). They contain a reasonable number of proteins, carbohydrates, minerals, fibers, and vitamins. Furthermore, they are low in calories, sodium, fats, and cholesterol (Ho et al., 2020; Ugbogu et al., 2020). Thus, the consumption of mushrooms can make a valuable addition to the unbalanced diets of people in developing countries (Marshall and Nair, 2009). However, there are numerous mushrooms that are not edible. '…many mushrooms are inedible or poisonous, such as the Amanita species that are used on arrows when local people hunt wild animals for meat' (Kadhila-Muandingi and Chimwamurombe, 2012). Other mushrooms are resourceful like ectomycorrhizas and are endowed with bioactive compounds and produce functional metabolites (Sridhar and Deshmukh, 2021).

Mushrooms are appreciated for their excellent sensory characteristics, flavor, and texture (Shivashankar and Premkumari, 2014). In the past, they were eaten mainly because of their distinctive taste (Das et al., 2021), but this has changed during the last few decades. Currently, mushrooms are also consumed for health reasons. They have been found to reduce the likelihood of cancer invasion and metastasis due to antitumor attributes, to have antibacterial properties, are antidiabetic, are immune system enhancers, and are cholesterol-lowering agents among other functions (Wasser, 2002; Valverde et al., 2015). More than 100 therapeutic effects are produced by mushrooms and fungi, of which the key medicinal uses are antioxidant, anticancer, antidiabetic, antiallergic, immunomodulating, cardiovascular protector, anticholesterolemic, antiviral, antibacterial, antimalarial, antiparasitic, antifungal, detoxification, and hepatoprotective effects; they also protect against tumor development and inflammatory processes (Stamets, 2005; Chang and Wasser, 2012; Valverde et al., 2015; Ho et al., 2020; Ugbogu et al., 2020). Therefore, mushrooms can help prevent life-threatening diseases (Prasad et al., 2015).

Furthermore, therapeutic effects of edible mushrooms have been reported in several preclinical and clinical studies, which indicates that they may be used in the adjuvant treatment of breast cancer (Novaes et al., 2011; Figueiredo and Régis, 2017; Berretta et al., 2022; Pathak et al., 2022). The use of mushrooms in conjunction with conventional cancer therapies has been associated with improvements in the well-being of patients due to reduced side effects, as well as with enhanced hematological and immunological parameters (Novaes et al., 2011). Mushrooms may also reduce the risk of developing cancer (Figueiredo and Régis, 2017). Certain proteins and antioxidants in mushrooms are said to be responsible for this by regulating cancerous cells and cytokine production (Pathak et al., 2022).

3.2.1 Mushroom Species Contributing to the Well-Being of Namibian Communities

Medicinal mushrooms in various Namibian ecosystems include *Ganoderma lucidum*—a genus of wood-inhabiting fungus within the Ganodermataceae family, which can improve health and livelihood of the communities (Figure 3.1). Other *Ganoderma* species found in Namibia include *Ganoderma tsugae*, *Ganoderma neo-japonicum*, and *Ganoderma applanatum* (Kadhila-Muandingi 2010). These species are used for medicinal purposes in the Oshana, Kavango, Zambezi (formerly known as Caprivi), and Ohangwena regions of Namibia, where they are predominantly distributed (Kadhila-Muandingi, 2010; Shikongo, 2012; Ekandjo and Chimwamurombe, 2021). These communities use *Ganoderma* mushrooms to strengthen babies' skull bones, improve the immune system of pregnant women, treat nose bleeds, for stress, shock, trauma, colds and flu, as well as head sores (ringworms) in children, and as a relaxant (Kadhila-Muandingi, 2010; Ekandjo and Chimwamurombe, 2021). The indigenous people in these regions also consume other mushrooms for medicinal purposes, treating both humans and animals, especially cattle, goats, and chickens. They use *Termitomyces* mushrooms, including pseudorrhiza, for medical applications such as prevention of miscarriages in pregnant women, severe diarrhea, sexually transmitted diseases, heart problems, cancer, and kidney problems, as well as healing of the umbilical cord for the newborn baby. Other mushrooms like truffles are used for eye treatments, while the poisonous mushrooms belonging to the Amanita species are used on arrows for hunting (Kadhila-Muandingi, 2010).

FIGURE 3.1   A wild indigenous *Ganoderma* mushroom (© Pauline Kadhila).

3.2.2 ETHNOPHARMACOLOGICAL EVIDENCE

A study by Mhanda, Kadhila-Muandingi, and Ueitele (2015) determined the minerals and trace elements in domesticated Namibian *Ganoderma* mushrooms. The results show that the fruiting bodies of the domesticated *G. lucidum* are a good source of protein, carbohydrates, crude fiber, and some nutritionally important minerals. This analysis found that the contents of moisture, crude protein, total lipids, available carbohydrates, dietary fiber, total ash, and calories agreed with the values reported in some previous studies done on *Ganoderma* species (Essien et al., 2013; Sharif et al., 2016).

3.2.3 MUSHROOM SPECIES CONTRIBUTING TO THE LIVELIHOOD OF NAMIBIAN COMMUNITIES

The use of wild mushrooms as food and a source of income is widespread in Namibia. It is very common to see local people, particularly young women, selling mushrooms by the roadside after the start of the rainy season (Figure 3.2). They stated that *Termitomyces schimperi*, *Terfezia pfeilii* (Kalahari Desert truffles) are some of the most sought after wild mushrooms in rural areas of Namibia. They are aware of wild edible mushrooms, and their importance to people who are generally poor and many of the mushrooms collected are for personal use (Kadhila-Muandingi and Chimwamurombe, 2012). A group of locals called Erari Mushroom Suppliers have received funding and training to produce oyster mushrooms. They produced about 10 tons for which they earned about N$ 3,000.00 collectively; in addition, 13 temporary employment opportunities were created (Admin, 2013).

3.2.4 SUSTAINABLE USE OF SELECTED MUSHROOM SPECIES

Mushroom cultivation plays an important role in the sustainable use of wild mushrooms that are sometimes overharvested by communities for use throughout the year. In addition, it can also

FIGURE 3.2   Wild harvested mushrooms *Kalaharituber pfeilii* and *Termitomyces schimperi are* on display for sale along the roads to generate income (© Pauline Kadhila).

support the local economy by contributing to subsistence food security, nutrition, and medicine, and therefore it can directly enhance livelihoods through financial, nutritional, and therapeutic advantages (Rautela et al., 2019). Since mushroom cultivation does not require access to land, it can be a viable and attractive activity for rural farmers and peri-urban dwellers in the country (Kadhila-Muanding and Mubiana, 2008). Mushroom cultivation activities can generate additional employment and income through local, regional, and national trade and offer processing enterprises opportunities. Indigenous medicinal mushrooms like *Ganoderma* and other types of mushrooms that have been identified as having potential for domestication can be cultivated through such opportunities. In fact, a study by Ueitele, Kadhila-Muanding, and Matundu (2014) established that *Ganoderma* mushrooms can be successfully domesticated and grown on corncobs and wood chips. As part of their community service, the University of Namibia, through the Zero Emissions Initiative (ZERI), is offering courses on mushroom cultivation. It can be a suitable and empowering income-generating option for women, as it can be combined with traditional domestic duties at home.

3.3   MANKETTI (*SCHINZIOPHYTON RAUTANENII*)

Schinziophyton rautanenii or Manketti, as it is commonly known in English, is a deciduous tree that grows up to 12 m tall and can provide good shade due to its large canopy (Graz, 2002). It has palmately compound leaves (Figure 3.3) and round, plum-sized drupe as fruit, which ripens once they have fallen to the ground. Locally, Manketti is known as Mangetti in Otjiherero, or Mungongo in Silozi, or Ugongo in Rukwangali. It is indigenous to Namibia, specifically in the Ohangwena, Kavango, and Zambezi regions, and other southern African countries, including Zambia, Botswana,

FIGURE 3.3   The palmately compound leaves of *Schinziophyton rautanenii* (© Iwanette du Preez-Bruwer).

Angola, Zimbabwe, Mozambique, and South Africa (Graz, 2002). Manketti has been used for centuries dating as far back as 7,000 years ago by San communities in the Kalahari Desert for food, medicine, energy, and cosmetic applications (Maroyi 2018).

3.3.1   MANKETTI CONTRIBUTING TO THE WELL-BEING OF NAMIBIAN COMMUNITIES

Manketti can be used medicinally; the bark is a remedy for abdominal complaints and diarrhea, and can also be used to treat morning sickness, which is when women experience nausea early in the day due to being pregnant (Leger, 1997; Quattrocchi, 2012). The roots are used to treat stomach pains and oil and skin ailments, which also have antibacterial properties (Maroyi, 2018). Perhaps the most interesting medicinal use is that of the leaves, which are used to treat sores on the skin that do not heal—a description similar to that of Kaposi sarcoma. Ethnobotanical surveys that were conducted in 2011 and 2013 in the Zambezi (then Caprivi) and Kavango regions, respectively, interviewing traditional knowledge holders to identify and document plants used within the regions for different infirmities, confirmed the medicinal applications for Manketti. The key informants indicated that alternative treatment from ethnomedicinal plant sources was mostly used for pain management in cancer patients. According to Dushimemaria (2014), another traditional use is to treat sores or surface wounds such as scrapes, abrasions, and scratches. Nutrients including phosphorous, magnesium, iron, sodium, zinc, and calcium, and even amino, and fatty acids were found in the fruit of Manketti (Gwatidzo et al., 2013; Maroyi, 2018).

3.3.2   ETHNOPHARMACOLOGICAL EVIDENCE

Analysis of the methanolic extracts of the root and bark of Manketti revealed the presence of anti-cancer classes of compounds such as alkaloids, anthraquinones, coumarins, flavonoids, and triterpenes (Dushimemaria, 2014). The same study also revealed the extracts to exhibit antioxidant and anti-protease activity. Furthermore, inhibition of the growth of cancer cells was also observed for the extracts. The extracts displayed *in vitro* anticancer activity against breast cancer MCF-7, renal cancer TK-10, and melanoma UACC-62 (Dushimemaria, 2014). This is a significant finding, because Manketti products may be utilized as an adjuvant or alternative cancer treatment, especially given the nation's rising cancer incidence. Breast cancer (14.5%), cervix uteri (10.7%), Kaposi sarcoma (12.6%), and prostate cancer (9.4%) are in the lead among the Namibian population (WHO, 2020).

3.3.3 Manketti as a Source of Income to Improve Livelihoods

Manketti plays an important role in rural areas regarding food security and income generation (Maroyi, 2018). Indigenous communities in Namibia, including the San, use different plant parts, especially nuts and fruits, as sources of food, beverages, oil, and medicines (Graz, 2002; Dushimemaria, 2014; Maroyi, 2018). While the kernels or the outer nutshell of Manketti can be eaten as a snack, raw or roasted (in the coals of a fire) (Ademe et al., 2014), they can also be crushed slightly and added in the preparation of certain food dishes as a thickening agent or for their unique flavor and oil (Ademe et al., 2014; Maroyi, 2018). On a larger scale, oil is produced for cosmetic applications such as skin and hair care (Ademe et al., 2014). The fruit pulp has been used to produce *kashipembe* (in Rumanyo, a Kavango dialect) or *ombike* (in Oshiwambo), which is a distilled alcoholic beverage (Misihairabgwi and Cheikhyoussef, 2017). The scale of household consumption (and home processing) of Manketti products in Namibia remains undocumented and are, therefore, entirely unknown in terms of volumes, economic importance, and food security contribution. However, the fruit is also used in exchange for other food sources, such as pearl millet or maize, with a 50 kg bag selling for N$ 50–100 (Mallet et al., 2015). A livelihood study of the San/ Vakwangali Natural Resource reported that a 1 l of kashipembe would cost N$ 12.50, about N$ 100 (U$ S5) today (Cole et al., 1998). Kernels have also been reported to fetch between N$ 15/kg and N$ 20/kg on the informal market. Manketti kernels are increasingly valued more widely for their outstanding nutritional content; the oil is known to be stable and a prized cosmetic ingredient. Between March 2005 and June 2006, 1.4 t of Manketti oil was sold in France for approximately N$ 80,000 through CRIAA-SADC (Mallet et al., 2015). Several companies have been formed around selling Manketti oil, such as Bio-Innovation Zimbabwe (BIZ), Kalahari Natural Oils Limited, Zambia, and Katutura Artisans Project Namibia. According to BIZ, the average yearly income is around USD 200 per producer (Masson, 2016). According to the authors, there is a lack of data for income from harvesting of Manketti for medicinal uses; traditional practitioners in Zambezi and Kavango regions mostly rely on agriculture for their income and not healing. This is because of the low value of the Manketti-based traditional medicines. However, research is showing potential for the development of therapeutics based on Manketti.

3.3.4 Sustainable Use of Manketti

In the Zambezi, Kavango, and Ohangwena regions, Manketti trees are found in large groves where they are dominant or co-dominant trees spaced as close as 20 m apart (Graz, 2002). There are success stories of propagation in Zambia and Namibia with the planting of truncheons (living fences) (Namibia Nature Foundation, 2017). However, this is limited by the requirements of the species, especially regarding soil needs. Commercialization could positively impact rural livelihoods through cash income generation, with little impact on the Manketti resources. Data on Manketti yields and purchases of nut kernels concluded positively on the feasibility of commercialization (Graz, 2002; Maroyi, 2018; Brendler, 2021).

3.4 DEVIL'S CLAW (*HARPAGOPHYTUM PROCUMBENS* AND *HARPAGOPHYTUM ZEYHERI*)

Harpagophytum—*Harpagophytum procumbens* and *Harpagophytum zeyheri* or Devil's claw, also known as grapple plant or wood spider (English), is commonly found in southern Africa in countries such as Namibia, Botswana, Zambia, South Africa, Angola, and Zimbabwe (Qi et al., 2006). Among communities in Namibia, especially the Nama/Damara and Afrikaans speaking,

FIGURE 3.4   *H. procumbens* or Devil's claw—on the left, the claw or hook-like fruit/seed pod after which the plant is named—growing wild in the Kalahari Desert (*Source*: https://www.cnseed.org/harpagophytum-procumbens-seed-devils-claw-seed.html).

it is called *gamagu* and or *Afrika aartappel*. It is also called *makakata* in Oshindonga, *omalyata* in Oshikwanyama, *otjihangatene* in Oshiherero, and *malamatwa* in Silozi. These names, together with 'Devil's claw' and 'grapple plant' and '*Harpagophytum*', the latter derived from the Greek word 'harpago' means 'grappling hook', stem from the appearance of the fruit that looks like a hook or claw (Figure 3.4) (Mncwangi et al., 2012).

Devil's claw is a perennial that grows in deep sandy soils. It has a lateral root (taproot) with several secondary storage tubers (Brendler, 2021). The taproot can grow 50 cm in length and is found up to 2 m deep, and secondary root tubers branch off horizontally (Strohbach and Cole, 2007). The latter can store up to 90% water and thus grows very well in dry conditions (Mowa and Maas, 2016). In Namibia, it grows well in the dry, harsh environment of the Kalahari Desert, which receives as little as 250 mm of rain per annum (Mncwangi et al., 2012).

3.4.1   Devil's Claw Contributing to the Well-Being of Namibian Communities

The secondary storage tubers are used medicinally for which they treat a number of infirmities, including arthritis, pain, fever, ulcers, boils, urinary tract infections, postpartum pain, sprains, and sores (Mncwangi et al., 2012). In addition to that, the indigenous communities in southern Africa specifically use them mostly for fever, arthritis, pain, ulcers, and boils and as a general health tonic (Mncwangi et al., 2012). Mcwangi (2012) further indicates that the San and Nama communities of Namibia use these plants for said aliments. This was confirmed by a survey that was conducted in 2017. It was found that the Karas and Hardap communities in the south of Namibia use the tubers of *H. procumbens* to alleviate or treat symptoms of indigestion, such as nausea, abdominal pain, diarrhea, bloating, and constipation (personal communication, 2017). They also use the tubers to treat fever, diabetes, eye problems, dysmenorrhea, burns, boils, wounds, sores, common cold, and swollen glands. Furthermore, the tubers are also taken as a form of cleansing or detox, especially after giving birth. The Khoi and San in South Africa also use Devil's claw in the treatment of fever, diabetes, blood diseases, and diarrhea (Mncwangi et al., 2012). The same groups also use it as a purgative for loss of appetite, menstrual pains, indigestion, as a bitter tonic, inflammation, syphilis, and in child birth (Gxaba and Manganyi, 2022). Devil's claw is used as complementary or alternative medicines in Germany, the United Kingdom, Holland, the United States, and the Far East

(Gxaba and Manganyi, 2022). In these countries, Devil's claw is being used for rheumatism and or arthritis, and indigestion, urinary tract infections, sores, fever, blood diseases, postpartum pain, sprains, ulcers, and as an appetite suppressant (Gxaba and Manganyi, 2022).

3.4.2 ETHNOPHARMACOLOGICAL EVIDENCE

In vitro studies on the biological activities of *H. procumbens* have been done, which demonstrated that extracts of *H. procumbens* exhibit antimalarial, analgesic, antioxidant, antidiabetic, antiepileptic, and antimicrobial activities (Mncwangi et al., 2012). On the other hand, clinical studies have shown Devil's claw to effectively treat ailments such as rheumatism and osteoarthritis (OA). The dosage of Devil's claw extract ranged from 60 to 3,000 mg and was shown to be effective in treating arthritis and rheumatism—an improvement in patients was observed, reduced stiffness and pain, as well as increased mobility (Akhtar and Haqqi, 2012). In addition, a study done by Chrubasik, Conradt, and Roufogalis (2004) indicated that 8 weeks post-treatment with Devil's claw extract at a dose of 60 mg daily, between 50% and 70% of patients suffering from non-specific low back pain or osteoarthritis of the knee or hip, reported improvement in pain, mobility, and flexibility. Safety information was reported by Brien, Lewith, and McGregor (2006). The extracts did not show any signs of toxicity; however, the data of many of these trials are doubtful mainly because of the methods used. However, *in vivo* studies in mice indicated the consumption of Devil's claw to be non-toxic; these mice were treated daily with *H. procumbens* capsules for up to 3 months (Al-Harbi et al., 2013). A more recent study also showed no toxicity of *H. procumbens*; in this case, rats and aqueous-ethanolic extracts of *H. procumbens* were used.

The medicinal properties of *H. procumbens* can be attributed to compounds such as harpagoside, harpagide, and procumbide; these are said to be the major active ingredients in Devil's claw (Akhtar and Haqqi, 2012; Manon et al., 2015; Gxaba and Manganyi, 2022). Harpagosides especially have been reported to reduce pain and inflammation in joints (Gxaba and Manganyi, 2022). The secondary tubers contain twice as many harpagosides as the taproot; however, whole-plant extracts have been reported to exhibit higher therapeutic effects than those prepared from individual plant parts (Raditic and Bartges, 2014). In addition, increased doses of the active ingredients are reported to have a wider therapeutic window (Chrubasik, Conradt, and Roufogalis, 2004; Raditic and Bartges, 2014). Chrubasik, Conradt, and Roufogalis (2004) also found that *H. procumbens* preparations containing 50–60 mg of harpagosides to relieve joint and lower back pain were more effective in alleviating pain and improving mobility than extracts with lower doses. Other compounds that are also found in the root include iridoid-glycosides, phytosterols; phenylpropanoids such as verbascoside; triterpenes, such as oleanolic acid, 3β-acetyloleanolic acid, and ursolic acid; flavonoids, such as kaempferol and luteolin; and unsaturated fatty acids, cinnamomic acid, chlorogenic acid, and stachyose (Brendler, 2021).

3.4.3 DEVIL'S CLAW CONTRIBUTING TO LIVELIHOODS

The resettlement farms in Namibia are relatively densely populated and heavily grazed. Thus, for settlers without livestock, Devil's claw harvesting often constitutes the only source of cash income (Kala et al., 2006). Local harvesters in rural areas of southern parts of the country who have access to the plant may sell the dried tubers to community members. Local harvesters in rural areas of southern parts of the country who have access to the plant may sell the dried tubers to other community members. Harvesting of Devils Claw is conducted as a subsistence activity at a low level and not as a business enterprise where harvesting is done on a large scale.

Since Namibia is the largest supplier of devil's claw (Mowa, 2008), *H. procumbens* contributes greatly to the economy of Namibia. Between 1992 and 2006, Namibia exported 95% of the raw material for Devil's claw, whereas South Africa and Botswana only contributed 2% and 3%, respectively (Mowa, 2008). In 2009, it was estimated that the sales value of *H. procumbens* was worth

1.06 million € (Mncwangi et al., 2012), equivalent to a little over 18 million NAD. Most of the time, local communities do not benefit from such trades. In 1997, the Sustainably Harvested Devil's Claw (SHDC) project was established to protect these communities by improving benefit-sharing opportunities.

The SHDC project suggests that improved benefit sharing for harvesters leads to better resource management and conservation by giving harvesters a financial incentive to ensure the tubers are not overharvested. However, the low price paid to harvesters is a major impediment to sustainable harvesting of the tubers. The project also shows the importance of traditional knowledge (e.g., the best time to harvest and how to avoid damaging the plants).

3.4.4   SUSTAINABLE USE OF DEVIL'S CLAW

Devil's claw is a protected plant species in Namibia. The main threat to Devil's claw is overharvesting for medicinal use or income-generating activities. According to Cole (2014), Devil's claw are usually harvested by subsistence farmers residing in resource-poor areas, where available resources are shared most of the time. The tubers are an important source of income for many of them. Wild harvesting of devil's claw tubers can be sustainable if only some of the tubers are taken and if enough is left behind for the plants to regenerate. However, the tubers have been overcollected in some areas, and the species are becoming rarer.

Although difficult to cultivate, it is possible and could solve the problem of sustainability. Cultivation on a commercial scale not only helps to reduce the pressure on wild plants but can also improve the quality of the harvested product. Cultivation also provides an opportunity to restore degraded land where unsustainable harvesting has occurred in the past. But some have pointed out the potential negative impact on the livelihoods of small-scale rural harvesters if large quantities of cultivated material dominated the export market and drove down prices. Micropropagation techniques such as making use of explants to rapidly multiply parent plants into many progeny plants with conserved medicinal properties can be used as a conservation strategy and may potentially provide an alternative to wild-harvested supplies for the herbal medicine.

The use of such techniques will also have other advantages such as a lower water requirement in the already dry southern parts of the country and the provision of nutrients for plant growth culture media to ensure viability of young plants. These approaches can contribute to the sustainability of the Devils Claw industry, which is delicately balanced demand for the plant exceeds its supply, hence trading in Devils Claw is controlled especially to overseas markets. Communities that trade in Devils claw have observed a decrease in the vegetation (medicinal plants) in their surrounding areas, and this is partly due to overharvesting and also changing rainfall patterns that are lower in volume and also due to longer spells of drought.

Devil's claw is classified as a protected species in Botswana, Namibia, and South Africa; permits are required for harvesting and exporting it. No part of the tubers or roots can be traded within the European Union without licensing. As Devil's claw is a restricted and scarce resource medicinal resource, communities that trade in this plant have developed practices that contribute to conserve the resource that include domestication and use of sustainable harvesting and pruning techniques. When harvesting tubers, they dig a hole away from the plant to avoid plant taproots and fill it in afterwards to allow the plants to continue growing.

The SHDC project contributed to the development of Namibia's policy on devil's claw and scientific research on the impact of harvest on growth rates to determine annual sustainable yields. Harvesters under the auspices of this project voluntarily use sustainable harvesting techniques. They are assisted through pre- and postharvest ecological surveys, to set local harvesting quotas, and to ensure good resource management practices such as not disturbing taproots, refilling holes, etc., are adhered to. The harvesters also strive to provide a high-quality product (Tubers are sliced with stainless steel knives to prevent discoloration and are dried on shade-net racks to avoid contamination by sand.). SHDC Harvesting groups are also equipped with scales and have access to secure storage facilities. This allows them to know exactly how much each harvester is supplying, how much the group is harvesting, and to collate commercially viable quantities of Devil's claw at central points where it can easily be collected by the exporter. In return, harvesters are paid a premium price directly by the exporter (at least 50%, and in some cases up to 1,000%, more than prices paid by informal-sector middlemen). Such approaches maintain the balance between harvesting greater yields of Devil's claw in a sustainable manner and securing greater economic returns for the harvesters.

3.5 GOVERNANCE/INSTITUTIONAL STRUCTURES TO ACCESS TRADITIONAL KNOWLEDGE AND/OR GENETIC RESOURCES

The Access to Biological and Genetic Resources and Associated Traditional Knowledge Act 2 of 2017 were enacted to protect local or traditional communities or maintain the country's rich biodiversity through the sustainable use of natural resources. This act is based on the ABS principles from the Nagoya Protocol established in 2010; this was adopted at the Convention on Biological Diversity (CBD) in 1992. ABS refers to how genetic resources such as medicinal and food plants and micro-organisms can be accessed, and how the benefits that result from their use are shared between the people or countries using the resources (users) and the people or countries that provide them (providers) (Sirakaya, 2019).

The ABS provisions of the CBD are designed to ensure that the physical access to genetic resources is facilitated, and that the benefits obtained from their use are shared equitably with the providers. This ensures that they are not misappropriated or misused to the disadvantage of knowledge holders or custodians of the genetic resource. This also includes valuable traditional knowledge associated with genetic resources. These agreements furthermore help governments to establish their national frameworks.

ABS is based on prior informed consent (PIC) being granted by a provider to a user and negotiations between parties to develop mutually agreed terms (MAT) to ensure the fair and equitable sharing of genetic resources and associated benefits. This policy can lead to more equitable sharing of benefits arising from the utilization of genetic resources. Benefits such sharing royalties when the resources are used to create a commercial product or non-monetary such as sharing research results, or developing research skills and knowledge within local communities. Users and providers must understand and respect institutional frameworks such as those outlined by the CBD and the Nagoya protocol. The framework in Namibia requires 'users' (researchers in this case) to apply for a research authorization permit before they can access communities and or genetic resources. The application process is done through the National Commission on Research, Science, and Technology (NCRST), which submits it to the relevant authorities such as the Ministry of Forestry, Environment, and Tourism. The ABS office under this ministry provides the necessary support to the users and issues the permit. This office also supervises and mediates the entire process.

3.6 CONCLUSION AND PROSPECTS

Traditional medicines play a significant part in Namibian primary health care as the first treatment option for most ailments. They are also an important source of income for Namibian communities in rural and urban areas; however, they are underutilized. For communities to benefit more, value addition to traditional medicine through research needs to be addressed. This highlights their beneficial properties and identifies novel uses of products. Sustainable utilization of these natural resources is encouraged, as well as cultivation should be done where applicable. Documentation of medicinal flora and associated traditional knowledge can contribute to their management and utilization to ensure sustainability. Furthermore, the increased use of medicinal mushrooms and plants can contribute to the sustainable development goals (SDGs) of the United Nations, including no poverty, zero hunger, good health, and well-being, decent work, and economic growth, sustainable cities, and communities, and responsible consumption, and production. Strengthening this sector may benefit and improve the living standard of poor people.

Namibia produces large quantities of agricultural waste; hence communities can venture into Agribusiness by cultivating mushrooms. Growing mushrooms is one of the few technologies that can produce food and medicine by managing waste. It is an ecofriendly process that does not require land and a lot of water as in the production of other vegetables. To unleash the potential of mushrooms as a tool for development, there is a need to enhance the communities' skills by educating them about mushrooms, both short-term and long-term skills. This can be beneficial to such communities in reducing poverty, creating employment, and improving livelihood.

Early-stage R&D has shown that Manketti extracts have the potential of being used as possible anticancer agents—a market worth billions of dollars globally. Further research should be conducted to identify a possible cure for cancer—a disease that may claim more deaths than HIV/AIDs. Devil's claw has been shown to have anti-inflammatory properties *in vitro* and *in vivo*. The authors suggest that further and more robust research should be done on Devil's Claw against disease models such as diabetes and immune-modulatory activities. This will increase the medicinal value of the plants and provide a basis for their integration into mainstream medicine in Namibia. However, a medicinal product or any medicinal applications from traditional medicines will require investment in research and development for any potential market to be realized.

Downstream activities such as collecting and processing plant material can contribute to the local green economy by employing local community members. Harvesting of the plant material can be done sustainably with minimal impact on the ecosystem, when alternative plant parts are used, e.g., using leaves or stems instead of roots and tubers. The development value chain is to be built from the ground up and be a test case for beneficiation of local communities while ushering development through utilizing Namibia's natural resources. However, the Access to Biological and Genetic Resources and Associated Traditional Knowledge Act, 2017 may challenge researchers or research institutions. Although it is good to have these procedures in place, the process may take months, if not years, to complete. However, a starting point could be the engagement of communities to develop an ABS-compliant value chain. These challenges need to be addressed to ensure that future generations can also enjoy the natural treasures and resources the country has to offer.

ACKNOWLEDGEMENTS

The authors would like to acknowledge the Multidisciplinary Research Services and UNAM Press of the University of Namibia who contributed to the writing of this chapter as well as the local communities who contributed to the knowledge we shared in this chapter.

REFERENCES

Ademe, Y., Axtell, B., Fellows, P., Gedi, L., Harcourt, D., Grenade, C.L., Lubowa, M., and Hounhouigan, J. (2014). *A Handbook for Setting Up and Running a Small-Scale Business Producing High-Value Foods.*, ACP-EU Technical Centre for Agricultural and Rural Cooperation (CTA), Wageningen.

Akhtar, N. and Haqqi, T.M. (2012). Current nutraceuticals in the management of osteoarthritis: a review. *Therapeutic Advances in Musculoskeletal Disease*, 4(3), 181–207.

Al-Harbi, N.O., Al-Ashban, R.M., and Shah, A. (2013). Toxicity studies on *Harpagophytum procumbens* (Devil's claw) capsules in mice. *Journal of Medicinal Plants Research*, 7, 3089–3097.

Berretta, M., Franceschi, F., Quagliariello, V., Montopoli, M., Cazzavillan, S., Rossi, P., and Zanello, P.P. (2022). The role of integrative and complementary medicine in the management of breast cancer patients on behalf of the integrative medicine research group (IMRG). *European Review for Medical and Pharmacological Sciences*, 26(3), 947–956.

Brendler, T. (2021). From bush medicine to modern phytopharmaceutical: a bibliographic review of Devil's claw (*Harpagophytum* spp.). *Pharmaceuticals*, 14(8), 726.

Brien, S., Lewith, G.T., and McGregor, G. (2006). Devil's claw (*Harpagophytum procumbens*) as a treatment for osteoarthritis: a review of efficacy and safety. *The Journal of Alternative and Complementary Medicine*, *12*(10), 981–993.

Chang, S.T. and Wasser, S.P. (2012). The role of culinary-medicinal mushrooms on human welfare with a pyramid model for human health. *International Journal of Medicinal Mushrooms*, *14*(2), 95–134.

Changxing, L., Galani, S., Hassan, F., Rashid, Z., Naveed, M., Fang, D., Ashraf, A., Qi, W., Arif, A., Saeed, M., Chishti, A.A., and Jianhua, L. (2020). Biotechnological approaches to the production of plant-derived promising anticancer agents: an update and overview. *Biomedicine and Pharmacotherapy*, *132*, 110918.

Cheikhyoussef, A., Mapaure, I., and Shapi, M. (2011). The use of some indigenous plants for medicinal and other purposes by local communities in Namibia with emphasis on Oshikoto region: a review. *Research Journal of Medicinal Plant*, *5*, 406–419.

Chrubasik, S., Conradt, C., and Roufogalis, B.D. (2004). Effectiveness of *Harpagophytum* extracts and clinical efficacy. *Phytotherapy Research: PTR*, *18*(2), 187–189.

Cole, D., Powell, N., !Aice, B., and #Oma, G. (1998). *San/Vakwangali Natural Resource and Livelihood Study, Mpungu and Kahenge Constituencies Kavango Region,-Namibia*. Centre for Research, Information, Action in Africa– Development and Consulting (Criaa-DC), Windhoek.

Das, A.K., Nanda, P.K., Dandapat, P., Bandyopadhyay, S., Gullón, P., Sivaraman, G.K., McClements, D.J., Gullón, B., and Lorenzo, J.M. (2021). Edible mushrooms as functional ingredients for development of healthier and more sustainable muscle foods: a flexitarian approach. *Molecules*, *26*(9), 2463.

Drexler, M. (2010). Prevention and treatment. *What You Need to Know About Infectious Disease*, National Academies Press (US). Washington (DC).

Dushimemaria, F. (2014). *An Investigation into the Antineoplastic Properties of* Schinziophyton Rautanenii *and* Colophospermum mopane. Ph.D Thesis, UNAM University of Namibia, Windhoek, Namibia.

Ekandjo, L. and Chimwamurombe, P.M. (2021). Genetic diversity of ganoderma species in the north eastern parts of Namibia using random amplified microsatellites (RAMS). *Journal of Pure and Applied Microbiology*, *6*(3), 1097–1104.

Ekor, M. (2014). The growing use of herbal medicines: issues relating to adverse reactions and challenges in monitoring safety. *Frontiers in Pharmacology*, *4*, 177.

Admin. Erari Mushroom Suppliers need funds for expansion (2013). Agriculture. *Namibia Economist*. https://economist.com.na/3771/agriculture/erari-mushroom-suppliers-need-funds-for-expansion/

Essien, E., Udoh, I. and Peter, N.-U. (2013). Proximate composition and anti-nutritive factors in some wild edible medicinal macrofungi. *Journal of Natural Product and Plant Resources*, *3*, 29–33.

Figueiredo, L. and Régis, W.C.B. (2017). Medicinal mushrooms in adjuvant cancer therapies: an approach to anticancer effects and presumed mechanisms of action. *Nutrire*, *42*(1), 28.

Graz, F.P. (2002). *Description and Ecology of* Schinziophyton rautanenii (Schinz) *Radcl.-Sm. in Namibia*. Namibian Scientific Society, vol. 27, pp. 19–35.

Gwatidzo, L., Botha, B., and Mccrindle, R. (2013). Determination of amino acid contents of manketti seeds (*Schinziophyton rautanenii*) by pre-column derivatisation with 6-aminoquinolyl-N-hydroxysuccinimi-dyl carbamate and RP-HPLC. *Food chemistry*, *141*, 2163–2169.

Gxaba, N. and Manganyi, M.C. (2022). The fight against infection and pain: devil's claw (Harpagophytum procumbens) a rich source of anti-inflammatory activity: 2011–2022. *Molecules*, *27*(11), 3637.

Ho, L.-H., Zulkifli, N.A., and Tan, T.-C. (2020). Edible mushroom: nutritional properties, potential nutraceutical values, and its utilisation in food product development. *An Introduction to Mushroom*, vol. 10, IntechOpen, UK.

Kadhila-Muanding, N.P. and Mubiana, F.S. (2008). Mushroom Cultivation: A Beginners Guide. University of Namibia, Namibia, Windhoek.

Kadhila-Muandingi, N.P. (2010). *The distribution, genetic diversity and uses of Ganoderma mushrooms in Oshana and Ohangwena regions of Northern Namibia*. Ph.D Thesis, Windhoek, Namibia. http://wwwisis.unam.na/theses/kadhila-muandingi2010.pdf.

Kadhila-Muandingi, N.P. and Chimwamurombe, P.M. (2012). Uses of ganoderma and other mushrooms as medicine in Oshana and Ohangwena regions of Northern Namibia. *Journal of Research in Agriculture*, *1*(2), 146–151.

Kala, C.P., Dhyani, P.P. and Sajwan, B.S. (2006). Developing the medicinal plants sector in Northern India: challenges and opportunities. *Journal of Ethnobiology and Ethnomedicine*, *2*, 32.

Leger, S. (1997). *The Hidden Gifts of Nature: A Description of Today's Use of Plants in West Bushmanland (Namibia)*. German Developing Service, Ministry of Environment and Tourism, Directorate of Forestry, Windhoek, Namibia.

Lichota, A. and Gwozdzinski, K. (2018). Anticancer activity of natural compounds from plant and marine environment. *International Journal of Molecular Sciences, 19*(11), 3533.

Mahomoodally, M.F. (2013). Traditional medicines in Africa: an appraisal of ten potent African medicinal plants. *Evidence-Based Complementary and Alternative Medicine, 2013*, 617459.

Mallet, M., Cole, D. and Davis, A.-L. (2015). *Establishing the status quo of and developing a strategic framework for further work on* Schinziophyton rautanenii *(manketti) in Namibia: Report for Deutsche Gesellschaft für Internationale Zusammenarbeit (GIZ) and the Indigenous Plant Task Team (IPTT).* Windhoek, Namibia.

Manon, L., Béatrice, B., Thierry, O., Jocelyne, P., Fathi, M., Evelyne, O., and Alain, B. (2015). Antimutagenic potential of harpagoside and *Harpagophytum procumbens* against 1-nitropyrene. *Pharmacognosy Magazine, 11*(Suppl 1), S29–S36.

Maroyi, A. (2018). Contribution of *Schinziophyton rautanenii* to sustainable diets, livelihood needs and environmental sustainability in Southern Africa. *Sustainability, 10*(3), 581.

Marshall, E. and Nair, N.G. (2009). *Make Money by Growing Mushrooms.* Food and .Agriculture Organization of the united Nation (FAO), Rome, Italy.

Masson, T., (2016). Mongongo nut or manketti tree. Bio-Innovation Zimbabwe. Harvey Brown Ave, Milton Park, Harare, Zimbabwe. https://www.bio-innovation.org/mongongo-or-manketti-tree-schinziophyton-rautanenii/

Mhanda, F.N., Kadhila-Muandingi, N.P. and Ueitele, I.S.E. (2015). Minerals and trace elements in domesticated Namibian Ganoderma species. *African Journal of Biotechnology, 14*(48), 3216–3218.

Ministry of Environment and Tourism, (2018). Sixth National Report to the Convention on Biological Diversity (2014–2018).CBD Sixth National Report - Namibia (English version)

https://www.cbd.int › doc › na-nr-06-en

Misihairabgwi, J. and Cheikhyoussef, A. (2017). Traditional fermented foods and beverages of Namibia. *Journal of Ethnic Foods, 4*(3), 145–153.

Mncwangi, N., Chen, W., Vermaak, I., Viljoen, A.M., and Gericke, N. (2012). Devil's Claw-a review of the ethnobotany, phytochemistry and biological activity of *Harpagophytum procumbens. Journal of Ethnopharmacology, 143*(3), 755–771.

Mowa, E., (2008). *Sustainable Management of* Harpagophytum procumbens *and the Effect of Effective Micro-organisms and Sulphuric Acid on its Seed Germination.* Ph.D Thesis, University of Namibia, Namibia. abstracts/mowa2008abs.pdf.

Mowa, E. and Maas, E. (2016). Influence of resting period on fruits and secondary tubers of *Harpagophytum procumbens* in Namibia. *International Science and Technology Journal of Namibia, 8*, 73–90.

Mudigonda, T. and Mudigonda, P. (2010). Palliative cancer care ethics: principles and challenges in the Indian setting. *Indian Journal of Palliative Care, 16*(3), 107–110.

Namibia Nature Foundation 2017. *Namibia Nature Foundation Annual Report 2015–2016.* Windhoek, Namibia.

Ngwira, P., Mbaiwa, J., and Kolawole, O.D. (2013). Community based natural resource management, tourism and poverty alleviation in Southern Africa: what works and what doesn't work. *Chinese Business Reviews*, David Publishing.

Novaes, M.R.C.G., Valadares, F., Reis, M.C., Gonçalves, D.R., and da Cunha Menezes, M. (2011). The effects of dietary supplementation with Agaricales mushrooms and other medicinal fungi on breast cancer: evidence-based medicine. *Clinics, 66*(12), 2133–2139.

Oyebode, O., Kandala, N.-B., Chilton, P.J., and Lilford, R.J. (2016). Use of traditional medicine in middle-income countries: a WHO-SAGE study. *Health Policy and Planning, 31*(8), 984–991.

Ozioma, E.-O.J. and Chinwe, O.A.N. (2019). Herbal Medicines in African Traditional Medicine. *Herbal Medicine, 10*, 191–214.

Pathak, M.P., Pathak, K., Saikia, R., Gogoi, U., Ahmad, M.Z., Patowary, P., and Das, A. (2022). Immunomodulatory effect of mushrooms and their bioactive compounds in cancer: a comprehensive review. *Biomedicine & Pharmacotherapy, 149*, 112901.

Prasad, S., Rathore, H., Sharma, S., and Yadav, A.S. (2015). Medicinal mushrooms as a source of novel functional food. *International Journal of Food Science, Nutrition and Dietetics, 4*(5), 221–225.

Qi, J., Chen, J.-J., Cheng, Z.-H., Zhou, J.-H., Yu, B.-Y., and Qiu, S.X. (2006). Iridoid glycosides from *Harpagophytum procumbens* D.C. (Devil's claw). *Phytochemistry, 67*(13), 1372–1377.

Quattrocchi, U. (2012). *CRC World Dictionary of Medicinal and Poisonous Plants: Common Names, Scientific Names, Eponyms, Synonyms, and Etymology,*1st edition. CRC Press, Boca Raton, FL.

Raditic, D.M. and Bartges, J.W. (2014). The role of chondroprotectants, nutraceuticals, and nutrition in rehabilitation. In: Millis, D. and Levine, D. (eds.), *Canine Rehabilitation and Physical Therapy* (2nd Edition). St. Louis, W.B. Saunders, pp. 254–276.

Rathore, H., Prasad, S., and Sharma, S. (2017). Mushroom nutraceuticals for improved nutrition and better human health: a review. *PharmaNutrition*, *5*(2), 35–46.

Rautela, I., Arora, H., Binjola, A., and Dheer, P. (2019). Potential and nutrition value of mushroom and its cultivation; an insight review, *International Journal of Engineering Science and Computing*, *9*(5), 22574–22582.

Sharif, D., Mustafa, G., Munir, H., Weaver, C., Jamil, Y., and Shahid, M. (2016). Proximate composition and micronutrient mineral profile of wild ganoderma lucidum and four commercial exotic mushrooms by ICP-OES and LIBS. *Journal of Food and Nutrition Research*, *4*, 703–708.

Shikongo, L.T. (2012). *Analysis of the Mycochemicals Components of the Indigenous Namibian Ganoderma Mushrooms*. Ph.D Thesis, University of Namibia, Namibia.

Shivashankar, M. and Premkumari, B. (2014). Preliminary qualitative phytochemical screening of edible mushroom *Hypsizygus ulmarius*. *Science, Technology and Arts Research Journal*, *3*, 122.

Sirakaya, A. (2019). Balanced options for access and benefit-sharing: stakeholder insights on provider country legislation. *Frontiers in Plant Science*, *10*, 1175.

Smith-Hall, C., Larsen, H.O., and Pouliot, M. (2012). People, plants and health: a conceptual framework for assessing changes in medicinal plant consumption. *Journal of Ethnobiology and Ethnomedicine*, *8*, 43.

Sridhar, K.R. and Deshmukh, S.K.(eds.) (2021). *Advances in Macrofungi: Pharmaceuticals and Cosmeceuticals*. CRC Press, NY.

Stamets, P. (2005). *Mycelium Running: How Mushrooms Can Help Save the World* (Illustrated edition). Ten Speed Press, Berkeley.

Strohbach, M. and Cole, D. (2007). Population dynamics and sustainable harvesting of the medicinal plant *Harpagophytum procumbens* in Namibia: result of the R+D project 800 86 005. Federal Agency for Nature Conservation, 53179 Bonn, Germany. http://www.bfn.de

Ueitele, I.S.E., Kadhila-Muanding, N.P., and Matundu, N. (2014). Evaluating the production of Ganoderma mushroom on corn cobs. *African Journal of Biotechnology*, *13*(22), 2215–2219.

Ugbogu, E.A., Emmanuel, O., Ude, V.C., Ijioma, S.N., Ugbogu, O.C., and Akubugwo, E.I. (2020). Nutritional composition and toxicity profile of *Cantharellus* species (Purple Mushroom) in rats. *Scientific African*, *8*, e00375.

Valverde, M.E., Hernández-Pérez, T., and Paredes-López, O. (2015). Edible mushrooms: improving human health and promoting quality life (online). *International Journal of Microbiology*, *2015*, 376387. Available from https://www.hindawi.com/journals/ijmicro/2015/376387/. Accessed 19 Apr 2018.

Wandati, T., Kenji, G., and Onguso, J. (2013). Phytochemicals in edible wild mushrooms from selected areas in Kenya. *Journal of Food Research*, *2*(3), 137.

Wasser, S.P. (2002). Medicinal mushrooms as a source of antitumor and immunomodulating polysaccharides. *Applied Microbiology and Biotechnology*, *60*(3), 258–274.

WHO, (2020). *WHO Report on Cancer: Setting Priorities, Investing Wisely and Providing Care for all*. World Health Organization, Geneva, Switzerland.

Wilson, J.W. and Primack, R.B. (eds.)(2020). Introduction to sub-saharan Africa. *Conservation Biology in Sub-Saharan Africa*, Open Book Publishers, Cambridge, pp. 23–60.

World Health Organization, (2013). WHO Traditional Medicine Strategy, 2014–2023. WHO Press, World Health Organization, 20 Avenue Appia, 1211 Geneva 27, Switzerland. https://apps.who.int/iris/handle/10665/92455

Zhang, J.-J., Li, Y., Zhou, T., Xu, D.-P., Zhang, P., Li, S., and Li, H.-B. (2016). Bioactivities and health benefits of mushrooms mainly from China. *Molecules*, *21*(7), 938.

Zlatić, N. and Stankovic, M. (2015). Medicinal plants in the treatment of cancer. *Communication*, *5*(3–4), 35–43.

4 Antiviral and Immune-Boosting Potentials of Four Common Edible Flowers

Gloria Aderonke Otunola and Anthony Jide Afolayan

CONTENTS

DOI: 10.1201/9781003206620-4

4.1   INTRODUCTION

Flowers, also referred to as blossoms or blooms, are the reproductive structures found in angiosperms (flowering plants), and their main function is for reproduction. Flowers (Figure 4.1) are characterized by visible, attractive, and functional floral structures, sometimes microscopic, though some grasses and cultivars only have rudimentary or vestigial flowers (Armstrong, 1998).

Flowers are made up of two main parts: (1) Vegetative—the perianth that consists of calyx, corolla, and *perigone* and (2) Reproductive or sexual parts made up of *Androecium* consisting of the male gametophytes and *Gynoecium* carrying the female organs (Figure 4.2).

In humans, the use of flowers for cooking and medicine can be traced as far back to Roman, Chinese, Middle Eastern, Greece, Thailand, Japanese, and Indian cuisine and cultures (Pires et al., 2019). Edible flowers are those that can be consumed safely, eaten as vegetables, herbs, or spices (Table 4.1).

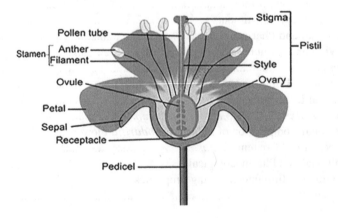

FIGURE 4.1   Flowers.

FIGURE 4.2   Parts of a flower.

TABLE 4.1

Some Edible Flowers

Scientific Name	Flavor	Color	Common Name
Abelmoschus esculentus	Vegetal	Medium-yellow	Okra
Allium schoenoprasum	Onion	Lavender-pink	Chives
Anethum graveolens	Herbal	Yellowish-green	Dill
Bauhinia purpurea	Sour	Purple	Purple bauhinia, Butterfly tree, Orchid tree
Borago officinalis	Anise	Lilac	Starflower
Brassica oleracea	Spicy	Green	Cabbage, etc.
Calendula officinalis	Slightly bitter	Yellow, orange	Marigold
Cichorium intybus	Herbal	Blue	Chicory
Foeniculum vulgare	Mildly anise	Yellow-green	Fennel
Hibiscus rosa-sinensis	Cranberry-like	Rose, red	Chinese hibiscus
Hibiscus sabdariffa	Sour	Red	Rossele, sorrel
Lavandula angustifolia	Sweet, perfumed	Lavender	Lavender, etc.
Matricaria recutita	Sweet apple	White	Chamomile
Mentha piperita	Minty	Purple	Mint, etc.
Musa paradisiaca	Sweet	Cream, Yellow	Banana, Plantain
Ocimum basilicum	Herbal	White, lavender	Basil
Phaseolus vulgaris	Vegetal	Purple	Common bean
Rosmarinus officinalis	Herbal	Blue	Rosemary
Salvia officinalis	Herbal	Purple-blue	Common sage
Syringa vulgaris	Varies	Lavender	Lilac
Tagetes tenuifolia	Spicy, herbal	Yellow	French marigold
Taraxacum officinale	Sweet, honey-like	Yellow	Common dandelion
Thymus vulgaris	Herbal	White	Thyme
Viola odorata	Sweet, perfumed	Purple, white	Common violet
Viola tricolor	Wintergreen	Purple and yellow	Heart's ease, etc.

Source: https://en.wikipedia.org/wiki/List_of_edible_flowers.

4.2 EDIBLE FLOWERS

Many cuisines all over the world use edible flowers in many different menus, to add flavor, aroma, and color to many dishes like sauces, soups, salads, cakes, beverages, and snacks.

Some foods are flowers or derived from flowers. Some examples include broccoli, artichokes, saffron, etc. (www.en.wikipedia.org/wiki/Edibleflower#cite_note-waterfields-2).

Some flowers have also been known to offer health benefits, including antioxidant, anti-inflammatory, antimicrobial, antiviral, anti-diabetes, anti-cancer, anti-obesity, anti-stress, immune boosting, hypotensive, and hypocholesterolemic among others (Figure 4.3).

Increased interest in natural and healthy foods has brought attention toward uncommon, unexplored, and underutilized edible flowers for alternative uses other than as food. The nutritional, nutraceutical, toxicology, bioactive contents, and health benefits of many edible flowers have been reported (Rop et al., 2012; Pires et al., 2019; Pinakin et al., 2020; Takahashi et al., 2020).

The COVID-19 global pandemic has led to the emergence of various treatment strategies involving known antiviral drugs and medications using varied mode of actions for treating affected patients. Various medicinal plants all over the world are also being explored for their antiviral potentials, especially against COVID-19.

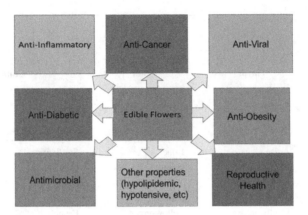

FIGURE 4.3   Health promoting functions of edible flowers.

This review aims to gather information on selected edible flowers having antiviral and immune-boosting properties as well as their nutrient value—bioactive properties along with the beneficial effects they may have on human health. These flowers were selected based on ethnobotanical, folkloric, preclinical (*in vitro* and *in vivo*), and clinical studies as available in the literature.

4.3   *MUSA X PARADISIACA* L.; *MUSA ACUMINATA* COLLA (BANANA INFLORESCENCE/FLOWER/BLOSSOM/HEART)

4.3.1   BOTANICAL DESCRIPTION

Family-Musaceae

Scientific name-*Musa x paradisiaca* L.; *Musa acuminata* Colla; English name-Banana

Banana flower or blossom grows on the end of the stem holding a cluster of bananas. It has a deep crimson or purple-skinned colored bracts and consists of tightly packed leaves or bracts that wrap around rows of thin-stemmed male flowers. It is the inner pale-colored bracts that are used for food.

Banana flowers (Figure 4.4) are arranged in clusters (called hands) on the peduncle, with female flowers located near the base and male ones at the distal end of the peduncle. Each hand of flower is enfolded by a bract that lifts at anthesis (period at which a flower is fully open and functional). Female banana flowers have a massive style, stigma, and stamen, which are usually reduced to staminodes that do not produce pollen, with male organs sometimes absent.

FIGURE 4.4   Banana Flower-Blossom-Heart.

Together, the ovary, style, and stigma make up the pistil or carpel. In a banana flower, three pistils fuse to produce a tri-pistillate ovary, style, and stigma (tip of the carpel, which receives pollen and on which the pollen grain germinates). Each flower is connected to a cushion of tissue on the peduncle by the pedicel. The female flowers reach anthesis before the male flowers.

In botanical references, cultivated banana is often listed as *Musa × paradisiaca* (Musaceae), although it is actually a complex hybrid derived from two diploid Asian species: *Musa acuminata* and *Musa balbisiana*. *M. acuminata* is a species native to the Malay Peninsula and adjacent regions and is thought to have given rise in total or in part to all edible banana varieties, while *M. balbisiana* is found from India eastwards to the tropical Pacific.

Hybridization probably occurred as *Musa acuminata* plants ($2n$ genome=AA) were increasingly cultivated over the distributional range of *Musa balbisiana* ($2n$ genome=BB). Although the *M. acuminata* cultivars were sterile because of being seedless, they did produce fertile pollen. Natural hybridization between *M. acuminata* (contributing to genome A) and *M. balbisiana* (contributing to genome B) led to the plantain and banana cultivars. All plantains and almost all-important bananas are triploid ($2n=3x=33$ chromosomes) and are monocotyledonous plants, belonging to the section Eumusa within the genus *Musa* of the family Musaceae, order Scitamineae.

Banana is the fourth largest crop in the world, widely grown in Latin America and the Caribbean, Africa, Asia, Europe, and Oceania (Arias, 2003; Salvador, 2012). Varieties of the genus *Musa* abound all over world, and up to 1,200 varieties of bananas have been identified, with India as the largest producer (21% of global production in 2000) of banana.

4.3.2 CULINARY USES

The leaf, pseudostem, flowers, fruit, or peel of the banana plant are useful nutritionally and medicinally traditionally in many parts of the world (Kumar et al., 2012).

Flowers from the banana (Musaecae) plant, also referred to as blossoms or heart are often consumed as vegetable in many Asian countries in curry, boiled, or deep fried, in salads, with rice and wheat bread (Wickramarachchi and Ranamukhaarachch, 2005; Liyanage et al., 2016). These flowers also possess very high nutritional value, are rich sources of roughage (dietary fiber), potassium, vitamins A, C, and E, minerals, fatty acids, myoinositol phosphates, and alpha-tocopherol. Table 4.2 shows some of the nutrient content per 100 g banana flower.

TABLE 4.2
Nutritional Composition of Banana Flower

Macronutrients	100 g Banana Flower
Protein (g)	1.6
Carbohydrates (g)	9.9
Fat (g)	0.6
Fiber (g)	5.7
Micronutrients	
Calcium (mg)	56.00
Phosphorous (mg)	73.3
Iron (mg)	56.4
Copper (mg)	13.00
Potassium (mg)	553.00
Magnesium (mg)	48.7
Vitamin E (mg)	1.07
Energy (kcal)	51.00

Source: Sheng et al. (2010).

4.3.3 Phytochemical Constituents

With regard to pytochemical constituents, Musaceae blossoms are valuable sources of flavonoids, saponin, essential, and non-essential amino acid, tannins, glycoside, and steroid. Other studies have reported that *Musa paradisiaca* flower contains tannins, saponins, reducing, and non-reducing sugars, sterols, saponins, and triterpenes; as well as hemiterpenoid glucoside (1,1-dimethylallyl alcohol), syringin, and benzyl alcohol glucoside (Martin et al., 2000; Ahmad et al., 2019). In addition, a tetracyclic triterpene ((24R)-4α-14α, 24-trimethyl-5-cholesta-8, 25 (27)-dien-3β-ol) has been isolated from the flowers, 3-rutinoside derivatives—delphinidin, pelargonidin, peonidin, and malvidin have been reported (Ketiku, 1973; Jang et al., 2002; Sharma et al., 2019). Umbelliferone and Lupeol, β- sitosterol, flavonoids, saponin, and other phenolic compounds such as catechin and isoquercetin, as well as arylpropanoid sucroses, triterpenes, benzofurans, stilbenes, and iridoids, are reported as the major bioactive components of banana flower (Swe, 2012; Ramu et al., 2017; Sandjo et al., 2019; Chiang et al., 2021; Amornlerdpison et al., 2021).

4.3.4 Traditional and Pharmacological Uses of Banana Flowers

Traditionally, banana flowers are used for treating bronchitis, diabetes, dysentery, and on ulcers, the sap for hemorrhoids, insect, and other stings, and bites, hysteria, epilepsy, leprosy, fevers, acute dysentery, and diarrhea. Young leaves are placed as poultices on burns and other skin afflictions; the astringent ashes of the unripe peel and of the leaves are taken for dysentery and diarrhea and are used for treating malignant ulcers; the roots are administered in digestive disorders, dysentery, and other ailments; banana seed mucilage is given in cases of diarrhea, and the peel and pulp of fully ripe bananas have antimicrobial activities (Imam and Akter, 2011; Kumar et al., 2012; Das et al., 2020; Mostafa, 2021).

The whole banana flower has been studied for its health benefits, which include food supplement for diabetics, anti-hyperglycemic, anti-hyperlipidemic, and antioxidant potentials, in addition to lowering cholesterol, anti-inflammatory, anti-cancer, and anti-aging (Bhaskar et al., 2012). Some other uses include the treatment of bronchitis, ulcers, hemorrhages, and hemorrhoids (Kumar et al., 2012; Mathew et al., 2017; Vu et al., 2018; Jouneghani et al., 2020). In addition to the afore-mentioned traditional uses, banana flowers are valued for their potential to treat liver, menstrual and uterine problems, pain, snakebite, wound-healing activities, and inhibition of HIV replication (Swanson et al., 2010; Panda et al., 2012; Pereira et al., 2015; Singh et al., 2016; Panda et al., 2020; Jouneghani et al., 2020).

4.3.5 Antiviral and Immune-Boosting Properties

The antiviral and immune-boosting properties of banana flowers have been collated and highlighted in this section.

Efficacy of the butanolic extract of *M. acuminata* Colla flower as antiviral for Acyclovir-resistant simple human Herpesvirus types 1 and 2, by inhibiting viral replication, has been reported by Martins et al., (2009). Extracts of flowers from Ambon and Klutuk bananas inhibited Tobamovirus at 86.34% and 91.22%, respectively (Nurviani et al., 2018). Sharma et al. (2019) reported the isolation of BanLec—a jacalin-related lectin, which has the capacity to bind with high mannose carbohydrate structures including, those found on viruses; an indication that BanLec could serve as an antivirucidal agent in the prevention of the sexual transmission of HIV-1.

The antiviral properties of banana flower could be attributed to Umbelliferone-a coumarin and Lupeol (Figure 4.5), which are the major bioactive constituents. Coumarins are therapeutic agents, occurring naturally as secondary metabolites in plants, bacteria, fungi, essential oils, and can also be chemically synthesized (Venugopala et al., 2013; Stefanach et al., 2018). UMB is a 7-hydroxycoumarin that is pharmacologically active, which because of its simple structure is generally accepted as the parent compound for the more complex coumarins, and is widely used as a synthon for a wider variety of coumarin heterocycles (Mazimba, 2017).

Current research has shown that coumarins are very effective and potent anti-viral agents against human immunodeficiency (HIV) virus, Enterovirus 71 (EV71), chikungunya, dengue, Ebola,

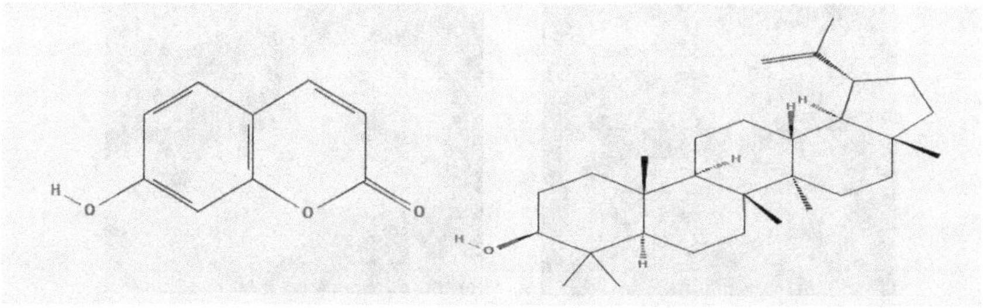

FIGURE 4.5   (left) Structures of Umbilliferone. (right) Structure of Lupeol.

coxsackievirus A16 (CVA16), hepatitis C virus (HCV), influenza, and many other viruses (Hassan et al., 2016; Wang et al., 2017; Gómez-calderón et al., 2017; Liu et al., 2019; Mishra et al., 2020).

Inhibition of proteins that are essential for viral penetration, replication, and infection, cellular pathway regulation including NF-κB (nuclear factor kappa-light-chain-enhancer of activated B cells), and anti-oxidative pathway including NrF-2 (The nuclear factor erythroid 2 (NFE2)-related factor 2) are some of the mechanisms for antiviral action (Liu et al., 2019; Mishra et al., 2020).

Lupeol—a pentacyclic triterpenoid is also pharmacologically active with anti-inflammatory, immune-modulatory, anti-cancer, and antiviral activities (Sharma et al., 2020).

Derivatives of lupane triterpenoids (lupeol) inhibit the HIV-1 RT-associated RNase H function (Flekhter et al., 2004). In another study, antiviral activity against the human immunodeficiency virus (HIV), herpes simplex virus (HSV), and Epstein–Barr virus (EBV) were reported for lupeol, betulin, and betulinic acid. Weak antiviral activities against influenza A, HSV, and high anti-HSV activity were also reported *in vitro* (Flekhter et al., 2004; Parvez et al., 2018). Also, Esposito et al. (2018) showed that Lupeol showed a very potent inhibition against HIV-1 IN, with an IC_{50} value of 17.7 µM (Chaniad et al., 2019).

A study by Mutai et al. (2012) showed that 50 µg/ml of Lupeol and 3-(E)-*trans* coumaroylbetulin reduced intracellular replication of the HSV-1 virus and disrupted completely (100%) the cell monolayer, as well as against HIV integrase (Chaniad et al., 2019). Computational docking also indicates that active compounds interacted with IN active sites-Asp64 and residues participating in 3-processing and Thr66, His67, and Lys159, strand-transfer reactions of the integration process.

With regards to immune-boosting activities, lupeol was shown to restore the protective immune responses and remarkable reduction of parasites against visceral leishmaniasis in a murine model (Kaur et al., 2019).

4.3.6 ASSESSMENT

There has been a long history of the use of banana blossoms for food, especially in Asian countries. The nutritional contents are quite high, and it contains high and diverse phytochemicals, especially lupeol and umbilliferone, credited with therapeutic properties. Traditionally, banana flower is used for anti-diabetic, anti-hypertensive, anti-snake, anti-obesity among others.

However, there has been a paucity regarding clinical and human studies. This should be a matter of urgent focus, especially with regard to its potential as a functional food.

4.4 *CALENDULA OFFICINALIS* LINN.(ASTERACEAE)

4.4.1 BOTANICAL DESCRIPTION

Kingdom: Plantae; Division: Magnoliophyta; Class: Magnoliopsida; Order: Asterales; Family: Asteraceae; Genus: *Calendula*; Species: *C. officinalis*.

Calendula also called pot marigold (African marigold) is used medicinally in several places all over the world and belongs to the Asteraceae (Sunflower or Aster) family.

FIGURE 4.6 *Calendula officinalis* Linn. Asteraceae.

Calendula officinalis (Figure 4.6) is an aromatic perennial herb with leaves lance-oblong, which can grow in height to 80 cm on branched erect stems. The disc florets are hairy on both the sides, tubular, hermaphrodite, 5–17 cm long, with margins entirely or occasionally waved or weakly toothed. The blossoms are yellow, with comprising a 4–7 cm thick head surrounded by two rows of hairy bracts. Under favorable conditions, the flowers may appear all year long with thorny curved achene fruit (Ashwlayan et al., 2018).

4.4.2 CULINARY USES

Calendula flower petals also called 'poor man's Saffron'—contains several nutrients and phytochemicals including carbohydrates, amino acids, lipids, fatty acids, carotenoids, terpenoids, flavonoids, quinones, coumarins, and other constituents (Jan et al., 2017). Its flowers have appetizing color and mild flavor, can be eaten fresh or dried, and have been used since the days of the Roman Empire to garnish and flavor food, added to soups, stews, salads, omelets, cakes, cookies, color cheese, and to replace the saffron color and flavor (Franzen et al., 2019).

4.4.3 NUTRITIONAL COMPOSITION

Traditionally, edible flowers have been used for human consumption in various cultures to the appearance, taste, and esthetics of food. Presently, however, consumers demand foods with beneficial health properties, plus the nutrients in them, for functional qualities like immune-boosting, antiviral, hypoglycemic, and hypolipidemic properties—the nutrients needed by the human body in health and disease states and the chemical/nutrient content of foods with nutritional potential, low calories, and affordability are of on-going research foci. *C. officinalis* has recently assumed the status of an edible flower, although it has been in used as medicine traditionally. Therefore, information on its nutrient composition (Table 4.3) is very important.

TABLE 4.3
Nutrient Composition of *Calendula officinalis*

Nutrients	(%)
Moisture	89.34a ± 0.100
Dry matter	10.66c ± 0.100
Ash	0.93b ± 0.005
Ether extract	1.32a ± 0.015
Protein	1.20c ± 0.014
Raw fiber	1.59c ± 0.105
Carbohydrate	5.62
Caloric value	45.52

Source: Adapted from Franzen et al. (2019).

TABLE 4.4

Phytochemical Constituents of *Calendula officinalis* Flower, Leaf, and Root

Plant Part	Phytochemical/ Bioactive Groups	Specific Bioactives
Flowers	Terpenoids	Lupeol, Ψ-taraxasteol, Erythrodiol, Calenduloside, Calendulaglycoside A, Calendulaglycoside B, Cornulacic acid acetate
	Coumarins	Esculetin, scopoletin, umbelliferone
	Flavonoids	Isoquercitrin, rutin, calendoflavoside, Quercetin, Isorhamnetin, Isorhamnetin-3-O-β-D glycoside, Narcissin
	Volatile oils	Cubenol, α-cadinol, oplopanone, methyllnoleate (73) Sabinene, limonene, α-pinene, p-cymene, non-anal, carvacrol, geraniol, nerolidol, T-muurolol, palustron
Leaves	Quinones	Phylloquinone, α-tocopherol, ubiquinone, plastoquinone
Roots	Terpenoids	Calenduloside B
Other constituents	Other phytochemicals	Loliolide (calendin), calendulin, n-paraffin, glucosides, calenduloside, flavonoids, volatile dyes, triterpenes, glycosides

4.4.4 PHYTOCHEMICAL CONSTITUENTS

The presence of different chemical compounds, including carbohydrates, amino acids, lipids, carotenoids, terpenoids, flavonoids, volatile oil, quinines, coumarins, and many other constituents have been reported in the different parts of *C. officinalis*, especially the flowers (Table 4.4).

4.4.5 TRADITIONAL AND PHARMACOLOGICAL USES

Traditionally, *Calendula* is used for wound healing, jaundice, blood purification, and as an antispasmodic. The pharmacological properties of *C. officinalis* (Figure 4.7) are many, diverse, and include diuretic, hemostatic, immune stimulant, anti-inflammatory, anti-pyretic, antiseptic, antispasmodic, antiviral, astringent, bitter, cardiotonic, carminative, dermagenic, and as a vasodilator.

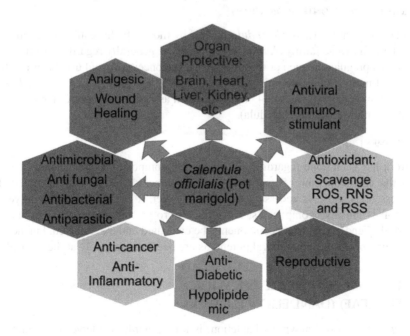

FIGURE 4.7 Pharmacological effects of *Calendula officinalis* Linn.

Other actions include antioxidant, wound healing, genotoxic, and antigenotoxic, anti-diabetic, hypo-glycemic, anticancerous, gastroprotective, and antiviral activities, and immune boosting among others and are not toxic (Baskaran, 2015; Jan et al., 2017).

4.4.6 ANTIVIRAL AND IMMUNE-BOOSTING PROPERTIES

Calendula flower has the ability to fight infections because of its antimicrobial properties, hence it is accredited to strengthen the immune system. The tea is often used for relief from coughs or nasal congestion. *Calendula* flower polysaccharide fractions with molecular weight in the range of 25,000–500,000 were reported to show significant immune-stimulating activity in granulocytes; other isolated polysaccharides stimulated phagocytosis of human granulocytes, but there is no direct mitogenic effect on human lymphocytes and thymocytes (European Medicines Agency—Committee on Herbal Medicinal Products—HMPC 2018).

Tincture of *C. officinalis* flower heads prevented *in vitro* replication of influenza A2 and influenza APR-8 and herpes simplex viruses, though aqueous extract was not active (Baskaran 2017). In addi-tion, the replication of HIV-1 in acutely infected lymphocytic MOLT-4 cells *in vitro* was prevented by chloroform extract of the flowers. Also, aqueous extract of *C. officinalis* flowers inhibited the replica-tion of encephalitis virus, and a chloroform extract inhibited HIV-1 reverse transcriptase activity in a dose-dependent intraperitoneal administration (Meenatchisundaram et al., 2009; Baskaran, 2017).

A dichloromethane–methanol (1:1) extract of *C. officinalis* flowers has potent anti-HIV activity *in vitro*, and the activity was attributed to the inhibition of HIV1-RT and suppression of the HIV-mediated fusion (Kalvatchev et al., 1997; Muley et al., 2009). The antiviral activity of the flower tincture was through suppression of the replication of influenza APR-8, influenza A2, and herpes simplex viruses (Arora et al., 2013).

According to the European Medicines Agency (Committee on Herbal Medicinal Products—HMPC 2018), a tincture of *Calendula* flower suppressed the replication of herpes simplex, influenza A2, and influenza APR-8 viruses *in vitro*, but the aqueous extract was not active; also, chloroform extracts inhibited replication of HIV Type I in acutely infected lymphocytic Molt-4 cells *in vitro* and HIV-I reverse transcriptase activity in a dose-dependent manner.

4.4.7 CAUTIONS AND CONTRAINDICATIONS

Care should be taken in the use of *Calendula* flowers by those who have allergic reactions to other members of the Asteraceae family. Although *Calendula* is generally regarded as on safe (GRAS) list and non-toxic, pregnant women or those who want to conceive should use it cautiously because of its ability to stimulate menstruation. Muley et al. (2009) reported that *calendula* flower extract caused allergy in nine patients out of 443 (2.03%), when assessed by patch testing method (https://www.herbrally.com/monographs/calendula).

4.4.8 ASSESSMENT

Not much work has been done regarding the nutritional composition of calendula flowers. Those reports show that calendula flower is nutritious and can be included in the daily human diet because they contain essential nutrients necessary for a healthy diet. It is, therefore, imperative that more investigation of its nutritive components, including macro and micronutrients, vitamins, amino/fatty acids profiles, and anti-nutrient components, be evaluated and documented. Further investiga-tion of its immune-boosting and antiviral potentials are needed, especially at the clinical levels.

4.5 TARAXACUM OFFICINALE F.H. WIGG.
(ASTERACEAE) (DANDELION FLOWER)

Taraxacum officinale, also known as dandelion, is a commonly available food plant that has a long history of human use, which is listed on the US Food and Drug Administration's 'generally recognized as safe' (GRAS) list for foods and supplements (Jędrejek, 2017; Fatima et al., 2018). Dandelion flowers are used to make salads, dandelion tea, wine, jam, syrup, or fritters.

Common dandelion has many pharmacological properties, although most reference to its use is in relation to its cultivation for rubber production (Venkatachalam et al., 2013).

4.5.1 BOTANICAL DESCRIPTION

Kingdom: Plantae; Subkingdom: Viridiplantae; Division: Tracheophyta; Subdivision: Spermatophytina; Class: Magnoliopsida; Order: Asterales; Family: Asteraceae; Genus: *Taraxacum F.H. Wigg.* Species: *T. officinale* F.H. Wigg. Common name: Dandelion.

Dandelion belongs to the genus *Taraxacum* of the Asteraceae family (Figure 4.8). It occurs in the tropics, cool highlands, and warm, sub-temperate, and temperate zones around the northern hemisphere, and can tolerate drought and frost. It is grown for food and medicine.

4.5.2 CULINARY USES

All plant parts of dandelion are edible, and its leaves, roots, and flowers are incorporated into different food products. Flowers are used to produce wines and desserts, and flavor components in food products such as dairy desserts, baked goods, gelatins, candy, puddings, and cheese (González-Castejón et al., 2012).

4.5.3 NUTRITIONAL COMPOSITION

There is a dearth in the literature regarding the nutrient content of dandelion flower. Most studies were on the phytochemical, antioxidant and biological potentials of the leaves, aerial parts, and roots, or in combination with other flowers (Lee et al., 2004; Lee and Hee-Kyung, 2015; Olas, 2022), not specifically on the flowers. Table 4.5 shows the nutritional content of dandelion flower.

4.5.4 PHYTOCHEMICAL CONSTITUENTS

Dandelion flower is a well-known phytomedicine with many diverse therapeutic uses. The most prominent bioactives in dandelion are sesquiterpene lactones, free sterols, triterpenoids, saponins, coumarins, polysaccharides, pectin, and resin (Fatima et al., 2018). Other constituents are luteolin-7-glucoside and two luteolin-7-diglucosides, hydroxycinnamic, chicoric, monocaffeyl tartaric, and

FIGURE 4.8 *Taraxacum officinale* L. Asteraceae.

TABLE 4.5

Nutritional Composition of *Taraxacum officinale* Flower

Macronutrients	%
Moisture	6.35 ± 0.34
Protein	14.9 ± 0.08
Fat	5.30 ± 0.34
Fiber	13.9 ± 0.8
Ash	9.8 ± 0.2
Carbohydrate	49.8 ± 1.1
Micronutrients	mg%
Calcium	120.1 ± 0.7
Phosphorus	445.3 ± 12.3
Iron	59.3 ± 3.3
Sodium	218.4 ± 2.2
Potassium	3016.6 ± 175.4
Magnesium	191.9 ± 3.6
Zinc	2.73 ± 0.21
Vitamins	mg%
Vitamin B_1 (mg%)	0.58 ± 0.13
Vitamin B_2 (mg%)	2.70 ± 0.15
Vitamin C (mg%)	134.1 ± 16.1
Niacin (mg%)	8.55 ± 0.80
Vitamin A (RE)	1935 ± 60.7

chlorogenic acids as well as vitamins C, D, and the B-complex, iron, silicon, and other major micronutrients (Fatima et al., 2018)

4.5.5 TRADITIONAL AND PHARMACOLOGICAL USES

An array of folkloric and traditional medicinal uses has been accredited to dandelion plant and its various parts. Dandelion flowers are reputed to have anodyne, cardiotonic, emollient, hepatic, and vulnerary activities, as well as the capacity to exhibit ameliorative effects on dyspepsia, bile duct disorders, spleen, and liver complaints, inflammatory diseases, and sedative properties.

The roots possess anti-inflammatory, antibacterial, anti-fungal, hepatic, hypnotic, purgative, and dyspepsia, bile duct disorders, spleen, liver complaints, and inflammatory diseases, while the leaves are credited with antacid, antioxidant, hypotensive, and restorative properties.

4.5.6 ANTIVIRAL AND IMMUNE-BOOSTING PROPERTIES

Immunomodulatory activities have been reported for polysaccharides from dandelion. It was found to restore suppressed immune functions through a dose-dependent pattern, in short-chain aldehyde reductase scalded mice (Luo et al., 1993; Venkatachalam et al., 2013).

A study (Shekarabi et al., 2021) reported that dietary supplementation with dandelion flower extract gave immune stimulation and resistance against *Streptococcus iniae* infection in rainbow trout (*Oncorhynchus mykiss*). Rainbow trout fingerlings were randomly assigned to 15 tanks and fed with 0 (control), 1–4 g/kg dandelion flower extract for 56 days. At the end of the trial, total leukocyte, lymphocyte counts, immunoglobulin M, total protein, and lysozyme were significantly ($P < 0.05$) enhanced in trout-fed DFE-added dandelion flower extract (3g/kg) fed groups compared to control. In addition, enzyme activities of skin mucus and protein (2 and 3g/kg), transcription levels of *interleukin-1β* and *interleukin-6* genes (3 and 4g/kg) were up-regulated in fish fed with dandelion flower extract. Finally, Interleukin-8 and lysozyme gene expression were elevated at 3g/kg, while mortality after *Streptococcus iniae* inoculation was significantly reduced.

Dandelion whole-plant extract was also shown to have potent inhibitory activity against HIV-1 replication and RT activity of the Human immunodeficiency virus type 1 (HIV-1) (Han et al., 2011). The leaf extract was also reported to prevent hepatitis C viral reproduction without damage to the human cell, as demonstrated through *in vitro* experiment (Han et al., 2011; Rehman et al., 2016).

Recently, aqueous extracts of dandelion were shown to block protein–protein interaction of spike S1 to the human ACE2 cell surface receptor for the original spike D614 and its mutant forms in human HEK293-hACE2 kidney and A549-hACE2-TMPRSS2 lung cell lines, as well as pseudo-typed lentivirus in SARS-COV-2 (Tran et al., 2021).

4.5.7 ASSESSMENT

This review revealed that dandelion is a well-known vegetable and medicinal plant. Most of the studies recorded were on the leaf and root, but not much has been done on the flowers. In addition, most of these studies were not clinical. It is, therefore, imperative that future studies should focus on the nutritional, pharmacological, and toxicity assessment of dandelion flower using animal and human models to validate its efficacy.

4.6 *HIBISCUS SABDARIFFA* LINNAEUS (MALVACEAE) (ROSELLE)

Hibiscus sabdariffa (roselle/sorrel) is a *Hibiscus* species and belongs to the family Malvaceae. The genus *Hibiscus* consists of more than 300 species of annual or perennial herbs, shrubs, or trees, but the most-studied species is *H. sabdariffa*, which is cultivated worldwide (López-Romero et al., 2018; Peredo Pozos et al., 2020).

4.6.1 BOTANICAL DESCRIPTION

Kingdom: Plantae; Phylum: Spermatophyta; Subphylum: Angiospermae; Class: Dicotyledonae; Order: Malvales; Family: Malvaceae (Mallow family); Genus: *Hibiscus*; Species: *Hibiscus sabdariffa* L. Common name: roselle.

H. sabdariffa (Figure 4.9) is an annual woody-based shrub that can grow 2–2.5 m in height. Leaves are 3–5 lobed, 8–15 cm long, and alternate on the stems. Flowers are white to pale yellow, with a dark red spot at the base of each petal, stout fleshy calyx, fleshy, and bright red as the fruit matures. Flowers are solitary, axial, nearly sessile, 5–7 cm in diameter, consisting of epicalyx—segments 8–12, distinct, lanceolate to linear, adnate at base of the calyx; calyx is thick, red, and fleshy, cup like, deeply parted, prominently 10 nerved; petals 5, yellow, twice as long as calyx. Stamens are numerous; the filaments are united into a staminal column; style is single, 5-branched near summit; stigma is capitate (Ross, 2003).

4.6.2 CULINARY USES

Flowers are used in food, beverage, and non-food medicinal applications in many countries of the world, as a source of nutrients and natural pigments to color foods (Pinela et al., 2019). The most used parts for food are the flowers and calyx, though other parts including seeds can be eaten raw or cooked to provide dietary fibers (Salem et al., 2021).

FIGURE 4.9   *Hibiscus sabdariffa* L. roselle/sorrel.

4.6.3   Nutritional Composition of *Hibiscus sabdariffa* Flower

H. sabdariffa flower is well known traditionally for its various pharmacological and nutritional properties. Currently, it is gaining prominence as a functional beverage, and several studies on its nutritional composition have been done and reported.

H. sabdariffa flower is reported to have high protein, carbohydrate, ash, carotene, and iron; other reports show high micronutrients and vitamins contents of the flower (Atta et al., 2013; Anel et al., 2016; Singh et al., 2017; Balarebe et al., 2019). An adapted proximate and micronutrient contents of *H. sabdariffa* flower are shown in Table 4.6.

TABLE 4.6
Nutritional Composition of *Hibiscus sabdariffa* Flower

Macronutrients	%
Moisture	10.50
Ash	11.67
Crude Fat	1.00
Crude Fiber	1.17
Protein	4.10
Carbohydrates	68.75
Micronutrients	mg/100g
Calcium	1.20
Magnesium	1.57
Phosphorus	5.48
Iron	833.00

(Continued)

TABLE 4.6 (*Continued*)
Nutritional Composition of *Hibiscus sabdariffa* Flower

Macronutrients	%
Potassium	160.50
Manganese	1.00
Sodium	15.33
Zinc	1.17
Copper	0.70
Vitamin C	53
Beta Carotene	285.29
Thiamin	0.05
Riboflavin	0.95
Niacin	0.06

Source: Atta et al. (2013); Anel et al. (2016); Singh et al. (2017); Balarebe et al. (2019).

4.6.4 PHYTOCHEMICAL CONTENT

The flower sepals of *H. sabdariffa* are the richest in bioactive compounds. The phytochemical constituents of roselle are diverse and include micronutrients, e.g., Ca, Fe, and Mg; soluble and insoluble dietary fibers; vitamins (e.g., ascorbic acid), anthocyanins (e.g., delphinidin-3-O-sambubioside and cyanidin-3-O-sambubioside); phenolics (e.g., chlorogenic acid and protocatechuic acid); flavonoids (e.g., quercetin, kaempferol, luteolin, and apigenin) and organic acids (e.g., citric acid and hibiscus acid) (Ali et al., 2005; Mojiminiyi et al., 2007; Salem et al., 2021).

4.6.5 TRADITIONAL AND PHARMACOLOGICAL USES

H. sabdariffa is credited with diverse pharmacological uses, including as an anti-hypertensive, anti-inflammatory, anti-pyretic, anti-nociceptive, anti-diabetic, anti-cancer, antispasmodic, anti-fungal, anti-parasitic, antimicrobial, anti-hypercholesterolemia, hepatic, nephroprotective, anti-obesity, and hypolipidemic properties (Akanbi et al., 2009; Da-Costa-Rocha et al., 2014; Zhang et al., 2014; Jabeur et al., 2017; Torky and Hossain, 2017; Salem et al., 2021).

4.6.6 ANTIVIRAL AND IMMUNE-BOOSTING PROPERTIES

Several studies have reported the *in vitro* and *in vivo* antiviral effects of extracts of *H. sabdariffa* flowers and calyx. Takeda et al. (2020) reported that the pH of hibiscus tea extract is acidic, which resulted in a rapid and potent antiviral activity driven largely by the acidic pH. However, although this was not effective *in vivo* in mice, hibiscus tea extract and protocatechuic acid—one of the major components of the extract—showed both potent acid-dependent antiviral activity and weak low-pH-independent activity suitable for medication and vaccination, as the activity was not affected by the neutral blood environment, and there was no loss to antigenicity of hemagglutinin. Another study (Hassan et al., 2017) revealed that aqueous extract of roselle has the capacity to inhibit replication of the HSV and thus can be used as a potential antiviral agent against HSV (Bhattacharya and Sharma, 2018). D'Souza et al. (2016) showed that aqueous extracts of *H. sabdariffa* successfully reduced Aichi virus (AiV) through the alteration of virus structure.

Hibiscus extracts were shown to have potent action against Herpes Simplex Virus-type 1 (HSV-1) attributed to the multiple bioactive components in the extract that reacted synergetically and effectively as virucidal and prophylactic agents (Torky and Hossain, 2017). Baatartsogt et al. (2016) reported that of 11 herbal tea extracts tested against H5N1 avian virus, only hibiscus tea showed *in vitro* high antiviral effects on the H5 subtypes of low and highly pathogenic avian influenza viruses H5N1 HPAIV.

4.6.7 ASSESSMENT

H. sabdariffa flower is a well-known and widely researched edible and therapeutic flower. It is mostly used as a tea tonic for the management of many health conditions, including hypertension, diabetes, obesity, and immune-boosting properties. Although used and consumed globally, hibiscus flower is also limited by lack of human clinical studies. This flower has a great potential for the formulation of functional food and beverage production, and source of natural, safe food phytomedicines, colorant, and dye.

4.7 CONCLUSIONS AND PROSPECTS

Dietary phytochemicals and bioactive agents have become a very relevant study area of nutrition and health. The focus is mainly directed at their therapeutic uses and use in the development of functional foods. Increasing scientific evidence suggests that edible flowers and their constituents are arsenals of bioactives and therapeutics for diverse ailments, and could be explored as leads for drug development, especially for their immune-boosting and antiviral properties. The recent COVID-19 pandemic, which caught the global world unprepared and unawares, calls for an urgent evaluation and constant readiness of the immune status of humans to combat such viral challenges. The edible flowers reviewed here, banana, *Calendula*, dandelion, and hibiscus flowers, although having great potentials for antiviral and immune modulation, are underutilized and need further investigations regarding nutritional, toxicological, and pharmacological potentials, especially human clinical studies to validate and establish these effects. In addition, the cultivation of these edible flowers and their recovery from waste during the processing of other main parts of the plants should be encouraged.

CONFLICT OF INTEREST

No conflicts declared.

ACKNOWLEDGEMENTS

This study was supported by the South Africa National Research Foundation, Research and Innovation Support and Advancement (NRF: RISA) CSURG: Grant No: 121264.

REFERENCES

Chaniad, P., Sudsai, T., Septama, A.W., Chukaew, A., and Tewtrakul, S. (2019). Evaluation of anti-HIV-1 integrase and anti-inflammatory activities of compounds from Betula alnoides buch-ham. *Advances in Pharmacological Sciences*, *2019*, 2573965.

Ahmad, B.A., Zakariyya, U.A., Abubakar, M., Sani, M.M., and Ahmad, M.A. (2019). Pharmacological activities of banana. *Banana Nutrition-Function and Processing Kinetics*. IntechOpen, London, UK.

Akanbi, W.B., Olaniyan, A.B., Togun, A.O., Ilupeju, A.E.O., and Olaniran, O.A. (2009). The effect of organic and inorganic fertilizer on growth, calyx yield and quality of Roselle (*Hibiscus sabdariffa* L.). *American-Eurasian Journal of Sustainable Agriculture*, *3*(4), 652–657.

Ali, B.H., Wabel, N.A., and Blunden, G. (2005). Phytochemical, pharmacological and toxicological aspects of *Hibiscus sabdariffa* L.: a review. *Phytotherapy Research: An International Journal Devoted to Pharmacological and Toxicological Evaluation of Natural Product Derivatives*, *19*(5), 369–375.

Amornlerdpison, D., Choommongkol, V., Narkprasom, K., and Yimyam, S. (2020). Bioactive compounds and antioxidant properties of banana inflorescence in a beverage for maternal breastfeeding. *Applied Sciences*, *11*(1), 343.

Anel, T.C., Thokchom, R., Subapriya, M.S., Thokchom, J., and Singh, S.S. (2016). *Hibiscus sabdariffa*-a natural micro nutrient source. *International Journal Advanced Research in Biological Science*, *3*(4), 243–248.

Arias, P. (2003). *The World Banana Economy, 1985–2002* (No. 1). Food and Agriculture Organization of the United Nations, Rome.

Ariffin, M.M., Khong, H.Y., Nyokat, N., Liew, G.M., Hamzah, A.S., and Boonpisuttinant, K. (2021). *In vitro* antibacterial, antioxidant, and cytotoxicity evaluations of musa paradisiaca cv. Sekaki florets from sarawak, Malaysia. *Journal of Applied Pharmaceutical Science*, *11*(5), 091â-099.DOI: 10.7324/JAPS.2021.110513.

Armstrong, W.P. (1998). The truth about cauliflory. *Zoonooz*, *71*, 20–23.

Arora, D., Rani, A., and Sharma, A. (2013). A review on phytochemistry and ethnopharmacological aspects of genus *Calendula*. *Pharmacognosy Reviews*, *7*(14), 179.

Arya Krishnan, S. and Sinija, V.R. (2016). Proximate composition and antioxidant activity of banana blossom of two cultivars in india. *International Journal of Agriculture and Food Science Technology*, *7*, 13–22.

Atta, S., Sarr, B., Diallo, A.B., Bakasso, Y., Lona, I., and Saadou, M. (2013). Nutrients composition of calyces and seeds of three Roselle (*Hibiscus sabdariffa* L.) ecotypes from Niger. *African Journal of Biotechnology*, *12*(26), 4174–4178.

Baatartsogt, T., Bui, V.N., Trinh, D.Q., Yamaguchi, E., Gronsang, D., Thampaisarn, R., Ogawa, H., and Imai, K. (2016). High antiviral effects of hibiscus tea extract on the H5 subtypes of low and highly pathogenic avian influenza viruses. *Journal of Veterinary Medical Science*, *78*(9), 16–0124.

Balarabe, M.A. (2019). Nutritional analysis of *Hibiscus sabdariffa* L.(Roselle) leaves and calyces. *Plant Journal*, *7*(4), 62–65.

Baskaran, K. (2017). Pharmacological activities of *Calendula officinalis*. *International Journal of Science and Research*, *6*(5), 43–47.

Bhaskar, J.J., Chilkunda, N.D., and Salimath, P.V. (2012). Banana (*Musa* sp. var. elakki bale) flower and pseudostem: dietary fiber and associated antioxidant capacity. *Journal of Agricultural and Food Chemistry*, *60*(1), 427–432.

Bhattacharya, B. and Sharma, C. (2018). Potential bioactive properties of *Hibiscus sabdariffa* (Roselle). *World Journal of Pharmacy and Pharmaceutical Science*, *7*(10), 718–734.

Cavanagh, H.M.A. and Wilkinson, J.M. (2002). Biological activities of lavender essential oil. *Phytotherapy Research*, *16*(4), 301–308.

Chiang, S.H., Yang, K.M., Lai, Y.C., and Chen, C.W. (2021). Evaluation of the *in vitro* biological activities of Banana flower and bract extracts and their bioactive compounds. *International Journal of Food Properties*, *24*(1), 1–16.

D'Souza, D.H., Dice, L., and Davidson, P.M., (2016). Aqueous extracts of *Hibiscus sabdariffa* calyces to control aichi virus. *Food and Environmental Virology*, *8*(2), 112–119.

Da-Costa-Rocha, I., Bonnlaender, B., Sievers, H., Pischel, I., and Heinrich, M. (2014). *Hibiscus sabdariffa* L.–A phytochemical and pharmacological review. *Food chemistry*, *165*, 424–443.

Das, A., Bindhu, J., Deepesh, P., Priya, G.S., and Soundariya, S., (2020). *In vitro* anticancer study of bioactive compound isolated from *Musa* extract (*Musa* acuminata). *Indian Journal Public Health Research and Development*, *11*, 340.DOI: 10.4314/tjpr.v8i5.48090.

Ejima, K., Brown, A.W., Schoeller, D.A., Heymsfield, S.B., Nelson, E.J., and Allison, D.B. (2019). Does exclusion of extreme reporters of energy intake (the 'Goldberg cutoffs') reliably reduce or eliminate bias in nutrition studies? Analysis with illustrative associations of energy intake with health outcomes. *The American Journal of Clinical Nutrition*, *110*(5), 1231–1239.

Esposito, F., Mandrone, M., Del Vecchio, C., Carli, I., Distinto, S., Corona, A., Lianza, M., Piano, D., Tacchini, M., Maccioni, E., and Cottiglia, F. (2017). Multi-target activity of Hemidesmus indicus decoction against innovative HIV-1 drug targets and characterization of Lupeol mode of action. *Pathogens and Disease*, *75*(6), p.ftx065.

Fatima, T., Bashir, O., Naseer, B., and Hussain, S.Z. (2018). Dandelion: phytochemistry and clinical potential. *Journal of Medicinal Plants Studies*, *6*(2), 198–202.

Flekhter, O.B., Boreko, E.I., Nigmatullina, L.R., Pavlova, N.I., Medvedeva, N.I., Nikolaeva, S.N., Tret'yakova, E.V., Savinova, O.V., Baltina, L.A., Karachurina, L.T., and Galin, F.Z. (2004). Synthesis and pharmacological activity of acylated betulonic acid oxides and 28-oxo-allobetulone. *Pharmaceutical Chemistry Journal*, *38*(3), 148–152.

Gómez-Calderón, C., Mesa-Castro, C., Robledo, S., Gómez, S., Bolivar-Avila, S., Diaz-Castillo, F., and Martínez-Gutierrez, M. (2017). Antiviral effect of compounds derived from the seeds of *Mammea americana* and *Tabernaemontana cymosa* on Dengue and Chikungunya virus infections. *BioMed Central Complementary and Alternative medicine, 17*(1), 1–12.

González-Castejón, M., Visioli, F., and Rodriguez-Casado, A. (2012). Diverse biological activities of dandelion. *Nutrition Reviews, 70*(9), 534–547. https://doi.org/10.1111/j.1753–4887.2012.00509.x

Gore, M.A. and Akolekar, D. (2003). Evaluation of banana leaf dressing for partial thickness burn wounds. *Burns, 29*(5), 487–492.

Guenova, E., Hoetzenecker, W., Kisuze, G., Teske, A., Heeg, P., Voykov, B., Hoetzenecker, K., Schippert, W., and Moehrle, M. (2013). Banana leaves as an alternative wound dressing. *Dermatologic Surgery, 39*(2), 290–297.

Han, H., He, W., Wang, W., and Gao, B. (2011). Inhibitory effect of aqueous dandelion extract on HIV-1 replication and reverse transcriptase activity. *BioMed Central Complementary and Alternative Medicine, 11*(1), 1–10.

Hasan, H.A. and Zahraa, A.E. (2020). Pharmacognostical and phytochemical study of *Calendula officinalis* L leaves cultivated in Baghdad. *American Institute of Physics Conference Proceedings, 2290*(1), 020019.

Hassan, M.Z., Osman, H., Ali, M.A., and Ahsan, M.J. (2016). Therapeutic potential of coumarins as antiviral agents. *European Journal of Medicinal Chemistry, 123*, 236–255.

Hassan, S.T., Švajdlenka, E., and Berchová-Bímová, K., (2017). *Hibiscus sabdariffa* L. and its bioactive constituents exhibit antiviral activity against HSV-2 and anti-enzymatic properties against urease by an ESI-MS based assay. *Molecules, 22*(5),722.

Hwu, J.R., Huang, W.C., Lin, S.Y., Tan, K.T., Hu, Y.C., Shieh, F.K., Bachurin, S.O., Ustyugov, A., and Tsay, S.C. (2019). Chikungunya virus inhibition by synthetic coumarin–guanosine conjugates. *European Journal of Medicinal Chemistry, 166*, 136–143.

Imam, M.Z. and Akter, S. (2011). *Musa paradisiaca* L. and *Musa sapientum* L.: a phytochemical and pharmacological review. *Journal of Applied Pharmaceutical Science, 01*(05), 14–20.

Jabeur, I., Pereira, E., Barros, L., Calhelha, R.C., Soković, M., Oliveira, M.B.P., and Ferreira, I.C. (2017). *Hibiscus sabdariffa* L. as a source of nutrients, bioactive compounds and colouring agents. *Food Research International, 100*, 717–723.

Jan, N., Andrabi, K.I., and John, R. (2017), December. *Calendula officinalis*-an important medicinal plant with potential biological properties. *Proceeding of the Indian National Science Academy, 83*(4), 769–787.

Jang, D.S., Park, E.J., Hawthorne, M.E., Vigo, J.S., Graham, J.G., Cabieses, F., Santarsiero, B.D., Mesecar, A.D., Fong, H.H., Mehta, R.G., and Pezzuto, J.M. (2002). Constituents of *Musa*×paradisiaca cultivar with the potential to induce the phase II enzyme, quinone reductase. *Journal of Agricultural and Food Chemistry, 50*(22), 6330–6334.

Jędrejek, D., Kontek, B., Lis, B., Stochmal, A., and Olas, B. (2017). Evaluation of antioxidant activity of phenolic fractions from the leaves and petals of dandelion in human plasma treated with H2O2 and H2O2/Fe. *Chemico-Biological Interactions, 262*, 29–37.

Jouneghani, R.S., Castro, A.H.F., Panda, S.K., Swennen, R., and Luyten, W. (2020). Antimicrobial activity of selected banana cultivars against important human pathogens, including candida biofilm. *Foods, 9*(4), 435.

Kalvatchev, Z., Walder, R., and Garzaro, D. (1997). Anti-HIV activity of extracts from *Calendula officinalis* flowers. *Biomedicine and Pharmacotherapy, 51*(4), 176–180.

Kalvatchev, Z., Walder, R., and Garzaro, D. (1997). Anti-HIV activity of extracts from *Calendula officinalis* flowers. *Biomedicine and Pharmacotherapy, 51*(4), 176–180.

Ketiku, A.O. (1973). Chemical composition of unripe (green) and ripe plantain (*Musa* paradisiaca). *Journal of the Science of Food and Agriculture, 24*(6), 703–707.

Kumar, K.S., Bhowmik, D., Duraivel, S., and Umadevi, M. (2012). Traditional and medicinal uses of banana. *Journal of Pharmacognosy and Phytochemistry, 1*(3), 51–63.

Kurkin, V.A. and Sharova, O.V. (2007). Flavonoids from *Calendula officinalis* flowers. *Chemistry of Natural Compounds, 43*(2), 216–217.

Lee, J.J. and Oh, H.K. (2015). Nutritional Composition and Antioxidative activity of different parts of Taraxacum coreanum and *Taraxacum officinale. Journal of the Korean Society of Food Culture, 30*(3), 362–369.

Lee, S.H., Rhie, S.G., Park, H.J., Han, G.J., and Cho, S.M. (2004). A Study of the nutritional composition of the dandelion by part (*Taraxacum officinale*). *The Korean Journal of Community Living Science, 15*(3), 57–61.

Liu, G., Wang, C., Wang, H., Zhu, L., Zhang, H., Wang, Y., Pei, C., and Liu, L. (2019). Antiviral efficiency of a coumarin derivative on spring viremia of carp virus in *vivo*. *Virus Research*, *268*, 11–17.

Liu, L., Hu, Y., Lu, J., and Wang, G., (2019). An imidazole coumarin derivative enhances the antiviral response to spring viremia of carp virus infection in zebrafish. *Virus Research*, *263*, 112–118.

Liyanage, R., Gunasegaram, S., Visvanathan, R., Jayathilake, C., Weththasinghe, P., Jayawardana, B.C., and Vidanarachchi, J.K. (2016). Banana blossom (*Musa* acuminate Colla) incorporated experimental diets modulate serum cholesterol and serum glucose level in Wistar Rats Fed with cholesterol. *Cholesterol*, *2016*, 9747412.

López-Romero, D., Izquierdo-Vega, J.A., Morales-González, J.A., Madrigal-Bujaidar, E., Chamorro-Cevallos, G., Sánchez-Gutiérrez, M., Betanzos-Cabrera, G., Alvarez-Gonzalez, I., Morales-González, Á., and Madrigal-Santillán, E. (2018). Evidence of some natural products with antigenotoxic effects. Part 2: plants, vegetables, and natural resin. *Nutrients*, *10*(12), 1954.

Luo, Z.H. (1993). The use of Chinese traditional medicines to improve impaired immune functions in scald mice. *Zhonghua zheng xing shao shang wai ke za zhi = Zhonghua zheng xing shao shang waikf (ie waike) zazhi = Chinese Journal of Plastic Surgery and Burns*, *9*(1), 56–58.

Martin, T.S., Ohtani, K., Kasai, R., and Yamasaki, K. (2000). A hemiterpenoid glucoside from *Musa paradisiaca*. *Natural Medicines*, *54*(4), 190–192.

Martins, F.O., Fingolo, C.E., Kuster, R.M., Kaplan, M.A.C., and Romanos, M.T.V. (2009). Atividade antiviral de *Musa acuminata* Colla, Musaceae. *Revista Brasileira de Farmacognosia*, *19*, 781–784.

Mazimba, O. (2017). Umbelliferone: sources, chemistry and bioactivities review. *Bulletin of Faculty of Pharmacy, Cairo University*, *55*(2), 223–232.

Meenatchisundaram, S., Parameswar, G., Subbraj, T., Suganya, T., and Michael, A. (2009. Note on pharmacological activities of *Calendula officinalis* L. *Ethnobotanical Leaflets*, *2009*(1), 5.

Mishra, S., Pandey, A., and Manvati, S. (2020). Coumarin: an emerging antiviral agent. *Heliyon*, *6*(1), 03217.

Mojiminiyi, F.B.O., Dikko, M., Muhammad, B.Y., Ojobor, P.D., Ajagbonna, O.P., Okolo, R.U., Igbokwe, U.V., Mojiminiyi, U.E., Fagbemi, M.A., Bello, S.O., and Anga, T.J. (2007). Antihypertensive effect of an aqueous extract of the calyx of *Hibiscus sabdariffa*. *Fitoterapia*, *78*(4), 292–297.

Mostafa, H.S. (2021). Banana plant as a source of valuable antimicrobial compounds and its current applications in the food sector. *Journal of Food Science*, *86*(9), 3778–3797.

Muley, B.P., Khadabadi, S.S., and Banarase, N.B. (2009). Phytochemical constituents and pharmacological activities of *Calendula officinalis* Linn (Asteraceae): a review. *Tropical Journal of Pharmaceutical Research*, *8*(5), 455–465. Mutai, C., Keter, L., Ngeny, L., and Jeruto, P. (2012). Effects of triterpenoids on Herpes simplex virus type1 (Hsv–1) *in vitro*. *Medicinal Aromatic Plants*, *1*, 106.

Nimisha, S.M. and Negi, P.S. (2017). Traditional uses, phytochemistry and pharmacology of wild Banana (*Musa acuminata* Colla): a review. *Journal of Ethnopharmacology*, *196*, 124–140.

Nurviani, N., Somowiyarjo, S., Sulandari, S., and Subandiyah, S. (2018). The inhibition of Tobamovirus by using the extract of banana flower. *Jurnal Perlindungan Tanaman Indonesia*, *22*(2),181–185.

Olas, B. (2022). New perspectives on the effect of dandelion, its food products and other preparations on the cardiovascular system and its diseases. *Nutrients*, *14*(7), 1350. https://doi.org/10.3390/nu14071350

Onyenekwe, P.C., Okereke, O.E., and Owolewa, S.O. (2013). Phytochemical screening and effect of *Musa paradisiaca* stem extrude on rat haematological parameters. *Current Research Journal of Biological Sciences*, *5*(1), 26–29.

Panda, S.K., Brahma, S., and Dutta, S.K. (2010). Selective antifungal action of crude extracts of *Cassia fistula* L.: a preliminary study on Candida and *Aspergillus* species. *Malaysian Journal of Microbiology*, *6*(1), 62–68.

Panda, S.K., Castro, A.H.F., Jouneghani, R.S., Leyssen, P., Neyts, J., Swennen, R., and Luyten, W. (2020). Antiviral and cytotoxic activity of different plant parts of banana (*Musa* spp.). *Viruses*, *12*(5), 549.

Panda, S.K., Patra, N., Sahoo, G., Bastia, A.K., and Dutta, S.K. (2012). Anti-diarrheal activities of medicinal plants of Similipal Biosphere Reserve, Odisha, India. *International Journal of Medicinal and Aromatic Plants*, *2*(1), 123–134.

Parvez, M.K., Alam, P., Arbab, A.H., Al-Dosari, M.S., Alhowiriny, T.A., and Alqasoumi, S.I. (2018). Analysis of antioxidative and antiviral biomarkers β-amyrin, β-sitosterol, lupeol, ursolic acid in Guiera senegalensis leaves extract by validated HPTLC methods. *Saudi Pharmaceutical Journal*, *26*(5), 685–693.

Peredo Pozos, G.I., Ruiz-López, M.A., Zamora Natera, J.F., Alvarez Moya, C., Barrientos Ramirez, L., Reynoso Silva, M., Rodriguez Macias, R., García-López, P.M., Gonzalez Cruz, R., Salcedo Perez, E., and Vargas Radillo, J.J. (2020). Antioxidant capacity and antigenotoxic effect of *Hibiscus sabdariffa* L. extracts obtained with ultrasound-assisted extraction process. *Applied Sciences*, *10*(2), 560.

Pereira, A. and Maraschin, M. (2015). Banana (*Musa* spp) from peel to pulp: ethnopharmacology, source of bioactive compounds and its relevance for human health. *Journal of Ethnopharmacology, 160,* 149–163.

Perusquía, M., Mendoza, S., Bye, R., Linares, E. and Mata, R., 1995. Vasoactive effects of aqueous extracts from five Mexican medicinal plants on isolated rat aorta. *Journal of Ethnopharmacology, 46*(1), 63–69.

Pinakin, D.J., Kumar, V., Suri, S., Sharma, R., and Kaushal, M. (2020). Nutraceutical potential of tree flowers: a comprehensive review on biochemical profile, health benefits, and utilization. *Food Research International, 127,* 108724.

Pinela, J., Prieto, M.A., Pereira, E., Jabeur, I., Barreiro, M.F., Barros, L., and Ferreira, I.C. (2019). Optimization of heat-and ultrasound-assisted extraction of anthocyanins from *Hibiscus sabdariffa* calyces for natural food colorants. *Food Chemistry, 275,* 309–321.

Pires, T.C., Barros, L., Santos-Buelga, C., and Ferreira, I.C. (2019). Edible flowers: emerging components in the diet. *Trends in Food Science & Technology, 93,* 244–258.

Ramu, R., Shirahatti, P., Naik, S., and Prasad, N. (2015). Impact of active compounds isolated from banana (*Musa* sp. var. Nanjangud rasabale) flower and pseudostem towards cytoprotective and DNA protection activities. *Education, 9*(10), 21–23.

Rehman, S., Ijaz, B., Fatima, N., Muhammad, S.A., and Riazuddin, S. (2016). Therapeutic potential of *Taraxacum officinale* against HCV NS5B polymerase: *in-vitro* and *In silico* study. *Biomedicine and Pharmacotherapy, 83,* 881–891.

Rop, O., Mlcek, J., Jurikova, T., Neugebauerova, J., and Vabkova, J. (2012). Edible flowers—a new promising source of mineral elements in human nutrition. *Molecules, 17*(6), 6672–6683.

Ross, I.A. (2003). *Hibiscus sabdariffa.* in *Medicinal Plants of the World,* Humana Press, Totowa, N J, pp. 267–275. https://doi.org/10.1007/978-1-59259-365-1_13

Salem, M.A., Zayed, A., Beshay, M.E., Abdel Mesih, M.M., Ben Khayal, R.F., George, F.A., and Ezzat, S.M. (2021). *Hibiscus sabdariffa* L.: phytoconstituents, nutritive, and pharmacological applications. *Advances in Traditional Medicine, 22*(23), 1–11.

Salvador, I.F. (2018). *Consumer Acceptability of Banana Blossom Sisig.* UNEJ e-Proceeding, ICAM, Jember, pp. 336–350.

Sandjo, L.P., dos Santos Nascimento, M.V., de H Moraes, M., Rodrigues, L.M., Dalmarco, E.M., Biavatti, M.W., and Steindel, M. (2019). NOx-, IL-1β-, TNF-α-, and IL-6-inhibiting effects and trypanocidal activity of banana (*Musa acuminata*) bracts and flowers: UPLC-HRESI-MS detection of phenylpropanoid sucrose esters. *Molecules, 24*(24), 4564.

Sarker, S.D. and Nahar, L. (2017). Progress in the chemistry of naturally occurring coumarins. *Progress in the Chemistry of Organic Natural Products, 106,* 241–304.

Sharma, N., Palia, P., Chaudhary, A., Verma, K., and Kumar, I. (2020). A review on pharmacological activities of lupeol and its triterpene derivatives. *Journal of Drug Delivery and Therapeutics, 10*(5), 325–332.

Shekarabi, S.P.H., Mostafavi, Z.S., Mehrgan, M.S., and Islami, H.R. (2021). Dietary supplementation with dandelion (*Taraxacum officinale*) flower extract provides immunostimulation and resistance against Streptococcus iniae infection in rainbow trout (*Oncorhynchus mykiss*). *Fish & Shellfish Immunology, 118,* 180–187.https://doi.org/10.1016/j.fsi.2021.09.004

Sheng, Z.W., Ma, W.H., Jin, Z.Q., Bi, Y., Sun, Z.G., Dou, H.T., Li, J.Y., and Han, L.N. (2010). Investigation of dietary fiber, protein, vitamin E and other nutritional compounds of banana flower of two cultivars grown in China. *African Journal of Biotechnology, 9*(25), 3888–3895.

Singh, B., Singh, J.P., Kaur, A., and Singh, N. (2016). Bioactive compounds in banana and their associated health benefits–A review. *Food Chemistry, 206,* 1–11.

Singh, P., Khan, M., and Hailemariam, H. (2017). Nutritional and health importance of *Hibiscus sabdariffa*: a review and indication for research needs. *Journal of Nutritional Health and Food Engineering, 6*(5), 00212.

Stefanachi, A., Leonetti, F., Pisani, L., Catto, M., and Carotti, A. (2018). Coumarin: a natural, privileged and versatile scaffold for bioactive compounds. *Molecules, 23*(2), 250.

Swanson, M.D., Winter, H.C., Goldstein, I.J., and Markovitz, D.M. (2010). A lectin isolated from bananas is a potent inhibitor of HIV replication. *Journal of Biological Chemistry, 285*(12), 8646–8655.

Swe, K.N.N. (2012) *Study on Phytochemicals and Nutritional Composition of Banana Flowers of Two Cultivars (Phee kyan and Thee hmwe).* Doctoral dissertation, MERAL Portal.

Takahashi, J.A., Rezende, F.A.G.G., Moura, M.A.F., Dominguete, L.C.B., and Sande, D. (2020). Edible flowers: bioactive profile and its potential to be used in food development. *Food Research International, 129,* 108868.

Takeda, Y., Okuyama, Y., Nakano, H., Yaoita, Y., Machida, K., Ogawa, H., and Imai, K. (2020). Antiviral activities of *Hibiscus sabdariffa* L. tea extract against human influenza A virus rely largely on acidic pH but partially on a low-pH-independent mechanism. *Food and Environmental Virology*, *12*(1), 9–19.

Torky, Z.A. and Hossain, M.M. (2017). Pharmacological evaluation of the *Hibiscus* herbal extract against Herpes simplex virus-type 1 as an antiviral drug in vitro. *International Journal of Virology*, *13*, 68–79.

Tran, H.T.T., Le, N.P.K., Gigl, M., Dawid, C., and Lamy, E. (2021). Common dandelion (*Taraxacum officinale*) efficiently blocks the interaction between ACE2 cell surface receptor and SARS-CoV-2 spike protein D614, mutants D614G, N501Y, K417N and E484K *in vitro*. *BioRxiv*, *14*(10), 1055.

Tsamo, C.V.P., Herent, M.F., Tomekpe, K., Emaga, T.H., Quetin-Leclercq, J., Rogez, H., Larondelle, Y., and Andre, C.M. (2015). Effect of boiling on phenolic profiles determined using HPLC/ESI-LTQ-Orbitrap-MS, physico-chemical parameters of six plantain banana cultivars (*Musa* sp). *Journal of Food Composition and Analysis*, *44*, 158–169.

Tsay, S.C., Lin, S.Y., Huang, W.C., Hsu, M.H., Hwang, K.C., Lin, C.C., Horng, J.C., Chen, I.C., Hwu, J.R., Shieh, F.K., and Leyssen, P. (2016). Synthesis and structure-activity relationships of imidazole-coumarin conjugates against hepatitis C virus. *Molecules*, *21*(2), 228.

Tschöp, M., Castaneda, T.R., Joost, H.G., Thöne-Reineke, C., Ortmann, S., Klaus, S., Hagan, M.M., Chandler, P.C., Oswald, K.D., Benoit, S.C., and Seeley, R.J. (2004). Does gut hormone PYY3–36 decrease food intake in rodents?. *Nature*, *430*(6996), 1–3.

Venkatachalam, P., Geetha, N., Sangeetha, P., and Thulaseedharan, A. (2013). Natural rubber producing plants: an overview. *African Journal of Biotechnology*, *12*(12).

Venugopala, K.N., Rashmi, V., and Odhav, B. (2013). Review on natural coumarin lead compounds for their pharmacological activity. *BioMed Research International*, *2013*.

Verma, P.K., Raina, R., Agarwal, S., and Kaur, H. (2018). Phytochemical ingredients and Pharmacological potential of *Calendula officinalis* Linn. *Pharmaceutical and Biomedical Research*, *4*(2), 1–17.

Voelkl, B., Vogt, L., Sena, E.S., and Würbel, H. (2018). Reproducibility of preclinical animal research improves with heterogeneity of study samples. *PLoS Biology*, *16*(2), 2003693.

Vu, H.T., Scarlett, C.J., and Vuong, Q.V. (2018). Phenolic compounds within banana peel and their potential uses: a review. *Journal of Functional Foods*, *40*, 238–248.

Wang, Y., Yan, W., Chen, Q., Huang, W., Yang, Z., Li, X., and Wang, X. (2017). Inhibition viral RNP and anti-inflammatory activity of coumarins against influenza virus. *Biomedicine and Pharmacotherapy*, *87*, 583–588.

Wickramarachchi, K.S. and Ranamukhaarachchi, S.L. (2005). Preservation of fiber-rich banana blossom as a dehydrated vegetable. *Science Asia*, *31*(3), 265.

Woronuk, G., Demissie, Z., Rheault, M., and Mahmoud, S. (2011). Biosynthesis and therapeutic properties of Lavandula essential oil constituents. *Planta Medica*, *77*(01), 7–15.

Zhang, B., Mao, G., Zheng, D., Zhao, T., Zou, Y., Qu, H., Li, F., Zhu, B., Yang, L., and Wu, X. (2014). Separation, identification, antioxidant, and anti-tumor activities of *Hibiscus sabdariffa* L. extracts. *Separation Science and Technology*, *49*(9), 1379–1388.

5 Characterization and Docking Studies of Immunomodulatory Active Compounds from *Rhododendron arboreum* Sm. Leaves

Nishant Rai, Pramod Rawat, Rakesh K. Bachheti,
Navin Kumar, Abdur Rauf, Umer Rashid, Vijay Jyoti Kumar,
Tanmay Sarkar and Raffaele Pezzani

CONTENTS

DOI: 10.1201/9781003206620-5

5.1   INTRODUCTION

Globally, there are more than 1,000 species of the *Rhododendron* genus known to be native to the temperate regions of Asia, North America, and Europe, while approximately 80 species with 10 subspecies and 14 varieties are known in the Indian subcontinent occurring from the western to eastern Himalayan region (Kumari et al., 2015). *Rhododendron arboreum* has a unique flower color characteristic. A pure, red colored flower is seen at lower altitudes, followed by pink color, and a white flower ascends at higher altitudes. The species name '*arboreum*' means tree like (Orwa et al., 2009). It grows at elevations from 1,370–3,200 m to 12–30 m. It is an evergreen heavily branched tree that reaches up normally to 14 m in height and 2.4 m in girth. In the Himachal Pradesh state of India, the plant is found at a higher altitude, and its flower is known for its medical and **ayurvedic** uses. Natives of Himachal Pradesh use '**Burans**' for making a sauce, especially in summer, to prevent **heatstroke, whereas juice of its flower** is consumed as an adaptogenic seasonal drink. *R. arboreum* leaves are reported to contain alkaloids, steroids, flavonoids, terpenoids, anthraquinones, glycosides, phlobatanins, phenols, saponins, and tannins (Madhvi et al., 2020; Paudel et al., 2020; Nisar et al., 2011; Kiruba et al., 2011; Saklani and Chandra 2015). *R. arboreum* leaves reportedly contain several active phytoconstituents, such as glucosides, gallic acid, quercetin, quercetin 3-O-beta-D-glucopyranosyl 1, L-rhamnopyranoside, 7,2'-dime, 2'-methoxy-4', 5-methylenedioxy flavone, epicatechin, rutin, ursolic acid, amyrin, epifriedelanol, friedelin, lupeol, β-amyrin, syringic acid, hyperoside, coumaric acid, arbutin, β-sitosterol, 22-stigmasten-3-one, linoleyl alcohol, ericolin, diethyl ester of terephthalic acid, and 1,2,3-propanetriyl ester. (Gill et al., 2015; Kumar et al., 2019; Painuli et al., 2016; Roy et al., 2014). This plant is equally reported to possess hepatoprotective, antioxidant, immunomodulatory, anti-inflammatory, antidiabetic, antinociceptive, antidiarrheal, oxytoxic, estrogenic, and CNS-depressant activities (Prakash et al., 2016; Sonar et al., 2012, 2013). It was observed that there was a significant influence of geographical location, climate, topography, etc., on the phytoconstituents and chemical and biological activities of *R. arboreum* (Thakur and Sidhu, 2014; Anand et al., 2014). The present study focuses on the preliminary screening of antioxidant properties in *R. arboreum* leaves and the isolation, purification and identification of bioactive compounds present in their methanolic extract using TLC, column chromatography, and HPLC.

5.2   MATERIALS AND METHODS

5.2.1   MATERIALS

Reagents and chemicals that include [3-(4,5-dimethylthiazol-2-yl)-2,5-diphenyltetrazolium bromide] (MTT) and trolox were procured from Hi-Media (Mumbai, Maharashtra, India). RAW 246.7 murine macrophage cell line) cell lines were procured from the National Centre for Cell Sciences, Pune,

Maharashtra, India. The cells were maintained and continuously subcultured in Dulbecco's modified Eagle's medium (DMEM) supplemented with 10% fetal bovine serum, amphotericin B (2.5 µg/mL), and streptomycin (50 µg/mL) at 37°C and 5% CO_2. All chemicals used in the present study, including standards, were procured from Sigma–Aldrich and Hi-Media, India. High-performance liquid chromatography (HPLC)-grade solvents and water were used in HPLC analysis. For the candidacidal assay, a culture of *Candida albicans* MTCC 3017 available in the Laboratory of Microbiology in Graphic Era was used.

5.2.2 COLLECTION OF SAMPLES

R. arboreum leaves were collected from the Ani-Jalori bypass district of Himachal Pradesh (31°28′28.9524″N, 77°25′21.3852″E), India. The collected leaves were identified (voucher specimen Accession No: 115589) by Dr. S. K. Srivastava, Scientist 'D', Botanical Survey of India (BSI), Dehradun India. The collected leaves were washed with fresh water and dried under the shed for 2 weeks. The dried leaves were then powdered and kept in sealed packets at 4°C until further use. In the present study, the extraction of the pulverized leaves was performed with a Soxhlet apparatus using increasing polarity bases (Figure 5.1).

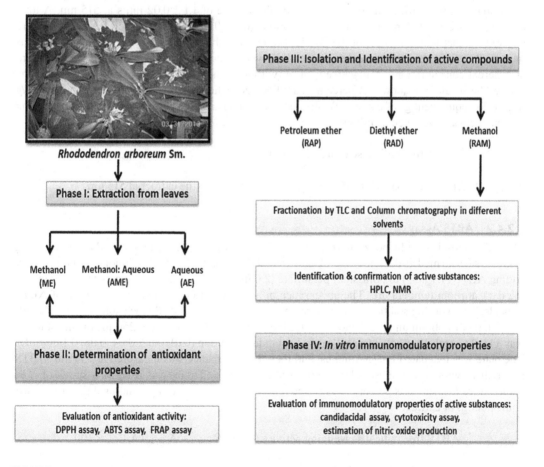

FIGURE 5.1   Flow chart of the study.

5.2.3 PREPARATION OF EXTRACTS

The pulverised *R. arboreum* leaf material (10 g each) was separately extracted in 100% methanol, 50% aqueous methanol, and water by the hot percolation method using a Soxhlet apparatus continuously for 8 hours. The respective extracts were named methanol extract (ME), aqueous methanol extract (AME), and aqueous extract (AE). All the extracts were filtered, collected, and concentrated using a vacuum rotary evaporator, dried and stored at −20°C separately (Painuli et al., 2018). These extracts were used to analyse their antioxidant activity. For the analysis of active phytoconstituents, *R. arboreum* air-dried leaves (130g) were separately crushed and consecutively extracted in different solvents with increasing polarity starting from the less polar solvent. Then, the extracts were dried carefully after each solvent for sequential extraction, *viz.*, petroleum ether (less polar) < diethyl ether < methanol (more polar) following the hot percolation method in a Soxhlet apparatus. The extracts were filtered and concentrated using a vacuum rotary evaporator, dried and stored at −20°C.

5.2.4 ANTIOXIDANT PROPERTIES

5.2.4.1 DPPH Assay

The 1,1-diphenyl-2-picrylhydrazyl (DPPH) assay was performed as suggested by Brand–Williams (Brand et al., 1995). The DPPH stock solution was prepared by dissolving 24 mg DPPH in 100 mL methanol and stored at −20°C until needed. The working solution was prepared by mixing a 10 mL stock solution in 45 mL methanol to obtain an absorbance of 1.1 ± 0.02 units at 515 nm. A total of 150 μL of extract (1 mg/mL) standard solutions (0.01, 0.02, 0.04, 0.08, and 0.1 mg/mL) were allowed to react with 2,850 μL of the DPPH solution for 2 hours in the dark. The absorbance was measured at 515 nm, and the scavenging activity was calculated. Trolox was used as a standard at different concentrations. A negative control (without standard/extract) was also included in the study. The results are shown as percent of DPPH-free radical scavenging activity* and μg TE/g (μg Trolox equivalent/g) of extract. Triplicate measurements were carried out, and the scavenging percentage was calculated as follows:

$$\text{*DPPH-free radical scavenging activity } (\%) = (A_{control} - A_{extract})/A_{control} \times 100$$

$A_{control}$: Absorbance of control at 515 nm; $A_{extract}$: Absorbance of the extract at 515 nm

5.2.4.2 ABTS Assay

The 2,2-azino-bis (3-ethylbenzothiazoline-6-sulfonic acid) diammonium salt (ABTS) scavenging assay was performed as described by Thaipong et al. (2006). ABTS stock solution was prepared by adding ABTS (7.4 mM) to potassium persulfate (2.6 mM) in a 1:1 ratio and incubated overnight in the dark at room temperature. The appearance of a blue color confirmed the formation of ABTS radicals. The working solution was prepared by mixing 1 mL of ABTS stock solution with 60 mL of methanol to obtain an absorbance of 1.1 ± 0.02 units at 734 nm. Then, 250 μL of this solution was added to each 10 μL of extract (1 mg/mL)/standard solution (0.01, 0.02, 0.04, 0.08, 0.1 mg/mL). The plate was incubated for 15 minutes, and then the absorbance was measured at 734 nm. ABTS and radical without extract/standard were used as controls. Trolox was used as a standard. The results were measured as percent ABTS free radical scavenging activity* and μg TE/g (μg Trolox equivalent/g) of extract.

$$\text{*ABTS free radical scavenging activity } (\%) = (A_{control} - A_{extract})/A_{control} \times 100$$

$A_{control}$: Absorbance of control at 73 4nm; $A_{extract}$: Absorbance of the extract at 734 nm

5.2.4.3 FRAP Assay

The ferric reducing antioxidant power (FRAP) assay was also measured through the total antioxidant power of the extract by the following method of Benzie and Strain (1999; 1996). The FRAP solution was freshly prepared by mixing 25 mL of 30 mM acetate buffer, 2.5 mL of 10 mM 2,4,6-tripyridyl-s-triazine (TPTZ), and 2.5 mL of 20 mM $FeCl_3.6H_2O$ solution. Then, 150 µL of extract (1 mg/mL) was allowed to react with 2,850 µL of the freshly prepared FRAP solution. The solution was incubated in the dark for 30 minutes. The absorbance was measured at 593 nm. The results were expressed in µg Trolox/g of extract using a standard curve prepared from the different concentrations of Trolox.

5.2.5 EXTRACTION AND PHYTOCONSTITUENT CHARACTERIZATION

5.2.5.1 Thin Layer Chromatography Analysis of the Methanolic Extract

R. arboreum methanolic leaf extract (RAM) was subjected to thin-layer chromatography (TLC) as per the conventional one-dimensional ascending method, as previously described (Swaroop et al., 2005). Aluminum TLC precoated plates (silica gel 60F-254, E. Merck) were subjected to the one-way ascending technique. Plates were developed in a twin trough chamber presaturated with multiple combinations of mobile phases comprising butanol (B), acetic acid (AA), water (W), chloroform (C), and methanol (M) in different ratios, such as B:AA:W::4:1:5, ACN:W::70:30, C:M::7:3,9:1,1:9,3:7,5:5,8:2, M:W::9:1, C:AA:W::10:9:1 and M:AA::6:4, 5:5, 7:3, 8:2, 4:6. The R_f value was calculated by standard procedures.

5.2.5.2 Isolation of Compounds by Column Chromatography

Column chromatography was performed to isolate and purify the constituents that showed dense and clear spots in the TLC of the RAM extract (Swaroop et al., 2005). RAM extract (10 mL) was introduced into the column, and then solvent mixtures (eluents) in different ratios were added into the column for each separation/elution. The solvents used in increasing polarity were chloroform:methanol: 9:1, 3:7, 5:5, 7:3, and 1:9 and methanol:water:1:9, 3:7, 5:5, 7:3, and 1:9, respectively. The fractions obtained were likewise named *R. arboreum* methanolic fractions 1, 2, and 3. They were concentrated, and their purity was determined using thin-layer chromatography.

5.2.5.3 TLC of Purified Fractions

Seven fractions of RAM extract isolated from column chromatography were again subjected to thin-layer chromatography of varying proportion of toluene:methanol in different ratios with increasing polarity (Madhvi et al., 2020; Sonar et al., 2012, 2013). The specificity of the method was confirmed by matching the color. Quercetin, kaempferol, ursolic acid, epigallocatechin gallate, and rutin were used as standards. The R_f value of each sample was evaluated using these standards. Out of the seven fractions, *R. arboreum* fraction 2 (**RAM fr2**) showed a prominent spot in toluene:methanol (8:1, v/v), and the further presence of active purified phytocompounds at spots 10 and 3 of **RAM fr2** was primitively confirmed, corresponding to the standards ursolic acid and kaempferol.

5.2.5.4 HPLC Analysis

For the HPLC analysis, the protocol suggested by Taralkar and Chattopadhyay (2012) was adopted. HPLC grade water, methanol, and acetonitrile were used for the analysis. Ursolic acid and kaempferol (Sigma–Aldrich) were used as standards. Isolated and purified fractions (**1**) and (**2**) were collected from TLC plates and dissolved in ethanol for HPLC analysis (Agilent 1220). An isocratic mobile phase of acetonitrile:methanol (80:20 v/v) and acetonitrile:methanol (70:30 v/v) with a flow rate of 0.5 and 1 mL/min was selected for identification. The column temperature was maintained at 35°C and 30 ± 0.1°C. The detection wavelength was set at 210 and 290 nm. Purified samples and standards were filtered using a Millipore syringe filter (0.2 µm). The volume of the standard and purified sample to be injected was 20 µL with a run time of 15 minutes.

5.2.5.5   IR and Mass Spectrometry

IR and mass spectrometry of isolated compounds were performed according to the previous works (Janakirman and Jeyaprakash, 2015; Keat et al., 2010).

5.2.6   IMMUNOMODULATORY PROPERTIES

To assess the immunomodulatory properties of *R. arboreum*, we used purified compounds (**1**) and (**2**) isolated from **RAM fr2**. Moreover, **RAM fr2**, which constituted both purified compounds (**1**) and (**2**), was used as a control. Compounds (**1**), (**2**), and **RAM fr2** were used at a 1,000 μg/mL concentration, including positive and negative controls for the determination of *in vitro* immuno-modulatory properties.

5.2.7   CANDIDACIDAL ASSAY

The method was performed using the protocol of Patil et al. (2010). A volume of 0.25 mL Hank's solution (control) and 0.25 mL of the mixture of human pooled sera (standard) were used. A suspension of leucocytes (7×10^6/mL) was prepared in 0.25 mL of Hank's solution. A volume of 0.25 mL of *C. albicans* suspension was added to the suspension and incubated at 37°C in a water bath for 60 minutes with shaking every 15 minutes. After 30 minutes, 100 μL solution in a sterile glass tube was taken for Giemsa staining. After 1 hour of incubation, 0.25 mL of 2.5% sodium deoxycholate was added. A volume of 4 mL of 0.01% methylene blue was added to each tube and mixed spun at 1,500 g at 4°C for 10 minutes. Approximately 300 *Candida* cells were counted using a hemocytometer. The proportions of dead cells, i.e., those that took up dye color, were determined.

5.2.8   CYTOTOXICITY ASSAY

The isolated purified extracts of *R. arboreum* leaves (**1**), (**2**) and control **RAM fr2** were assessed for cytotoxicity analysis (Kwon et al., 2016).

5.2.8.1   Cytotoxicity of H_2O_2

A cytotoxicity assay was performed in RAW 246.7 macrophages exposed to H_2O_2 (at different concentrations, data not shown). Significant toxicity was demonstrated in RAW macrophage cells at 18 hours after 30 minutes of exposure to H_2O_2 at concentrations as low as 0.5 mM. The dose–response relationship in terms of toxicity in RAW264.7 cells treated with H_2O_2 was determined. Reduction of H_2O_2 concentration in the culture medium in the presence of RAW cells was also measured. The antiproliferative activity was evaluated by MTT reduction assay (Kwon et al., 2016). RAW cells (1×10^5) were seeded in 96-well microtiter plates. The plate was then incubated for 24 hours at 37°C with 5% CO_2 to obtain 80%–90% cell confluence. Final concentrations of extracts (**1**) and (**2**) and control **RAM fr2** were maintained in decreasing concentrations (80 μg/mL–1.25 μg/mL) and for H_2O_2 (2 mM–0.03125 mM). The plates were incubated for 24 hours at 37°C with 5% CO_2. After 18 hours of incubation, the cells were washed with phosphate-buffered saline (PBS). MTT solution (5 mg/mL in PBS) was added to each well and incubated for another 3 hours at 37°C. The medium was removed, and dimethyl sulfoxide (DMSO) was added to dissolve the formazan dye. The optical density was measured at 570 nm. All the experiments were performed in triplicate and repeated three times, and the results are shown in the form of percent inhibition of the cell line using the formula.

$$\% \text{ Cell Inhibition} = 100 - \% \text{ Cell Viability}$$

$$\% \text{ Cell Viability*} = \text{A570 of treated cells/A570 of untreated cells} \times 100$$

The mean value of the percent inhibition $\pm$ SD was used for plotting the bar graphs at each concentration.

5.2.8.2 H₂O₂ Dose Curve of Purified Compounds

The dose curve of purified compounds **(1)**, **(2)** and **RAM fr2** was obtained using their different doses (80 µg/mL–1.25 µg/mL) to determine their maximum tolerated dose for RAW 246.7 cells by MTT assay. After dose determination, RAW264.7 cells were treated with compounds pre, post, and concomitantly with the administration of H_2O_2. In pretreated samples, compounds were added 30 minutes before H_2O_2 treatment, while in the post-treated sample, extracts were added 30 minutes after H_2O_2 treatment, and concomitantly, the compounds and H_2O_2 were added simultaneously to RAW 246.7 cells. After the determination of pre, post, and concomitant treatment results of the respective compounds **(1)**, **(2)** and **RAM fr2**, RAW 246.7 cells were treated with their maximum-tolerated dose 30 minutes prior to the addition of H_2O_2 in RAW 246.7 cells followed by 18 hours of incubation. After completion of 18 hours of treatment following H_2O_2 addition, media from each plate were collected and used to determine the NO concentration by performing a Griess assay. Then, 100 µl of MTT was added (0.5 mg/5 mL) to each well, and the cells were incubated in the dark at 37°C for 4 hours. The medium was removed after 4 hours, the crystals were dissolved in 100 µl of dimethyl sulfoxide, and the absorbance was recorded at 570 nm using an ELISA plate reader. Endotoxin-activated plasma (LPS) was used as a positive control, while PBS was used as a negative control (Alley et al., 1988).

5.2.8.3 Estimation of Nitric Oxide (NO) Production

RAW 264.7 cells were seeded in 96-well plates at a density of 5×10^5 cells well^{-1} and incubated for 24 hours. After incubation for 18 hours, 50 µl of cell culture medium supernatant was mixed with an equal volume of Griess reagent (1% sulfanilamide in 5% phosphoric acid, 0.1% *N*-(1-naphthyl) ethylenediamine in H_2O) and incubated at room temperature for 15 minutes. The absorbance was measured at 540 nm. The nitrite concentrations were determined by extrapolation from a standard sodium nitrite curve (Lopez et al., 2015; Othman et al., 2015; Wang et al., 2013).

5.2.9 Docking Studies

Docking studies were performed on proinflammatory cytokines, i.e., tumor necrosis factor-α (TNF-α) and interleukin-1β (IL-1β), using Molecular Operating Environment (MOE 2016) software. Three-dimensional crystal structures of the proteins were obtained from the RCSB Protein Data Bank (PDB). The accession numbers of the obtained proteins are 2AZ5 (co-crystallized with a small-molecule inhibitor 307) for TNF-α and 3O4O for IL-1β. The docking procedure was validated by re-docking of the native ligands. For 2AZ5, the native co-crystallized ligand was extracted and prepared in a manner comparable to that for the test compounds. For 3O4O, the binding site was predicted by using the MOE site finder option and dummies were created. Docking was carried out using the triangle matcher algorithm (placement stage) and scored by the London dG scoring function. Energy minimization of the ligand, preparation of structures of the downloaded enzymes, and active site identification were carried out according to the previously reported procedure. The assessment of the docking results and investigation of their surface with graphical representations were carried out using MOE and a Discovery Studio visualizer.

5.2.10 Statistical Analysis

Statistical analysis was expressed as the mean ± SEM (standard error of the mean). The results were analysed by using one-way analysis of variance (ANOVA) followed by Dunnett's 't'-test to determine statistical significance.

5.3 RESULTS

The percent yield of *R. arboreum* leaves with increasing order of polarity were 3.55%, 3.66%, and 12.97%, for petroleum ether (RAP), diethyl ether (RAD), and methanol (RAM), respectively, as evaluated by Soxhlet extraction.

5.3.1 ANTIOXIDANT ASSAY

5.3.1.1 DPPH Assay

Different dilutions of Trolox were used to plot a calibration curve, resulting in the standard linear equation $y=-2.861x+0.963$ with the regression coefficient $R^2=0.979$. The aqueous and 50% methanolic extracts showed DPPH scavenging activities of $87.32 \pm 0.007\%$ and $90.02 \pm 0.003\%$ at a 1 mg/mL concentration, respectively, while the methanolic extract showed the highest DPPH-free radical scavenging activity ($90.32 \pm 0.005\%$) at a 1 mg/mL concentration (Suppl. Figure 5.1(a), Suppl. Table 5.1A).

5.3.1.2 ABTS Assay

The calibration graph for the percent ABTS activity of Trolox at different concentrations showed a linear equation $y=-2.993x+1.272$ and $R^2=0.989$. The aqueous and 50% methanolic extracts showed percent of free radical scavenging activities of $86.74 \pm 0.05\%$ and $94.35 \pm 0.05\%$, respectively, while the methanolic extract showed the maximum percent scavenging activity ($94.50 \pm 0.024\%$) (Suppl. Figure 5.1(b), Suppl. Table 5.1B).

5.3.1.3 FRAP Assay

FRAP activities of different dilutions of Trolox were used to plot a standard calibration graph, which resulted in the linear equation $y=8.4367x+0.155$ and correlation coefficient $R^2=0.979$. The absorbance at 593 nm of each extract (1 mg/mL) was measured, and FRAP activity was calculated in terms of Trolox equivalent μg/gm of the extract. The FRAP activities of the *R. arboreum* methanolic, aqueous, and 50% methanolic extracts were found to be 567 ± 0.15 Trolox mg/gm, 648 ± 0.03 Trolox mg/gm, and 451.5 ± 0.11 Trolox mg/gm, respectively (Suppl. Figure 5.1(c), Suppl. Table 5.1C). Several works have shown that *R. arboreum* extracts possess high antioxidant potential, including high free radical scavenging activity (Bhandari and Rajbhandari, 2014; Gill et al., 2015).

5.3.1.4 Thin-Layer Chromatography of the Methanolic Extract

Out of the different solvent mixtures used in varying ratios, the C:M:9:1 and M: W:9:1 mobile phases showed the best separation, which was further used in column chromatography (Suppl. Figure 5.2).

5.3.2 EXTRACTION AND PHYTOCONSTITUENT CHARACTERIZATION

5.3.2.1 Thin-Layer Chromatography of Purified Compounds

Out of the seven fractions, **RAM fr2** showed a prominent spot in the toluene:methanol (8:1, v/v) solvent system. The R_f value of **RAM, fr2,** spot 10, i.e., compound (**1**), was 0.92, corresponding to the standard ursolic acid at 0.93. The presence of ursolic acid in *Rhododendron arboreum* leaves was also reported in other works (Madhvi et al., 2020; Sonar et al., 2012, 2013). Additionally, **RAM fr2** showed a prominent spot in the chloroform:methanol (11:3) development system. The R_f value of standard kaempferol was evaluated to be 0.33, corresponding to **RAM fr2** spot 3, i.e., compound (**2**) at 0.32 (Suppl. Figures 5.3–5.4).

5.3.2.2 HPLC Analysis

Quantification of Active Components: The HPLC calibration curve revealed the presence of ursolic acid and kaempferol based on the retention time of the identified peaks compared to standards. The concentration of purified compound (**1**) was estimated to be in the range of 3.59 ± 2.66 mg ursolic acid/g of extract. The concentration of the purified compound (**2**) was estimated to be in the range of 0.743 ± 1.52 mg kaempferol/g of extract (Figure 5.2).

5.3.2.3 Structural Characterization

Compound TMS-10: crystallized from methanol. The EI–MS spectrum showed a molecular ion peak at *m/z* 457 that corresponds to the molecular formula $C_{30}H_{48}O_3$. IR (v_{max}^{kBr}): 3,435, 2,941,

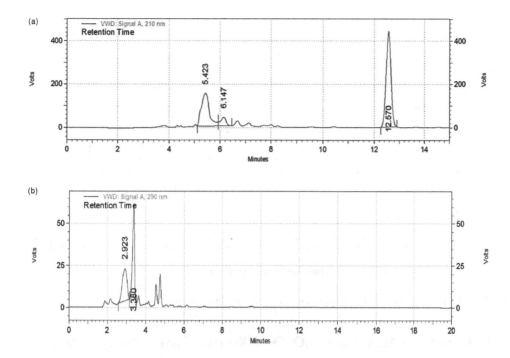

FIGURE 5.2 HPLC chromatograms of isolated compounds **1** (a) and **2** (b).

1,704, 1,033, and 755. EI–MS: (DART-MS) m/z 457 [M]$^+$, other peaks for the fragment ions were obtained at 455, 439, 249, 191, and 130 m/z. Scan: 13 times: 0.7–0.9 minutes frag = 150.0 V, ISOCRATIC.m.

NMR data were similar to Aziz and Saha (2020). Thus, the compound TMS-10 was identified as ursolic acid.

CMS-3 compound: M.P. 222–223°C Analysis of CMS-3 corresponded to the molecular formula $C_{15}H_{10}O_6$, which was confirmed by the presence of a molecular ion peak [M]$^+$ at m/z 287 in its EI-mass spectrum.

IR (v_{max}^{kBr}): cm^{-1} 3,414, 2,945, 1,601, 1,494, 1,379, 1,306, 1,183, 816, 674, etc.
EI–MS: DART-MS m/z 287 [M]$^+$, 325, 317, 309, 303, and 282.
Scan: 37 times: 0.5–1.1 minutes, frag = 150.0 V

The previous study identified this compound as a flavonoid (Tien et al., 2016). Based on this work and on combining and comparing all the NMR data, compound CMS-3 was identified as kaempferol. The molecular formula was established as $C_{15}H_{10}O_6$ based on a molecular ion peak at m/z 287 [M$^+$H]$^+$.

5.3.3 CANDIDACIDAL ASSAY

R. arboretum-purified compounds (**1**) (csm-3), (**2**) (tms-10), and **RAM fr2** showed significant candidacidal activity equal to 33%, 41.7%, and 31%, respectively, at a concentration of 1 mg/mL. The positive control (i.e., pooled sera) showed 39% candidacidal activity, while the normal control (Hank's solution) showed 27% candidacidal activity at the same concentration (Figure 5.3).

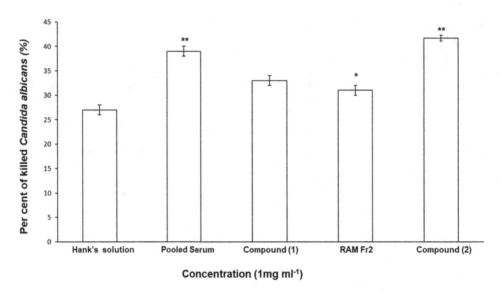

FIGURE 5.3   Percent of killed *C. albicans* evaluated by candidacidal assay.

5.3.4   Cytotoxicity Activity and H_2O_2 Dose–Response Curve of the Extracts

The role of H_2O_2 on cell viability was checked at different concentrations from 2 to 0.03125 mM for 18 hours. The previous work by Piao et al. revealed that RAW 264.7 macrophages treated with 12 hours of exposure to 0.5 mM H_2O_2 induced apoptosis rather than necrosis (Kwon et al., 2016). Based on these data, we showed that H_2O_2 significantly affected cell viability at a dose of up to 500 µM (the sublethal dose used for experiments) compared to the untreated control. The calculated LC_{50} was approximately 250 µM H_2O_2 (see Suppl. Table 5.2).

5.3.4.1   Dose Curve of a Purified Isolated Compound

The viability assay performed using different doses of isolated compounds **(1)**, **(2)**, and **RAM fr2** extract showed an increase in the cell viability of RAW 246.7 cells (cells concomitantly treated with 0.5 mM H_2O_2) (Figure 5.4 and Table 5.1). Compounds **(1)**, **(2)**, and **RAM fr2** extract exerted protective effects against H_2O_2-induced cytotoxicity, especially at 20 and 40 µg/mL, compared to the control (untreated cells).

5.3.4.2   Effect of Purified Isolated Compounds on RAW264.7 Cells at Different Time Points

Compounds **(1)**, **(2)**, and **RAM fr2** (40 µg/mL each) were added at three different times (i.e., pre, simultaneously, and post addition of H_2O_2 to RAW 246.7 cells). The effects of each compound on RAW264.7 cells at different times were recorded as percent cell viability (Figure 5.5). Cell viability was found to be the maximal when the extracts were added before H_2O_2 treatment.

5.3.4.3   Estimation of Nitric Oxide (NO) Production

Compounds **(1)**, **(2)**, and **RAM fr2** stimulated RAW264.7 cells to produce iNOS. The percent of stimulation in pretreatments showed maximum iNOS production, i.e., 644%, 802%, and 460% for compounds **(1)**, **(2)**, and control **RAM fr2**, respectively (Figure 5.6).

5.3.5   Docking Studies

Docking studies were performed on TNF-α and IL-1β by using Molecular Operating Environment (MOE 2016) software. Three-dimensional crystal structures of the proteins were obtained from the

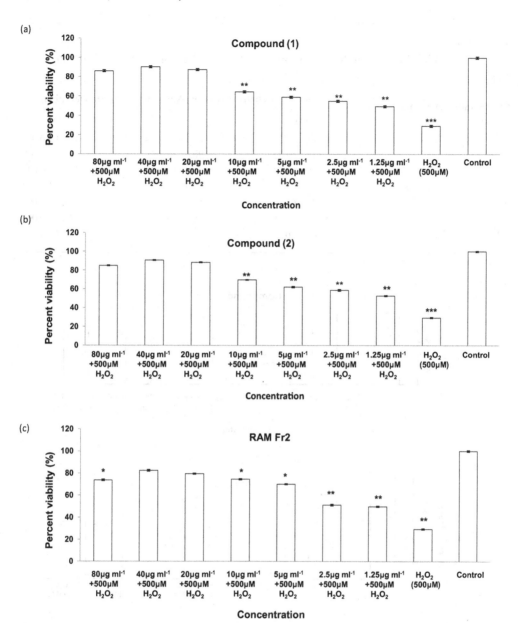

FIGURE 5.4 RAW 264.7 cells were pretreated with purified compound (**1**) (a), purified compound (**2**) (b), and isolated fraction RAM Fr2 (c).

RCSB Protein Data Bank (PDB). The accession numbers of the obtained proteins are 2AZ5 (co-crystallized with a small-molecule inhibitor 307) for TNF-α and 3O4O for IL-1β. Three-dimensional interaction plots of the identified compounds kaempferol and ursolic acid at the binding site of TNF-α are shown in Figure 5.7. Kaempferol forms two hydrogen bond interactions and three π–π stacking interactions with the amino acid residues of TNF-α. Ser60 forms a hydrogen bond interaction with the hydroxyl groups at 2.87 Å. Tyr151 forms hydrogen bond interactions with the carbonyl oxygen of the chromen-4-one ring. Tyr59 forms bifurcated π–π stacking interactions with the chromen-4-one ring. Leu120 forms an amide–π interaction with the phenyl ring (Figure 5.7a). Ursolic acid forms hydrogen bond interactions with Leu120 and Tyr151. Tyr59 stabilizes the ligand–enzyme complex by forming a π–alkyl type of hydrophobic interaction (Figure 5.7b). The computed binding energy values for

TABLE 5.1

Percent Cell Viability of Isolated Compounds (1), (2) and RAM Fr2 in the RAW 264.7-Cell Line

Concentration of Extracts (μg/mL)	Percent of Cell Viability (Mean Value with Standard Deviation)		
	(1)	(2)	RAM Fr2
80	86.17 ± 0.026	84.98 ± 0.044	73.73 ± 0.045
40	90.37 ± 0.032	90.69 ± 0.031	82.44 ± 0.14
20	87.55 ± 0.072	88.43 ± 0.048	79.45 ± 0.034
10	64.28 ± 0.072	69.72 ± 0.027	74.51 ± 0.124
5	58.85 ± 0.067	61.93 ± 0.073	70.09 ± 0.092
2.5	54.84 ± 0.056	58.52 ± 0.087	51.15 ± 0.069
1.25	49.45 ± 0.0123	52.5 ± 0.017	49.77 ± 0.045

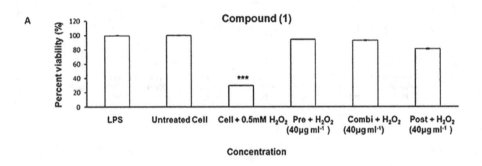

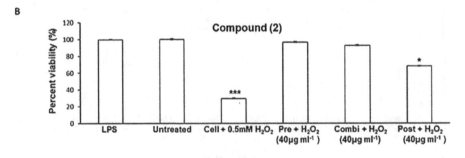

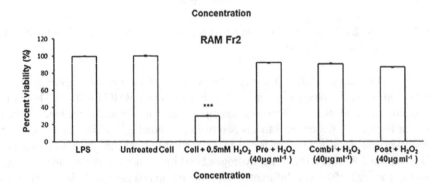

FIGURE 5.5   RAW 264.7 cells treated at different times with purified compound 1 (A), purified compound 2 (B), and isolated fraction RAM Fr2 (C) and evaluation of per cent cell viability..

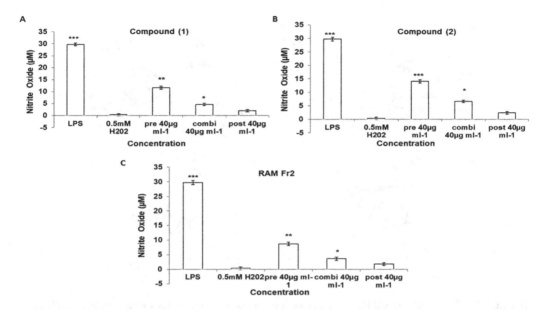

FIGURE 5.6 RAW 264.7 cells treated at different times with purified compound 1 (A), purified compound 2 (B), and isolated fraction RAM Fr2 (C) and evaluation of nitric oxide production.

kaempferol and ursolic acid at the binding site of TNF-α are −5.6430 kcal/mol and −5.6090 kcal/mol, respectively.

Three-dimensional interaction plots of the identified compounds at the binding site of IL-1β are shown in Figure 5.7. Kaempferol forms five hydrogen bond interactions with the amino acid residues present in the binding site of IL-1β. Gln14 forms hydrogen bond interactions with the hydroxyl group at a distance of 2.43 Å. Asn166 forms a bifurcated hydrogen bond interaction with the chromen-4-one ring carbonyl and hydrogen–oxygen atom at distances of 2.89 Å and 3.05 Å, respectively, while Phe167 also establishes bifurcated hydrogen bond interactions with the hydroxyl group of the chromen-4-one ring at distances of 2.24 Å and 2.80 Å (Figure 5.7c). In ursolic acid, the carboxylic group forms two hydrogen bond interactions with Phe167 and Val170. His178 also stabilizes the ligand–enzyme complex by forming a π–alkyl type of hydrophobic interaction (Figure 5.7d). The computed binding energy values for kaempferol and ursolic acid at the binding site of IL-1β are −6.7754 kcal/mol and 5.7513 kcal/mol, respectively.

5.4 DISCUSSION

R. arboreum is considered a rich source of secondary metabolites[6] and has multiple pharmacological properties (Tewari et al., 2018; Prakash et al., 2016; Raza et al., 2015; Roy et al., 2014; Sonar et al., 2012, 2013). *R. arboreum* aqueous extracts showed excellent free-radical scavenging activity against DPPH, ABTS, and FRAP; indeed, the previous studies reported strong antioxidant activity of *R. arboreum* leaf extracts. Hydromethanolic leaf extracts of *R. arboreum* prepared through maceration have been analysed for antioxidant activity following DPPH and FRAP assays (Prakash et al., 2016, 2007). Our data revealed that the *R. arboretum* methanolic extract showed maximum scavenging activity by DPPH, ABTS, and FRAP assays, in accordance with the previous work in which the methanolic leaf extract retained 78.60% antioxidant activity at 500 µg/mL (Kumar et al., 2014). Aziz and Saha (2020) revealed that uroslic acid was completely dissolved in chloroform and methanol, and its mass spectrum is m/z 456. Its IR spectrum absorption band at 3,421 cm⁻¹ due to the O–H stretching vibration, 284 and 1,171 cm⁻¹ due to the C–O stretching vibrations, 2,925 and 2,871 cm⁻¹ for saturated C–H stretching vibrations, C=O stretching vibrations was indicated by the absorption band

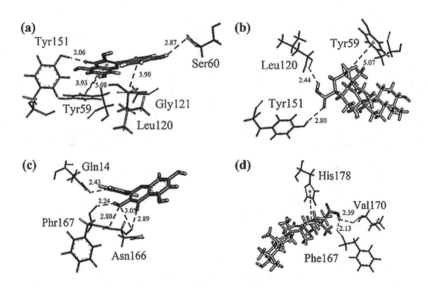

FIGURE 5.7 Three-dimensional interaction plots of the identified compounds kaempferol and ursolic acid at the binding site of TNF-α (a and b) and IL-1β (c and d).

found at 1,687 cm^{-1}, and bands at 1,456 and 1,376 cm^{-1} due to C–H bending vibrations, respectively. Our study showed the EI–MS spectrum showed a molecular ion peak at m/z 457 that corresponds to the molecular formula $C_{30}H_{48}O_3$. IR (v_{max}^{kBr}): 3435, 2941, 1704, 1033, and 755. EI–MS: (DART-MS) m/z 457 [M]$^+$, other peaks for the fragment ions were obtained at 455, 439, 249, 191, and 130 m/z. Scan: 13 times: 0.7–0.9 minutes frag = 150.0 V, ISOCRATIC.

The presence of in the 1H NMR spectrum, the one-proton unresolved singlet at δ 5.12 showed the presence of an olefinic proton at C-12 in the molecule. This was supported by peaks at δ 138.1 125.4 in the 13C NMR spectrum, which indicated the presence of an olefinic bond. The triplet at δ 3.08 1H, J=7.2 Hz) clearly showed the presence of >CHOH group, which was further supported by the signal at δ 78.8 in the 13C NMR spectrum, indicating the carbon at position-3 in the structure. The presence of seven methyl groups in the compound was ascertained by the five 3H singlets at δ 0.97, 0.86, 0.80, 0.69, 0.66, and two 3H doublets at δ 0.83 and 0.75 in the 1H NMR spectrum. The ^{13}C NMR spectrum showed 28 signals for 30 carbons. The two intensified peaks at δ 38.6 and 27.9 were obtained due to four carbons. The carbon of carboxyl group (C-28) was indicated by the signal at δ 180.7. In our studies, the signal appearing at 11.02 δ indicates the presence of a —COOH group at 28th position in the structure. The signal appearing at 3.3 δ indicates the presence of –CH—OH group at three positions—the –OH group accounting for the peak of 4.28 δ. The presence of signal at 5.12 δ signifies the presence of an unsaturated double bound. The seven –CH$_3$ groups appeared between 0.6 and 1.3. The shift position of the isolated compound was almost identical with that of the authentic sample of uroslic acid.

The previous study suggested a strong relationship between the antioxidant activity of herbal extracts and their immunomodulatory potential (Roy et al., 2014). It has been observed that such immunomodulatory properties of plant extracts could be imputable to phenolic and flavonoid compounds (Kumar et al., 2014). For example, Belapurkar et al., showed potent antioxidant effects of a polyherbal preparation (Belapurkar et al., 2014). These extracts possess immunomodulatory potential, given their role in neutralizing reactive oxygen species (ROS) and scavenging free radicals. *In vitro* and *in vivo* studies have shown that phenolic compounds induce macrophage modulation and inflammatory mediator secretion from macrophages and other leucocytes (Sonar et al., 2013; Gomez et al., 2010). In our preliminary study, the *R. arboreum* methanolic extract demonstrated the presence of phenolic and flavonoid compounds, which suggested the presence of significant antioxidant properties and immunomodulatory activity of the *R. arboreum* extract (Rawat et al., 2020). Similarly, the immunomodulatory activity of a methanolic extract of fruit and bark of *Ficus glomerata* Roxb has been evaluated using a nitroblue tetrazolium (NBT) assay, phagocytosis of killed *C. albicans* assay, and chemotaxis assay (Srikumar et al., 2005). Macrophages are the first line of defense in innate immunity against microbial infection, engulfing, digesting, and presenting antigens to trigger adaptive immune responses. For this reason, and for its availability and experimental reproducibility, we performed an *in vitro* study on RAW 264.7 macrophages (This cell line is frequently used to study immune cellular targets, especially those related to the inflammatory process.). Our data showed that compound 2 possessed great candidacidal effects, suggesting antifungal potential that needs to be corroborated in the future by other assays. Moreover, we demonstrated that compounds (**1**) and (**2**) and isolated fraction **RAM Fr2** at 40 µg/mL had the best protective activity against H_2O_2 cytotoxicity (Figure 5.4). *R. arboreum* extracts seem to not affect cell viability and thus cannot be considered toxic to the cells, even if other studies are necessary to confirm these data. Moreover, the protective effects of the extracts were more evident when the pretreatment was performed (Figure 5.5). This result is in line with the previous work in which taurine chloramine protected RAW 264.7 cells from apoptosis caused by H_2O_2. Indeed, macrophages treated with 0.4 mM H_2O_2 underwent apoptosis without showing immediate signs of necrosis, and the cells pretreated with taurine chloramine were protected from such insult (Piao et al., 2011). The pretreatment of cells with *R. arboreum* extracts can result in the transport of phytoconstituents within the cell, where they can exert scavenger activity against reactive oxygen species (ROS) generated within and outside the cell.

Previous studies revealed that NO levels in LPS-induced RAW264.7 macrophages could be suppressed by plant extracts with high polyphenol content (Kwon et al., 2016; Alley et al., 1988). We showed that pretreatment with *R. arboreum* extract demonstrated immunomodulatory activity on RAW 264.7 cells, increasing significantly with NO production (Figure 5.6). The production of NO is strictly related to the immune response, and this increase observed in pretreated RAW 264.7 cells suggests a potential therapeutic tool for NO and a quantitative index of macrophage activation. According to He et al., (2019), *R. arboreum* extracts significantly attenuated cell damage, which can be impaired by upregulating iNOS expression (He et al., 2019). Nonetheless, such hypotheses should be confirmed by quantitative analysis, such as protein or gene expression assays (i.e., real-time PCR or Western blot). Moreover, NO production, as demonstrated by pretreatment with *R. arboreum* extracts, can be associated with cell viability, again emphasizing that this plant can have a role in the macrophage immune response.

Based on *in vitro* results on the RAW 264.7 macrophage cell line and literature work, we performed docking studies on proinflammatory mediators such as TNF-α and IL-1β (Shawky et al., 2020). Several studies revealed that these targets are involved in inflammation and can be easily studied by RAW 264.7-cell lines (Liu et al., 2018; Pan et al., 2017). Indeed, like our data, a recent work reported the effects of novel peptides from *Porphyridium purpureum* in RAW 264.7 cells (Kavitha et al., 2019). The two peptides reduced the secretion of TNF-α and IL-6 in RAW 264.7 cells at noncytotoxic concentrations, and, by a docking study, they were shown to target the scavenging receptors CD36 and SRA1 and the Map Kinase p38. In another study, proinflammatory cytokines (IL-6 and TNF-α) were inhibited by extracts and isolated compounds derived from *Origanum vulgare* (Mir et al., 2020). Our molecular docking data revealed that the two identified compounds could inhibit the two cytokine mediators with high affinity, as suggested by the computed binding energy values for compound 1 and compound 2 at the binding sites of IL-1β (−6.7754 kcal/mol and 5.7513 kcal/mol, respectively) and TNF-α (−5.6430 kcal/mol and −5.6090 kcal/mol, respectively) (Figure 5.7).

These results suggest that NO cell viability was not altered, and that pretreatment with *R. arboreum* extracts can modulate the immune response by influencing IL-1β and TNF-α cytokines. More data are needed to reinforce such hypotheses.

5.5 CONCLUSION AND PROSPECTS

The methanolic leaf extract of *R. arboreum* exhibited free radical scavenging activity and reducing potential, suggesting high antioxidant properties. The preparation of *R. arboreum* extracts with the increased polarity of organic solvents using Soxhlet extraction has shown a rich profile of phyto-constituents, especially phenolics and flavonoids of pharmacological relevance. The identification of purified compounds confirmed the presence of ursolic acid and kaempferol (Tien et al., 2016; Sathyadevi and Subramanian 2015).

The viability of macrophage cells was found to be the maximal when the purified compounds ursolic acid and kaempferol were added before H_2O_2 treatment, which clearly demonstrated that pre-treated cells could face H_2O_2 induced stress. Thus, prior treatment of cells with isolated *R. arboreum* compounds probably resulted in the transport of phytochemicals inside the cell, suggesting the ability to scavenge ROS generated in the cell. Therefore, isolated compounds (ursolic acid and kaempferol) can be considered the active constituents of *R. arboreum* exhibiting potential anti-inflammatory and antioxidant properties. However, further investigations and toxicological assays are needed to support these preliminary results on the immunomodulatory activity *of R. arboreum* extracts.

ACKNOWLEDEGEMENTS

The authors are thankful to the Department of Biotechnology, Graphic Era University, Dehradun, India for providing all laboratory facilities, CDRI, Lucknow, India for providing instrumentation facilities and Syed Mohsin Waheed for arranging technical support related to cell culture work at JNU, New Delhi, India.

REFERENCES

Alley, M.C., Scudiero, D.A., Monks, A., Hursey, M.L., Czerwinski, M.J., Fine, D.L., Abbott, B.J., Mayo, J.G., Shoemaker, R.H., and Boyd, M.R. (1988). Feasibility of drug screening with panels of human tumor cell lines using a microculture tetrazolium assay. *Cancer Research, 48*, 589–601.

Anand, J., Upadhyaya, B., Rawat, P., and Rai, N. (2014). Biochemical characterization and pharmacognostic evaluation of purified catechins in green tea (*Camellia sinensis*) cultivars of India. *Three Biotech, 5*(3), 285–294.

Aziz, S. and Saha, K. (2020). Isolation and structural studies on chemical constituents from *Catharanthus roseus* leaf ethyl acetate extract. *Scholars Academic Journal of Bioscience, 8*, 411–416.

Belapurkar, P., Goyal, P., and Tiwari-Barua, P. (2014). Immunomodulatory effects of triphala and its individual constituents: a review. *Indian Journal of Pharmaceutical Science, 76*(6), 467–475.

Benzie, I.F. and Strain, J.J. (1996). The ferric reducing ability of plasma (FRAP) as a measure of antioxidant power: the FRAP assay. *Analytical Biochemistry, 239*, 70–76.

Benzie, I.F.F. and Strain, J.J (1999). Ferric reducing/antioxidant power assay: direct measure of total antioxidant activity of biological fluids and modified version for simultaneous measurement of total antioxidant power and ascorbic acid concentration. *Methods of Enzymology, 299*, 15–27.

Bhandari, L. and Rajbhandari, M. (2014). Isolation of quercetin from flower petals, estimation of total phenolic, total flavonoid and antioxidant activity of the different parts of *Rhododendron arboreum* Smith, *Scientific World ,.12*(12), 34–40.

Brand-Williams, W., Cuvelier, M.E., and Berset, C. (1995). Use of a free radical method to evaluate antioxidant activity. *Lebensmittel-Wissenschaft Technologie-Food Science and Technology, 28*, 25–30.

Gill, S., Panthari, P., and Kharkwal, H. (2015). Phytochemical investigation of high altitude medicinal plants *Cinnamomum tamala* (Buch-ham) Nees and Eberm and *Rhododendron arboreum* Smith. *American Journal of Phytomedicine and Clinical Therapeutics, 3*, 512–528.

Gomez-Flores, R., Hernández-Martínez, H., Tamez-Guerra, P., Tamez-Guerra, R., Quintanilla-Licea, R., and Monreal-Cuevas, E. (2010). Antitumor and immunomodulating potential of *Coriandrum sativum, Piper nigrum,* and *Cinnamomum zeylanicum. Journal of Natural Products, 3,* 54–63.

He, M.T., Lee, A.Y., Park, C.H., and Cho. E.J. (2019). Protective effect of *Cordyceps militaris* against hydrogen peroxide-induced oxidative stress *in vitro. Nutrition Research and Practice, 13,* 279.

Janakiraman, M. and Jeyaprakash. K. (2015). Evaluation of phytochemical compounds in leaf extract of *Vitex negundo* L. using TLC, UV–VIS and FTIR analysis. *International Journal of Health Science and Research, 5,* 289–295.

Kavitha, M.D., Gouda, K.G.M., Rao, S.J.A., Shilpa, T.S., Shetty, N.P., and Sarada, R. (2019). Atheroprotective effect of novel peptides from *Porphyridium purpureum* in RAW 264.7 macrophage cell line and its molecular docking study. *Biotechnology Letters, 41,* 91–106.

Keat, N.B., Umar, R.U., Lajis, N.H., Chen, T.Y., Li, T.Y., Rahmani, M., and Sukari, M.A. (2010). Chemical constituents from two weed species of Spermacoce (Rubiaceae). *Malaysian Journal of Analytical Science, 14,* 6–11.

Kiruba, S., Mahesh, M., Nisha, S., Paul, Z.M., and Jeeva, S. (2011). Phytochemical analysis of the flower extracts of *Rhododendron arboreum* Sm. ssp. *nilagiricum* (Zenker) Tagg. *Asian Pacific Journal of Tropical Biomedicine, 1*(2), S284–S286.

Kumar, D., Arora, S., Kumar, M., Thakur, M.K., and Singh, A.P. (2014). Membrane stabilizing activity and antioxidant effect of *Sida cordata* (Burm F.) bioss and *Rhododendron arboreum* Sm. leaves. *International Journal of Pharmaceutical Sciences and Research, 5,* 165–170.

Kumar, V., Suri, S., Prasad, R., Gat, Y., Sangma, C., Jakhu, H., and Sharma, M. (2019). Bioactive compounds, health benefits and utilization of *Rhododendron*: a comprehensive review. *Agriculture and Food Security, 8*(1), 6.

Kumari, K., Srivastva, H., and Mudgal, V. (2015). Medicinal importance and utilization of *Rhododendron*-a review. *Research in Environment and Life Science, 8*(4), 761–766.

Kwon, D.H., Cheon, J.M., Choi, E.O., Jeong, J.W., Lee, K.W., Kim, K.Y., Kim, S.G., Kim, S., Hong, S.H., Park, C., Hwang, H.J., and Choi, Y.H. (2016). The immunomodulatory activity of *Mori folium*, the leaf of *Morus alba* L., in RAW 264.7 macrophages *in vitro. Journal of Cancer Prevention, 21,* 144–151.

Liu, S., Jia, H., Hou, S., Xin, T., Guo, X., Zhang, G., Gao, X., Li, M., Zhu, W., and Zhu, H. (2018). Recombinant Mtb9.8 of *Mycobacterium bovis* stimulates TNF-alpha and IL-1beta secretion by RAW264.7 macrophages through activation of NF-kappaB pathway via TLR2. *Scientific Reports, 8,* 1928.

Lopez-Garcia, S., Castaneda-Sanchez, J., Jimenez-Arellanes, A., Dominguez-Lopez, L., Castro-Mussot, M., Hernandez-Sanchez, J., and Luna-Herrera, J. (2015). Macrophage activation by Ursolic and Oleanolic acids during mycobacterial infection. *Molecules, 20,* 14348–14364.

Madhvi, S.K., Sharma, M., Iqbal, J., Younis, M., and Sheikh, R. (2020). Phytochemical analysis, total flavonoid, phenolic contents and antioxidant activity of extracts from the leaves of *Rhododendron arboreum. Research Journal of Pharmacy and Technology, 13,* 1701–1706.

Mir, R.H., Sawhney, G., Verma, R., Ahmad, B., Kumar, P., Ranjana, S., Bhagat, A., Madishetti, S., Ahmed, Z., Jachak, S.M., Choi, S., and Masoodi, M.H. (2020). *Oreganum Vulgare*: *In vitro* assessment of cytotoxicity, molecular docking studies, antioxidant, and evaluation of anti-inflammatory activity in LPS stimulated RAW 264.7 cells. *Medicinal Chemistry, 17*(9), 983–993.

Nisar, M., Ali, S., and Qaisar, M. (2011). Preliminary phytochemical screening of flowers, leaves, bark, stem and roots of *Rhododendron arboreum. Middle-East Journal of Scientific Research, 10,* 472–476.

Orwa, C., Mutua, A., Kindt, R., Jamnadass, R., and Simons, A. (2009). *Agroforestree Database: A Tree Reference and Selection Guide, version 4.* World Agroforestry Centre, Nairobi. In *Agroforestree Database: a tree reference and selection guide. Version 4.*

Othman, A.R., Abdullah, N., Ahmad, S., Ismail, I.S., and Zakaria, M.P. (2015). Elucidation of *in vitro* anti-inflammatory bioactive compounds isolated from *Jatropha curcas* L. plant root. *BioMed Central complementary and alternative medicine, 15*(1), 1–10.

Painuli, S., Joshi, S., Bhardwaj, A., Meena, R.C., Misra, K., Rai, N., and Kumar, N. (2018). *In vitro* antioxidant and anticancer activities of leaf extracts of *Rhododendron arboreum* and *Rhododendron* campanulatum from Uttarakhand region of India. *Pharmacognosy Magazine, 14,* 294.

Painuli, S., Rai, N., and Kumar. N. (2016). Gas chromatography and mass spectrometry analysis of a methanolic extract of leaves of *Rhododendron arboreum. Asian Journal of Pharmaceutical and Clinical Research, 9,* 101–104.

Pan, G., Xie, Z., Huang, S., Tai, Y., Cai, Q., Jiang, W., Sun, J., and Yuan, Y. (2017). Immune-enhancing effects of polysaccharides extracted from *Lilium lancifolium* thunb. *International Immunopharmacology, 52,* 119–126.

Patil, J., Jalalpure, S., Hamid, S., and Ahirrao, R. (2010). *In vitro* immunomodulatory activity of extracts of Bauhinia vareigata Linn bark on human neutrophils. *Iranian Journal of Pharmacology and Therapeutics*, 9, 41–46.

Paudel, A., Panthee, S., Shakya, S., Amatya, S., Shrestha, T.M., and Amatya, M.P. (2011). Phytochemical and antibacterial properties of *Rhododendron campanulatum* from Nepal. *Asian Journal of Traditional Medicines*, 6, 252–258.

Piao, S., Cha, Y.N., and Kim, C. (2011). Taurine chloramine protects RAW 264.7 macrophages against hydrogen peroxide-induced apoptosis by increasing antioxidants. *Journal Clinical Biochemistry and Nutrition*, 49, 50–56.

Prakash, D., Upadhyay, G., Singh, B., Dhakarey, R., Kumar, S., and Singh, K. (2007). Free-radical scavenging activities of Himalayan *Rhododendrons*. *Current Science*, 92, 526–532.

Prakash, V., Rana, S.,and Sagar, A. (2016). Studies on antibacterial activity of leaf extracts of *Rhododendron arboreum* and *Rhododendron campanulatum*. *International Journal of Current Microbiology and Applied Sciences*, 5, 315–322.

Rawat, P., Anand, J., Kumar, N., and Rai, N. (2020). Biochemical characterization of *Rhododendron arboreum* leaves from Himachal Pradesh region. *Biochemical and Cellular Archives*, 20, 425–428.

Raza, R., Ilyas, Z., Sajid, A., Nisar, M., Khokhar, M.Y., and Iqbal, J. (2015). Identification of highly potent and selective α-Glucosidase inhibitors with antiglycation potential, isolated from *Rhododendron arboreum*. *Records of Natural Products*, 9, 262.

Roy, J., Handique, A., Barua, C., Talukdar, A., Ahmed, F., and Barua, I. (2014). Evaluation of phytoconstituents and assessment of adaptogenic activity *in vivo* in various extracts of *Rhododendron arboreum* (leaves). *Indian Journal of Pharmaceutical and Biological Research*, 2, 49–56.

Sathyadevi, M. and Subramanian, S. (2015). Extraction, isolation and characterization of bioactive flavonoids from the fruits of *Physalis peruviana* Linn extract. *Asian Journal of Pharmaceutical and Clinical Research*, 8, 152–157.

Shawky, E., Nada, A.A., and Ibrahim, R.S. (2020). Potential role of medicinal plants and their constituents in the mitigation of SARS-CoV-2: identifying related therapeutic targets using network pharmacology and molecular docking analyses. *Royal Society of Chemistry Advances*, 10, 27961–27983.

Srikumar, R., Parthasarathy, N.J., and Devi, R.S. (2005). Immunomodulatory activity of triphala on neutrophil functions. *Biological and Pharmaceutical Bulletin*, 28, 1398–1403.

Swaroop, A., Gupta, A.P., and Sinha, A.K. (2005). Simultaneous determination of quercetin, rutin and coumaric acid in flower of *Rhododendron arboreum* by HPTLC. *Chromatographia*, 62, 649–652.

Taralkar, S. and Chattopadhyay, S. (2012). An HPLC method for determination of ursolic acid and betulinic acids from their methanolic extracts of *Vitex Negundo* linn. *Journal of Analytical Bioanalytical Techniques*, 3, 1–6.

Tewari, D., Sah, A.N., and Bawari, S. (2018). Pharmacognostical evaluation of *Rhododendron arboreum* Sm. from uUttarakhand. *Pharmacognosy Journal*, 10(3), 527–532.

Thaipong, K., Boonprakob, U., Crosby, K., Cisneros-Zevallos, L., and Byrne, D.H. (2006). Comparison of ABTS, DPPH, FRAP, and ORAC assays for estimating antioxidant activity from guava fruit extracts. *Journal of Food Composition and Analysis*, 19, 669–675.

Thakur, S. and Sidhu, M. (2014). Phytochemical screening of some traditional medicinal plants. *Research Journal of Pharmaceutical, Biological and Chemical Sciences*, 5, 1088–1097.

Tien, V.N., Duc, L.V., and Thanh, T.B. (2016). Isolated compounds and cardiotonic effect on the isolated rabbit heart of methanolic flower extract of *Nerium oleander* L. *Research Journal of Phytochemistry*, 10, 21–29.

Wang, M.L., Hou, Y.Y., Chiu, Y.S., and Chen, Y.H. (2013). Immunomodulatory activities of *Gelidium amansii* gel extracts on murine RAW 264.7 macrophages. *Journal of Food and Drug Analysis*, 21(4), 397–403.

Supplementary Materials are provided online at: https://www.routledge.com/9781032071732.

6 Herbal and Traditional Medicine Practices against Viral Infections

Perspectives against COVID-19

Satyajit Tripathy, Innocensia Mangoato, Barsha Dassarma,
Abdur Rauf, Syed Uzair Ali Shah, Hlupheka Chabalala,
Ishaq N. Khan and Motlalepula Gilbert Matsabisa

CONTENTS

6.1 INTRODUCTION

Since the outburst of a very contagious novel Severe Acute Respiratory Syndrome Coronavirus 2 (SARS-CoV-2) in late December 2019, the disease, subsequently named COVID-19, has been causing unprecedented havoc at a global scale. COVID-19 was originally reported in Wuhan, China in late 2019 (WHO, 2020; Onat Kadioglu, 2020). On 13th January 2020, the full genome sequencing of the virus revealed that it was a novel coronavirus (GenBank No. MN908947). SARS-CoV-2 is the given official name. The disease caused by SARS-CoV-2 has been designated Coronavirus Disease 2019 (COVID-19) (Kadioglu et al., 2020), and the World Health Organization has declared it a global pandemic (WHO). COVID-19—a modern form of coronavirus that is circulating exponentially at a global level has infected people in more than 180 countries and continues to make headline news coverage worldwide. While the world pushes for accelerated approvals for vaccine development, which is time-consuming, preventative, and unlikely to be a cure, physicians and country leaders are considering repurposing old drugs like Chloroquine (CQ)/Hydroxychloroquine (HCQ), Remdesivir, and other antimalarials as treatment agents. A Brazilian study report warns that the high-dose hydroxychloroquine-group patients had more severe QT prolongation (syndrome of heart rhythm disorder, causes arrhythmias) and with a tendency near higher lethality compared to the low dose. According to the authors of this study, the proportion of mortality rate was overlapping with

DOI: 10.1201/9781003206620-6

patients who did not take CQ (Borba et al., 2020). Thus, work to establish successful therapeutics and inexpensive diagnosis against COVID-19 are desperately required.

In the past two decades, the scientific world has been paying tremendous interest in conventional medicine forms of traditional medicine disciplines, now named as 'Complementary and Alternative Medicine (CAM)', CAM also opened a significant category of systematic, holistic patient care medicine, and this has made private industry change on how they view medicine in large. Plant-derived compounds have become the backbone of herbal medicine and are now a significant component of conventional pharmaceuticals. In fact, drugs extracted from natural products resources, especially those in the plant kingdom, are of tremendous importance in terms of their possible applications for handling a multiplicity of human disorders (Surh, 2011; Patwardhan and Vaidya, 2010). The use of natural products as antiviral, anti-pyretic, with immunomodulating activity, and anti-infectives (Tripathy et al., 2020) were considered with available literature on plants with antiviral activity, and the role of natural products as possible interventions for COVID-19 or what is known on their effects on SARS-COV-2 virus explored will be highlighted in this review.

6.2 METHODOLOGY

This review was based on the study of selected publications on plants, isolated natural compounds, and traditional medicine practices against virus infections and especially COVID-19. To prepare the review, we looked for published scientific publications, reviews, and reports between March 2020 and mid-February 2021. We included 20 studies, which had discussed experimental results on plants and/or the use of traditional medicine against COVID-19. PubMed (http://www.pubmed.com), Scopus (http://www.scopus.com), Google Scholar (http://www.scholar.google.com), Web of Science (www.webofscience.com), and Science Direct were systematically searched using keywords like 'Traditional medicine', 'COVID-19', 'Natural products', 'SARS-COV-2', 'herbal medicine', and 'antivirus'. Furthermore, relevant textbooks, patents, digital documents, and published bulletins from World Health Organizations (WHO) were considered to gather all information on herbal medicine, plant, natural compound used against viral infection in past. The results from the literature review have been organized into different sections.

6.3 RESULTS

6.3.1 Mechanism of Infection by SARS-CoV-2

Coronaviruses are single-stranded plus-sense RNA viruses that belong to the Coronaviridae family and Nidovirales order (Vellingiri et al., 2020). There are four types of coronaviruses namely; α-, β-, δ- and γ-coronaviruses. To date, scientists have identified seven coronaviruses that can cause diseases in human beings, and these include β-coronaviruses such as Severe Acute Respiratory Syndrome Coronavirus (SARS-CoV) and the other being the Middle East respiratory syndrome coronavirus (MERS-CoV). These viruses are susceptible to zoonotic transmissions with a broad infection rate in mammalian and avian species (Vellingiri et al., 2020; Millet and Whittaker, 2018; Wang et al., 2020; Malik et al., 2020). SARS-CoV-2 is a zoonotic virus that belongs to the subgenus Sarbecovirus and shares 96.2% sequence homology with a bat coronavirus. It is thought to have been transmitted to humans via an unidentified intermediary animal vector from which it has spread contagiously from human to human (Vellingiri et al., 2020; Malik et al., 2020).

SARS-CoV-2 infection causes chronic diarrhea and vomiting, as well as a flu-like sickness with nausea, cough, sore throat, and exhaustion. A small percentage of patients, such as the elderly and immunocompromised people, develop acute respiratory distress syndrome (ARDS), septic shock, and multi-organ failure, all of which lead to death. The severe morbidity and mortality associated with SAR-CoV-2 infections are due to the host's excessive and persistent immune response and reaction against infection (excessive Host Reaction Process) (Shetty et al., 2020). The presence of the virus, which has been detected by the immune system, causes the milder symptoms that occur during the viral response. The SARS-CoV-2 envelope spike glycoprotein (S-protein) attaches to its host cell receptor angiotensin-converting enzyme 2 (ACE2) on the cell surface. The viral genome is released once the virus particle has entered a host cell. A viral polymerase is used to clone mRNA, producing many different variants of viral mRNA (Du et al., 2009). The translation into the endoplasmic reticulum (ER) membrane of the translated proteins (spiked protein, membrane, and envelope protein) is achieved by translating these RNA copies through an endoplastal ribosome attachment and released finally by exocytosis. Cytoplasmic ribosomes translate the nucleocapsid proteins into mRNA. The nucleocapside is internalized into the vesicular ER/Golgi (Alanagreh et al., 2020).

The SARS-COV-2 virus replicates and mutates, making vaccine development difficult. COVID-19 does not integrate into the host genome of infected cells, and past research has shown that humans produce a strong immune response to coronaviruses. However, it should be recalled that attempts to develop a vaccine against the SARS-COV-1 of the 2003 outbreak were not successful. Even though the experimental vaccines have shown to have some efficacy in animals, the vaccines exhibited severe immunopathology—a severe hyperactive immune response that caused greater damage to the animals. Despite current evidence, there could be a chance that a vaccine may be developed against the SARS-COV-2. Considering the inadequacies of viral vaccines for the prevention or treatment of viral pandemics, chemotherapeutics still offer hope in the fight for any emerging viral pandemics (Li, 2020).

6.3.2 MEDICINAL PLANTS AND THEIR ROLE IN VIRAL INFECTIONS

Many plants have been used in traditional medicines to treat several illnesses such as fevers, pneumonia, coughs, colds, upper respiratory problems, asthma, headaches, loss of appetite, influenza, lung inflammation, etc. (Thring and Weitz, 2006; Liu NQ et al., 2009; Clarkson et al., 2004). Some plants have been found to have antiviral properties against the herpes simplex virus, while others are effective against the SARS virus. The Star anise, *Illicum verum* Hook. f, has been one of the successful traditional medicinal plant to have been developed into the proprietary drug Tamiflu (oseltamivir) from the pods of the plant. Now Tamiflu is an antiviral indicated for influenza A and B, Swine flu, H1N1 (WHO, 2011), H3N2 (Thorlund, 2011) H5N1, the Avian influenza or 'bird flu' (McKimm-Breschkin, 2013), and the H7N9 Avian flu (Hay, 2013).

The combination of herbs and their identification was based on the American eclectic medicine with evidence from 222 Eclectic physicians in 1919 who provided top herbs for influenza and pneumonia. From these 11 botanical remedies; the following endorsements were recorded (influenza; pneumonia) Table 6.1 (Brinker, 2007). *Echinacea pupurea* is now considered a standard herbal remedy for the flu. Modern research now promotes the use of extracts of *E. purpurea* root, whole plant,

---

TABLE 6.1

Eleven Botanical Remedies Endorsed by Physicians during the Spanish Flu

S.No.	Plant Name & Family	S. No.	Plant Name & Family
01	*Aconitum napellus*, Ranunculaceae	07	*Asclepias tuberose*, Asclepiadaceae
02	*Actaea racemosa*, syn: *Cimicifuga racemosa*, Ranunculaceae	08	*Sanguinaria canadensis*, Papaveraceae
03	*Lobelia inflata*, Campanulaceae	09	*Bryonia alba*, Curcubitaceae
04	*Atropa belladonna*, Solanaceae	10	*Eupatorium perfoliatum*, Asteraceae
05	*Gelsemium sempervirens*, Loganiaceae	11	*Cephalis ipecacuanha*, Rubiaceae
06	*Veratrum viride*, Liliaceae		

Source: Abascal (2006).

or aerial plant juice as more effective in viral respiratory infections (Abascal, 2006). Several clinical studies show that a combination of herbs that include Echinacea in a proprietary product formulation and *E. angustifolia* root in combination with other herbs have been employed successfully for both preventing and treating colds and flu (Abascal, 2006).

6.3.3 Natural Compounds Against Viral Infection

Secondary metabolites produced by plants serve an important role in plant physiology and have a wide range of potential applications, including antioxidants, anti-inflammatory, anti-cancer, antimicrobial, and antiviral properties (Panche et al., 2016; Kumar and Pandey, 2013). Various flavonoids, such as flavanols and flavonols, have been studied extensively for their antiviral properties (Lalani and Poh, 2020; Rane et al., 2020). Few key plants are used in the Indian System of Traditional Medicine for antiviral activities and for conditions with similar symptoms to COVID-19.

Curcumin: Turmeric from *Curcuma longa* L. root is used in the Indian traditional system of medicine for the treatment of diseases of the skin, digestive problems, and pain. It is indeed a staple medicine ingredient in Ayurvedic medicine—a traditional method of healing. Curcumin, the key chemical element in turmeric, has been discovered as having improved antiviral action against viruses such as dengue virus (serotype 2), herpes simplex virus, human immunodeficiency viruses, Zika, and Chikungunya (Rane et al., 2020; Chiang et al., 2005; Patwardhan et al., 2020).

Apigenin: Apigenin is a compound isolated from sweet basil (*Ocimum basilicum L.*) and has proven as an effective anti-hepatitis B virus, anti-adenovirus, as well as against in vitro studies of the African swine fever virus and other RNA viruses (Chiang et al., 2005; Hakobyan et al., 2016).

Pterostilbene: One of the plant secondary metabolites primarily found in the *Vaccunium corymbosum* (Blueberries) and *Pterocarpus marsupium* Roxb. Pterostilbene appears to have a variety of preventative and therapeutic characteristics, including preventing the reproduction of a variety of viruses, including herpes simplex (HSV) 1 and 2, influenza viruses, and others (Rane et al., 2020; McFadden 2013).

Luteolin: It is present in several plants, including broccoli (*Brassica olearacea* var *italic*), pepper (*Piper nigrum L.*), thyme (*Thymus vulgaris L.*), and celery (*Apium graveolus*). Studies have demonstrated that luteolin has beneficial neuroprotective effects. It also has immunomodulatory and antioxidant effects. Both the reactivations of the cell HIV-1 and inhibition of the Epstein–Barr virus are inhibited by the highly appreciated flavones. It has antiviral effects on Chikungunya, Japanese encephalitis viruses (Rane et al., 2020; Fan et al., 2016), and acute coronaviral syndrome (SARS-CoV), in addition to these antivirals (Liu et al., 2020).

Quercetin: Quercetin is graded as one of the six-flavonoid sub-classes. For respiratory syncytial virus type 1, HSV-1, and HSV-2, quercetin shows dose-dependent antiviral action. It has also been suggested that it might be used as an Ebola virus preventive medicine (Fanunza et al., 2020; Nile, et al., 2020).

Kaempferol: Kaempferol is one of the most common glycoside aglyconic flavonoids. Kaempferol, derived from Ficus benjamina L. (Weeping fig) leaves, has been shown to suppress HCMV, HSV-1, HSV-2, and influenza A. (Mani et al., 2020; Zakaryan et al., 2017).

Fisetin: Fisetin is a bioactive molecule of the flavonol group of compounds present in fruits and vegetables like strawberry (*Fragaria ananassa*), apple (*Malus domestica*), persimmon (*Diospyros kaki*), grape (*Vitis vinifera*), onion (*Allium cepa*), and cucumber (*Cucumis sativus*). Fisetin, a flavonol, has been shown that it inhibits viral entry and fusion of virus cells to prevent CHIKV as well as HIV-1 infection (Zakaryan et al., 2017).

Scientific findings are revealing that resveratrol and pterostilbene have antiviral activity in viruses like HIV-1(Rane et al., 2020). Resveratrol has been demonstrated to inhibit the poliovirus receptor (PVR) and the Coronavirus (MERS-CoV), as well as reduce MERS-CoV nucleocapside (N) protein formation (Ghildiyal and Gabrani, 2020; Farshi et al., 2020). Resveratrol has also been demonstrated to suppress PVR, which prevents virus multiplication (Rane and colleagues, 2020).

6.3.4 Traditional Therapy Against COVID-19

As with the many other previous global epidemics associated with coronaviruses, there are currently no scientifically proven and validated western or traditional therapies for the treatment of SARS-CoV-2. The current treatment strategy for COVID-19 is supportive care, which is supplemented by the combination of broad-spectrum antibiotics, antivirals, antimalarials, anti-inflammatory, corticosteroids, and convalescent plasma aimed at treating COVID-19 symptoms (Vellingiri et al., 2020; Yang et al., 2020). Researchers are continually working hard to develop COVID-19 diagnostics, as well as medicines and vaccines for the treatment of SARS-CoV-2. To date, treatment involves the use of drugs such as Remdesivir, Favipiravir (T-705), and Arbitol that have been shown to be effective in *in vitro* studies. Additionally, the use of chloroquine was supported by the drug's efficacy in both *in vitro* and *in vivo* studies (Wang et al., 2020).

Even though there is no specific treatment or standard of care for coronavirus infections, certain methods have been tried to manage the disease. In the Indian system of medicine, these approaches may be categorized to include complementary approaches such as Allopathic, Unani, and Homeopathic treatments (Vellingiri et al., 2020; Ali and Alharbi, 2020). Additionally, there are also claims from various quarters including the use of Traditional Chinese Medicine (TCM) and Korean Oriental Medicine (KOM) for the use of herbal traditional medicines in the treatment of COVID-19 (Muthappan and Ponnaiah, 2020). There is little to no verifiable facts on the use of African traditional remedies to prevent or treat COVID-19. However, there is anecdotal and empirical evidence that African traditional remedies can be used to treat disorders caused by a viral infection and other infectious agents.

COVID-19 has been reported to be treated using Chinese Traditional Herbal Medicine (TCM), either alone or in combination with western treatments. TCM has been shown to cure SARS, which was caused by the SARS-CoV epidemic in 2002, albeit the mechanism of action by which these preparations achieve their pharmacological impact is still unknown (Wang et al., 2020; Yang et al., 2020). Furthermore, there is now even strong evidence to identify multiple herbal formulae and chemical entities contained in TCM with anti-SARS-CoV activity (Yang et al., 2020).

In China, there are now 303 clinical trials examining the efficacy and safety of therapies for COVID-19 patients. There are 50 clinical trials (16.5%) on the use of TCM, with 14 (4.6%) of these research exploring TCM and Western medicine therapies in combination (Yang et al., 2020). The effect of self-made herbal preparations was explored in 22 TCM clinical trials (7.3%), while commercially accessible TCM items were studied in 14 TCM trials (4.6%) (Wang et al., 2020; Yang et al., 2020). Based on these studies, generally used Chinese herbs against COVID-19 include *Astragalus membranaceus, Saposhnikovia divaricata* (Turcz. ex Ledeb.) Schischk., *Glycyrrhiza uralensis* Fisch. ex DC., *Lonicera japonica* Thunb., *Atractylodes macrocephala* Koidz., *Forsythia suspensa* (Thunb.) Vahl, *Platycodon grandiflorus* (Jacq.) A. DC., *Atractylodes lancea* (Thunb.) DC., *Cyrtomium fortunei* J.Sm., *and Agastache rugosa* (Fisch. & C. A. Mey.) Kuntze, (Yang et al., 2020).

A traditional TCM injection made from a preparation documented by Chinese physicians in the 1830s, during the Qing Dynasty, made from a decoction of herbs, *Carthamus tinctorius* L. (Honghua), *Paeonia lactiflora* Pall. (Chishao), Chuanxiong (*Chuanxiong rhizhome*), *Salvia miltiorrhiza* Bunge (Dashen), and *Angelica sinensis* (Oliv.) Diels (Danggui). This formula called Xuebijing, in an injectable form, is found to ameliorate inflammatory reactions, respiratory distress, and hypoxia said to be caused by COVID-19 infection (Yu-Liang Zhang et al, 2020). It is reported that the mechanism (Figure 6.1) through which the Xuebijing works is via the ACE2 and some signaling pathways of hypoxia, factor-1, P13K-Akt and NF-kB through the regulation of PK-1, VEGF-A, B-cell lymphoma-2, TNF, and other undefined targets (Zhang L et al., 2020).

Another TCM herb, *Citrus maxima* (Burm.) Merr. (Huanjuhong), with its isolated chemical compound—naringin, is reported to alleviate multiple respiratory diseases, and its potential application in the prevention and treatment of COVID-19 is proposed (Su et al., 2020).

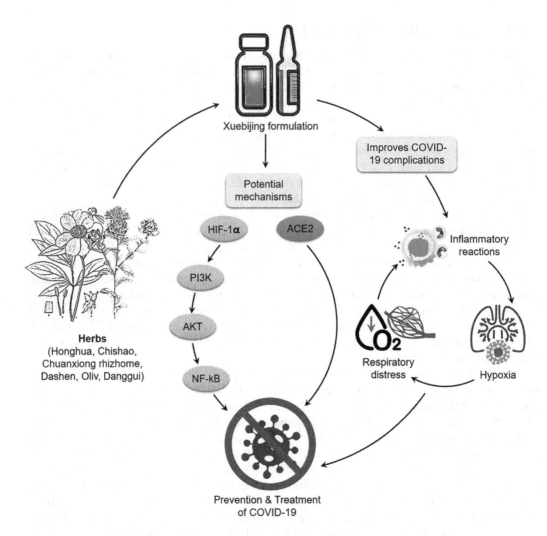

FIGURE 6.1 Schematic presentation on mode of action of Xuebijing against inflammation, respiratory distress, and hypoxia, caused by COVID-19 infection via HIF-1α, ACE2 signaling pathway.

Naringin is said to be anti-tussive, an expectorant, lung function enhancer, lung injury modulator, pulmonary fibrosis attenuator, and antiviral immune response enhancer. A study of a proprietary TCM, Shuanghuanglian oral liquid, by the Shanghai Institute of Materia Medica, Chinese Academy of Science, jointly with the Wuhan Virus Institute, found that the medicine inhibited the coronavirus (http://3g.163.com/news/article/F49EQ5QP055004XG.html). This study is being further continued clinically at the Shanghai Public Health Clinical Centre and the Tongji Hospital, Huangzhong University of Science and Technology (Li et al., 2020); TCM has been very successful in treating other viral diseases such as SAR-COV-2, H1N1, Influenza A H7N9, and Ebola (Cui, 2020), and TCM is thought that it could play a significant role in the cure, management, and treatment of COVID-19. Zhang et al. (2020) presented three TCM proprietary medicines and three TCM decoctions under clinical evaluations (Tables 6.2 and 6.3 respectively). Based on the reviews conducted on TCM formulae and their composition of medicinal plants, it is observed that there are certain key plants (Table 6.4) commonly used in TCM preparations.

Traditional Indian medicine systems are one of the oldest medical systems in human history, and they play a significant role in meeting global health care needs (Jaiswal and Williams, 2017). Ayurveda, Siddha, Unani, Yoga, Naturopathy, and Homeopathy (AYUSH) are traditional Indian

TABLE 6.2

Treatment Efficacy of Three Proprietary TCM Medicines for Coronavirus Pneumonia and New Approved Indications

Drug Name	Site of Study% (Number of Enrolled Participants)	Results	New Indications Proposed for Drug
Xuebijing injection	33 Hospitals (710)	Reduced 28-day mortality by 8.8% in patients with severe pneumonia. Shortened time in mechanical ventilation and reduced length of hospitalization	In severe/critical COVID-19 pneumonia can be indicated for inflammatory response syndrome and multiple organ failure
Jinhua Qinggan granule	Wuhan Frontline Hospitals (102)	Reduced ratio of patients who progressed to severe disease condition by 2/3, shortened fever by 1.5 days and improved WBC, neutrophil and lymphocyte counts	Now for routine treatment of fever, cough, and fatigue in mild-to-moderate cases of coronavirus pneumonia
	Beijing Youan Hospital (80)	Improved on time to convert viral nuclei acid by 2.5 days. Increased WBC and lymphocyte counts	Indicated for mild-to-moderate coronavirus pneumonia
Lianhua Qingwen capsule	23 Hospitals (284)	Improved lung imaging from 64.1% to 83.8%. Clinical cure rate improved from 66.2% to 78.9%. The progression of mild-to-severe cases reduced by 50%.	New indication is for fever, cough, and fatigue in mild-to-moderate cases of coronavirus pneumonia

therapeutic techniques that have been effectively used to treat a variety of ailments (Vellingiri et al., 2020). In Ayurveda, *Rasayana* is a therapy described to enhance the qualities of rasa (body fluids), to give a person to attain health qualities such as longevity, freedom from health disorders, intelligence and enhanced memory power, youthfulness, and optimum development of the physique and sense organs. *Rasayana* medicines are said to increase resistance to various diseases perhaps due to their properties of immunomodulation and antioxidant (Chandresh, 2017). Many *Rasayana* plants are said to exhibit pharmacological qualities like immunomodulation, antioxidation, anti-inflammatory, hypocholesterolaemia, antiasthmatic, hepatoprotective, anti-arrhythmic, cardiotonic, antifungal, antiviral, diuretic, and many other medicinal activities (Pattanayak, 2020; Kumar et al., 2012). The Ministry for Complementary and Alternative Medicine in India has released ayurvedic guidelines on how to improve immunity and self-care for managing COVID-19 (Golechha, 2020). A comprehensive AYUSH medicine strategy for the D-19 focuses on prevention through behavioral improvement, food control, mental recovery, immune-enhancement prophylaxis, and quick treatments (Vellingiri et al., 2020). In mild cases of COVID-19, these AYUSH guidelines recommend drinking hot water, eating hot food, and drinking herbal decoctions, gargling medicated water, inhaling steam, and using local therapies. These steps are claimed to help with the symptomatic alleviation of minor ailments (Tillu et al., 2020). Fresh hot vegetable soups made with *Raphanus sativus* (radish), *Trigonella foenu-graecum* Linn. (trigonella) leaves, drum stick vegetable pods, and pulses (lentils, green gram/mung beans, chickpeas) seasoned with spices like *Zingiber officinale* (ginger), *Allium sativum* (garlic), *Cuminum cyminum* (cumin seeds), and *Brassica nigra* ((L.) K. Koch (black mustard) seeds (Tillu et al., 2020).

Although scientific evidence on coronavirus using Indian medicinal plants is still sparse, one study found that a few Indian plants utilized in the Tamil Nadu traditional medicine system have anti-mouse corona viral activity (a surrogate of SARS-CoV) (Table 6.5). *Vitex trifolia* L. and *Sphaeranthus indicus* L., for example, have been discovered to lower inflammatory cytokines via the NF-kB pathway, which has been linked to respiratory distress in SARS-CoV, while other plants have been shown to have antiviral potential (Vellingiri et al., 2020). Although multiple medicinal

TABLE 6.3
The Clinical Studies of Three TCM Decoctions and Their Efficacy in Treatment of Novel Coronavirus Pneumonia

TCM Decoction	Composition	Clinical Site (Number of Participants in Study)	Treatment Results
Qingfei Paidu	*Ephedrae Herba, Glycyrrhizae Radix et Rhizoma Praeparata cum Melle, Armeniacae Semen Amarum, Gypsum Fibrosum, Cinnamomi Ramulus, Alismatis Rhizoma, Polyporus Artactylodis Macrocephalae Rhizoma, Poria, Bupleuri Radix, Scutellariae Radix, Pinelliae Rhizoma Praeparatum cum Zingibere et Alumine, Zingiberis Rhizoma Recens, Osmundae Rhizoma, Farfarae Flos, Belamcandae Rhizoma, Asari Radix et Rhizoma, Dioscoreae Rhizoma, Aurantii Fructus Immaturus, Citri Reticulatae Pericarpium, Pogostemonis Herba.*	66 Designated medical institutions (1 262)	1,214 cases cured (96.1%) of severe cases 73.7% cured and none progressed to severe cases. Qingfei Paidu two courses over six days reduced lung lesions by 90%
Xuanfei Baidu	*Ephedrae Herba, Pogostemonis Herba, Gypsum Fibrosum, Armeniacae Semen Amarum, Pinelliae Rhizoma Praeparatum, Magnoliae Officinalis Cortex, Artactylodis Rhizoma, Tsaoko Fructus, Poria, Astragali Radix, Paeoniae Radix Rubra, Descurainiae Semen Lepidii Semen, Rhei Radix et Rhizoma, Glycyrrhizae Radix et Rhizoma.*	Wuhan hospital of traditional chinese medicine, hubei provincial hospital of integrated Chinese and western medicine (70 cases in test arm and 50 in control arm	Significant effect on inflammation was seen on TCM decoction with increased lymphocyte count increased by 17.0% c.f. control arm. Clinical cure rate increased by 22.0%
		First affiliated hospital of Henan university of Chinese medicine (40 mild and moderate cases)	Average negative conversion time for nuclei acid was 9.7 days. None of the cases progressed to severe or critical stages. CT diagnosis showed a 85.0% improvement rate
		Wuhan hospital of traditional Chinese medicine, Hubei provincial hospital of integrated traditional Chinese and western medicine and jiangxia makeshift cabin hospital (500 cases)	For mild and moderate cases, Xuanfei significantly alleviated fever, cough, and fatigue. CT diagnosis showed significant improvement. None of the cases progressed to severe form.
Huashi baidu	*Ephedrae Herba, Armeniacae Semen Amarum, Gypsum Fibrosum, Coicis Semen, Atractylodis Rhizoma, Pogostemonis Herba, Artemisiae Annuae Herba, Polygoni Cuspidati Rhizoma et Radix, Verbenae Herba, Imperatae Rhizoma, Descurainiae Semen Lepidii Semen, Citri Grandis Exocarpium, Glycyrrhizae Radix et Rhizoma.*	Wuhan jinyintan hospital (75 severe cases)	CT diagnosis showed marked pulmonary inflammation improvement as well as clinical symptoms. Time to nuclei acid conversion and hospital stay were reduced by 3 days.
		Street Health Center at Jiangjun Road (124 moderate cases)	There was a significant difference in time for nucleic acid negative conversion and in symptoms
		Dongxihu Makeshift Cabin hospital (894 cases of mild and moderate cases, 452 in TCM arm)	There was a significant difference in time for nucleic acid negative conversion and in symptoms

TABLE 6.4
Herbs Commonly Used in TCM Preparations and Recommended for Use in COVID

Atractylodis Rhizoma	*Ephedrae Herba*	
Pogostemonis Herba	*Glycyrrhizae Radix et Rhizoma*	*Pinelliae Rhizoma Praeparatum*
Gypsum Fibrosum	*Descurainiae Semen Lepidii Semen*	*Rhizoma Atractylodis Macrocephalae*
Armeniacae Semen Amarum	*Coicis Semen*	*Poria cocos*

INDEX	1x	2x	3x	4x

Number of times the plant appears in different formulations of TCM* in this review.

TABLE 6.5
Indian Medicinal Plants with Anti-Mouse Corona Viral Activity

S.No.	Plant Name	S.No.	Plant Name
01	*Indigofera tinctoria (AO),*	07	*Gymnema sylvestre*
02	*Leucas aspera*	08	*Sphaeranthus indicus*
03	*Clerodendrum inerme Gaertn*	09	*Evolvulus alsinoides*
04	*Vitex trifolia*	10	*Abutilon indicum*
05	*Cassia alata*	11	*Clitoria ternatea*
06	*Pergularia daemi*		

plants have been identified to either ameliorate the symptoms of COVID-19 or as potential antiviral agents, they will have to be further investigated and validated as possible antiviral agents in the treatment of SARS-CoV-2.

In a recent *in-silico* study, the extracts of Indian traditional medicinal plants for instance, *Nyctanthes arbor-tristis* L. (harsingar*), Tinospora cordifolia* (Willd.) *Miers* (giloy), *Aloe barbadensis miller* (aloe vera), *Curcuma longa* (turmeric), *Azadirachta indica* A. Juss. (neem), *Withania somnifera* (ashwagandha), *Zingiber officinale* (ginger), *Allium cepa* L. (red onion), *Ocimum sanctum* (tulsi), *Cannabis sativa* L. (cannabis), and *Piper nigrum* (black pepper) have been found to inhibit the COVID-19 protease enzyme. The results revealed that all these plants possess COVID-19 protease inhibitor capabilities to some degree. *Nyctanthes arbor-tristis* (harsingar), *Aloe barbadensis miller* (aloe vera), and *Tinospora cordifolia* (giloy) were shown to be the most powerful COVID-19 protease inhibitors of the 11 plants investigated, based on binding affinity, log P, and log S values. *Curcuma longa* (turmeric), *Azadirachta indica* (neem), *Withania somnifera* (ashwagandha), and *Zingiber officinale* are all potential COVID-19 protease inhibitors (ginger). All these plant extracts are reported to have higher inhibitory potentials than chloroquine and hydroxychloroquine (Srivastava et al., 2020).

Unani and Ayurvedic methods of treatment are mainly based on the use of plant botanicals (Revathy et al., 2012). *Glycyrrhiza glabra, Allium cepa, Allium sativum, Ocimum sanctum, Ocimum tenuiflorum, Piper nigrum, Cinnamomum verum, Daucus maritimus, Curcuma longa,* and other plants have been recorded to be utilized in Unani and Ayurvedic medicine. These plants' aqueous extracts, along with lemon juice and honey, have been demonstrated to be beneficial against flu and common cold viral infections. These botanical formulations may hold promise in the prevention, treatment, and management of COVID-19 (Ali and Alharbi, 2020), and they are worth additional investigation in controlled preclinical and clinical research to establish their efficacy and safety.

One study investigated the potential of *Cannabis sativa* L. extracts with a high CBD content in modulating ACE2 expressions in COVID-19 gateway tissues. Extracts of *Cannabis sativa* L. were made to see how they affected gene expression and molecular cascades that drive inflammation and other important cellular functions. Extracts of novel efficacious *Cannabis sativa* L. types may become a helpful addition armament in the treatment of COVID-19 and a superb GRAS (Generally Recognized as Safe) adjunct therapy, according to the findings of this study (Wang et al., 2020). Furthermore, these findings could lead to the development of *cannabis*-based COVID-19 prophylactic treatments that target reduced viral entry via the oral cavity, including as mouthwashes and throat gargles, and can be used in clinical practice as well as at home for self-medication (Wang et al., 2020).

Several compounds have been reported to demonstrate antiviral bioactivities, such as flavonoids, in medical plants (Khaerunnisa et al., 2020). The possible inhibitors of COVID-19 M^{pro} are kaempferol, quercetin, luteolin-7-glucoside, desmethoxycurcumin, naringin, apigenin-7-glucoside, oleuropine, curcumin, catechin, epicatechin gallate, zingerol, gingerol, and allicin, which on the whole are antioxidants. An empirical silicon analysis showed that the drugs are identical in efficacy to nelfinavir, known as Viracept—an antiretroviral drug used to treat Human Immunodeficiency Virus (HIV). Nelfinavir is a class of medication compounds called protease inhibitors (PIs) and is usually used along with other antiretroviral drugs. Much research has documented the presence and abundance in nature of these phenolic compounds, especially medicinal plants (Khaerunnisa et al., 2020).

Sweet wormwood, or *Artemisia annua* L., is a member of the Asteraceae family. It is native to China, but it has long been used as a pressed juice and tea to cure malaria, and its symptoms in Africa and Asia (Efferth, 2017; Lang et al., 2019). Extensive scientific research shows that *A. annua* has antiviral, anti-malarial, anti-diabetic, and anti-cancer properties. An *in vitro* study by Li et al. (2005) reported that compounds extracted from *A. annua* exerted good anti-SARS-CoV activity with 50% effective concentration among 200 herbal extracts tested (Islam et al., 2020; Shahrajabian et al., 2020). These findings serve as evidence that *A. annua* can, therefore, be explored for possible antiviral activity against SARS-CoV-2. *A. annua* has also been investigated for its potential antiviral activity against multitudinous haemorrhagic fevers; including yellow fever, congo fever, Hanta virus fever, typhoid fever, meningococcal infection, and Ebola virus to name a few (Saleem et al., 2020).

Chloroform extracts of *A. annua* have been shown to relax the airway smooth muscle in a mouse model *in vivo*. The relaxing effect was linked to limiting voltage-dependent Ca^{2+} channels-mediated Ca^{2+} influx and increasing BK-mediated K^+ conductance, according to this study. These findings support the development of *A. annua* as a new bronchodilator for the treatment of obstructive airway disorders (Huang et al., 2017). Of particular interest is the use of *A. annua* together with a concoction of other medicinal plants in Madagascar for a product— COVID-Organics (CVO) and claimed to be a cure for COVID-19. Very little is known of the composition of CVO other than that one of the ingredients is *A. annua*. The studies conducted to support this indication for CVO are also very scanty. Although WHO has distanced its support for the Madagascar product, many African countries have shown support including procuring the CVO for their countries. WHO Afro is setting up technical and advisory committees on COVID-19 to help stimulate research on COVID-19 in the continent, and to establish collaborating centers with appropriate facilities and expertise to guide the continental research on COVID-19.

The anti-malarial drug artemisinin isolated from plant *A. annua* L and the plant itself are under investigations against COVID-19 infections in a clinical trial study in Madagascar (https://www.sciencemag.org/news/2020/05/unproven-herbal-remedy-against-covid-19-could-fuel-drug-resistant-malaria-scientists). WHO acknowledges that conventional, complementary, and alternative medicines have many advantages, and that Africa has a long history of conventional and traditional medicine use, which play significant roles in the disease treatments of the continent's population. As alternative COVID-19 therapies, medicinal plants such as *A. annua* L. must be considered, and their effectiveness and adverse side effects must be checked (https://www.afro.who.int/news/who-supports-scientifically-proven-traditional-medicine) before they can be recommended for inclusion in the national drug policies and recommended for new treatment. Similarly, these traditional therapies should be researched and tested before again they could be outrightly rejected by resolutions of the World Health Assembly (WHA) on traditional medicines (WHA56.3 2003; WHA44.34 1991; WHA30.49 1977; WHA40.33 1987; and WHA42.43 1989)

African traditional medicines, although they have been used for centuries, have not been coded and documented nor have their philosophies been described. There are many extemporaneous preparations and formulations used in traditional medicines, and these are neither documented nor systematically tested. There is, however, a lot of literature on the use of African medicinal plants for different diseases. Many medicinal plants have been reported for use in respiratory problems, asthma, fevers, anti-infectives, pneumonia, and as antivirals (Beuscher et al., 1994; Mehrbod et al., 2018; Ogbole et al., 2018; Silva et al., 1997; Konéa et al., 2007; More et al., 2012).

There has been an increased demand for *A. afra* for its use in COVID-19 in South Africa and neighboring countries. The environmental status of *A. afra* will be stressed as the demand for the plants and its unsustainable harvesting continues uncontrolled. The plant is used for coughs, whooping cough, influenza, asthma, fever, hepatitis, loss of appetite, colic, headache, intestinal worms, and malaria (van Wyk, 2008). Other uses of the plant include in stomach and intestinal upset, measles, jaundice, and parasitic infections such as roundworms, pinworms, tapeworms, hookworms, and flukes. The use of *A. afra* in COVID-19 is unknown and not documented. So, the use of the plant against COVID-19 is taken with care, as appropriate preclinical and clinical studies for this indication are needed.

6.4 CONCLUSION AND PROSPECTS

COVID-19 is a new pandemic and very contagious. WHO has described COVID-19 as a pandemic whose research and development of interventions are in their infancy (Wang, 2020). While all attempts and funding have been redirected to find the cure or prevention for COVID-19 and acknowledging the costs of developing the preventative vaccines and chemotherapy, the cost of which may be prohibitory to many patients that really need the medications and that such interventions may take months to years to be made available, the use of traditional medicines should be explored and not forgotten.

TCM and the Indian Systems of Traditional Medicines have been systematized and formalized over the years. Traditional formulations that have been utilized for the prevention and treatment of old and new diseases have been documented in these medical systems. The use of the Chinese wormwood plant, qinghao, *A. annua* led to the isolation of artemisinin and its derivatives artesunate for complicated malaria (Liao, 2009; Tu, 2011), and the Chinese anise plant from which Tamiflu or oseltamivir, used for the treatment of viral infections (Hay, 2013), are giving hope for the continued and sustained research in traditional medicines based on indigenous knowledge for search for new compounds for new and emerging diseases as well as the old diseases such as malaria that have plagued especially the developing and third world.

India, through the AYUSH ministry, has seen the institutionalization of traditional Indian medicines into the national health care system, thus giving birth to internationally accepted complementary medicines disciplines such as, Ayurveda, Siddha, Unani, Yoga, and homeopathy. These medicines disciplines have formalized and standardized traditional practices and formulae without changing them and their chemical extractions to find active chemical moieties.

The African traditional medicine system, despite its century's old use, is being disadvantaged by its lack of systematization, philosophy, Materia Medica, formulae, and general documentation. Its principles and theorems have not been systematically documented. However, the medicines continue to gain popularity, and this is an advantage for the African continent to research and document its medicines and medicinal plants. There is now research emerging on the screening and pharmacological testing of African medicines where some plants that have been known to be used for disease and symptoms of colds, flu, fever, and respiratory infections are beginning to emerge (Mehrbod, 2018; Ogbole, 2018; Silva et al., 1997; Konéa, 2007).

Current research is being driven by the pharmacology of medicinal drugs to determine the active chemicals in the plants, and standardize the dose and understand their pharmacokinetics and pharmacodynamics. Some natural compounds like kaempferol, quercetin, luteolin, curcumin, gingerol, allicin etc., should be studied to find new prophylactic and treatments for the COVID-19 disease. Researchers in China, India, and Africa are also looking at the traditional formulae and repurposing them through clinical trials for new indications for COVID-19 (Zhang, 2020). As per the WHO request, there is a need to continue the preclinical and clinical research and development of traditional medicines. Traditional medicines continue to give evidence of their potential to contribute to finding a prevention or cure treatment for COVID-19. Research in any traditional medicine that has preclinical data or shows promise of safety and efficacy and the WHO should be approached for solidarity support for peer review and joint multicenter clinical trials. We hope that this review will help the researcher as a reference point for the future studies. There is a need to develop research strategies to address emergency research and pandemics using traditional medicines and medicinal plants.

ACKNOWLEDGEMENTS

The authors express gratitude to the Department of Pharmacology University of the Free State, Bloemfontein, Republic of South Africa for providing the opportunity and time to prepare this review. Figure 6.1. Schematic presentation on mode of action of Xuebijing against inflammation, respiratory distress, and hypoxia, caused by COVID-19 infection via HIF-1α, ACE2 signaling pathway.

REFERENCES

Abascal, K. (2006). *Herbs & Influenza: How Herbs Used in the 1918 Flu Pandemic can be Effective Today.* Tigana Press, Montgomery, Illinois.

Alanagreh, L.A., Alzoughool, F., and Atoum, M. (2020). The human coronavirus disease COVID-19: its origin, characteristics, and insights into potential drugs and its mechanisms. *Pathogens, 9*(5), 331.

Ali, I. and Alharbi, O.M. (2020). COVID-19: disease, management, treatment, and social impact. *Science of the Total Environment, 728,* 138861.

Beuscher, N., Bodinet, C., Neumann-Haefelin, D., Marston, A., and Hostettmann, K. (1994). Antiviral activity of African medicinal plants. *Journal of Ethnopharmacology, 42*, 101–109.

Borba, M., de Almeida Val, F., Sampaio, V.S., Alexandre, M.A., Melo, G.C., Brito, M., Mourao, M., Sousa, J.D.B., Guerra, M.V.F., Hajjar, L., and Pinto, R.C. (2020). Chloroquine diphosphate in two different dosages as adjunctive therapy of hospitalized patients with severe respiratory syndrome in the context of coronavirus (SARS-CoV-2) infection: preliminary safety results of a randomized, double-blinded, Phase IIb clinical trial (Cloro Covid-19 Study). *MedRxiv*, JAMA Netw Open, *3* (4): e208857. doi:10.1001/jamanetworkopen.2020.8857.

Brinker, F. (2007). Herbs & Influenza: how herbs used in the 1918 Flu Pandemic can be effective today, American Botanical Council. *Herbal Gram, 75*, 69. https://www.herbalgram.org/resources/herbalgram/issues/75/table-of-contents/article3153/

Chandresh, R., Kaundal, M., and Srivastava, R. (2017). Role of Rasayana herbs as immunomodulator. *International Ayurvedic Medical Journal, 5*, 3643–3648.

Chiang, L.C., Ng, L.T., Cheng, P.W., Chiang, W., and Lin, C.C. (2005). Antiviral activities of extracts and selected pure constituents of *Ocimum basilicum*. *Clinical and Experimental Pharmacology and Physiology, 32*(10), 811–816.

Clarkson, C., Maharaj, V.J., Crouch, N.R., Grace, O.M., Pillay, P., Matsabisa, M.G., Bhagwandin, N., Smith, P.J., and Folb, P.I. (2004). *In vitro* antiplasmodial activity of medicinal plants native to or naturalized in South Africa. *Journal of Ethnopharmacology, 92*(2–3), 177–191.

Cui, H.T., Li, Y.T., Guo, L.Y., Liu, X.G., Wang, L.S., Jia, J.W., Liao, J.B., Miao, J., Zhang, Z.Y., Wang, L., and Wang, H.W. (2020). Traditional Chinese medicine for treatment of coronavirus disease 2019: a review. *Traditional Medicine Research, 5*(2), 65–73.

Du, L., He, Y., Zhou, Y., Liu, S., Zheng, B.J., and Jiang, S. (2009). The spike protein of SARS-CoV-a target for vaccine and therapeutic development. *Nature Reviews Microbiology 7*(3), 226–236.

Efferth, T. (2017). From ancient herb to modern drug: artemisia *annua* and artemisinin for cancer therapy. *Seminars in Cancer Biology, 46*, 65–83.

Fan, W., Qian, S., Qian, P., and Li, X. (2016). Antiviral activity of luteolin against Japanese encephalitis virus. *Virus Research, 15*, 112–116.

Fanunza, E., Iampietro, M., Distinto, S., Corona, A., Quartu, M., Maccioni, E., Horvat, B., and Tramontano, E. (2020). Quercetin blocks Ebola virus infection by counteracting the VP24 interferon inhibitory function. *Antimicrobial Agents and Chemotherapy, 64*(7), 00530–20.

Farshi, P., Kaya, E.C., Hashempour-Baltork, F., and Khosravi-Darani, K. (2020). A comprehensive review on the effect of plant metabolites on coronaviruses: focusing on their molecular docking score and IC50 values. *Mini-Reviews in Medicinal Chemistry, 22*(3), 457–483. Epub ahead of print.

Ghildiyal, R. and Gabrani, R. (2020). Antiviral therapeutics for chikungunya virus. *Expert Opinion on Therapeutic Patents, 30*(6), 467–480.

Golechha, M. (2020). Time to realise the true potential of Ayurveda against COVID-19. *Brain, Behavior, and Immunity, 87*, 130–131.

Hakobyan, A., Arabyan, E., Avetisyan, A., Abroyan, L., Hakobyan, L., and Zakaryan, H. (2016). Apigenin inhibits African swine fever virus infection in vitro. *Archives of Virology, 161*(12), 3445–3453.

Hay, A.J. and Hayden, F.G. (2013). Oseltamivir resistance during treatment of H7N9 infection. *The Lancet, 381*(9885), 2230–2232.

Huang, J., Ma, L.Q., Yang, Y., Wen, N., Zhou, W., Cai, C., Liu, Q.H., and Shen, J. (2017). Chloroform extract of *Artemisia annua* L. Relaxes mouse airway smooth muscle. *Evidence-Based Complementary and Alternative Medicine, 2017*, 9870414.

Islam, M.T., Sarkar, C., El-Kersh, D.M., Jamaddar, S., Uddin, S.J., Shilpi, J.A., and Mubarak, M.S. (2020). Natural products and their derivatives against coronavirus: a review of the non-clinical and pre-clinical data. *Phytotherapy Research, 34*(10), 2471–2492.

Jaiswal, Y.S. and Williams, L.L. (2017). A glimpse of Ayurveda–The forgotten history and principles of Indian traditional medicine. *Journal of Traditional and Complementary Medicine, 7*(1), 50–53.

Kadioglu, O., Saeed, M., Johannes, G.H., and Efferth, T. (2020). Identification of novel compounds against three targets of SARS CoV-2 coronavirus by combined virtual screening and supervised machine learning. *Computers in Biology and Medicine, 133*, 104359.

Khaerunnisa, S., Kurniawan, H., Awaluddin, R., Suhartati, S., and Soetjipto, S. (2020). Potential inhibitor of COVID-19 main protease (Mpro) from several medicinal plant compounds by molecular docking study. *Preprints, 2020*, 2020030226.

Koné, W.M., Atindehou, K.K., Kacou-N, A., and Dosso, M. (2007). Evaluation of 17 medicinal plants from Northern Cote d'Ivoire for their *in vitro* activity against *Streptococcus pneumoniae*. *African Journal of Traditional, Complementary and Alternative Medicines, 4*(1), 17–22.

Kumar, D., Arya, V.,Kaur, R., Bhat, Z.A., Gupta, V.K., and Kumar, V. (2012). A review of immunomodulators in the Indian traditional health care system. *Journal of Microbiology, Immunology and Infection, 45*, 165–184.

Kumar, S. and Pandey, A.K. (2013). Chemistry and biological activities of flavonoids: an overview. *The Scientific World Journal, 2013*, 162750.

Lalani, S. and Poh, C.L. (2020). Flavonoids as antiviral agents for enterovirus A71 (EV-A71). *Viruses, 12*(2), 184.

Lang, S.J., Schmiech, M., Hafner, S., Paetz, C., Steinborn, C., Huber, R., El Gaafary, M., Werner, K., Schmidt, C.Q., Syrovets, T., and Simmet, T. (2019). Antitumor activity of an *Artemisia annua* herbal preparation and identification of active ingredients. *Phytomedicine, 62*, 152962.

Li, H., Zhou, Y., Zhang, M., Wang, H., Zhao, Q., and Liu, J. (2020). Updated approaches against SARS-CoV-2. *Antimicrobial Agents and Chemotherapy, 64*(6), 00483–20.

Li, M., Yang, X., Li, K., and Xie, Y.Q. (2020). Traditional Chinese medicine for novel coronavirus pneumonia treatment: main force or supplement?. *Traditional Medicine Research, 5*(2), 62–4.

Liao, F. (2009). Discovery of Artemisinin (qinghaosu). *Molecules, 14*(2), 5362–5366.

Liu, M., Yu, Q., Yi, Y., Xiao, H., Putra, D.F., Ke, K., Zhang, Q., and Li, P. (2020). Antiviral activities of *Lonicera japonica* Thunb. Components against grouper iridovirus *in vitro* and *in vivo*. *Aquaculture, 519*, 734882.

Liu, N.Q., Van der Kooy, F., and Verpoorte, R. (2009). *Artemisia afra*: a potential flagship for african medicinal plants. *South African Journal of Botany, 75*(2), 185–195.

Malik, Y.S., Sircar, S., Bhat, S., Vinodhkumar, O.R., Tiwari, R., Sah, R., Rabaan, A.A., Rodriguez-Morales, A.J., and Dhama, K. (2020). Emerging coronavirus disease (COVID-19), a pandemic public health emergency with animal linkages: current status update. *Preprints,* 2020030343. doi: 10.20944/preprints202003.0343.v1.

Mani, J.S., Johnson, J.B., Steel, J.C., Broszczak, D.A., Neilsen, P.M., Walsh, K.B., and Naiker, M. (2020). Natural product-derived phytochemicals as potential agents against coronaviruses: a review. *Virus Research, 30,* 197989.

McFadden, D. (2013). A review of pterostilbene antioxidant activity and disease modification. *Oxidative Medicine and Cellular Longevity, 2013,* 575482.

McKimm-Breschkin, J.L. (2013). Influenza neuraminidase inhibitors: antiviral action and mechanisms of resistance. *Influenza and Other Respiratory Viruses, 7,* 25–36.

Mehrbod, P, Abdalla, M.A., Njoya, E.M., Ahmed, A.S., Fotouhi, F., Farahmand, B., Gado, D.A., Tabatabaian, M., Fasanmi, O.G., Eloff, J.N., and McGaw, L.J. (2018). South African medicinal plant extracts active against influenza A virus. *BioMed Central Complementary and Alternative Medicine, 18*(1), 112.

Millet, J. K. and Whittaker. G.R. 2018. Physiological and molecular triggers for SARS-CoV membrane fusion and entry into host cells. *Virology, 517,* 3–8.

More, G., Lall, N., Hussein, A., and Tshikalange, T.E. (2012). Antimicrobial constituents of *Artemisia afra* Jacq. ex Willd. against periodontal pathogens. *Evidence Based Complementary and Alternative Medicine, 2012,* 252758.

Muthappan, S. and Ponnaiah, M. (2020). Time to tread cautiously during public health emergencies: reactions from traditional and complementary/alternative medical systems to ongoing Coronavirus (COVID-19) outbreak. *Journal of Ayurveda and Integrative Medicine, 13*(1), 100315.Ahead of print.

Nile, S.H., Kim, D.H., Nile, A., Park, G.S., Gansukh, E., and Kai, G. (2020). Probing the effect of quercetin 3-glucoside from *Dianthus superbus* L against influenza virus infection- *In vitro* and *in silico* biochemical and toxicological screening. *Food and Chemical Toxicology, 135,* 110985.

Ogbole, O.O., Akinleye, T.E., Segun, P.A., Faleye, T.C., and Adeniji, A.J. (2018). *In vitro* antiviral activity of twenty-seven medicinal plant extracts from Southwest Nigeria against three serotypes of echoviruses. *Virology Journal, 15*(1), 110.

Panche, A.N., Diwan, A.D., and Chandra, S.R. (2016). Flavonoids: an overview. *Journal of Nutritional Science, 5,* 1–15.

Pattanayak, S. (2020). *Use of succulent biomedicines to control COVID-19.* Calcutta Block and Print, Sikdar Bagan street, Kolkata, India.

Patwardhan, B. and Vaidya, A.D. (2010). Natural products drug discovery: accelerating the clinical candidate development using reverse pharmacology approaches. *Indian Journal of Experimental Biology, 48*(3), 220–7.

Patwardhan, M., Morgan, M.T., Dia, V., and D'Souza, D.H. (2020). Heat sensitization of hepatitis A virus and Tulane virus using grape seed extract, gingerol and curcumin. *Food Microbiology, 90,* 103461.

Rane, J.S., Chatterjee, A., Kumar, A., and Ray, S. (2020). Targeting SARS-CoV-2 spike protein of COVID-19 with naturally occurring phytochemicals: an *in silco* study for drug development. *Journal of Biomolecular Structure & Dynamics, 39*(16), 6306–6316.

Revathy, S.S., Rathinamala, R., and Murugesan, M. (2012). Authentication methods for drugs used in Ayurveda, Siddha and Unani Systems of medicine: an overview. *International Journal of Pharmaceutical Sciences and Research, 3*(8), 2352.

Saleem, M., Tanvir, M., Akhtar, M.F., and Saleem, A. (2020). Crimean-Congo hemorrhagic fever: etiology, diagnosis, management and potential alternative therapy. *Asian Pacific Journal of Tropical Medicine, 13*(4), 143.

Shahrajabian, M.H., Sun, W., Shen, H., and Cheng, Q. (2020). Chinese herbal medicine for SARS and SARS-CoV-2 treatment and prevention, encouraging using herbal medicine for COVID-19 outbreak. *Acta Agriculturae Scandinavica, Section B—Soil and Plant Science, 70*(5), 437–443.

Shetty, R., Ghosh, A., Honavar, S.G., Khamar, P., and Sethu, S. (2020). Therapeutic opportunities to manage COVID-19/SARS-CoV-2 infection: present and future. *Indian Journal of Ophthalmology, 68*(5), 693.

Silva, O., Barbosa, S., Diniz, A., Valdeira, M.L., and Gomes, E. (1997). Plant extracts antiviral activity against *Herpes simplex* virus type 1 and African swine fever virus. *International Journal of Pharmacognosy, 35*(1), 12–16.

Srivastava, A.K., Kumar, A., and Misra, N. (2020). On the Inhibition of COVID-19 protease by Indian herbal plants: an *in-silico* investigation. arXiv preprint arXiv: 2004.03411.

Su, W.W., Wang, Y.G., Li, P.B., Wu, H., Zeng, X., Shi, R., Zheng, Y.Y., Li, P.L., and Peng, W. (2020). The potential application of the traditional Chinese herb *Exocarpium Citri grandis* in the prevention and treatment of COVID-19. *Traditional Medicine Research, 5*(3), 160–166.

Surh, Y.J. 2011. Reverse pharmacology applicable for botanical drug development–inspiration from the legacy of traditional wisdom. *Journal of Traditional and Complementary Medicine, 1*(1), 5.

Thorlund, K., Awad, T., Boivin, G., and Thabane, L. (2011). Systematic review of influenza resistance to the neuraminidase inhibitors. *BioMed Central Infectious Diseases, 11*(1), 1–13.

Thring, T.S. and Weitz, F.M. (2006). Medicinal plant use in the Bredasdorp/Elim region of the Southern Overberg in the Western Cape Province of South Africa. *Journal of Ethnopharmacology, 103*(2), 261–275.

Tillu, G., Chaturvedi, S., Chopra, A., and Patwardhan, B. (2020). Public health approach of Ayurveda and Yoga for COVID-19 prophylaxis. *Journal of Alternative and Complementary Medicine, 26*(5), 360–364.

Tripathy, S., Dassarma, B., Roy, S., Chabalala, H., and Matsabisa, M.G. (2020). A review on possible modes of actions of Chloroquine/ Hydroxychloroquine: repurposing to mitigate SAR-COV-2 (COVID 19) pandemic. *International Journal of Antimicrobial Agents, 56*(2), 106028.

Tu, Y. (2011). The discovery of artemisinin (qinghaosu) and gifts from Chinese medicine. *Nature Medicine, 17*(10), 1217–1220.

Van Wyk, B.E. (2008). A broad review of commercially important southern African medicinal plants. *Journal of Ethnopharmacology, 119*(3), 342–355.

Vellingiri, B., Jayaramayya, K., Iyer, M., Narayanasamy, A., Govindasamy, V., Giridharan, B., Ganesan, S., Venugopal, A., Venkatesan, D., Ganesan, H., and Rajagopalan, K. (2020). COVID-19: a promising cure for the global panic. *Science Total Environment, 4*, 138277.

Wang, B., Kovalchuk, A., Li, D., Ilnytskyy, Y., Kovalchuk, I., and Kovalchuk, O. (2020). In search of preventative strategies: novel anti-inflammatory high-CBD *Cannabis Sativa* extracts modulate ACE2 expression in COVID-19 gateway tissues. *Preprints*, 2020040315.

Wang, L.S., Wang, Y.R., Ye, D.W., and Liu, Q.Q. (2020). A review of the 2019 novel Coronavirus (Sars-CoV-2) based on current evidence. *International Journal of Antimicrobial Agents, 19*, 105948.

World Health Organization. (1977). Resolution WHA30. 49 on Promotion and development of training and research in traditional medicine. World Health Organization, Geneva.

World Health Organization. (1987). Resolutions and decisions of regional interest adopted by the fortieth-second World Health Assembly and by the Executive Board at its seventy-ninth and eightieth sessions (No. EM/RC34/4).

World Health Organization. (1989). Resolutions and decisions of regional interest adopted by the forty-second World Health Assembly (42.43) and by the Executive Board at its eighty-third and eighty-fourth sessions (No. EM/RC36/4).

World Health Organization. (2011). Donor Report Pandemic Influenza A, 1 March 2011, WHO, Geneva. http://whqlibdoc.who.int/publications/2012/9789241564427_eng.pdf.

World Health Organization. (2020). Coronavirus disease (COVID-2019) situation reports. Accessed on February 23, 2020. https://www.who.int/emergencies/diseases/novel-coronavirus-2019/situation-reports.

World Health Organization. (1991). SEA/RC44/13-Consideration of resolutions of regional interest adopted by the world health assembly and the executive board (No. SEA/RC44/13). WHO Regional Office for South-East Asia.

World Health Organization. (2003). Resolutions and decisions of regional interest adopted by the fifty-sixth World Health Assembly and by the Executive Board at its 111th and 112th sessions (No. EM/RC50/3).

Yang, Y., Islam, M.S., Wang, J., Li, Y., and Chen, X. (2020). Traditional Chinese medicine in the treatment of patients infected with 2019-new coronavirus (SARS-CoV-2): a review and perspective. *International Journal of Biological Sciences, 16*(10), 1708.

Zakaryan, H., Arabyan, E., Oo, A., and Zandi, K. (2017). Flavonoids: promising natural compounds against viral infections. *Archives of Virology, 162*(9), 2539–2951.

Zhang, L., Chen, J., Ke, C., Zhang, H., Zhang, S., Tang, W., Liu, C., Liu, G., Chen, S., Hu, A., and Sun, W. (2020). Ethanol extract of *Caesalpinia decapetala* inhibits influenza virus infection *in vitro* and *in vivo*. *Viruses, 12*(5), 557.

Zhang, Y.L., Cui, Q., Zhang, D., Ma, X., and Zhang. G.W, (2020). Efficacy of Xuebijing injection for the treatment of coronavirus disease 2019 via network pharmacology. *Traditional Medicine Research, 5*(4), 201–215.

Zhang, Y.L., Zhang, W.Y., Zhao, X.Z., Xiong, J.M. and Zhang, G.W. (2020). Treating Covid-19 by traditional Chinese medicine: a charming strategy? *Traditional Medicine Research, 5*, 178–181.

7 Traditional Herbal Remedies Used in Management of Acute Bronchitis, Common Cold, and Severe Acute Respiratory Syndrome Coronavirus-2 in Southern Africa

Makomborero Nyoni, Tatenda Clive Murashiki,
Farisai Caroline Chibage, Moira Amanda Mubani,
Reward Muzerengwa, Cyprian Mahuni,
Praise Chirilele, Tafadzwa Phyllis Maranjisi,
Caroline Peninah Kwenda, Progress Dube,
Prisca Nonceba Ncube, Munyaradzi Mativavarira
and Deckster Tonny Savadye

CONTENTS

7.1 INTRODUCTION

The use of herbal medicines in the management of communicable and non-communicable diseases is widely accepted and has a very long history of practice (Mahomoodally et al., 2013; James et al., 2018; Saini et al., 2022). Globalisation has resulted in the use of traditional herbal medicines as complementary or alternative medicines (CAM), which are easily accessible and cost-effective therapy options. The ease of access and cost-effectiveness of traditional herbal medicines has resulted in widespread use mainly as CAM (Welz et al., 2019) in regions of the world including Southern Africa.

Populations of Southern Africa use traditional herbal medicines in management of respiratory conditions, including acute bronchitis, common cold (Mahomoodally et al., 2013), and severe acute respiratory syndrome coronavirus (SARS-CoV-2) (Mathew et al., 2020; Attah et al., 2021; Mfengu et al., 2021). In recognition of increasing popularity of traditional herbal medicines, Southern African countries including Zimbabwe and South Africa have developed policies to promote incorporation of traditional herbal medicines and other CAM in local health systems (National Health Strategy for Zimbabwe 2016–2020; James et al., 2018; Abrams et al., 2020). The incorporation of traditional herbal medicines and other CAM in local health systems has led to increased research and efforts to further develop CAM in the management of common diseases, including acute bronchitis, common cold, and SARS-CoV-2.

In Southern Africa, respiratory conditions are commonly treated with remedies prepared from plants including *Achyranthes aspera* Linn., commonly known as *isinama* in Zulu; *moxato* in Tswana; *bhomane* in Sotho, and *udombo* in Ndebele, which contains phytochemicals with weak antiviral activity but high antioxidant, antihistamine, and anti-inflammatory properties. *Achyranthes aspera* acts as mast cell stabiliser and possesses broncho-protecting effects (Younis et al., 2018). In addition, *A. aspera* is rich in ascorbic acid conferring antioxidant properties and boosts immunity against SARS-CoV (Adeleye et al., 2021) and common cold. The anti-inflammatory properties of *Achyranthes aspera* may alleviate inflammation caused by SARS-CoV and may be effective in the management of SARS-CoV-2 (Adeleye et al., 2021). *Achyranthes aspera* is added to other excipients in Hoopinil syrup, which is taken at a dosage of 10 mL consumed six times per day during treatment of common cold and cough (Younis et al., 2018).

A decoction prepared from leaves and roots of *Abrus precatorius* climbers commonly known as rosary pea in English or *amabope* in Ndebele is a potential remedy for management of SARS-CoV-2 (Adeleye et al., 2021). The leaf and root of *Abrus precatorius* contain glycyrrhizin—a phytochemical that possesses anti-SARS-CoV-2 effects (Adeleye et al., 2021). *Allium sativum,* commonly known as garlic, possess significant antiviral activity against coronaviruses (Keyaerts et al., 2007). *Psidium guajava* Linn., commonly known as guava in English or koejawal in Afrikaans, with multipurpose use in management of malaria and bacterial infections was reported to possess anti-SARS-CoV activity (Fukumoto, Goto and Hayashi, 2010; Adeleye et al., 2021). A decoction of *Psidium guajava* Linn. tree leaves possess antiproliferative, antimicrobial (Maroyi, 2018), anti-inflammatory, and immune-system stimulatory activities (Adeleye et al., 2021) and contains phytochemicals with potential to break down coronavirus proteins (Laksmiani et al., 2020; Tungadi et al., 2020; Erlina et al., 2020).

Zingiber officinale, commonly known as ginger, possesses high antioxidant, antimicrobial, and anti-inflammatory activities (Adeleye et al., 2021). Ginger has potential for inhibition of infections from coronaviruses (Laksmiani et al., 2020; Rathinavel et al., 2020; Mbadiko et al., 2020) and is recommended as a component of remedies for the management of SARS-CoV-2 (Prasad, Muthamilarasan and Prasad, 2020; Cinatl et al., 2003). A concoction comprising leaves of *Eucalyptus globulus* and *Psidium guajava* Linn. in water may be taken orally during management of acute bronchitis or common cold (Maroyi, 2018).

Traditional herbal medicines have potential for management of acute respiratory infections; however, the long- and short-term risks of use should be established in order to safeguard human health. This is of importance because traditional herbal medicines containing toxic phytochemicals and secondary metabolites that pose adverse health effects or interact with other drugs taken by patients were reported (Gamaniel, 2000; Auerbach et al., 2012; Fasinu, Bouic and Rosenkranz, 2012). Safety concerns are also linked to adulteration (Newmaster et al., 2013) and poor-quality control during preparation of herbal remedies (Chang et al., 2017). Despite concerns on safety, awareness of the public on potential toxicity of herbal medicines may be poor, and the general perception may be that herbal medicines are always safe. Misinformation on safety of traditional herbal remedies may expose patients to risk of adverse health effects and poisoning through consumption of harmful doses of phytochemicals. Traditional herbal medicines should, therefore, be analysed and documented for safety and efficacy.

In this chapter, we highlight and provide a consolidated and critical coverage of traditional herbal medicines used in the management of acute bronchitis, common colds, and SARS-CoV-2 in

Southern Africa. Information on part of plant used and mode of preparation for traditional herbal remedies are also provided.

7.1.1 Symptoms and Management of Acute Bronchitis

Acute bronchitis is mainly caused by infection with viruses (Clark et al., 2014; Gencay et al., 2010), including rhinovirus, enterovirus, influenza A and B, parainfluenza, coronavirus, human metapneumovirus, and respiratory syncytial virus (Clark et al., 2014; Singh et al., 2021). Bacteria are detected in 1%–10% of cases of acute bronchitis (Clark et al., 2014; Gencay et al., 2010; Macfarlane et al., 2001), and atypical bacteria including *Mycoplasma pneumoniae, Chlamydophila pneumonia*, and *Bordetella pertussis* are rare causes of the disease (Kinkade and Long, 2016).

Acute bronchitis is a temporary inflammatory response to an infection of the bronchi epithelium that arises after a complex chain of events post-infection. A predominant and defining symptom of acute bronchitis is cough (Kinkade and Long, 2016). The primary diagnostic consideration in patients with suspected acute bronchitis is ruling out more serious causes of cough including asthma, exacerbation of chronic obstructive pulmonary disease, heart failure, or pneumonia (Kinkade and Long, 2016). Supportive care and management of symptoms is the mainstay of treatment for acute bronchitis. The role of antibiotics is limited, and over-the-counter medications are often recommended as first line of treatment for acute bronchitis (Kinkade and Long, 2016).

Complementary or alternative medicines are commonly used in management of acute bronchitis (Kinkade and Long, 2016). Rhizomes and tubers of *Pelargonium sidoides*—a plant native to South Africa are used in formulation of remedies for management of acute bronchitis (Nowicki and Murray, 2020). The plant possesses mild antibacterial, antiviral, and mucolytic properties (Hart, 2014).

7.1.2 Symptoms and Management of Common Cold

Acute upper respiratory tract infection, also known as common cold, is a self-limiting infection of the upper respiratory tract mainly caused by rhinoviruses, coronaviruses, and adenoviruses (DeGeorge, Ring and Dalrymple, 2019; Schellack et al., 2020). The viruses often infect adults and children up to three and eight times a year, respectively, mainly through inhalation of airborne particles, direct contact with hands, or objects carrying the pathogens (Pappas, 2018; Cock and Van Vuuren, 2020; Schellack et al., 2020). Symptoms of common cold often last up to ten days and include cough, sneezing, runny nose, sore throat, congested airways, difficulty breathing, increased mucus production, headache, muscular pain, and fever (Ismail and Schellack, 2017; Pappas, 2018; Cock and Van Vuuren, 2020, Schellack et al., 2020).

Effective treatments for symptoms of the common cold in adults are limited to intranasal ipratropium, over-the-counter analgesics, decongestants with or without antihistamines, and zinc. Nasal saline irrigation, analgesics, and time are the mainstays of treatment for common cold in children (DeGeorge, Ring and Dalrymple, 2019).

In Southern Africa, over-the-counter medicines and herbal medicines are often taken concurrently during management of common cold (Schellack et al., 2020). Traditional herbal remedies have gained popularity in management of common cold due to ease of use, low cost, safety, and efficacy. Traditional remedies used in management of common cold typically contain extracts from *Cussonia spicata*, commonly known as *mufenje* or *musheme* in Zimbabwe, *Artemisia afra*, ginger, eucalyptus, and peppermint preparations (Liu, Van der Kooy and Verpoorte, 2009; Merhbod et al., 2018).

7.1.3 Symptoms and Management of SARS-CoV-2

Majority of coronaviruses typically cause mild infections of the respiratory tract in humans; however, coronaviruses that are associated with severe acute respiratory syndrome (SARS) and Middle East respiratory syndrome (MERS) cause severe symptoms and may be lethal (Cock and Van Vuuren, 2020). Cases of SARS-CoV were first reported in 2003, and new cases were reported in Wuhan city of China in December of year 2019, marking the beginning of a new coronavirus epidemic caused by SARS-CoV-2. In March of 2020, the World Health Organization declared SARS-CoV-2 a global pandemic (Lambelet et al., 2020).

The most common symptoms of SARS-CoV-2 are fever, tiredness, and dry cough, which may be accompanied by aches and pains, nasal congestion, runny nose, sore throat, or diarrhoea (Orisakwe, Orish, and Nwanaforo, 2020). Presently, effective treatments modalities that target SARS-CoV-2 remain elusive and largely unknown (Orisakwe, Orish, and Nwanaforo, 2020). Due to the broad clinical spectrum of SARS-CoV-2, there are no specific therapeutic agents available at present. Measures to manage the disease are critical, and a significant population of Southern Africa resorted to use of traditional herbal remedies (Adeleye et al., 2021). Several herbal extracts or their derivatives have shown potential antiviral efficacy (Mirzaie et al., 2020). A variety of plants are currently being used in formulation of remedies for management of symptoms of SARS-CoV-2 disease (Adeleye et al., 2021). The most used herbal remedy for management of SARS-CoV-2 is *Lippia javanica*, commonly known as "lemon bush" in English, "*zinziniba*" in Xhosa, and "*zumbani*" in Shona (Mathew et al., 2020; Mfengu et al., 2021). During preparation, the leaves, roots, or twigs are boiled in water, and infusion may be taken orally, or steam from the preparation may be inhaled (Maroyi, 2017).

7.2 TRADITIONAL HERBAL REMEDIES FOR MANAGEMENT OF COMMON COLD AND ACUTE BRONCHITIS

Traditional herbal remedies are often prepared to alleviate symptoms instead of specifically targeting the disease. Traditional herbal remedies used in management of respiratory conditions may contain extracts from one plant; however, a single plant may not be a solution to a viral respiratory infection, and a range of plants containing phytochemicals of diverse modes of action may be required. Increased efficacy of herbal remedies can, therefore, be achieved by a combination of two or more plants (Yarnell, 2018).

Decoctions and infusions are the most used remedies in management of pathogenic diseases (Cock and Van Vuuren, 2018, 2019), including common cold and acute bronchitis (Cock and Van Vuuren, 2020). The burning of plant volatiles during management of respiratory diseases is widely practiced throughout the world (Mohagheghzadeh et al., 2006). In Southern Africa, burning of parts of a plant and inhalation of the smoke is commonly practiced (Cock and Van Vuuren, 2020). Moreover, boiling parts of a plant in water and inhalation of the steam, smoking parts of a plant, and grinding or pounding plant materials and sniffing the powder are remedies commonly used in management of common cold and acute bronchitis (Cock and Van Vuuren, 2020).

About 257 plant species in Southern Africa are used in formulation of traditional herbal remedies for management of viral respiratory conditions, including acute bronchitis and common cold (Cock and Van Vuuren, 2020). Plant species used in formulation of traditional herbal remedies in Southern Africa are listed in Table 7.1. The plant species come from families including Amaryllidaceae, Anaradriaceae, Apiaceae, Apocynaceae, Asphodelaceae, Aspleniaceae, Asteraceae, Celestraceae, Combretaceae, Fabaceae, Geraniaceae, Lamiaceae, Malvaceae, Myrtaceae, Poaceae, Polypodiaceae, Pteridaceae, Rananculaceae, Rosaceae, Rutaceae, Scrophulariaceae, Solanaceae, Verbenaceae, and Xanthorrhoeaceae (Cock and Van Vuuren, 2020). Only a few of these species have had their inhibitory activity validated against viral respiratory pathogens via *in vitro* testing. Most species have only had their use to treat viral acute bronchitis, and common cold documented in ethnobotanical

TABLE 7.1

Traditional Herbal Remedies Used in the Management of Viral Bronchitis and Common Colds in Southern Africa

Plant Species Used in Remedy	Family	Common Name(s)	Part of Plant Used in Preparation of Remedy	Use, Preparation, and Application of Remedy	References
Acacia arenaria Schinz	Fabaceae	Sand acacia (English), sanddoring (Afrikaans)	Root	A decoction is consumed during management of common colds.	Von Koenen (2001)
Acaciadecurrens Willd.	Fabaceae	Blackwattle, early green wattle (English)	Trunk exudate/gum	Used during management of bronchitis.	Watt and Breyer-Brandwijk (1962)
Acacia hebeclada DC.	Fabaceae	Cattle pod acacia (English), Kersdoringboom (Afrikaans)	Root	A decoction is consumed during management of common colds.	Von Koenen (2001)
Acacia karoo Hayne	Fabaceae	Sweet thorn (English), soetdoring (Afrikaans), mooka (Tswana), umuNga (Zulu, Xhosa)	Leaves and bark	Used during management of common colds. Preparation and application are not specified.	Watt and Breyer-Brandwijk (1962); Hutchings et al. (1996) Van Wyk et al. (2009)
Acacia mearnsii De Wild.	Fabaceae	Uwatela (Zulu)	Not specified	Used during management of common cold and influenza. Preparation and application not specified.	Mhlongo and Van Wyk (2019)
Acacia mellifera (Vahl.) Benth.	Fabaceae	Hook thorn, black hook (English), swarthaak (Afrikaans)	Roots	Chewed during management of common colds.	Von Koenen (2001)
Acacia nilotica (L.) Delile	Fabaceae	Red heart, scented thorn (English), lekkerreulpeul (Afrikaans)	Bark and leaves	Used during management of common colds.	Hutchings et al., (1996); Von Koenen (2001)
Acacia Senegal (L.) Willd.	Fabaceae	Threethorntree (English), driedoringakasia (Afrikaans)	Gum (trunk exudate)	Used as an expectorant during management of common colds and influenza.	Von Koenen (2001)
Acacia sieberana var. *Woodii* (Burtt Davy) Keayb & Brenan	Fabaceae	Paperbark thorn (English), Papierbasdoring (Afrikaans)	Leaf and bark	Decoctions are used as an expectorant during management of common colds and influenza.	Von Koenen (2001)
Achyranthes aspera L.	Amaranthaceae	Devil horse whip (English), Langplitskafblom (Afrikaans)	Roots	A decoction used during management of bronchitis and common colds.	Von Koenen (2001) Watt and Breyer-Brandwijk (1962)

(Continued)

TABLE 7.1 (Continued)
Traditional Herbal Remedies Used in the Management of Viral Bronchitis and Common Colds in Southern Africa

Plant Species Used in Remedy	Family	Common Name(s)	Part of Plant Used in Preparation of Remedy	Use, Preparation, and Application of Remedy	References
Acokanthera oppositifolia (Lam.) Codd.	Apocynaceae	Bushman poison (English), boesmansgif (Afrikaans), inhlungunyembe (Zulu), intlungunyembe (Xhosa)	Leaves	Leaf decoction used during management of common cold; however, this species is highly toxic, and caution is required.	Watt and Breyer-Brandwijk (1962); Van Wyk et al. (2009) Philander (2011)
Acorus calamus L.	Acoraceae	Indaluqwatha, indawolucwatha, uzulucwatha (Zulu)	Not specified	Used during management of common colds and influenza.	Mhlongo and Van Wyk (2019)
Adenopodia spicata (E. Mey.) C.Presl	Fabaceae	Spiny splinter bean (English), stekelsplinterboontjie (Afrikaans), ibobo, ubobo, umbambangwe (Zulu)	Bark	A decoction used during management of common colds.	Watt and Breyer-Brandwijk (1962)
Adiantum chilense var. *sulphureum* (Kaulf.) Kuntze× Hicken	Pteridaceae	Common maidenhairfern (English)	Leaves	Leaves are smoked during management of common cold.	Watt and Breyer-Brandwijk (1962) Moffett (2010)
Adiantum capillus-veneris L.	Pteridaceae	Maiden hair fern(English), venushaar (Afrikaans)	Leaves	Leaves are smoked during management of common cold.	Watt and Breyer-Brandwijk (1962); Hutchings et al. (1996); Von Koenen (2001); Moffett (2010)
Adansonia digitata L.	Malvaceae	Baobab (English), kremetartboom (afrikaans)	Fruit	Boiled and consumed during management of common cold.	Von Koenen (2001)
Aeollanthus buchnerianus Briq.	Lamiaceae	Rocksage (English), klipsalie (Afrikaans)	Leaves	Leaves are smoked during management of common cold.	Watt and Breyer-Brandwijk (1962); Moffett (2010)
Agathosma betulina (P.J. Bergius) Pillans	Rutaceae	Bucgh (English), boegoe, letuling (Afrikaans)	Leaves	An infusion is drunk during management of common cold.	De Beer and Van Wyk (2011)
Albizia antunesiana Harms	Fabaceae	Purple-leaved false thorn (English), persblaarvalsdoring (Afrikaans)	Bark	Chewed during management of common cold.	Von Koenen (2001)

(Continued)

TABLE 7.1 (Continued)

Traditional Herbal Remedies Used in the Management of Viral Bronchitis and Common Colds in Southern Africa

Plant Species Used in Remedy	Family	Common Name(s)	Part of Plant Used in Preparation of Remedy	Use, Preparation, and Application of Remedy	References
Alepidea amatymbica Eckl &. Zeyh.	Apiaceae	Kalmoes (Afrikaans), lesoko (Sotho), iqwili (Xhosa), ikhathazo (Zulu)	Roots and stem	Stems are burnt and smoke is inhaled for management of common cold. A root decoction used for management of common cold and influenza.	Watt and Breyer-Brandwijk (1962); Hutchings et al. (1996); Van Wyk et al. (2009); Moffett (2010); Philander (2011)
Allium cepa L.	Amaryllidaceae	Onion (English)	Bulb	Used for management of influenza.	Watt and Breyer-Brandwijk (1962)
Allium sativum L.	Amaryllidaceae	Garlic (English)	Bulb	Used for management of influenza.	Watt and Breyer-Brandwijk (1962)
Aloe arborescens Mill.	Xanthorrhoeaceae	Krantzaloe (English), kransaalwyn (Afrikaans), ikalene (Xhosa), inkalane, umhlabana (Zulu)	Not specified	Used for management of common cold and influenza. Preparation and application not specified.	Mhlongo and Van Wyk (2019)
Aloe maculata All.	Xanthorrhoeaceae	Soapaloe, zebraloe (English)	Leaves	A decoction used for management of common cold and influenza.	Watt and Breyer-Brandwijk (1962); Kose et al. (2015)
Anemone caffra Harv.	Ranunculaceae	Unknown	Roots	Powdered roots taken as a snuff for management of common cold.	Watt and Breyer-Brandwijk (1962)
Anemone vesicatoria (L.f.) Prantl	Ranunculaceae	Blisterleaf (English), brandblaar, katjiedrieblaar (Afrikaans)	Leaves and roots	Root decoctions are used for management of common cold and influenza. Application not specified for the leaves.	Van Wyk et al. (2009) Watt and Breyer-Brandwijk (1962)
Anisodontea triloba (Thunb.) D.M. Bates	Malvaceae	Wildsalie (Afrikaans)	Leaves	A leaf infusion is consumed for management of common cold.	De Beer and Van Wyk (2011)
Antiphiona pinnatisecta (S. Moore) Merxm.	Asteraceae	Unknown	Roots	The Sanusean infusion used for management of influenza.	Von Koenen (2001)
Aptosimum indivisum Burch. ex. Benth.	Scrophulariaceae	Kinkhoesbos, agtdaegeneesbos (Afrikaans)	Not specified	Used for management of common cold.	Hulley and Van Wyk (2019)

(Continued)

TABLE 7.1 (Continued)
Traditional Herbal Remedies Used in the Management of Viral Bronchitis and Common Colds in Southern Africa

Plant Species Used in Remedy	Family	Common Name(s)	Part of Plant Used in Preparation of Remedy	Use, Preparation, and Application of Remedy	References
Artemisia afra Jacq Ex. Willd.	Asteraceae	African wormwood (English), als, alsem, wildeals (Afrikaans), lengana (Sotho, Tswana), umhlonyane (Xhosa, Zulu)	Leaves	Infusions and decoctions are used for management of common cold, coughs, influenza, and bronchitis. Fresh leaves may also be inserted directly into the nose.	Smith (1888) Watt and Breyer-Brandwijk (1962) Van Wyk et al. (2009); Moffett. (2010) Kose et al. (2015)
Asplenium cordatum (Thunb.) Sw.	Aspleniaceae	Resurrection fern (English), skubcaring (Afrikaans)	Leaves and rhizome	Leaves are smoked for management of common cold. A rhizome decoction is drunk for management of common cold and relieves sore throat.	Watt and Breyer-Brandwijk (1962); Moffett (2010)
Asplenium monanthes L.	Aspleniaceae	Unknown	Leaves	Smoked for management of common cold.	Watt and Breyer-Brandwijk (1962); Moffett (2010)
Asplenium trichomanes L.	Aspleniaceae	Maidenhair spleenwort (English)	Leaves	Smoked for management of common cold.	Watt and Breyer-Brandwijk (1962); Moffett (2010)
Baccharoides adoensis var. *mossambiquensis*. (Steetz) "Isawumi, El- Ghazaly & B.Nord."	Asteraceae	Unknown	Not specified	Infusions used for management of influenza.	Smith (1888) Watt and Breyer-Brandwijk (1962)
Ballota africana (L.) Benth.	Lamiaceae	kattekruid, kattekruie (Afrikaans)	Whole plant	Infusion is used for management of common cold and influenza.	Watt and Breyer-Brandwijk (1962); Van Wyk et al. (2009); Hulley and Van Wyk (2019)
Bauhinia petersiana Bolle	Fabaceae	Camel foot (English), koffiebos Afrikaans)	Leaves	Boiled and the steam inhaled for management of common cold and influenza.	Von Koenen (2001)

(Continued)

TABLE 7.1 (Continued)

Traditional Herbal Remedies Used in the Management of Viral Bronchitis and Common Colds in Southern Africa

Plant Species Used in Remedy	Family	Common Name(s)	Part of Plant Used in Preparation of Remedy	Use, Preparation, and Application of Remedy	References
Berkheya setifera DC.	Asteraceae	Leleme la khomo (Sotho)	Roots and leaves	Used for management of common cold. Preparation and application not specified.	Kose et al. (2015)
Bidens pilosa L.	Asteraceae	Blackjack, beggar ticks, cobblerspegs, sticky beaks (English), mutsine, tsine (Shona)	Not specified	Used for management of common cold and influenza. Preparation and application not specified.	Mhlongo and Van Wyk (2019)
Blepharis espinosa Phillips	Acanthaceae	Unknown	Not specified	Used for management of common cold.	Moffett (2010)
Buddle saligna Willd.	Scrophulariaceae	False olive (English), witolien (Afrikaans), lelothwane (Sotho), ungqeba (Xhosa), igqeba- elimhlophe (Zulu)	Leaves	A leaf decoction used for management of common cold.	Watt and Breyer-Brandwijk (1962)
Berkheya setifera DC.	Asteraceae	Buffalo-tongue thistle (English), rasperdissel (Afrikaans), ikhakhasi (Zulu), indlebe-lenkomo (Xhosa), lelelemia-khomo, nsoantsane (Sotho)	Roots	A decoction prepared and consumed for management of common cold and coughs.	Watt and Breyer-Brandwijk (1962)
Brunsviga griandiflora Lindl.	Amaryllidaceae	Giant candellabria (English), reusekandelaarblom (Afrikaans)	Bulb	A decoction prepared from the crushed bulb is consumed for management of common cold and coughs.	Watt and Breyer-Brandwijk (1962)
Bulbine frutescens Willd.	Xanthorrhoeaceae	Unknown	Leaves	The dried leaves are smoked for management of common cold.	Watt and Breyer-Brandwijk (1962)
Bulbine narcissifolia Salm-Dyck	Xanthorrhoeaceae	Khomo ea balisa	Bulbs or roots.	Used for management of common cold. Preparation and application not specified.	Kose et al. (2015)

(Continued)

TABLE 7.1 (Continued)
Traditional Herbal Remedies Used in the Management of Viral Bronchitis and Common Colds in Southern Africa

Plant Species Used in Remedy	Family	Common Name(s)	Part of Plant Used in Preparation of Remedy	Use, Preparation, and Application of Remedy	References
Carpobrotus edulis (L.) N.E.Br.	Aizoaceae	Sour fig, Cape fig, Hottentot fig (English), vyerank, ghaukum, ghoenavy, hotnotsvye, Kaapvy, perdevy, rankvy (Afrikaans), ikhambi-lamabulawo, umgongozi (Zulu)	Leaves	Juice from squeezed leaves is diluted in water and used to relieve sore throat associated with colds.	Hutchings et al. (1996); Felhaber and Mayeng (1997)
Carissa bispinosa (L.) Defs. ex Brenan.	Apocynaceae	Amathungulu (Zulu)	Not specified	Used for management of common cold and influenza. Preparation and application not specified.	Mhlongo and Van Wyk (2019)
Cassine transvaalensis (Burtt Davy) Codd.	Celestraceae	Transvaalsaffron (English), lepel, lepelhout, waterboom (Afrikaans)	Leaves	Adults chew the leaves and swallow the sap for management of influenza. An infusion is given to children for the same purpose.	Hutchings et al. (1996); Von Koenen (2001)
Catha edulis (Vahl) Endl.	Celestraceae	Bushman tea (English), boesmanstee (Afrikaans), khat (Arabic)	Leaves	Leaves used for management of influenza.	Watt and Breyer-Brandwijk (1962); Van Wyk et al. (2009)
Chamarea capensis (Thunb) Eckl. &.Zeyh.	Apiaceae	Vinkelbol, vinkelwortel (Afrikaans)	Root	The infusions are consumed for management of common cold.	Hulley and Van Wyk (2019)
Cheilanthes eckloniana Met.	Pteridaceae	Ecklonlipfern, resurrection fer (English)	Root and leaves	A decoction is consumed for management of common cold. The leaves are also smoked for management of common cold.	Moffett (2010); Watt and Breyer-Brandwijk (1962)
Cheilanthes hirta Sw.	Pteridaceae	Mamauoaneng (Sotho)	Root	A decoction is consumed for management of common cold.	Watt and Breyer-Brandwijk (1962); Hutchings et al. (1996); Kose et al. (2015) *(Continued)*

TABLE 7.1 (Continued)

Traditional Herbal Remedies Used in the Management of Viral Bronchitis and Common Colds in Southern Africa

Plant Species Used in Remedy	Family	Common Name(s)	Part of Plant Used in Preparation of Remedy	Use, Preparation, and Application of Remedy	References
Cheilanthes involuta var. *obscura* (N.C.Anthony) N.C.Anthony	Pteridaceae	Unknown	Root	A decoction is consumed for management of common cold.	Moffett (2010)
Chaenostoma floribundum Benth.	Scrophulariaceae	Unknown	Root	A decoction used for management of common cold in children.	Watt and Breyer-Brandwijk (1962); Moffett (2010)
Chloris virgata Sw	Poaceae	Feather finger grass (English)	Not specified	Used for management of common cold.	Moffett (2010)
Chrysocoma ciliata L.	Asteraceae	Beesbos (Afrikaans)	Leaves	A decoction is consumed for management of common cold.	De Beer Van Wyk (2011)
Cinnamomum camphora (L.) J.Presl.	Lauraceae	Camphor Tree (English) , kanferboom (Afrikaans), uroselina (Zulu)	Essential oil distilled from the wood	Used for management of common cold.	Van Wyk et al. (2009); Philander (2011)
Clematis brachiata Thunb.	Ranunculaceae	Morara oa thaba (Sotho)	Not specified	Used for management of common cold.	Hutchings et al .(1996); Moffett (2010); Kose et al (2015)
Cleome angustifolia Forssk.	Cleomaceae	Unknown	Roots	The steam from boiling roots inhaled for management of influenza.	Von Koenen (2001)
Cliffortia ilicifolia L.	Rosaceae	Unknown	Not specified	Used as an expectorant that relieves the symptoms of common cold.	Watt and Breyer-Brandwijk (1962)
Cliffortia odorata L.f.	Rosaceae	Wildewingerd (Afrikaans)	Not specified	An infusion is consumed for management of common cold.	Watt and Breyer-Brandwijk (1962); Philander (2011)
Conyza scabrida DC.	Asteraceae	Bakhos, oondbos (Afrikaans), isavu (Xhosa)	Leaves	Infusions are consumed or powdered leaves taken as snuff for management of common cold and influenza.	Hutchings et al. (1996); Van Wyk et al. (2009); De Beer and Van Wyk (2011); Philander (2011); Hulley and Van Wyk (2019)

(*Continued*)

TABLE 7.1 (Continued)
Traditional Herbal Remedies Used in the Management of Viral Bronchitis and Common Colds in Southern Africa

Plant Species Used in Remedy	Family	Common Name(s)	Part of Plant Used in Preparation of Remedy	Use, Preparation, and Application of Remedy	References
Chloris virgata Sw.	Poaceae	Feather-fingergrass, Rhodesgrass (English)	Roots	A decoction is added to the bath for management of common cold.	Watt and Breyer-Brandwijk (1962)
Chrysanthemum morifolium Ramat.	Asteraceae	Unknown	Leaves and flowers	People with cold sleep on a pillow filled with leaves and flowers to relieve the symptoms.	Watt and Breyer-Brandwijk (1962)
Clematis brachiata Thunb.	Ranunculaceae	Ihlonzoleziduli, inhlongo, umdloza, umfufuna (Zulu)	Stem	Bruised stem to taken as snuff for management of common cold.	Smith (1888); Watt and Breyer-Brandwijk (1962); York et al. (2011)
Combretum molle. R.Br. ex G. Don	Combretaceae	Umbondo, umbondwe (Zulu)	Leaves	An infusion is consumed for management of common cold.	York et al.(2011)
Corymbia gummifera (Gaertn) K. D. Hill & L. A.S. Johnson	Myrtaceae	Red bloodwood (English)	Leaves	Essential oils inhaled for management of common cold.	Watt and Breyer-Brandwijk (1962)
Citrus limon (L.) Osbeck	Rutaceae	Lemon (English)	Fruit Juice	Consumed for management of common cold.	Watt and Breyer-Brandwijk (1962); York et al. (2011)
Crassula alba Forssk.	Crassulaceae	Unknown	Juice	Juice mixed with water is used as nasal cleanser during management of influenza.	Watt and Breyer-Brandwijk (1962); Hutchings et al. (1996)
Crinum bulbispermum (Burm.f) Milne. -Redh. & Schweik.	Amaryllidaceae	Orange River lily, Vaal River lily (English), Oranjerivier (Afrikaans), anduze (Zulu)	Bulb	A decoction is prepared from the crushed bulb and consumed for management of common cold.	Watt and Breyer-Brandwijk (1962); Moffett (2010)
Crinum macowanii Baker	Amaryllidaceae	umduze (Zulu)	Bulbs	A decoction is prepared from the crushed bulb and consumed for management of common cold.	Van Wyk et al. (2009)

(Continued)

TABLE 7.1 (Continued)
Traditional Herbal Remedies Used in the Management of Viral Bronchitis and Common Colds in Southern Africa

Plant Species Used in Remedy	Family	Common Name(s)	Part of Plant Used in Preparation of Remedy	Use, Preparation, and Application of Remedy	References
Cymbopogon nardus (L.) Rendle	Poaceae	Citronella grass (English)	Whole plant	Used for management of common cold. Treatment and application not specified.	Watt and Breyer-Brandwijk (1962); Hutchings et al. (1996)
Cyperus longus L.	Cyperaceae	Galingale (English)	Tuber	Powdered tuber taken as a snuff for management of common cold.	Watt and Breyer-Brandwijk (1962)
Datura spp.	Solanaceae	Various common names for different species.	Not specified	Used for management of common cold.	Moffett (2010)
Dicerothamnus rhinocerotis (L.f.) Koek.	Asteraceae	Rhinoceros bush (English)	Not specified	Used for management of influenza. Preparation and application methods were not specified.	Watt and Breyer-Brandwijk (1962)
Dicoma anomala Sond.	Asteraceae	Fever bush, stomach bush (English), maagbitterwortel, kalwerbossie, koorsbossie, gryshout, maagbossie (Afrikaans), inyongana (Xhosa), isihlabamakhondlwane, umuna (Zulu)	Roots	Powdered root bark is taken as a snuff for management of common cold. A decoction may also be consumed for the same purpose.	Watt and Breyer-Brandwijk (1962); Hutchings et al. (1996) Von Koenen (2001); Moffett (2010); Kose et al. (2015)
Dicoma capensis Less.	Asteraceae	Wilde karmedi; koorsbossie, (Afrikaans)	Roots	Powdered root bark is taken as a snuff for management of common cold. A decoction may also be consumed for the same purpose.	Hutchings et al. (1996); VanWyk et al. (2009); De Beer Van Wyk (2011)
Digitaria sanguinalis (L.) Scop.	Poaceae	Crab grass (English)	Whole plant	A decoction is consumed for management of common cold.	Smith (1888)
Dioscorea dregeana (Kunth) D. Durand & Schinz.	Dioscoreaceae	Ingevu, intanaebovu, udakawa (Zulu)	Not specified	Used for management of common cold and influenza. Preparation and application not specified.	Mhlongo and Van Wyk (2019)

(Continued)

TABLE 7.1 (Continued)
Traditional Herbal Remedies Used in the Management of Viral Bronchitis and Common Colds in Southern Africa

Plant Species Used in Remedy	Family	Common Name(s)	Part of Plant Used in Preparation of Remedy	Use, Preparation, and Application of Remedy	References
Dodonaea viscosa (L.) Jacq.	Sapindaceae	Sandolive (English), sandolien, ysterbos (Afrikaans), mutata-vhana (Venda)	Leaves and twigs	Used for management of common cold and influenza.	Nortje and Van Wyk (2015); Van Wyk et al. (2009); De Beer and VanWyk (2011); Hulley and Van Wyk (2019)
Dolichotherix ericoides (Lam.) Hilliard & B.L. Burtt.	Asteraceae	Klipanostber, ganoster (Afrikaans)	Not specified	Infusions are consumed for management of common cold.	Hulley and Van Wyk (2019)
Drimia altissima (L.f.) Ker Gawl.	Asparagaceae	Tall whites quill (English)	Bulb	The powdered bulb is ingested for management of influenza and bronchitis.	Watt and Breyer-Brandwijk (1962)
Dysphania ambrosioides (L.) Mosyakin & Clemants	Amaranthaceae	Wormseed, Jesuittea (English), Ikhambileslumo, isinuka, isinukamasimb, umanxiweni, (Zulu)	Not specified	Used for management of common cold and influenza.	Hulley and Van Wyk (2019); Mhlongo and Van Wyk (2019); Watt and Breyer-Brandwijk (1962); Hutchings et al (1996)
Elaphoglossum conforme (Sw.) Schott	Pteridaceae	Unknown	Rhizome	A decoction is consumed for management of common cold.	Moffett (2010)
Elaphoglossum petiolatum (Sw.) Urb.	Pteridaceae	Graceful Tongue Fern (English)	Rhizome	A decoction is consumed for management of common cold.	Watt and Breyer-Brandwijk (1962); Moffett (2010)
Empleur umunicapsulare (L.f) Skeels.	Rutaceae	Bergboegoe, bokboegoe, langblaarboegoe (Afrikaans)	Not specified	Used for management of common cold and influenza. Preparation and application not specified	Hulley and Van Wyk (2019)
Equisetum ramosissimum Desf.	Equisetaceae	Branched horsetail (English)	Rhizome	A decoction of the rhizome used for management of common cold.	Watt and Breyer-Brandwijk (1962); Moffett (2010)

(Continued)

TABLE 7.1 (Continued)
Traditional Herbal Remedies Used in the Management of Viral Bronchitis and Common Colds in Southern Africa

Plant Species Used in Remedy	Family	Common Name(s)	Part of Plant Used in Preparation of Remedy	Use, Preparation, and Application of Remedy	References
Eriocephalus africanus L.	Asteraceae	Kapokbos, kaapkaroo (Afrikaans)	Not specified	Infusions are used for management of common cold.	Hulley and Van Wyk (2019)
Eriocephalus punctulatus DC.	Asteraceae	Wild Rosemary, Capesnowbush (English), kapokbos (Afrikaans)	Not specified	Used to fumigate huts inhabited by common cold patients.	Watt and Breyer-Brandwijk (1962)
Erigeron bonariensis L.	Asteraceae	Unknown	Leaves	A leaf infusion is consumed for management of common cold.	Watt and Breyer-Brandwijk (1962); Hutchings et al .(1996)
Eriosema cordatum E. Mey.	Fabaceae	Ugwayana, umuthi wamadoda, umvusandoda (Zulu)	Not specified	Used for management of common cold and influenza. Preparation and application not specified.	Mhlongo and Van Wyk (2019)
Eriosema distinctum. N.E. Br.	Fabaceae	Ugqomfane, umvusandoda, uqonsi (Zulu)	Not specified	Used for management of common cold and influenza. Preparation and application not specified.	Mhlongo and Van Wyk (2019)
Erythrophleum suaveolens (Guill & Perr.) Brenan	Fabaceae	Woodland Waterbury, water pear (English), waterpeer (Afrikaans)	Bark	Powdered bark taken as snuff for management of common cold.	Watt and Breyer-Brandwijk (1962)
Eucalyptus camaldulensis Dehnh.	Myrtaceae	Umgamthrini (Zulu)	Gum (trunk exudate)	The gum is dissolved in water and used for management of common cold.	Watt and Breyer-Brandwijk (1962); Mhlongo and Van Wyk (2019)
Eucalyptus globulis Labill.	Myrtaceae	Bluegum (English)	Leaves	Decoctions and infusions are used for management of common cold and influenza.	Smith (1888) Watt and Breyer-Brandwijk (1962)
Eucomis autumnalis (Mill.) Chitt.	Asparagaceae	Autumn pineapple lily, pineapple flower (English), wilde pynappel, krulkoppie (Afrikaans), ubuhlungu becanti, isithothobala mthunzi (Xhosa), umathunga, ukhokho, umakhandakantsele (Zulu)	Not specified	Used for management of common cold and influenza. Preparation and application not specified.	Mhlongo and Van Wyk (2019)

(Continued)

TABLE 7.1 (Continued)
Traditional Herbal Remedies Used in the Management of Viral Bronchitis and Common Colds in Southern Africa

Plant Species Used in Remedy	Family	Common Name(s)	Part of Plant Used in Preparation of Remedy	Use, Preparation, and Application of Remedy	References
Euryops spp.	Asteraceae	Different Names for Different Species	Not specified	Ingested for management of influenza.	Watt and Breyer-Brandwijk (1962)
Foeniculum vulgare Mill.	Apiaceae	Vinkel, takvinkel Afrikaans)	Not specified	Infusions are drunk for management of common cold.	Hulley and Van Wyk (2019)
Geranium incanum Burn. f.	Geraniaceae	vroue boosie, bergtee, amarabossie (Afrikaans), gope-sethsohtlaka, (Sotho)	Leaves	Leaf infusions are used for management of bronchitis.	Van Wyk et al .(2009)
Geranium ornithopodon Eckl. & Zeyh.	Geraniaceae	Unknown	Leaves	A leaf decoction is consumed for management of common cold.	Moffett (2010)
Gerbera ambigua (Cass.) Sch. Bip.	Asteraceae	Sebok (Sotho)	Roots	A root infusion is consumed for management of severe colds.	Kose et al. (2015)
Gerbera piloselloides (L.) Cass.	Asteraceae	Tsebeapela (Sotho)	Not specified	Used for fumigation of huts inhabited by common cold patients.	Watt and Breyer-Brandwijk (1962); Hutchings et al. (1996); Moffett (2010); Kose et al. (2015)
Gerbera viridifolia (DC.) Sch.Bip.	Asteraceae	Blushing barberton daisy (English), griquateebossie (Afrikaans), lyeza lamazi (Xhosa)	Not specified	Smoke from the burning plant is inhaled for management of common cold. An infusion is consumed for the same purpose.	Watt and Breyer-Brandwijk (1962); Moffett (2010)
Gladiolus dalenii Van Geel	Iridaceae	Parrotgladiolus, Natalily (English), papegaai-gladiowildeswaardlelieus, (Afrikaans), umnunge (Xhosa), udwendweni, uhlakahle (Zulu), khahla- e-kholo (Sotho)	Corm	A decoction is used for management of common cold. Smoke From Burning corms is inhaled for the same purpose.	Hutchings et al. (1996); Von Koenen (2001); Moffett (2010); Watt and Breyer-Brandwijk (1962)

(Continued)

TABLE 7.1 (Continued)

Traditional Herbal Remedies Used in the Management of Viral Bronchitis and Common Colds in Southern Africa

Plant Species Used in Remedy	Family	Common Name(s)	Part of Plant Used in Preparation of Remedy	Use, Preparation, and Application of Remedy	References
Glycyrrhiza glabra L.	Fabaceae	Liquorice licorice (English)	Rhizome	Decoctions and infusions are used for management of common cold, influenza and bronchitis.	Watt and Breyer-Brandwijk (1962)
Gnidia anthylloides (L.f.) Gilg.	Thymelaeaceae	Brandbossie (Afrikaans), indolo, into zwane (Zulu)	Roots	Used for management of influenza.	Hutchings et al. (1996); Moffett (2010); Watt and Breyer-Brandwijk (1962)
Gnidia kraussiana Meisn.	Thymelaeaceae	Imprevu, umsila wenger, umahedeni (Zulu)	Not specified	Used for management of common cold and influenza. Preparation and application specified.	Mhlongo and Van Wyk (2019)
Gomphocarpus fruticosus (L.) W.T. Aiton	Apocynaceae	Milkweed (English), melkbos, tonteldoos (Afrikaans), begana, lerete-le-ntja (Sotho), modimolle (Sotho), msinga-isalukazi (Zulu)	Not specified	Used for management of common cold.	Von Koenen (2001); Moffett (2010)
Grangea maderaspatana (L.) Poir.	Asteraceae	Unknown	Not specified	A decoction is used for management of common cold and influenza. The crushed leaves may also be directly inserted into the nostrils for the same purpose.	Watt and Breyer-Brandwijk (1962)
Gunnera perpensa L.	Gunneraceae	Wild Rhubarb, river pumpkin (English), wilderamenas, ravi pampoen (Afrikaans), qobo (Sotho), rambola-vhadzimu (Venda), iphuzighobo, (Xhosa), ugobho (Zulu)	Leaves and roots	Used for management of common cold.	Watt and Breyer-Brandwijk (1962)

(Continued)

TABLE 7.1 (Continued)

Traditional Herbal Remedies Used in the Management of Viral Bronchitis and Common Colds in Southern Africa

Plant Species Used in Remedy	Family	Common Name(s)	Part of Plant Used in Preparation of Remedy	Use, Preparation, and Application of Remedy	References
Gymnosporia buxifolia Szynszyl.	Celestraceae	Lemoen Doring, wondering, pendoringbos (Afrikaans)	Roots	Root decoctions are consumed for management of common cold.	Hulley and Van Wyk (2019); Watt and Breyer-Brandwijk (1962)
Gymnosporia heterophylla (Eckl &. Zeyh.) Loes.	Celestraceae	Gewone Pendoring (Afrikaans)	Leaves	An infusion is consumed for management of common cold and influenza.	Von Koenen (2001); Hutchings et al. (1996)
Haplocarpha scaposa Harv.	Asteraceae	Papetloana (Sotho)	Root	A decoction is consumed for management of common cold.	Kose et al. (2015); Watt and Breyer-Brandwijk (1962)
Helichrysum appendiculatum (L.f.) Less.	Asteraceae	Sheep ears everlasting (English), skaapoorbossie (Afrikaans), senkotoana (Sotho), ibode, indlebeyemvu (Zulu)	Leaves	The leaves are eaten raw for management of common cold.	Watt and Breyer-Brandwijk (1962); Smith (1888)
Helichrysum caespiticium (DC.) Sond. Ex Harv.	Asteraceae	Phat Mangaka (Sotho)	Leaves	Smoke from burning leaves is inhaled for management of common cold.	Kose et al. (2015); Moffett (2010); Watt and Breyer-Brandwijk (1962)
Helichrysum cymosum (L.) D. Don.	Asteraceae	Kooibos (Afrikaans)	Not specified	Used for management of common cold. Preparation and application not specified.	Hulley and Van Wyk (2019)
Helichrysum dregeanum Sond. & Harv.	Asteraceae	Bergankeerkaroo, vaalberganker (Afrikaans)	Leaves	Leaves smoked to relieve common cold.	Watt and Breyer-Brandwijk (1962); Moffett (2010)
Helichrysum cochleariforme DC.	Asteraceae	Unknown	Not specified	Used for management of bronchitis. Preparation and application not specified.	Watt and Breyer-Brandwijk (1962)
Helichrysum luteoalbum (L.) Rchb.	Asteraceae	Impepho, in old lane (Zulu)	Not specified	Used for management of common cold and influenza. Preparation and application not specified.	Mhlongo and Van Wyk (2019)

(Continued)

TABLE 7.1 (Continued)

Traditional Herbal Remedies Used in the Management of Viral Bronchitis and Common Colds in Southern Africa

Plant Species Used in Remedy	Family	Common Name(s)	Part of Plant Used in Preparation of Remedy	Use, Preparation, and Application of Remedy	References
Helichrysum nudifolium Less.	Asteraceae	Everlastings (English), hottentot steeb kossie, oil goed (Afrikaans), ice, indlebe bhokwe, undleni (Xhosa), icholocholo, imphepho (Zulu)	Leaves and roots	Leaves consumed as remedy for common cold. Root decoctions are used for the same purpose.	Watt and Breyer-Brandwijk (1962); Hutchings et al. (1996); Van Wyk et al. (2009)
Helichrysum odoratissimum (L.) Sweet	Asteraceae	Everlastings (English), kooigoed (Afrikaans) imphepho (Zulu)	Roots	Decoctions are consumed during management of common colds.	Hulley and Van Wyk (2019); Kose et al. (2015); Moffett (2010) Van Wyk et al. (2009); Hutchings et al.(1996); Watt And Breyer-Brandwijk (1962)
Helichrysum peduncular Hilliard & B.L. Burtt.	Asteraceae	Unknown	Roots	Root decoction consumed during management of coughs and common cold.	Watt and Breyer-Brandwijk (1962)
Helichrysum rugulosum Less.	Asteraceae	Marotole, motlosa-ngaka, moto antonyan (Sotho)	Not specified	Used for fumigation of huts of common cold patients.	Watt and Breyer-Brandwijk (1962); Moffett (2010)
Hermannia cuneifolia Jacq.	Malvaceae	Wildeheuningeneesbossie, pleisterbossie (Afrikaans)	Leaves	A leaf infusion is used for management of common cold.	De Beer Van Wyk (2011)
Hermannia salviifolia L.f.	Sterculiaceae	Katjie Drie Blaar (Afrikaans)	Not specified	Used during management of common cold. Preparation and application not specified	Hulley and Van Wyk, 2019
Hernia hirsute L.	Caryophyllaceae	Hairy Rupturewort (English)	Root	A root decoction is used during management of common cold.	Watt and Breyer-Brandwijk, (1962)
Hoodia gordonii (Masson) Sweet ex Decne.	Apocynaceae	Hoodia, ghaap, kakimas (Afrikaans) Yellowstar, starlily, starflower	Fleshy stems	Used during management of common cold. Preparation And Application not specified	Philander (2011)

(Continued)

TABLE 7.1 (Continued)
Traditional Herbal Remedies Used in the Management of Viral Bronchitis and Common Colds in Southern Africa

Plant Species Used in Remedy	Family	Common Name(s)	Part of Plant Used in Preparation of Remedy	Use, Preparation, and Application of Remedy	References
Hypoxis hemerocallidea Fisch., C.A. Mey. & Ave-Lall.	Hypoxidaceae	Yellow star, star lily, star flower (English), sterblom, geel sterretje, gifbol (Afrikaans), moli kharatsa, lotsane (Sotho); inkomfe, inkomfe enkulu (Zulu), inongwe, ilabatheka, ixhalanxa, ikhubalo lezithunzela (Xhosa), tshuka (Tswana)	Not specified	Used during management of common cold and influenza. Preparation and application not specified.	Mhlongo and Van Wyk (2019)
Ilex mitis (L.) Radlk.	Aquifoliaceae	Cape holly, African holly, water tree (English), waterboom, waterhout (Afrikaans), nonaname (Northern Sotho), ipuphuma (Zulu), unduma (Xhosa), phukgu, phukgile (Sotho), mutaganzwa-khameel (Venda)	Leaves and bark	The leaves and bark are pound to produce a lather which is used to wash the bodies of influenza patients.	Watt and Breyer-Brandwijk (1962)
Imperata cylindrica (L.) Raeusch	Poaceae	Kunai grass, cogongrass (English)	Roots	Roots used during management of common cold.	Moffett (2010); Watt and Breyer-Brandwijk (1962)
Kalanchoe paniculata Harv.	Crassulaceae	Hasieoor (Afrikaans), indabulaluvalo (Zulu), sehlakwahlakwane (Sotho)	Roots	The fresh root is chewed, or the powdered root is taken as a snuff during management of common cold.	Watt and Breyer-Brandwijk (1962)
Laggera decurrens (Vahl.) Merxm.	Asteraceae	Wolbos (Afrikaans)	Leaves and roots	Boiled and the steam is inhaled during management of common cold.	Von Koenen (2001)
Lantana camara L.	Verbenaceae	Tickberry (English)	Leaves	Used during management of common cold. Preparation and application were not specified.	Watt and Breyer-Brandwijk (1962); Von Koenen (2001)
Lantana rugosa Thunb.	Verbenaceae	Bird's brandy (English)	Leaves	Leaf extracts used during management of common cold.	Von Koenen (2001)

(Continued)

TABLE 7.1 (Continued)

Traditional Herbal Remedies Used in the Management of Viral Bronchitis and Common Colds in Southern Africa

Plant Species Used in Remedy	Family	Common Name(s)	Part of Plant Used in Preparation of Remedy	Use, Preparation, and Application of Remedy	References
Lebeckia sericea Thunb.	Fabaceae	Silverpea (English), bloufluitjiesbos, vaalertjiebos (Afrikaans)	Not specified	Plant used as a remedy for common cold.	Watt and Breyer-Brandwijk (1962)
Leonitis ocymifolia (Burm.f) Iwarsson	Lamiaceae	Wild dagga (English), wild dagga (Afrikaans)	Leaves and stems	Used during management of common cold. Preparation and application not specified.	Hulley and Van Wyk (2019); Watt and Breyer-Brandwijk (1962)
Leonotis ocymifolia var. *schinzii* (Gürke) Iwarsson	Lamiaceae	Unknown	Leaves and stems	Decoction used for coughs and common cold.	Watt and Breyer-Brandwijk (1962)
Leonotis leonurus (L.) R. Br.	Lamiaceae	Wild dagga (English), wild dagga, duiwels tabak (Afrikaans), mvovo (Xhosa), uyshwala-bezinyoni (Zulu)	Leaves and stems	A decoction of the leaves and stems used during management of coughs, common cold and influenza.	Moffett (2010); Van Wyk et al. (2009); Smith (1888); Watt and Breyer-Brandwijk (1962); Hutchings et al.(1996); Van Wyk et al. (2009); Moffett (2010); Hulley and Van Wyk (2019)
Lessertia frutescens subsp. *frutescens* Goldblatt & J.C. Manning	Fabaceae	Keurtjie, beeskeurtiebos, kankerbos (Afrikaans)	Not specified	An infusion is consumed during management of common cold.	Hulley and Van Wyk (2019)
Leucas martinicensis (Jacq.) R.Br.	Lamiaceae	Tumbleweed (English), tolbossie (Afrikaans)	Leaves	An infusion is used during management of common cold and influenza.	Von Koenen (2001)
Leucas lavandulifolia (Sm.) Raf.	Lamiaceae	Umagumede (Zulu)	Not specified	Used during management of common cold and influenza. Preparation and application not specified.	Mhlongo and Van Wyk (2019)

(Continued)

TABLE 7.1 (Continued)
Traditional Herbal Remedies Used in the Management of Viral Bronchitis and Common Colds in Southern Africa

Plant Species Used in Remedy	Family	Common Name(s)	Part of Plant Used in Preparation of Remedy	Use, Preparation, and Application of Remedy	References
Leucas pechuelii (Kuntze) Baker	Lamiaceae	Unknown	Entire plant	A decoction is made from the entire plant and taken as snuff during management of influenza.	Von Koenen (2001)
Lichtensteinia interrupta E. Mey.	Apiaceae	Umkhalaphanga (Zulu)	Roots	A root decoction is consumed during management of common cold.	Smith (1888); Watt and Breyer-Brandwijk (1962); Hutchings et al. (1996); Moffett (2010)
Lippia javanica (Burm F.) Spreng	Verbenaceae	Fever tea, lemon bush (English), koorbossie, beukesbossie, lemoenbossie (Afrikaans), inzinzinba (Xhosa), umsuzwane, umswazi (Zulu), zumbani (Shona)	Leaves and stems	A decoction used during management of common cold, influenza and bronchitis.	Watt and Breyer-Brandwijk (1962); York et al. (2011); Mhlongo and Van Wyk (2019)
Lonchocarpus nelsii (Schinz) Heering & Grimme	Fabaceae	Apple leaf (English), appelblaar (Afrikaans)	Roots	Smoke from the burning roots is inhaled for management of common cold.	Von Koenen (2001)
Lycopodium clavatum L.	Lycopodiaceae	Clubmoss, stag-horn clubmoss, running clubmoss, ground pine (English)	Whole plant	Used for management of common cold.	Moffett (2010)
Melolobium candicans (E. Mey.) Eckl. & Zeyh.	Fabaceae	Wild dagga (English), wild dagga (Afrikaans)	Leaves and stems	A decoction of the leaves and stems is drunk for management of common cold.	De Beer and Van Wyk (2011)
Mentha aquatica L.	Lamiaceae	Water mint (English)	Bark	A decoction is used during management of common cold.	Watt and Breyer-Brandwijk (1962); Hutchings et al. (1996)

(Continued)

TABLE 7.1 (Continued)

Traditional Herbal Remedies Used in the Management of Viral Bronchitis and Common Colds in Southern Africa

Plant Species Used in Remedy	Family	Common Name(s)	Part of Plant Used in Preparation of Remedy	Use, Preparation, and Application of Remedy	References
Mentha longifolia (L.)	Lamiaceae	Wild mint (English), kruisement, balderjan (Afrikaans), koena-ya-thabo (Sotho), inixina, inzinziniba (Xhosa), ufuthana, lomhlanga (Zulu)	Leaves, roots, and stems	A decoction is taken, or the leaves are added to milk and consumed during management of common cold, influenza, and bronchitis.	Watt and Breyer-Brandwijk (1962); Hutchings et al. (1996); Van Wyk et al. (2009); Moffett (2010); De Beer and Van Wyk (2011); Hulley and Van Wyk (2019)
Mentha spicata L.	Lamiaceae	Spearmint (English), imboza (Xhosa)	Leaves	A decoction is used during management of common cold.	Hulley and Van Wyk (2019)
Metalasia densa (Lam.) P. O.Karis	Asteraceae	Tee (Sotho)	Not specified	Used to fumigate huts of people suffering from common cold.	Watt and Breyer-Brandwijk (1962); Moffett (2010)
Microglossa mespilifolia (Less.) B.L. Rob.	Asteraceae	Inkhambi elimhlophe, umazambezi (Zulu)	Not specified	Used during management of common cold and influenza. Preparation and application not specified.	Mhlongo and Van Wyk (2019)
Millettia grandis (E.Mey.) Skeels	Fabaceae	Ubobolwehlathi (Zulu)	Not specified	Used during management of common cold and influenza. Preparation and application not specified.	Mhlongo and Van Wyk (2019)
Mohria caffrorum (L.) Desv	Anemiaceae	Carrot fern, scented fern (English), brandbossie (Afrikaans)	Leaves	The leaves are smoked during the management of a common cold.	Watt and Breyer-Brandwijk (1962); Moffett (2010)
Mikania natalensis DC.	Asteraceae	Ihlozi (Zulu)	Not specified	Used during management of common cold and influenza. Preparation and application not specified.	Mhlongo and Van Wyk (2019)
Monsonia burkeana Planch. ex Harv	Geraniaceae	Crane's bill (English), angelbossie, keitabossie, naaldbossie, teebos (Afrikaans)	Leaves and roots	A decoction is used during management of common cold.	Watt and Breyer-Brandwijk (1962)

(Continued)

TABLE 7.1 (Continued)

Traditional Herbal Remedies Used in the Management of Viral Bronchitis and Common Colds in Southern Africa

Plant Species Used in Remedy	Family	Common Name(s)	Part of Plant Used in Preparation of Remedy	Use, Preparation, and Application of Remedy	References
Monsonia emarginata L'Her.	Geraniaceae	Dysentery herb (English), geita, geitabossie, keitabossie, naaldbossie (Afrikaans)	Leaves and roots	A decoction is used during management of common cold.	Smith (1888); Watt and Breyer-Brandwijk (1962);
Montinia caryophyllacea Thunb.	Montiniaceae	Pepperbush, wild clove-bush (English), berg klapper, peperbos (Afrikaans)	Leaves	Pulverised leaves are taken as a snuff during management of common colds.	Von Koenen (2001)
Morella serrata (Lam.) Killick	Myricaceae	Mountain waxberry, lance-leaved waxberry (English), smalblarwasbessie, berg wasbessie, waterolier (Afrikaans), isibhara, umaluleka (Xhosa), lyethi, ulethi, umakhuthula (Zulu)	Not specified	Used during management of common cold.	Moffett (2010)
Myrothamnus flabellifolia Welw	Myrothamnaceae	Resurrection plant (English), bergboegoe (Afrikaans), uvukwabafile (Zulu)	Leaves	Leaf infusions are used during management of common cold.	Watt and Breyer-Brandwijk (1962); Von Koenen (2001); Van Wyk et al. (2009)
Nicotiana glauca Graham	Solanaceae	Mustard tree, tree tobacco (English), tabakboom, wildetabak, volstruisgifboom (Afrikaans), mohlafotha (Sotho)	Leaves	Powdered leaves are taken as a snuff during management of common cold.	Moffett (2010)
Nicotiana rustica L	Solanaceae	Strong tobacco, Aztec tobacco (English)	Leaves	Powdered leaves are taken as a snuff during management of common colds.	Watt and Breyer-Brandwijk (1962)
Notobubon tenuifolium (Thunb.) Magee	Apiaceae	Wildekoelsaad, wilde vinkel (Afrikaans)	Not specified	Used during management of common cold. Preparation and application not specified.	Hulley and Van Wyk (2019)
Ocimum americanum L.	Lamiaceae	Hoary basil (English), wilde basielkruid (Afrikaans)	Leaves	The leaves are burnt, and the smoke inhaled during management of common cold.	Von Koenen (2001)

(Continued)

TABLE 7.1 (Continued)

Traditional Herbal Remedies Used in the Management of Viral Bronchitis and Common Colds in Southern Africa

Plant Species Used in Remedy	Family	Common Name(s)	Part of Plant Used in Preparation of Remedy	Use, Preparation, and Application of Remedy	References
Oncosiphon piluliferum L. f. Kallersj € o	Asteraceae	Stinkkruid (Afrikaans)	Not specified	Used during management of common cold and bronchitis. Preparation and application not specified.	Hulley and Van Wyk (2019)
Oncosiphon suffruticosum L. Kallersjo	Asteraceae	Stinkkruid, wirmkruid (Afrikaans)	Whole plant	Used during management of influenza.	Van Wyk et al. (2009); Nortje and Van Wyk (2015); Hulley and Van Wyk (2019)
Osmitopsis asteriscoides Less	Asteraceae	Bels, belskruie (Afrikaans)	Leaves	Used during management of influenza.	Van Wyk et al. (2009); Philander (2011)
Otholobium polystictum (Harv.) C.H.Stirt	Fabaceae	Kite hook-leaved pea (English), vlieebos (Afrikaans), mohlonecha,	Root	Used during management of common cold.	Moffett (2010); Watt and Breyer-Brandwijk (1962)
Pachycarpus rigidus E. Mey. ex Eckl. & Zeyh.	Apocynaceae	Ishongwe (Zulu)	Not specified	Used during management of common cold.	Moffett (2010)
Packera heterophylla (Fisch.) E. Wiebe	Asteraceae	Unknown	Leaves	Smoke from burning leaves is inhaled during management of common cold.	Watt and Breyer-Brandwijk (1962); Moffett (2010)
Pechuel-Loeschea leubnitziae (Kuntze) O. Hoffm	Asteraceae	Bitterbos (mbiguous)	Leaves	Smoke from burning leaves is inhaled during management of common cold.	Von Koenen (2001)
Pegolettia baccharidifolia Less.	Asteraceae	Ghwarrieson, heuningdou (Afrikaans)	Not specified	Used during management of common cold. Preparation and application not specified.	Hulley and Van Wyk (2019)
Pellaea calomelanos (Sw.) Link.	Pteridaceae	Hardfern (English), lehorometso (Sotho), inkomankomo (Zulu)	Leaves	Leaves are smoked for management of common cold.	Watt and Breyer-Brandwijk (1962); Van Wyk et al. (2009) Moffett (2010); Philander (2011); Smith (1888)

(Continued)

TABLE 7.1 (Continued)
Traditional Herbal Remedies Used in the Management of Viral Bronchitis and Common Colds in Southern Africa

Plant Species Used in Remedy	Family	Common Name(s)	Part of Plant Used in Preparation of Remedy	Use, Preparation, and Application of Remedy	References
Pellaea ambiguous (Sw.) Baker	Pteridaceae	Unknown	Leaves	Leaves are smoked for management of common cold.	Watt and Breyer-Brandwijk (1962)
Pentzincana (Thunb.) Kuntze	Asteraceae	Skaapkaroobos, ankerkaroo, kleinkaap karoobos (Afrikaans)	Not specified	Used for management of common cold and influenza Preparation and application not specified.	Hulley and Van Wyk (2019)
Passiflora suberosa L.	Passifloraceae	Unyawo lenkukhu (Zulu)	Not specified	Used for management of common cold and influenza. Preparation and application not specified.	Mhlongo and Van Wyk (2019)
Pelargonium abrotanifolium Jacq.	Geraniaceae	Bergsalie (Afrikaans)	Leaves	A leaf infusion is usedfor management of common cold and influenza.	De Beer Van Wyk (2011)
Pelargonium graveolens L'Her.	Geraniaceae	Rose geranium (English), wildemalva (Afrikaans)	Leaves	Leaves are steamed and vapours are inhaled for management of common cold.	Van Wyk et al. (2009)
Pelargonium ramosissimum Willd.	Geraniaceae	Dassieboegoe, dassiebos (Afrikaans)	Leaves	Decoctions or tinctures used for management of common cold.	Smith (1888); Watt and Breyer-Brandwijk (1962); De Beer Van Wyk (2011)
Pentanisia prunelloides (Koltzsch) Walp.	Rubiaceae	Wild Verbena (English), sooibrandbossie (Afrikaans), setimamollo (Sotho), icimamlilo (Zulu)	Roots	Used for management of common cold and influenza Preparation and application were not specified	Van Wyk et al. (2009); Watt and Breyer-Brandwijk (1962); Hutchings et al. (1996); Kose et al. (2015)
Persicaria lapathifolia (L.) Delarbre	Polygalaceae	Uxhaphoxana, uxhaphozi (Zulu)	Not specified	Used for management of common cold and influenza. Preparation and application not specified.	Mhlongo and van Wyk (2019)

(Continued)

TABLE 7.1 (Continued)
Traditional Herbal Remedies Used in the Management of Viral Bronchitis and Common Colds in Southern Africa

Plant Species Used in Remedy	Family	Common Name(s)	Part of Plant Used in Preparation of Remedy	Use, Preparation, and Application of Remedy	References
Phyla scaberrima (Juss. Ex Pers.) Moldenke	Verbenaceae	Unknown	Leaves	A decoction of the leaves is used for management of common cold.	Smith (1888)
Plectranthus ambiguous (Bolus) Codd.	Lamiaceae	Iboza, imbatatane (Zulu)	Not specified	Used for management of common cold and influenza. Preparation and application not specified.	Mhlongo and Van Wyk (2019)
Plectranthus laxiflorus Benth.	Lamiaceae	Citronellaspurflower (English), sitrinellaspoorsalie (Afrikaans), ubebebe (Xhosa)	Leaves	A decoction made from powdered leaves used as an enema for management of influenza.	Watt and Breyer-Brandwijk (1962); Hutchings et al. (1996); Ngwenya et al. (2003)
Polygala schinziana Chodat	Polygalaceae	Kanjengena (Afrikaans)	Roots	A decoction is consumed for management of common cold and cough.	Hutchings et al. (1996); Von Koenen (2001)
Pleopeltis macrocarpa (Bory ex Willd) Kaulf.	Polypodiaceae	Lance leaf polypody (English)	Not specified	A decoction is consumed for management of common cold.	Watt and Breyer-Brandwijk (1962); Moffett (2010)
Protea repens L.	Proteaceae	Sugarbush (English), suikerbos (Afrikaans)	Flowers	Syrup prepared from the flower nectar is used by people with colds and influenza as a cough mixture.	Van Wyk et al. (2009)
Pteronia incana (Brum.) DC.	Asteraceae	Skieterbos, keurtjiebos, kraakbos (Afrikaans)	Not specified	Used for management of common cold and influenza. Preparation and application not specified.	Hulley and Van Wyk (2019)
Pulicaria scabra (Thunb.) Druce	Asteraceae	Fleabane (English), aambeibos (Afrikaans), ithaphuka (Zulu)	Leaves	The powdered leaves are used for management of common cold.	Watt and Breyer-Brandwijk (1962)

(Continued)

TABLE 7.1 (*Continued*)
Traditional Herbal Remedies Used in the Management of Viral Bronchitis and Common Colds in Southern Africa

Plant Species Used in Remedy	Family	Common Name(s)	Part of Plant Used in Preparation of Remedy	Use, Preparation, and Application of Remedy	References
Pycreus nitidus (Lam.) J. Raynal	Cyperaceae	Unknown	Rhizome	Used for management of common cold. Preparation and application not specified.	Watt and Breyer-Brandwijk (1962)
Ranunculus capensis Thunb.	Rananculaceae	Blistering Leaves (English), blandblare, katjiedrieblaar (Afrikaans)	Roots	A root infusion is consumed for management of common cold.	Smith (1888)
Ranunculus multifidus Forssk.	Rananculaceae	Buttercupflower (English), botterblom, brandblaar, geelbotterbom, kankerblaar (Afrikaans), hlapi (Sotho), uxhaphozi, ishashakazane (Zulu)	Leaf	Leaves used for management of common cold.	Smith (1888); Watt and Breyer-Brandwijk (1962); Hutchings et al. (1996); Moffett (2010)
Rhigozum trichotomum Burch.	Bignoniaceae	Driedoring (Afrikaans)	Stems	Stems are chewed during management of common cold.	Von Koenen (2001)
Rhoicissus tridentata (L.f.) Willd. & R.B. Drumm.	Vitaceae	Northern Bushman grape, bitter gape (English), noordelikeboesmandruif, bitterdruif, droog-my-keel (Afrikaans), isaqoni, umnxeba, ulatile (Xhosa), isinwazi, umthwazi (Zulu), morara-oa-thaba (Sotho), murumbula- mbudzana (Venda)	Not specified	Used for management of common cold.	Moffett (2010)
Rhus divaricata Eckl & Zeyh.	Anacardiaceae	Fire thorn Karee, rusty leaved currant, Mountain kuni-bush (English)	Roots	A root decoction is used for management of common cold, influenza and bronchitis.	Watt and Breyer-Brandwijk (1962)
Rubu sludwigii Eckl & Zeyh.	Rosaceae	Wild raspberry, Silver bamble (English), braambos, wildebraam (Afrikaans), itshalo, unomhloshane (Zulu), monoko-metsi (Sotho)	Roots	A root decoction is used for management of common cold.	Watt and Breyer-Brandwijk (1962)

(*Continued*)

TABLE 7.1 (Continued)

Traditional Herbal Remedies Used in the Management of Viral Bronchitis and Common Colds in Southern Africa

Plant Species Used in Remedy	Family	Common Name(s)	Part of Plant Used in Preparation of Remedy	Use, Preparation, and Application of Remedy	References
Rumex lanceolatus Thunb.	Polygalaceae	Tongblaar (Afrikaans)	Leaves	Fresh leaves are ground and taken as snuff during management of common cold.	Von Koenen (2001)
Ruta graveolens L.	Rutaceae	Wynruit (Afrikaans)	Leaves	Leaf infusions are consumed for management of common cold and influenza.	De Beer Van Wyk (2011); Nortje and Van Wyk (2015); Hulley and VanWyk (2019)
Salix hirsute Thunb.	Salicaceae	Cape Willow (English), vaalwilger, wilgerboom (Afrikaans)	Bark	Used for management of common cold and influenza.	Von Koenen (2001); Hulley and VanWyk (2019)
Salvia africana-lutea L.	Lamiaceae	Bloebloomsalie (Afrikaans)	Not specified	A decoction is used for management of common cold.	Watt and Breyer-Brandwijk (1962); Philander (2011)
Salvia chamelaeagnea Berg.	Lamiaceae	Bloublomsalie (Afrikaans)	Leaves and flowers	A decoction is used for management of common cold.	Watt and Breyer-Brandwijk (1962); Hulley and Van Wyk (2019)
Salvia dentata Aiton	Lamiaceae	Bergsalie (Afrikaans)	Leaves	Leaf decoctions are consumed for management of common cold.	De Beer Van Wyk (2011); Nortje and VanWyk (2015)
Salvia microphylla Kunth.	Lamiaceae	Rooisalie, rooiblomsalie (Afrikaans)	Not specified	Infusions are consumed for management of common cold.	Hulley and Van Wyk (2019)
Salvia stenophylla Burch. Ex Benth.	Lamiaceae	Maagpynbos, fynblaarsalie (Afrikaans)	Leaves	A leaf infusion is consumed for management of common cold and influenza.	Von Koenen (2001)
Schistostephium flabelliforme Less.	Asteraceae	Unknown	Not specified	An infusion is consumed for management of common cold.	Smith (1888)
Schkuhria Pinnata (Lam.) Kuntze. ex. Thell.	Asteraceae	Kleinkakiebos, klein-gousblom (Afrikaans)	Leaves	Powdered leaves are swallowed with water for management of common cold and influenza.	Watt and Breyer-Brandwijk (1962); Von Koenen (2001)

(Continued)

TABLE 7.1 (Continued)
Traditional Herbal Remedies Used in the Management of Viral Bronchitis and Common Colds in Southern Africa

Plant Species Used in Remedy	Family	Common Name(s)	Part of Plant Used in Preparation of Remedy	Use, Preparation, and Application of Remedy	References
Schotia brachypetala Sond.	Fabaceae	Weeping Boer Bean, tree fuchsia, African walnut (English hullboerboon), (Afrikaans), umfofofo, umgxam (Xhosa), ihluze, umgxamu, uvovovo (Zulu)	Not specified	Used for management of common cold and influenza. Preparation and application not specified.	Mhlongo and Van Wyk (2019)
Securidaca longipedunculata Fresen.	Polygalaceae	Violet tree (English), krinkhout (Afrikaans)	Roots	A root infusion is used for management of common cold.	Von Koenen (2001)
Searsia Divaricata (Eckl. & Zeyh.) Moffett	Anacardiaceae	Mountainkuni-bush (English)	Not specified	Used during management of common cold.	Moffett (2010)
Searsia erosa (Thunb.) Moffett	Anacardiaceae	Broomkaree, besembos (English)	Not specified	Used for management of common cold.	Moffett (2010)
Searsia lancea (L.f.) F.A. Barkley	Anacardiaceae	African Sumac, willowrhus (English)	Leaves	Used for management of common cold. Preparation and application not specified.	Mulaudzi et al. (2012)
Searsia natalensis (Bernh. ex C. Krauss) F.A. Barkley	Anacardiaceae	Natal rhus (English)	Roots	A decoction is used during management of influenza.	Watt and Breyer-Brandwijk (1962)
Searsia undulata (Jacq.) T.S. Yi, A.J.Mill & J. Wen.	Anacardiaceae	Kuni-bush (English), koeniebos, garrabos (Afrikaans), t (Khoi)'kuni	Leaves, bark and roots	Leaves chewed to relieve colds. A leaf decoction is consumed during management of common cold. Alternatively, fresh leaves are chewed for the same purpose. No preparation is provided for the bark and roots.	Van Wyk et al (2009) Hulley. and Breyer-Brandwijk (1962); VanWyk (2019); Watt and Breyer-Brandwijk (1962)

(Continued)

TABLE 7.1 (*Continued*)

Traditional Herbal Remedies Used in the Management of Viral Bronchitis and Common Colds in Southern Africa

Plant Species Used in Remedy	Family	Common Name(s)	Part of Plant Used in Preparation of Remedy	Use, Preparation, and Application of Remedy	References
Selaginella caffrorum (Milde) Hieron	Selaginellaceae	Resurrection plant (English)	Not specified	Used during management of common cold.	Moffett (2010); Kose et al (2015).
Senecio asperulus DC.	Asteraceae	Moferefere (Sotho)	Not specified	Decoction used during management of common cold and influenza.	Watt and Breyer-Brandwijk (1962); Moffett (2010); Kose et al. (2015)
Senecio brachypodus DC.	Asteraceae	Forest senecio (English)	Roots	An infusion is consumed during management of common cold.	Watt and Breyer-Brandwijk (1962)
Senecio dregeanus DC.	Asteraceae	Unknown	Roots	A root decoction is consumed during management of common cold.	Watt and Breyer-Brandwijk (1962)
Senecio rhyncholaenus DC.	Asteraceae	Unknown	Leaves	Smoke from burning leaves inhaled during management of common cold.	Watt and Breyer-Brandwijk (1962); Moffett (2010)
Senna italica Mill.	Fabaceae	Italian senna (English), elandsertjie (Afrikaans)	Root	The root is pounded, soaked in milk and consumed during management of influenza.	Watt and Breyer-Brandwijk (1962); Hutchings et al. (1996); Von Koenen (2001)
Seriphium plumosum L.	Asteraceae	Khoi-kooigoed, slangbos, slangbossie (Afrikaans)	Leaves	Leaves are boiled, and the steam is inhaled during management of influenza.	Ngwenya et al. (2003)
Siphonochilus aethiopicus (Schweif.) B.L. Burt.	Zingiberaceae	African ginger (English), isiphephetho, indungulo (Zulu)	Roots and rhizomes	Chewed during management of influenza.	Hutchings et al. (1996); Van Wyk et al. (2009); Philander (2011)
Solanum capense L.	Solanaceae	Nightshade (English)	Bark and roots	A decoction is consumed during management of common cold.	Smith (1888); Hutchings et al. (1996)

(*Continued*)

TABLE 7.1 (*Continued*)
Traditional Herbal Remedies Used in the Management of Viral Bronchitis and Common Colds in Southern Africa

Plant Species Used in Remedy	Family	Common Name(s)	Part of Plant Used in Preparation of Remedy	Use, Preparation, and Application of Remedy	References
Solanum americanum Mill.	Solanaceae	Black nightshade (English)	Fruit	Ripe fruits taken with honey during management of common cold.	Von Koenen (2001)
Solanum refractum Hook. & Arn.	Solanaceae	Nastergal, nasgal (Afrikaans)	Not specified	Infusions are consumed to treat common cold.	Hulley and Van Wyk (2019)
Spilanthes mauritiana (A. Rich ex Pres) DC.	Asteraceae	Isishoshokazane, isisinini (Zulu)	Not specified	Used during management of common cold and influenza. Preparation and application not specified.	Mhlongo and Van Wyk (2019)
Spirostachys africana Sond.	Euphorbiaceae	African mahogany, African sandalwood, headache tree, jumping beantree, tamboti (English) agelhout, angelhout, gifboomelkhout, dandaleenhout, tambootie (Afrikaans), muonze (Venda),morekuri (Northern Sotho), umtamboti (Zulu), umthombothi (Xhosa)	Bark	Used during management of influenza. Preparation and application not specified.	Mulaudzi et al. (2012)
Stachys hyssopoides Burch ex. Benth.	Lamiaceae	Pinksalie (Afrikaans), selaoane (Sotho)	Not specified	Used during management of common cold.	Moffett (2010)
Sutherlandia frutescens (L.) R.Br.	Fabaceae	Cancer bush (English), kankerbos (Afrikaans), 'musa-motlepelopelo, (Sotho), insiswa, unwele (Xhosa, Zulu)	Leaves	A decoction is used during management of influenza.	VanWyk et al. (2009); De Beer and Van Wyk (2011); Nortje and Van Wyk (2015)
Sutherlandia humilis E. Phillips & R.A. Dyer	Fabaceae	Unknown	Leaves	A decoction is used during management of influenza.	Watt and Breyer-Brandwijk (1962)
Sutherlandia microphylla Burch.	Fabaceae	Unknown	Leaves	Infusions or decoctions are used during management of influenza.	Watt and Breyer-Brandwijk (1962)

(Continued)

TABLE 7.1 (*Continued*)
Traditional Herbal Remedies Used in the Management of Viral Bronchitis and Common Colds in Southern Africa

Plant Species Used in Remedy	Family	Common Name(s)	Part of Plant Used in Preparation of Remedy	Use, Preparation, and Application of Remedy	References
Syzygium cordatum Hochst ex Krauss.	Myrtaceae	Waterberry (English), water bessie, waterboom (Afrikaans), undoni (Zulu), umswi, umjomi (Xhosa), mawthoo (Sotho), motlho (Northern Sotho), mutu (Venda)	Leaves	Used during management of common cold. Preparation and application not specified.	York et al (2011) Mulaudzi et al. (2012); Mhlongo and Van Wyk (2019)
Tagetes minuta L.	Asteraceae	Ikhambilempaka, usangwana (Zulu)	Not specified	Used during management of common cold and influenza. Preparation and application not specified.	Mhlongo and Van Wyk (2019)
Tarchonanthus camphoratus L.	Asteraceae	Wild camphor bush (English), wildekanferbos, vaalbos (Afrikaans), sefahla (Sotho), mofahlana (Sotho), mathola (Xhosa), iggeba- elimhlophe (Zulu)	Leaves and twigs	Infusions or tinctures are taken during management of bronchitis.	Von Koenen (2001); Van Wyk et al. (2009)
Tecoma capensis (Thunb.) Lindl.	Bignoniaceae	Umunyane, uthswalabenyoni (Zulu)	Not specified	Used during management of common cold and influenza. Preparation and application not specified.	Mhlongo and Van Wyk (2019)
Terminalia prunioide M. A. Lawson	Combretaceae	Purple-pod Terminalia (English), deurmekaar, sterkbos (Afrikaans)	Roots	A root infusion is consumed during management of common cold.	Von Koenen (2001)
Terminalia sericea Burch. ex DC.	Combretaceae	Silver cluster leaf (English), vaalboom (Afrikaans), mususu (Venda)	Roots and leaves	Infusions are consumed during management of common cold.	Hutchings et al. (1996)

(Continued)

TABLE 7.1 (Continued)
Traditional Herbal Remedies Used in the Management of Viral Bronchitis and Common Colds in Southern Africa

Plant Species Used in Remedy	Family	Common Name(s)	Part of Plant Used in Preparation of Remedy	Use, Preparation, and Application of Remedy	References
Tetradenia riparia (Hochst.) Codd.	Lamiaceae	Misty plume bush, gingerbush (English), gemerbos, watersalie (Afrikaans), iboza, ibozane (Zulu)	Leaves	An infusion is used during management of blocked and runny nose among people with common cold and influenza.	Felhaber and Mayeng (1997); Van Wyk et al (2009); York et al. (2011); Mhlongo and Van Wyk (2019)
Teucrium africanum Thunb.	Lamiaceae	Drievingertee (Afrikaans)	Not specified	Decoction used during management of influenza.	Watt and Breyer-Brandwijk (1962); Hulley and Van Wyk (2019)
Thamnosma africana Engl.	Rutaceae	Fleabush (English)	Leaves and twigs	An infusion is consumed during management of common cold and influenza.	Von Koenen (2001)
Thesium costatum A.W. Hill	Santalaceae	Kleinswartstorm (Afrikaans), marakalle (Sotho)	Nots specified	Multiple *Thesium* spp. used during management of common cold.	Watt and Breyer-Brandwijk (1962); Kose et al. (2015)
Toddalia asiatica (L.) Lam.	Rutaceae	Cockspurorange, climbing orange (English)	Root	Root decoction used during management of influenza.	Watt and Breyer-Brandwijk (1962)
Trephrosia semiglabra Sond.	Fabaceae	Unknown	Roots	Root decoction used during management of common cold.	Watt and Breyer-Brandwijk (1962)
Tropaeolum majus L.	Tropaeolaceae	Kappertjie (Afrikaans)	Not specified	Used during management of common cold. Preparation and application not specified.	Hulley and Van Wyk (2019)
Tulbaghia acutiloba Harv.	Amaryllidaceae	Wild Garlic (English), wildeknoffel (Afrikaans), konofolo, setotha-fotha (Sotho)	Whole plant	Used during management of common cold. Preparation and application not specified.	Kose et al. (2015)

(Continued)

TABLE 7.1 (Continued)

Traditional Herbal Remedies Used in the Management of Viral Bronchitis and Common Colds in Southern Africa

Plant Species Used in Remedy	Family	Common Name(s)	Part of Plant Used in Preparation of Remedy	Use, Preparation, and Application of Remedy	References
Tulbaghia violacea Harv.	Amaryllidaceae	Wild Garlic (English), wildeknoffel (Afrikaans), ihaqa (Zulu)	Bulbs and leaves	Used during management of common cold.	Van Wyk et al (2009); Hulley And VanWyk (2019); Mhlongo and Van Wyk (2019)
Valeriana capensis Thunb.	Caprifoliaceae	Capevalerian (English), wildebalderjan (Afrikaans)	Rhizome	Used during management of bronchitis.	Van Wyk et al. (2009)
Vangueria infausta Burch.	Rubiaceae	Wild medlar (English), wildemispel (Afrikaans)	Leaves and roots	Infusions are consumed during management of common cold.	Hutchings et al. (1996) Von Koenen (2001)
Vepris lanceolata G. Don.	Rutaceae	Whiteironwood (English), witysterhout (Afrikaans), muhondwa (Venda), umzane (Xhosa), umozana (Zulu)	Root	Powdered root used during management of influenza.	Smith (1888); Watt and Breyer-Brandwijk (1962)
Viscum capense L.f.	Santalaceae	Capemistletoe (English), lidjiestee, voelent (Afrikaans)	Whole plant	An infusion is used during management of bronchitis.	Van Wyk et al (2009); Nortje and VanWyk (2015)
Volkameria glabra (E. Mey.) Mabb. & Y.W. Yuan	Lamiaceae	Umqoqongo (Zulu)	Not specified	Used during management of common cold and influenza. Preparation and application not specified.	Mhlongo and Van Wyk (2019)
Waltheria indica L.	Malvaceae	Unknown	Roots	Roots are chewed during management of common cold and influenza.	Von Koenen (2001)
Warburgia salutaris (G. Bertol.) Chiov.	Canellaceae	Pepper-batreek (English), peperbasboom (Afrikaans), mulanga, manaka (Venda), isibhaha (Zulu)	Bark	Used during management of cough and common cold.	Hutchings et al. (1996); Van Wyk et al. (2009)

(Continued)

TABLE 7.1 (*Continued*)
Traditional Herbal Remedies Used in the Management of Viral Bronchitis and Common Colds in Southern Africa

Plant Species Used in Remedy	Family	Common Name(s)	Part of Plant Used in Preparation of Remedy	Use, Preparation, and Application of Remedy	References
Withania somnifera (L.) Dunal	Solanaceae	Indian ginseng, poison gooseberry, wintercherry (English), bitterappelliefie, koorshout (Afrikaans), ubuvuma (Xhosa), ubuvimbha (Zulu)	Root	Decoction consumed during management of common cold and influenza.	Smith (1888); Watt and Breyer-Brandwijk (1962); Moffett (2010); Kose et al. (2015)
Xysmalobium undulatum (L.) W.T. Aiton.	Apocynaceae	Wildcotton (English), melkbos (Afrikaans)	Root	Decoction is applied on the chest during management of common cold.	Von Koenen (2001); Watt and Breyer-Brandwijk (1962); Von Koenen (2001)
Zanthoxylum capense (Thunb.) Harv.	Rutaceae	Knobwood (English)	Bark	Used during management of common cold. Preparation and application not specified.	Philander (2011)
Zanthoxylum davyi Waterm.	Rutaceae	Forest knobwood (English)	Bark	Powdered bark used during management of common cold and severe coughs.	Watt and Breyer-Brandwijk (1962)
Zingiber officinale Roscoe	Zingiberaceae	Ginger (English), gemmer (Afrikaans)	Root	Used during management of common cold, influenza, and sinusitis.	Hulley and Van Wyk (2019)

surveys and laboratory-based studies are required to validate their traditional use (Cock and Van Vuuren, 2020).

The growth forms of plants used in formulation of remedies include ferns, herbs, trees, and shrubs. In Southern Africa, there is a prominent use of herbaceous plants in remedies that may be attributed to high number of ethnobotanical studies conducted, widespread availability, and vastness of traditional knowledge. Herbaceous plants grow rapidly, and prominent use of herbaceous plants has an advantage of sustainable production and use of remedies. Use of different plant growth forms in formulation of remedies may be attributed to differences in socio-cultural beliefs, ecological status, and variations in practice by formulators (Cock and Van Vuuren, 2020).

Leaves are the most frequently used part in formulation of remedies. Roots, bulbs, rhizomes, stems, twigs, flowers, corms, and whole plants were also used in remedies (Cock and Van Vuuren, 2020). Tree bark is also used in formulation of remedies; however, uncontrolled harvesting of bark may damage or kill trees. Tree bark should be used sparingly to limit tree damage and ensure sustainable production of remedies. Harvesting parts of plants without inflicting serious damage is preferred, particularly, for endangered and threatened species (Cock and Van Vuuren, 2020).

In Southern Africa, ethnobotanical records for most plant species used in formulation of remedies for management of acute bronchitis and common cold are incomplete, and information on parts of plant used in formulation of remedies is not available for some species. Further ethnobotanical studies are required to provide the important data.

In Southern Africa, few studies have been conducted to isolate bioactive compounds and determine efficacy, dosage, and safety of the herbal formulations. Only one species, *C. glabrum*, was screened for ability to block a single strain of H1N1 influenza (Mehrbod et al., 2018), and evaluation of antiviral activity against other strains of influenza and other respiratory viruses is yet to be conducted (Cock and Van Vuuren, 2020). Other plant species used traditionally in formulation of remedies should be evaluated to isolate bioactive compounds and determine efficacy, dosage, and safety.

7.3 HERBAL REMEDIES WITH POTENTIAL FOR MANAGEMENT OF SARS-COV-2

Herbal remedies played an important role in providing primary healthcare needs in Southern Africa during the on-going SARS-CoV-2 pandemic. Readily available herbal remedies were widely used in management of SARS-CoV-2 (Mathew et al., 2020; Attah et al., 2021; Mfengu et al., 2021), although there is limited published data to support the efficacy of the various remedies. At present, fast tracked attempts are being made to discover, repurpose, or develop preventive and treatment options for SARS-CoV-2 from the wealth of indigenous knowledge on use of medicinal plants of Southern Africa.

In year 2020, Madagascar authorised use of an indigenous herbal remedy produced from a species under the *Artemisia* genus for the prevention and management of SARS-CoV-2. World Health Organisation (WHO) carefully discouraged the official positioning of the remedy as a cure for the disease, and emphasised the need for evidence for satisfactory efficacy and safety from clinical trials (Attah et al., 2021). World Health Organisation and Africa Centre for Disease Control are supporting and cooperating with Madagascar in the design and implementation of clinical trials to validate the efficacy and possible adverse effects of the remedy (Attah et al., 2021; WHO, 2020, 2020a).

In silico analyses are scientific experiments or research conducted or produced by means of computer modelling or simulation. *In silico* analyses including the mining of databases containing chemicals originating from plants and computer-based drug design are increasingly being applied in drug discovery. The emerging technologies are regularly applied in the rapid identification of potential compounds during fast tracked drug discovery in emergency situations, including the on-going SARS-CoV-2 pandemic (Attah et al., 2021).

An advantage of *in silico* analyses is that there is a significant reduction in the time frame for the identification of potential active compounds, the plants they originate from, and the analysis of their suitability in combating pathogenic diseases and, therefore, speed up drug discovery (Terstappen and Reggiani, 2001; Pascolutti and Quinn, 2014; Ubani et al., 2020). Documented potential active compounds that have demonstrated *in silico* inhibitory activity against SARS-CoV-2 pathogenic proteins and isolated from plants available in Southern Africa include amaranthin (134) (*Amaranthus tricolor* L.-Amaranthaceae), myricitrin (32) (Myrica cerifera (L.) Small—Myricaceae), isoflavones (30) (Psorothamnus arborescens (A.Gray) Barneby—Fabaceae), nigellicine (21), nigellidine (22), nigellone (28), carvacrol (24), hederin (25), thymol (26), thymoquinone (27), thymohyroquinone (29) (Nigella sativa L.), Calceolarioside B (135) (Fraxinus sieboldiana Blume—Oleaceae), Licoleafol (137) (Glycyrrhiza uralensis Fisch. ex DC—Fabaceae), methyl rosmarinate (31) (*H. atrorubens Poit*), and myricetin 3-O-beta-D glucopyranoside (136) (Camellia sinensis L. Kuntze—Theaceae) (Attah et al., 2021). Although *in silico* findings provide limited evidence, results should form the basis for future in-depth *in vitro, in vivo,* and clinical studies.

Medicinal plants commonly used in Southern Africa that have been evaluated using the molecular docking method for use in management of SARS-CoV-2 include *Acacia senegal. Sutherlandia frutescens, Hypoxis hemerocallidea*, and *Xysmalobium undulantum* (Dwarka et al., 2020). The bioactive compounds arabic acid and L-canavanine found in *Acacia senegal* and *Sutherlandia frutescens*, respectively, exhibited favourable binding modes with strong interactions in the active site of 3C-like main protease—an enzyme which plays a major role in viral replication. Hypoxoside and uzarin extracted from *Hypoxis hemerocallidea* and *Xysmalobium undulatum*, respectively, showed favourable binding orientations characterised by strong interactions within the inhibitor-binding site of SARS-CoV-2 receptor binding domain and SARS-CoV-2 RNA-dependent polymerase. Further evaluation of hypoxoside, uzarin, arabic acid, and L-canavanine may serve as a starting point in the discovery of novel SARS-CoV-2 therapeutic (Dwarka et al., 2020).

7.4   CONCLUSION AND PROSPECTS

There is widespread use of traditional herbal remedies in management of acute bronchitis, common cold, and SARS-CoV-2 in Southern Africa. Majority of populations in Southern Africa are using herbal remedies with limited pharmacological evaluation. Despite extensive ethnobotanical records and the relatively large number of remedies formulated from Southern African plant species, only a single herbal formulation has been tested for ability to inhibit activity of a respiratory virus. Studies should be conducted to evaluate other herbal formulations for ability to inhibit viral respiratory pathogens that they are used against. Evaluation of herbal remedies should be conducted to isolate the bioactive compounds and determine efficacy, dosage, and safety in management of viral respiratory conditions. We encourage increased efforts in cultivation of medicinal plants, creation of new markets for various herbal products, enforcement of regulations for conserving ecosystems, and engagement of communities in natural resource management and value addition of products derived from medicinal plants. Establishment of nurseries and botanical gardens accompanied by local community awareness and involvement in the protection of medicinal plant species are advocated.

REFERENCES

Abrams, A.L, Falkenberg, T., Rautenbach, C., Moshabela, M., Shezi, B., van Ellewee, S. and Street, R. (2020). Legislative landscape for traditional health practitioners in Southern African development community countries: a scoping review. *BMJ open, 10*(1), e029958

Adeleye, O.A., Femi-Oyewo, M.N., Bamiro, O.A., Bakre, L.G., Alabi, A., Ashidi, J.S., Balogun-Agbaje, O.A., Hassan, O.M., and Fakoya, G. (2021). Ethnomedicinal herbs in African traditional medicine with potential activity for the prevention, treatment, and management of coronavirus disease 2019. *Future Journal of Pharmaceutical Sciences, 7*(1), 72. https://doi.org/10.1186/s43094-021-00223–5

Attah, A.F., Fagbemi, A.A., Olubiyi, O., Dada-Adegbola, H., Oluwadotun, A., Elujoba, A., and Babalola, C. P. (2021). Therapeutic potentials of antiviral plants used in traditional African medicine with COVID-19 in focus: a Nigerian perspective. *Frontiers in Pharmacology*, *12*, 596855. https://doi.org/10.3389/fphar.2021.596855

Auerbach, B.J., Reynolds, S.J., Lamorde, M., Merry, C., Kukunda-Byobona, C., Ocama, P., Semeere, A.S., Ndyanabo, A., Boaz, I., Kiggundu, V., Nalugoda, F., Gray, R.H., Wawer, M.J., Thomas, D.L., Kirk, G.D., Quinn, T.C., Stabinski, L., and Rakai Health Sciences Program (2012). Traditional herbal medicine use associated with liver fibrosis in rural Rakai, Uganda. *Public Library of Science One*, *7*(11), e41737. https://doi.org/10.1371/journal.pone.0041737

Chang, S.Y., Voellinger, J.L., Van Ness, K.P., Chapron, B., Shaffer, R.M., Neumann, T., White, C.C., Kavanagh, T.J., Kelly, E.J., and Eaton, D.L. (2017). Characterization of rat or human hepatocytes cultured in microphysiological systems (MPS) to identify hepatotoxicity. *Toxicology In Vitro: An International Journal Published in Association with BIBRA*, *40*, 170–183. https://doi.org/10.1016/j.tiv.2017.01.007

Cinatl, J., Morgenstern, B., Bauer, G., Chandra, P., Rabenau, H., and Doerr, H.W. (2003). Glycyrrhizin, an active component of liquorice roots, and replication of SARS-associated coronavirus. *Lancet (London, England)*, *361*(9374), 2045–2046. https://doi.org/10.1016/s0140-6736(03)13615-x

Clark, T.W., Medina, M.J., Batham, S., Curran, M.D., Parmar, S., and Nicholson, K.G. 2014. Adults hospitalised with acute respiratory illness rarely have detectable bacteria in the absence of COPD or pneumonia; viral infection predominates in a large prospective UK sample. *The Journal of Infection*, *69*(5), 507–515. https://doi.org/10.1016/j.jinf.2014.07.023

Cock, I.E., and Van Vuuren, S.F. 2020. The traditional use of southern African medicinal plants in the treatment of viral respiratory diseases: a review of the ethnobotany and scientific evaluations. *Journal of Ethnopharmacology*, *262*, 113194. https://doi.org/10.1016/j.jep.2020.113194

Cock, I.E., Selesho, I., Van Vuuren, S.F. (2019). A review of the traditional use of South African medicinal plants for the treatment of malaria. *Journal of Ethnopharmacology*, *245*, 112176. https:// doi.org/10.1016/j.jep.2019.112176.

Cock, I.E., Selesho, M.I., and Van Vuuren, S.F. (2018). A review of the traditional use of Southern African medicinal plants for the treatment of selected parasite infections affecting humans. *Journal of Ethnopharmacology*, *220*, 250–264

De Beer, J.J.J., and Van Wyk, B.-E. (2011). An ethnobotanical survey of the Agter-Hantam, Northern Cape Province, South Africa. *South African Journal of Botany*, *77*, 741–754.

DeGeorge, K.C., Ring, D. J. and Dalrymple, S.N. (2019). Treatment of the common cold. *American Family Physician*, *100*(5), 281–289.

Dwarka, D., Clement, A., Mellem, J., Soliman, J., Mahmoud, E., and Himansu, B. (2020). Identification of potential SARS-CoV-2 inhibitors from South African medicinal plant extracts using molecular modelling approaches. *South African Journal of Botany*, *133*, 273–284. https://doi.org/10.1016/j.sajb.2020.07.035.

Erlina, L., Paramita, R.I., Kusuma, W.A., Fadilah, F., Tedjo, A., and Pratomo, I.P. et al. (2020).Virtual screening on Indonesian herbal compounds as COVID-19 supportive therapy: machine learning and pharmacophore modeling approaches. *BMC Medical Informatics and Decision Making*,*1*(6), 2–35. https://doi.org/10.21203/rs.3.rs-29119/v1.

Fasinu, P.S., Bouic, P.J., and Rosenkranz, B. (2012). An overview of the evidence and mechanisms of herb-drug interactions. *Frontiers in Pharmacology*, *3*, 69. https://doi.org/10.3389/fphar.2012.00069.

Felhaber, T. and Mayeng, I. (1997). *South African Traditional Healers' Primary Health Care Handbook*, Kagiso Publishers, Cape Town, South Africa.

Fukumoto, S., Goto, T., and Hayashi, S. (2010). *Anti-SARS Coronavirus Agent, and Product Containing Anti-SARS Coronavirus Agent*. Taiwan Patent TW201018473A.

Gamaniel, K.S. (2000). Toxicity from medicinal plants and their products. *Nigerian Journal of Natural Products and Medicine*, *4*, 4–8.

Gencay, M., Roth, M., Christ-Crain, M., Mueller, B., Tamm, M., and Stolz, D. (2010). Single and multiple viral infections in lower respiratory tract infection. *Respiration*, *80*(6), 560–567.

Hart, A.M. (2014). Evidence-based diagnosis and management of acute bronchitis. *The Nurse Practitioner*, *39*(9), 32–40. https://doi.org/10.1097/01.NPR.0000452978.99676.2b

Hulley, I.M., Van Wyk, B.-E. (2019). Quantitative medicinal ethnobotany of Kannaland (western Little Karoo, South Africa): non-homogeneity amongst villages. *Journal of Botany*, *122*, 225–265. https://doi.org/10.1016/j.sajb.2018.03.014

Hutchings, A., Scott, A.H., Lewis, G., Cunningham, B. (1996). *Zulu Medicinal Plants: An Inventory* (1st edition), University of Natal Press, Pietermaritzburg.

Ismail, H. and Schellack, N. (2017). Colds and flu–an overview of the management. *South African Family Practice*, *59*(3), 5–12

James, P.B., Wardle, J., Steel, A., and Adams, J. (2018). Traditional, complementary and alternative medicine use in sub-saharan Africa: a systematic review. *BMJ global health*, *3*(5), e000895. https://doi.org/10.1136/bmjgh-2018-000895

Keyaerts, E., Vijgen, L., Pannecouque, C., Van Damme, E., Peumans, W., Egberink, H., Balzarini, J., and Van Ranst, M. (2007). Plant lectins are potent inhibitors of coronaviruses by interfering with two targets in the viral replication cycle. *Antiviral Research*, *75*(3), 179–187. https://doi.org/10.1016/j.antiviral.2007.03.003

Kinkade, S. and Long, N.A. (2016). Acute bronchitis. *American Family Physician*, *94*(7), 560–565.

Kose, L.S., Moteetee, A., and Van Vuuren, S. (2015). Ethnobotanical survey of medicinal plants used in the Maseru district of Lesotho. *Journal of Ethnopharmacology*, *170*, 184–200.

Laksmiani, N.L., Larasanty, L.P., Santika, A.J., Prayoga, P.A., Dewi, A.K., Dewi, N.K. (2020). Active compounds activity from the medicinal plants against SARS-CoV-2 using *in silico* assay. *Biomedical Pharmacology Journal*, *13*(2), 873–881. https://doi.org/10.13005/bpj/1953.

Lambelet, V., Vouga, M., Pomar, L., Favre, G., Gerbier, E., Panchaud, A., and Baud, D. (2020). SARS-CoV-2 in the context of past coronaviruses epidemics: consideration for prenatal care. *Prenatal Diagnosis*, *40*(13), 1641–1654. https://doi.org/10.1002/pd.5759

Liu, N.Q., Van der Kooy, F., andVerpoorte, R. (2009). *Artemisia afra:* a potential flagship for African medicinal plants? *South African Journal of Botany*, *75*(2), 185–195.

Macfarlane, J., Holmes, W., Gard, P., Macfarlane, R., Rose, D., Weston, V., Leinonen, M., Saikku, P., and Myint, S. (2001). Prospective study of the incidence, aetiology and outcome of adult lower respiratory tract illness in the community. *Thorax*, *56*(2), 109–114. https://doi.org/10.1136/thorax.56.2.109

Mahomoodally, F.M. (2013). Traditional medicines in Africa: an appraisal of ten potent African medicinal plants. *Evidence-Based Complementary and Alternative Medicine*, 1–14. https://doi.org/10.1155/2013/617459.

Maroyi, A. (2017). *Lippia javanica* (Burm.f.) Spreng.: traditional and commercial uses and phytochemical and pharmacological significance in the African and Indian Subcontinent. *Evidence-Based Complementary and Alternative Medicine*, 1–34. https://doi.org/10.1155/2017/6746071.

Maroyi, A. (2018). *Ethnomedicinal Uses of Exotic Plant Species in South-Central Zimbabwe*, NISCAIR-CSIR, India.

Matthew M., Chingono, F., Mangezi, S., Mare, A., and Mbazangi, S. (2020). Hidden variables to COVID-19: Zimbabwe. *Cambridge Open Engage*, 3–18. https://doi.org/10.33774/coe-2020-1mqnz

Mbadiko, C.M., Inkoto, C.L., Gbolo, B.Z., Lengbiye, E.M., Kilembe, J.T., Matondo, A., Mwanangombo, D.T., Ngoyi, E.M., Bongo, G.N., Falanga, C.M., Tshibangu, D.S.T., Tshilanda, D.D., Ngbolua, K.T.N., and Mpiana, P.T. (2020). A mini review on the phytochemistry, toxicology and antiviral activity of some medically interesting zingiberaceae species. *Journal of Complementary and Alternative Medical Research*, *9*(4), 44–56. https://doi.org/10.9734/jocamr/2020/v9i430150.

Mehrbod, P., Abdalla, M.A., Njoya, E.M., Ahmed, A.S., Fotouhi, F., Farahmand, B., Gado, D.A., Tabatabaian, M., Fasanmi, O.G., Eloff, J.N., McGaw, L.J., and Fasina, F.O. (2018). South African medicinal plant extracts active against influenza A virus. *BMC Complementary and Alternative Medicine*, *18*(1), 112. https://doi.org/10.1186/s12906-018-2184-y

Mfengu, M., Shauli, M., Engwa, G.A., Musarurwa, H.T., and Sewani-Rusike, C.R. (2021). *Lippia javanica* (Zumbani) herbal tea infusion attenuates allergic airway inflammation via inhibition of Th2 cell activation and suppression of oxidative stress. *BMC Complementary Medicine and Therapies*, *21*(1), 192. https://doi.org/10.1186/s12906-021-03361-8

Mhlongo, L.S., and Van Wyk, B.E. (2019). Zulu medicinal ethnobotany: new records from the Amandawe area of KwaZulu-Natal. *Journal of Botany*, *122*, 266–290.

Mirzaie, A., Halaji, M., Dehkordi, F.S., Ranjbar, R., and Noorbazargan, H. (2020). A narrative literature review on traditional medicine options for treatment of corona virus disease 2019 (COVID-19). *Complementary Therapies in Clinical Practice*, *40*, 101214. https://doi.org/10.1016/j.ctcp.2020.101214

Moffett, R. (2010). *Sesotho Plant and Animal Names and Plants Used by the Basotho*. Sun Press, Free State, South Africa.

Mohagheghzadeh, A., Faridi, P., Shams-Ardakani, M., Ghasemi, Y. (2006). Medicinal smokes. *Journal of Ethnopharmacology*, *108*(2), 161–184.

Mulaudzi, R.B., Ndhlala, A.R., Kulkarni, M.G., Van Staden, J. (2012). Pharmacological properties and protein binding capacity of phenolic extracts of some Venda medicinal plants used against cough and fever. *Journal of Ethnopharmacology*, *143*, 185–193.

Newmaster, S.G., Grguric, M., Shanmughanandhan, D., Ramalingam, S., and Ragupathy, S. (2013). DNA barcoding detects contamination and substitution in North American herbal products. *BMC Medicine*, *11*, 222. https://doi.org/10.1186/1741-7015-11-222

Ngwenya, M.A., Koopman, A., and Williams, R. (2003). *Zulu Botanical Knowledge*. National Botanical Institute, Durban, South Africa.

Nortje, J.M., and Van Wyk, B.-E. (2015). Medicinal plants of the Kamiesberg, Namaqualand, South Africa. *Journal of Ethnopharmacology*, *171*, 205–222.

Nowicki, J. and Murray, M.T. (2020). *Therapeutic Considerations: In Textbook of Natural Medicine* (5th edition). Churchill Livingstone.

Orisakwe, O.E., Orish, C.N., and Nwanaforo, E.O. (2020). Coronavirus disease (COVID-19) and Africa: acclaimed home remedies. *Scientific African*, *10*, e00620. https://doi.org/10.1016/j.sciaf.2020.e00620

Pappas, D.E. (2018). The common cold. *Principles and Practice of Pediatric Infectious Diseases*, Elsevier, *199–202*, e1. https://doi.org/10.1016/B978-0-323-40181-4.00026-8.

Pascolutti, M. and Quinn. R.J. (2014). Natural products as lead structures: chemical transformations to create lead-like libraries. *Drug Discovery Today*, *19*(3), 215–221.

Philander, L.A. (2011). An ethnobotany of Western Cape Rasta bush medicine. *Journal of Ethnopharmacology*, *138*(2), 578–594. https://doi.org/10.1016/j.jep.2011.10.004

Prasad, A., Muthamilarasan, M., and Prasad, M. (2020). Synergistic antiviral effects against SARS-CoV-2 by plant-based molecules. *Plant Cell Reports*, *39*(9), 1109–1114. https://doi.org/10.1007/s00299-020-02560-w.

Rathinavel, T., Palanisamy, M., Palanisamy, S., Subramanian, A., and Thangaswamy, S. (2020). Phytochemical 6-Gingerol – a promising drug of choice for COVID-19. *International Journal on Advanced Science and Engineering*, *6*(4), 1482–1489. https://doi.org/10.29294/IJASE.6.4.2020.1482-1489.

Saini, R., Sharma, N., Oladeji, O.S., Sourirajan, A., Dev, K., Zengin, G., El-Shazly, M., and Kumar, V. (2022). Traditional uses, bioactive composition, pharmacology, and toxicology of *Phyllanthus emblica* fruits: a comprehensive review. *Journal of Ethnopharmacology*, *282*, 114570. https://doi.org/10.1016/j.jep.2021.114570

Schellack, N., Schellack, G., and Ismail, H. (2020). An overview of cold and flu management. *South African Pharmaceutical Journal*, *87*(3), 38–42.

Singh, A., Avula, A., Zahn, A. (2021). *Acute Bronchitis*. StatPearls Publishing, Treasure Island, FL.

Smith, A. (1888). *A Contribution to South African Materia Medica*, (2nd edition). Lovedale Institution Press, South Africa.

Terstappen, G.C. and Reggiani, A. (2001). *In silico* research in drug discovery. *Trends in Pharmacological Sciences*, *22*(1), 23–26.

Tungadi, R., Tuloli, T.S., Abdulkadir, W., Thomas, N., Hasan, A.M., Sapiun, Z. et al. (2020). COVID-19: clinical characteristics and molecular levels of candidate compounds of prospective herbal and modern drugs in Indonesia. *Journal of Pharmaceutical Science*, *26*, S12–S23. https://doi.org/10.34172/PS.2020.49.

Ubani, A., Agwom, F., Morenikeji, O.R., Shehu, N., Luka, P., Umera, A., Umar, U., Omale, S., Nnadi, N.E., and Aguiyi, J.C. (2020). Molecular docking analysis of some phytochemicals on two SARS-CoV-2 targets: potential lead compounds against two target sites of SARS-CoV-2 obtained from plants. *BioRxiv*, 1–21 https://doi.org/10.1101/2020.03.31.017657.

Van Wyk, B.-E., van Oudtshoorn, B., Gericke, N. (2009). *Medicinal Plants of South Africa* (2nd edition), Briza Publications, Pretoria, South Africa.

Von Koenen, E. (2001). *Medicinal, Poisonous and Edible Plants in Namibia* (4th edition), Klaus Hess Publishers, Windhoek, Namibia.

Watt, J.M. and Breyer-Brandwijk, M.G. (1962). *The Medicinal and Poisonous Plants of Southern and Eastern Africa* (2nd edition), E & S Livingstone Ltd, Edinburg and London.

Welz, A.N., Emberger-Klein, A., and Menrad, K. (2019). The importance of herbal medicine use in the German health-care system: prevalence, usage pattern, and influencing factors. *BMC Health Services Research*, *19*(1), 1–11.

WHO (2020). WHO Supports Scientifically-Proven Traditional Medicine. Available at: https://www.afro.who.int/news/who-supports-scientifically-proven-traditional-medicine?gclid=CjwKCAjwjLD4BRAiEiwAg5NBFlOWbdSg5OgzIsNBICCwbaCndOvz_Nk8onOJzRLqZw9YhMVHMhRsbxoC9_wQAvD_BwE.

World Health Organisation (2020a). Expert Panel Endorses Protocol for COVID-19 Herbal Medicine Clinical Trials. Available at: https://www.afro.who.int/news/expert-panel-endorses-protocol-COVID-19-herbal-medicine-clinical-trials. Accessed 23 Sep 2021.

Yarnell, E. (2018). Herbs for viral respiratory infections. *Alternative and Complementary Therapies*, Mary Ann Lierbert Inc., vol. 24, 35–43. https://doi.org/10.1089/act.2017.29150.eya

York, T., De Wet, H., and Van Vuuren, S.F. (2011). Plants used for treating respiratory infections in rural Maputaland, KwaZulu-Natal, South Africa. *Journal of Ethnopharmacology*, *135*, 696–710.

Younis, W., Asif, H., Sharif, A., Riaz, H., Bukhari, I.A., and Assiri, A.M. (2018). Traditional medicinal plants used for respiratory disorders in Pakistan: a review of the ethno-medicinal and pharmacological evidence. *Chinese Medicine*, *13*(1), 1–29

8 Solanum nigrum Seed Viability and Germination, and Soil Modulation Effect on Seedling Emergence

*Adijat Funke Ogundola, Anthony Jide Afolayan
and Callistus Bvenura*

CONTENTS

8.1   INTRODUCTION

Seed viability is an important factor in seed germination, as it determines the success of crop production (Perry, 1980). Seeds possess some properties that are necessary for crop establishment that leads to profitable crop production—a goal of every professional crop producer. This goal is driven by production of good quality seed lots; after all achieving and maintaining high-quality seed leads to seed/crop yield increase. Additional factors of successful crop production involve seed handling (mechanisms), seed moisture content, and the prevailing environmental factors that determine crop establishment. However, seed germination performance depends on genetic make-up—properties that enhance its germination, but the environmental influence matters in seed vigour (ISTA, 2015). Early emergence, on the other hand, is described as a trait that describes the crop more competitive and, therefore, escapes the shading effect of their competitors in the field (Beheshtian-mesgaran et al., 2006). It is also understandable that uniform seedling emergence encourages large-scale seed production that enhances marketability and profitability.

Seed production is one of the strategies adopted globally to conserve wild plant species of interest to both scientists and agriculturalists (Modi et al., 2006). However, much importance, priority is placed on seed regeneration of wild plants with potential nutraceutical properties and, therefore, valuable health benefits (Powell et al., 2013). Among some wild and useful plants that have gained the attention of researchers is *Solanum nigrum*—an annual herbaceous plant belonging to the family Solanaceae, commonly known as Black Nightshade. It grows up to 0.6 m in wooded and disturbed habitats. The leaves are green ovate to heart shaped with white flowers, which, in turn, produce green matured fruits that turn purple or black when ripe (Edmond and Chweya, 1997).

DOI: 10.1201/9781003206620-8

This plant grows in the wild in most parts of the world, including the Eastern Cape Province, South Africa. *S. nigrum* is a nutraceutical plant and has been extensively explored by traditional users to pharmaceutical applications in the treatment of different diseases and ailments (Odhav et al., 2007). The use of this plant as an immune booster and a reservoir of micronutrients have attracted research interests on its seed regeneration (Bvenura and Afolayan, 2013). This step ensures a continuous and dependable source for the plant. However, the cultivation of *S. nigrum* is currently restricted to the peasant farmers who scavenge for the plant in the wild on occasional needs for folkloric use (Modi et al., 2006).

S. nigrum, being a wild vegetable, needs sustainable edaphic conditions for maximum seedling production (Flyman and Afolayan, 2006). Suitable soils with unfavourable soil depth may not be adequate for profitable seedling production. However, optimum sowing depth on a suitable soil is required to improve crop seedling emergence, which is a desired goal for crop establishment (Finch-Savage and Basel, 2016). Seedling emergence may be determined by the synergetic influence of two or more factors at a time or singly by soil type, sowing depth, and sources of the seeds among others, as reported by Forcella et al. (2000). The synergetic effects of sowing depth and temperature, and planting depth and temperature on seedling emergence of *S. nigrum* have been reported (Cousens and Mortimer, 1995; Chauhan et al., 2006). However, no study has examined the synergetic influence of soil types and sowing depth on seedling emergence of the plant.

Therefore, the current study aimed to investigate the viability, germination, and soil modulation effect and sowing depths on *S. nigrum* seedling emergence.

8.2 MATERIALS AND METHODS

Seeds of *S. nigrum* were extracted from the pulp of ripe mature berries collected in Alice (32046'47"S and 26050' 5"E and 524 m a.s.l) in the Eastern Cape, South Africa. The plant was identified by Prof Griesson of Department of Botany, University of Fort Hare, South Africa and with Voucher number (BVE11/017), and was deposited at the Giffen herbarium of the same University.

A total 100 seeds in triplicates were dried at 110°C to a constant weight in a pre-set incubator, after which the differences in weight before and after drying were used to calculate percentage moisture content.

Seed viability tests followed the Tetrazolium technique of Grabe (1970), as described for Solanaceae by Peters (2000). The seeds were immersed in water overnight at temperatures between 21°C and 26°C. The seeds were carefully cut longitudinally along the margin to prevent embryo damage. The dissected seeds were soaked in 0.1% solution of 2, 3, 5-triphenyltetrazolium chloride at 20–25°C for 24 hours in the dark (Etejere and Okoko, 1989). The seeds were removed and washed thoroughly in distilled water, soaked in 95% ethanol, and then observed under the light stereo microscope. Viable seeds appeared reddish in colour.

Germination tests included eight temperature regimes (5°C, 10°C, 15°C, 20°C, 25°C, 30°C, 35°C, and 40°C). Seeds of *S. nigrum* in petri dishes were germinated separately under different temperatures, as listed in pre-set incubators.

Continuous light, 12 hours/12 hours light/dark photoperiods, and continuous darkness were carried out in separate Conviron growth chambers (Model E15; Controlled Environment Limited, Winnipeg, Manitoba, Canada).

Seeds of *S. nigrum* were scarified mechanically by scratching the seed's radicle end with a 100-grit sandpaper of 10 cm×10 cm, holding grip between the left and right palms, and gently rubbing against each other for 15 minutes (Belel, 2016). About 2 g of soil was mixed with 100 seeds and rubbed for 10–15 minutes within palms (van Reinsburg et al., 2012).

TABLE 8.1

First and Last Days of Germination of *S. nigrum* in Different Seed Treatments

Treatments			Photoperiodism					Temperatures (°C)						Scarification	
Germination treatments	Control	Continuous light	Total darkness	Alt. Light and darkness	5	10	15	20	25	30	35	40	Sand paper	sand
First day of germination	4	3	8	3	6	3	3	3	3	4	4	3	3	3
Last day of germination	9	8	9	8	9	9	8	8	8	8	4	3	8	9

For each of the listed seed germination experiment—three replicates of 50 treated seeds, each were placed on the laboratory work bench in sterile petri dishes, lined with 9 cm discs of Whatman No. 1 filter paper moistened with distilled water. The seeds were examined daily for 12 days and considered germinated when the radicle was visible (De Villalobos et al., 2002).

Soil was collected from a fallow land at Research farm, University of Fort Hare, Alice, Eastern Cape, South Africa. The soil was air-dried, ground, and sieved into sand, silt, and clay particles using a Vibracion S.L modelo FTLVH-0200 sieving machine. Relative combinations of sand, silt, and clay in different ratios to formulate soil types followed the method of Kellogg (1993) as follows: sandy clay loam (ST_1)-60%, 30%, and 10%, silty clay loam (ST_2)-10%, 60%, and 30%, clay loam (ST_3)-36%, 30%, and 34% loam (ST_4)-40%, 40%, and 20% and sand clay loam (ST_0), which is the control. Three soil samples were taken from each of the soil types and analysed for their chemical properties, as shown in Table 8.1.

This experiment was conducted for 12 days in the glasshouse at temperatures ranging between 28°C and 30°C. The five different soil types (ST_0–ST_4) were filled into 65 cm×33 cm nursery trays with 200 holes at depths of 0–4 cm. Each treatment had three replicates, and each replicate had ten experimental units.

8.3 STATISTICAL ANALYSIS

All recorded data from germination and seedling emergence were statistically analysed using MINITAB Release 17. A one-way analysis of variance was used to compare the means of various germination and seedling emergence treatments, and a two-way analysis of variance was used to determine the interaction between seedling emergence and different sowing depths. Means were segregated using Fisher's Least Significant Difference (LSD) paired wise comparison. The means were treated as significantly different at $p < 0.05$.

8.4 RESULTS

S. nigrum seeds recorded an average moisture content of 6.12%. And the seeds were 74.3% viable. The petri-dish revealed the highest mean seed germination percentage (94.67%) under 15°C, 20°C, and 25°C temperature regimes and scarification with sandpaper. However, the results for the 10°C temperature treatment were not significantly different from the former (Figure 8.1). The First Day of Germination (FDG) was on day 3 and Last Day of Germination (LDG) on day 8, as shown in Table 8.1.

This indicates that the seeds quickly germinate if exposed to the right conditions. In the present study, continuous and alternating light-induced germination was significantly ($p < 0.05$) lower than the control (Figure 8.2). The FDG reported for continuous darkness (day 8) indicates the important role light plays in the germination of *S. nigrum* seeds. In this study, sandpaper recorded significantly

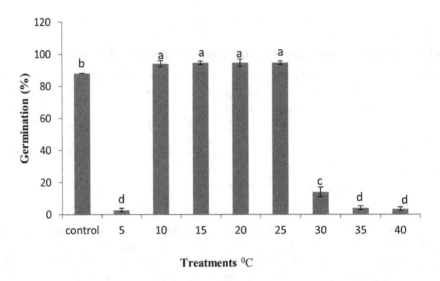

FIGURE 8.1   The effect of temperature on *S. nigrum* seed germination.

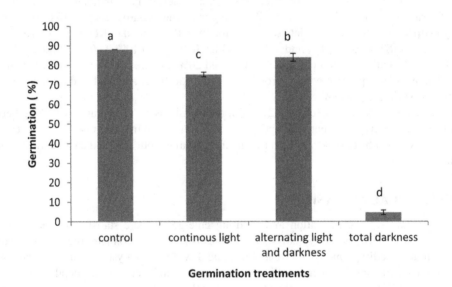

FIGURE 8.2   The effect of light on *S. nigrum* seed germination.

high germination percentage (94.67%), while sand recorded lower results (75%) relative to the control (Figure 8.3).

The effects of soil types on *S. nigrum* seedling emergence are shown in Figures 8.4–8.8. The result of experimental soil analysis was presented in Table 8.2.

Silty clay loam soil recorded the highest seedling emergence at 2 cm depth, while 0 cm and 4 cm recorded the lowest. Statistical analysis established an interaction between sowing depth and seedling emergence, and between soil types and seedling emergence. The current results show that sowing depth had a significant ($p < 0.01$) effect on seedling emergence, while soil types had a significant ($p < 0.05$) effect on seedling emergence. Silty clay loam soil recorded the highest seedling emergence at 2 cm depth, while 0 cm and 4 cm recorded the lowest. Statistical analysis established an interaction between sowing depth and seedling emergence and between soil types and seedling

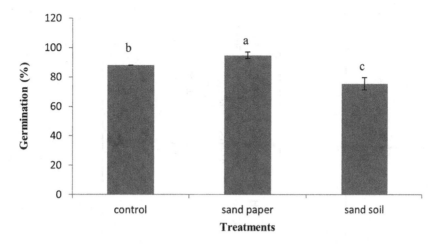

FIGURE 8.3 The effect of scarification on *S. nigrum* seed germination.

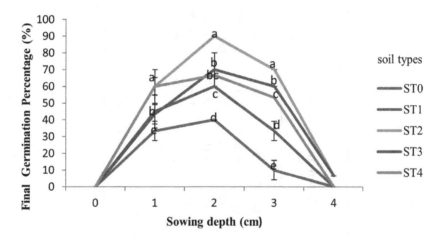

FIGURE 8.4 Effect of soil types on final germination percentage (FGP) of *S. nigrum*.

emergence. The current results show that sowing depth and soil types had a significant effect on seedling emergence.

The highest FGP was recorded in silty clay loam soil (ST_2) at 2 cm sowing depth, while the lowest was observed in the control (ST_0), as shown in Figure 8.4.

No seedling emergence was recorded in the control and sandy loam soils at all sowing depths. The higher the seedling emergence, the higher the FGP. The results of the current study indicate that seedling emergence is inversely related to sand composition. The higher the EG value, the greater the potential of the seed lot to germinate. The results of the present study indicate that soil types and sowing depth significantly affected the EG. As shown in Figure 8.5, The EG was high in silty clay loam (ST_2) at all sowing depths. The lowest EG was recorded in clay loam at 3 cm and 4 cm sowing depths, and simultaneously in sandy clay loam and the control at 1 cm sowing depth.

The CVG gives an indication of the rapidity of germination. Therefore, the higher the CVG value, the faster the time the seeds take to germinate. In the present study, the CVG values in different soil types and sowing depths are shown in Figure 8.6. The lowest CVG was recorded in silty

TABLE 8.2

Physical Properties and Elemental Composition of Soil before the Experiment

mg/kg

Soil types	P	K	N	Ca	Mg	Zn	Mn	Cu	Org. cont%	pH	Clay%	Sand%	Silt %
ST_0	60 ± 0.7^d	524 ± 0.7^a	3.4 ± 0.7^c	1389 ± 0.5^a	332 ± 0.1^b	6.0 ± 0.3^b	29 ± 0.7^d	10.5 ± 0.5^c	4 ± 0.5^c	6.22 ± 0.7^a	10 ± 0.7^c	60 ± 0.7^a	30 ± 0.5^c
ST_1	62 ± 0^c	467 ± 0.7^d	3.6 ± 0^c	1357 ± 0.5^b	347 ± 0.5^a	8.1 ± 0.5^a	66 ± 0.7^a	15.6 ± 0.7^a	3.9 ± 0.5^c	$5.70.7^b$	21 ± 0.5^b	66 ± 0.7^a	13 ± 0.7^d
ST_2	68 ± 0.7^a	524 ± 0.5^a	5 ± 0.5^a	1278 ± 0.5^e	316 ± 0.0^d	5.9 ± 0^{bc}	45 ± 0^b	10.6 ± 0^c	5 ± 0.7^a	5.7 ± 0.5^b	30 ± 0.5^a	10 ± 0.5^c	60 ± 0^a
ST_3	60 ± 0.5^b	519 ± 0^b	4.8 ± 0^b	1318 ± 0^c	321 ± 0.0^c	5.3 ± 0.5^c	38 ± 0^c	10.3 ± 0.5^c	4.5 ± 0^b	$5.630.^b$	34 ± 0^a	36 ± 0.5^c	30 ± 0^c
$S.T_4$	63 ± 0.7^b	482 ± 0.5^c	4.6 ± 0^b	1290 ± 0.5^d	330 ± 0.5^b	6.2 ± 0.7^b	45 ± 0^b	11.4 ± 0.7^b	4.8 ± 0.7^a	5.63 ± 0.5^b	20 ± 0.7^b	40 ± 0^b	40 ± 0.5^b

Values shown are mean±SD different.

Letters down each column represents significant differences at $p<0.05$.

ST0, sandy loam (control); ST1, sandy loam; ST2, silty clay loam; ST3, clay loam; and ST4, loam

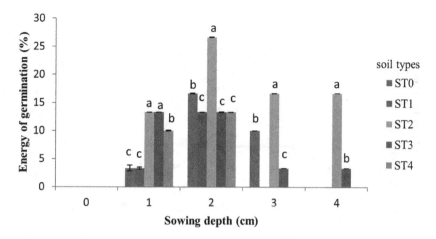

FIGURE 8.5   Effect of soil types on energy of germination (EG) of *S. nigrum*.

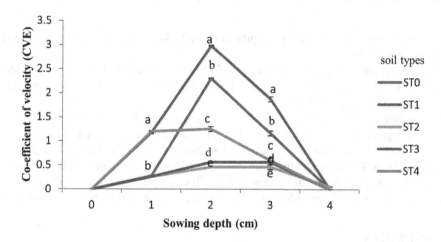

FIGURE 8.6   Effect of soil types on Co-efficient of velocity (CVE) of germination of *S. nigrum*.

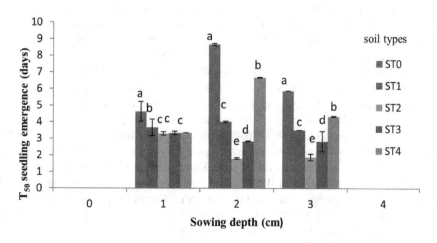

FIGURE 8.7   Effect of soil types on 50% seedling emergence of *S. nigrum*.

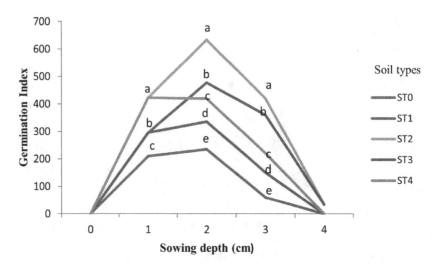

FIGURE 8.8 Effect of soil types on Germination Index (GI) of *S. nigrum*.

clay loam soil at 2 cm and 3 cm sowing depths, and the highest was recorded in sandy clay loam (ST$_1$) at 2 cm sowing depth.

The T$_{50}$ of the current seed population was significantly affected by the soil types, as shown in Figure 8.7. The lowest T$_{50}$ was obtained from silty clay loam (ST$_2$) at 2 cm sowing depth, while the highest was in the control soil (ST$_0$).

The GI emphasises both the percentage of germination and its speed. A higher GI value denotes a higher percentage and rate of germination. Soil types in this study affected the GI in the order: ST$_2$ > ST$_3$ > ST$_4$ > ST$_1$ > ST$_0$, as shown in Figure 8.8. The highest GI was recorded in silty clay loam (ST$_2$) at 2 cm sowing depth. This indicates that seeds planted on this soil type had both higher germination percentage as well as rate of germination. However, the GI was significantly high in all the soil types in comparison with the control. Each of the soil types recorded a significant GI level at 2 cm sowing depth.

8.5 DISCUSSION

Moisture content 6.12% for *S. nigrum* in this study presumably indicates the ability of *S. nigrum* seeds to be stored for long periods of time without compromising their viability. The result is in accordance with an earlier study, which indicated that *S. nigrum* seeds had a moisture composition between 8.19% and 8.36% (Bvenura and Afolayan, 2014). Luna and Wilkinson (2009) reported that most orthodox seeds require moisture content between 6% and 10% for them to retain viability after long storage in which the result of this study falls within. These findings are also within the range reported by Suthar et al. (2009).

The 74.3% viability of *S. nigrum* seed lot in this study is an indication that the seeds are capable to produce high germination percentage results. The high viability percentage may be responsible for the ability of the plant to retain viable in the soil for long time and its consequently germination at later stage, causing problems to crops (Edmond and Chweya, 1997).

The high germination percentage 88%–96.4% from control and different treatments in this study showed that the seed lot is of good quality and capable of germinating under normal conditions. These results are in close agreement with those of Kambizi et al. (2006), who conducted a study and reported that 80% germination on *Withania somnifera* belongs to the same solanaceae family. The high germination percentages from seed treated with different temperatures are in close agreement with those of Suthar et al. (2009), who reported 100% germination after incubating seeds of *S. nigrum* at 27°C and the lowest at 5°C as in this study as well. Also, high germination percentages at 20°C and 25°C corroborate those of Larina (2008), who reported high germination at 20°C and 25°C. However, lower germination percentage (2.67%) recorded at 5°C deviates from the report of Brandel (2005), who concluded that breakages in dormancy occur at low temperatures. Lower germination percentage at low temperature indicates relatively low metabolic activities at lower temperatures (Okuzanya, 1980).

High seed germination percentages 84.75% reported in continuous light and alternating light and darkness agree with the report of Gairola et al. (2011). The study highlighted light as one of the most important environmental factors that promote seed germination. Failure of seeds to germinate in continuous darkness in this study agrees with the report of Beningno-Gonzalez et al. (2009) that continuous darkness at constant temperature inhibits seed germination. In addition, Maloof et al. (2000) reported that optimal growth and development are directly proportional to light received, and the results are in close agreement with those of the present study.

The high germination percentage (96.4%) from the sandpaper treatment is higher than 78% reported by Suthar et al. (2009). The present study indicates that the seed lot has a high germination potential; however, dormancy may be the inhibiting factor.

In this study, no seedling emergence was observed at 0 cm and 4 cm sowing depth; the result compares favourably with those of Chachalis and Reddy (2000). The highest FPSE obtained at 2 cm depth on the five soil types in our study agrees with the observation of seedling emergence at 2 cm depth in a study carried out by Zhou et al. (2005). Failure of seedling emergence at 0 cm and few seedling emergences at 1 cm sowing depths may be attributed to exposure of the seeds to harsh weather elements such as direct sunlight, which in turn could have led to loss of moisture, subsequent seed dryness, and failure to break dormancy (Desbiolles, 2002; Suthar et al., 2009).

In this study, observation of seedling emergence evaluated on the fourth day after sowing was high in all soil types in 1 cm and 2 cm sowing depths showed that the seed is not prone to delay emergence. However, the highest seedling percentage observed in silty clay loam soil at 2 cm sowing depth distinguished it as the best treatment. The EG indicates the capacity of seed lot germinating on the fourth day of the experiment, and this is a critical day in seed germination (Farooq et al., 2005).

High germination index in all the soil treatments at 1 cm and 2 cm sowing depths in this study showed the high potential of the seeds of *S. nigrum* at these sowing depths. However, significantly higher result on silty clay loam at 2 cm sowing depth is an indication that *S. nigrum* seed lot possessed higher potential to produce healthy seedlings on these treatments. In addition, the result indicates that *S. nigrum* can produce healthy seedlings within a short period, no matter the various environmental conditions (Beena and Jayaran, 2010).

8.6 CONCLUSIONS

The present study indicates that *S. nigrum* seeds are sufficiently viable, but seed dormancy may be the factor hindering germination, and this can be overcome in several ways. Sandpaper, 15°C, 20°C, and 25°C temperature treatments resulted in high germination of up to 94.6%. Temperatures less than 5°C, higher than 35°C, as well as continuous darkness hindered germination. Germination was first and last observed on the third and ninth day after sowing, respectively. Sowing depth significantly affected seedling emergence. Silty clay with the highest seedling emergence record at 2 cm sowing depth indicates that there is an influence of soil modulation on *S. nigrum* seeding emergence.

ACKNOWLEDGEMENTS

This work was supported by the National Research Foundation of South Africa, Cape Peninsula University of Technology, Tertiary Educational Trust Fund, LAUTECH, Ogbomoso, Nigeria.

REFERENCES

Beena, K.A. and Jayaram,.K.M. (2010). Effect of temperature on seed water content and viability of green pea (*Pisum sativum* L.) and Soybean (*Glycine max* L. Merr.) seeds. *International Journal of Botany, 6*(2), 122–12.

Beheshtian-Mesgaran, M.Z.E., Nassiri, M., and Rahimian, H. (2006). Improvement of Iranian wheat cultivars bred during 1956–1995 in relation to wild oat competition. *Iranian Journal of Weed Science, 2,* 32–52.

Belel, M. (2016). Effect of different scarification methods on the germination of Petai Belalang (*Leuceana leucocephala*) seed. *Global Advanced Journal of Agricultural Sciences, 5*(1), 28–32.

Benigno-Gonzalez, R., Mulualem, T., Guillermo Castro, M., and Per Christer, O. (2009). Seed germination and seedling establishment of Neotropical dry forest species in response to temperature and light conditions. *Journal of Forestry Research, 20*(2), 99–104.

Brandel, M. (2005). The effect of stratification temperatures on the level of dormancy in primary and secondary dormant seeds of two Carex species. *Plant Ecology, 178,* 163–169.

Bvenura, C. and Afolayan, A.J. (2013). Growth and physiological response to organic and/or inorganic fertilisers of wild *Solanum nigrum* L. cultivated under field conditions in Eastern Cape Province, South Africa. *Acta Agriculturae Scandinavica, Section B - Soil and Plant Science, 63*(8), 683–693.

Bvenura, C. and Afolayan, A.J. (2014). Maturity effects on mineral concentration and uptake in *Solanum nigrum* L. *Journal of Applied Botany and Food Quality, 87,* 220–226.

Chachalis, D. and Reddy, K.N. (2000). Factors affecting Campsis radicans seed germination and seedling emergence. *Weed Science, 48*(2), 212–216.

Chauhan, B.S., Gill, G., and Preston, C. (2006). Seed germination and seedling emergence of three horn bedstraw (*Galium tricornutum*). *Weed Science, 54,* 867–872.

Cousens, R. and Mortimer, M. (1995). *Dynamics of Weed Populations.* Cambridge University Press, Cambridge.

Desbiolles, J. (2002). Optimising Seedling Depth in the Paddock. Internet: http://www.unisa.edu.au/amrdc/Areas/Proj/SeedTrials/Seedingdeptharticle Kerribee.pdf.

Edmonds, J.M. and Chweya, J.A. (1997). *Black Nightshade.* Solanum nigrum L. *and Related Species.* Promoting the conservation and use of underutilised and neglected crops. 15. Institute of Plant Genetics and Crop Plant Research, Gatersleben and International Plant Genetic Resources Institute, Rome, Italy.

Etejere, E.O. and Okoko, T.A. (1989). Seed production, germination and emergence of *Euphorbia heterophylla* L. *Nigerian Journal of Botany, 2,* 143–147.

Farooq, M., Basra, S.M.A., Hafeez, K., and Ahmad, N. (2005). Thermal hardening: a new seed vigor enhancement tool in rice. *Journal of Integrated Plant Biology, 47,* 187–193.

Finch-Savage, W.E. and Bassel, G.W. (2016). Seed vigour and crop establishment: extending performance beyond adaptation. *Journal of Experimental Botany, 67*(3), 567–591.

Flyman, M.V. and Afolayan, A.J. (2006). The suitability of wild vegetables for alleviating human dietary deficiencies. *South African Journal of Botany, 72,* 49–497.

Forcella, F., Benech-Arnold, R.L., Sanchez, R., and Ghersa, C.M. (2000). Modeling seedling emergence. *Field Crops Research, 67,* 123–139.

Gairola, K.C., Nautiyal, A.R., and Dwived, A.K. (2011). Effect of temperature and germination media on seed germination of *Jatropha curcas* Linn. *Advances in Bioresearch, 2*(2), 66–71.

Grabe, D.F. (1970). *Tetrazolium testing handbook for agricultural seeds.* Contribution no. 29 to the handbook of seed testing. Association of Official Seed Analysts, *29,* 1–62.

ISTA. (2015). International rules for seed testing.International Seed Testing Association. Basserdorf, Switzerland: Jansen Van Rensburg, W.S., Van Averbeke, W., Slabbert, R., Faber, M.,Van Jaarsveld, P., Van Heerden, I., Wenhold, F., and Oelofse, A. (2002). African leafy vegetables in South Africa. *Water South Africa, 33,* 317–326.

Kambizi, L., Adebola, P.O., and Afolayan, A.J. (2006). Effects of temperature, prechilling and light on seed germination of *Withania somnifera,* a high value medicinal plant. *South African Journal of Botany, 72,* 11–14.

Kellogg, C.E. (1993). *Soil Survey Division Staff: Soil Survey Manual.* United States Department of Agriculture, Washington, USA.

Larina, S.Y. (2008). *Solanum nigrum.* In: Afonin, A.N., Greene, S.L,. Dzyubenko, N.I., and Frolov, A.N. (eds.,). *Interactive Agricultural Ecological Atlas of Russia and Neighboring Countries.* Economic plants and their diseases, pests, and weeds. http://www.agroatlas.spb.ru/en/content/weeds/Solanum_nigrum/.

Luna, T. and Wilkinson, K.M. (2009). Collecting, processing, and storing seeds. In: Dumroese, R.K, Luna, K. and Landis, T.D. (eds.,). Nursery Manual for Native Plants. a Guide for Tribal Nurseries. Vol I, Nursery Management. *Interior, Environment, and Related Agencies Appropriations for 2010: Hearings Before a Subcommittee of the Committee on Appropriations*, House of Representatives, One Hundred Eleventh Congress, First Session, Part 3.

Maloof, J.N., Borevitz, J.O., Weigel, D., and Choryn J. (2000). Natural variation in phytochrome signaling. *Seminars in Cell and Developmental Biology, 11*, 523–530.

Modi, M., Modi, A.T., and Hendricks, S. (2006). Potential role for wild vegetables in household food security: a preliminary case study in Kwazulu-Natal, South Africa. *African Journal of Food, Agriculture, Nutrition and Development, 6*, 1–13.

Odhav, B., Beekrum, S., Akula, U., and Baijnath, H. (2007). Pre-liminary assessment of nutritional value of traditional leafy vegetables in KwaZulu-Natal. *South African Journal of Food Composition and Analysis, 20*, 430–435.

Okuzanya, O.T. (1980). Germination and growth of *cristata* L., under Light and temperature regimes. *American Journal of Botany, 67*, 854–858.

Perry, D.A. (1980). The concept of seed vigour and its relevance to seed production techniques. In: Hebblethwaite P.D. (ed.), *Seed Production*, Butterworths, London, pp. 585–591.

Peters, J. (2000). Tetrazonium testing handbook. Contribution no 29, The Handbook on Seed Testing. Prepared by the Tetrazolium subcommittee of the association of official seed analysts, Part 2. Lincoln, Nesbraka.

Powell, B., Maundu, P., Kuhnlein, H.V., and Johns, T. (2013). Wild foods from farm and forest in the East Usamba mountains, Tanzania. *Ecologia Food Nutrition, 52*, 451–478.

Suthar, A.C., Naik, V.R., and Mulani, R.M. (2009). Seed and seed germination in *Solanum nigrum* Linn. *American-Eurasian Journal of Agricultural and Environmental Science, 5*(2), 179–183.

Van Rensburg, W.S.J., Van Averbeke, W., Beletse, G., and Slabbert, M.M. (2012). *Black Nightshade (*Solanum retroflexum *Dun.), Production Guidelines for African Leavy Vegetables*. Water Research Commission, Pretoria, 31–32.

Villalobos, A.E., Peláez, D.V., Bóo, R.M., Mayor, M.D., and Elia, O.R. (2002). Effect of high temperatures on seed germination of *Prosopis caldenia* Burk. *Journal of Arid Environments, 52*, 371–378.

Zhou, J., Deckard, E.L., and Ahrens, W.H., (2005). Factors affecting germination of hairy nightshade (*Solanum sarrachoides*) seeds. *Weed Sciences, 531*, 41–45.

9 Indian Medicinal Plants and Their Potential Against Coronaviruses

Manisha Nigam and Abhay Prakash Mishra

CONTENTS

9.1 INTRODUCTION

Coronaviruses (CoVs) have the biggest genome of any RNA virus with such a high rate of recombination that this category of viruses is continually producing new variants capable of infecting various hosts (Wyganowska-Swiatkowska et al., 2020). In humans, they induct respiratory infirmities that encompass mild symptoms to severe complications leading to pneumonia and other respiratory ailments. Acute lower respiratory tract infectious diseases are one of the leading causes of morbidity and mortality worldwide, accounting for over 4 million fatalities each year (Cock and Van Vuuren, 2020).

In the Indian traditional system of medicine, there are peculiar and acknowledged systems, which are Ayurveda, Yoga and Naturopathy, Unani, Siddha, and Homoeopathy (AYUSH) (Prasad, 2002). People in India rely on indigenous natural sources to cure various respiratory ailments due to a lack of healthcare facilities. Across many regions of the world, herbal medicines for the treatment of respiratory diseases are commonplace (Alamgeer et al., 2018). There is mounting evidence that herbal medications are efficient against viral infections do not acquire resistance, aid in the stimulation of the host immune system, and can affect viruses in a precise way (Wyganowska-Swiatkowska et al., 2020). Phytotherapeutic compounds have been used for disease management since prehistoric times, but their use has skyrocketed in the recent decade. Plants are a significant source of natural chemicals with a wide range of biological and pharmacological effects (Mishra et al., 2018).

The ongoing outbreak of coronavirus disease 2019 (COVID-19) caused by severe acute respiratory syndrome-coronavirus-2 (SARS-CoV-2) has demonstrated the necessity for novel vaccine and therapeutic candidates' development. Although there is currently not any specific cure for

DOI: 10.1201/9781003206620-9

COVID-19, except to adopt preventative procedures such as maintaining social isolation, hygiene, and improving immunity. As a result, new therapy solutions for the current deadly epidemic can be identified by reclaiming the ancient applications of Indian medicinal herbs and formulations. Moreover, because of the ingrained side effects of conventional drugs, a substantial populace has moved over to traditional medications as a substitute to conventional therapies owing to numerous causes comprising lower cost, easy accessibility, and reduced or no side effects. Many traditional AYUSH (Ministry of AYUSH, Government of India) formulations that are very well immune modulators have been used in respiratory illnesses and allergy illnesses for centuries. AYUSH has identified and advised these inventions for their usage as a preventative strategy. Most of these are being tested for clinical trials in COVID-19-infected subjects (Ahmad et al., 2021). This chapter emphasizes the antiviral activities from several potent Indian folklore plants against coronaviruses with emphasis on SARS-CoV-2.

9.2 CORONAVIRUSES: A GENERAL OVERVIEW

Coronaviruses represent a distinct group of huge, enveloped positive-sense RNA viruses belonging to the order Nidovirales, comprising four families, namely Coronaviridae, Mesoniviridae, Roniviridae, and Arteriviridae. Coronaviridae consists of two sub-families: Torovirinae and Coronavirinae. Coronavirinae sub-family again differed into four genera, namely, α -, β-, γ-, and δ-coronaviruses (Vellingiri et al., 2020), of which α- and β-coronaviruses are only infectious to mammals, whereas γ- and δ-coronaviruses infect birds mostly, though some infect mammals too (Vellingiri et al., 2020). In the 1960s, coronaviruses were discovered for the first time in humans (Wyganowska-Swiatkowska et al., 2020). Coronaviruses have previously been viewed as generally innocuous respiratory microorganisms to people. Among the coronaviruses identified to infect humans, severe acute respiratory syndrome coronavirus (SARS-CoV), Middle East respiratory syndrome coronavirus (MERS-CoV), and SARS-CoV-2 possess the capacity to cause severe maladies, whereas HKU1, NL63, 229E, and OC43 exhibit mild signs (Su et al., 2016). SARS-CoV-2—a coronavirus reported in China (In Wuhan City December 2019) (Zhu et al., 2020) is seventh among the coronaviruses that infect humans and has provoked current global fear (Cock and Van Vuuren, 2020).

9.3 PATHOGENIC EFFECT OF COVID-19

In terms of genomic structure, SARS-CoV-2 shares 75%–80% of the characteristics of SARS-CoV and 50% of those of MERS-COV. It is already known that SARS, MERS, and COVID-19 patients commonly develop lymphopenia and have a higher likelihood of developing pneumonia and progressing to respiratory failure (Li et al., 2021). According to clinical and experimental findings from studies of COVID-19, the immune inflammatory response is crucial in SARS-CoV-2 infection. Several studies have suggested that an abnormal immune response (rather than a direct viral cytopathic effect) plays a much greater role in the effects of COVID-19. Similarly, COVID-19 is thought to cause pathological changes to vital organs directly in response to the cytopathic effect of SARS-CoV-2 as well as indirectly by damaging immune responses triggered by SARS-CoV-2. During the early stage of COVID-19, patients were observed to have the highest viral load (To et al., 2020). In addition, SARS-CoV and SARS-CoV-2 bind to the same receptor—Angiotensin-Converting Enzyme 2 (ACE-2), suggesting that both viruses enter the body via similar pathways. ACE-2 is widely expressed in vascular endothelial cells and smooth muscle cells in all organs, which might cause extensive vascular endothelial cell injury and possibly contribute to the development of multiple organ lesions in patients with COVID-19 infection (He et al., 2006; Liang et al., 2020). Recent research suggests that cardiac cells expressing high amounts of ACE-2 might serve as target cells for SARS-CoV-2 in patients with basic heart failure disease (Chen et al., 2020b). Similarly, there has been a report of cardiac injury in 7%–23% of patients with COVID-19, which leads to increased mortality (Shi et al., 2020).

Studies using COVID-19 models suggest that the common type of damage caused by SARS-CoV-2 infection also occurs within the immune system, and spleen and lymphoid atrophy are associated with cytokine activation, suggesting that SARS-Co-V2 might directly damage immune cells (Chen et al., 2020a; Chan et al., 2020; C. Huang et al., 2020; Bermejo-Martin et al., 2020). SARS-CoV-2 can activate pattern recognition receptors (PRRs) in the body, which leads to the antiviral innate immune response as well as organ dysfunction and damage to cells (Li et al., 2021). It has been observed that both damage-associated molecular patterns (DAMPs) and pathogen-associated molecular patterns (PAMPs) may contribute to the systemic dysregulation of the innate immune response and play a role in the development of multiple organ dysfunction syndrome (MODS).

Furthermore, Diao et al. found that COVID-19 not only reduced the number of T cells but also functionally exhausted the surviving T cells (Diao et al., 2020). Additionally, T-cell subpopulation differentiation and functional imbalances play major roles in some inflammatory conditions (Li et al., 2021).

COVID-19 also causes both pulmonary and systemic inflammation, which can lead to MODS in high-risk patients (Chen et al., 2020c). According to the number of studies, the primary diagnostic criterion for severe or critical SARS-Co-V2 is organ dysfunction, including acute respiratory distress syndrome (ARDS), shock, acute myocardial injury, liver injury, renal injury, and MODS being the most common. (Xie et al., 2020a; Wang et al., 2020; Yang et al., 2020; Bermejo-Martin et al., 2020; Li et al., 2021). However, it is widely assumed that the core pathophysiology of critical COVID-19 is severe ARDS (Yang et al., 2020). The primary cause of ARDS appears to be the extravagant and unrestrained inflammation apparently caused by SARS-CoV-2 infection. In addition, in both severe SARS and COVID-19, hypercytokinemia plays a crucial role in pathogenic inflammation. In 2019, hypercytokinemia and lymphopenia were found in COVID-19 patients, suggesting that there is a specific dysregulated immunological phenotype linked to remarkably high severity (Li et al., 2021).

MERS-CoV and SARS-CoV infect and reproduce rapidly in human macrophages and dendritic cells, causing abnormal release of proinflammatory cytokines and chemokines (Yoshikawa et al., 2009; Li et al., 2021). Viroporin 3a has also been reported to activate the nucleotide oligomerization domain-like receptor protein three inflammasome and secrete IL1β in macrophages during SARS-CoV infection, suggesting that PAMP-PRR signaling in macrophages may result in the production of proinflammatory cytokines in COVID-19 (Fu et al., 2020).

Several studies confirmed that lymphopenia (defective acquired immunity) is a prevalent trait in COVID-19 patients, and it has been linked to severity of illness and death (Xie et al., 2020a; Wang et al., 2020; Guan et al., 2020; Chen and Li, 2020). Immunosuppression might make it difficult to get rid of the virus and can lead to secondary infections. In individuals with severe COVID-19 (5–15.5%), hospital-acquired secondary infection is common. Secondary bacterial infection was found in 14.3% of COVID-19 patients, according to a recent meta-analysis, which included 3,448 patients from 28 trials.

9.4 ANTIVIRAL ACTIVITY OF INDIAN MEDICINAL PLANTS AGAINST CORONAVIRUSES

In this situation in which preventive and therapeutic agent has not been established or recommended yet for coronaviruses, medicinal plants are a boon of God for curative, supplementary, and recuperative patients. Owing to their efficacy, cost-effectiveness, eco-friendliness, feasibility, lesser or no side effects herbal plants, and their formulations need to be explored further for therapeutic use against viruses with emphasis on coronaviruses. Indian medicinal plants had offered numerous novel leads in history (Mukherjee et al., 2007) and possess brilliant potential in providing future new chemical entities (NCE) for developing brilliant therapeutic representatives.

The following section detailed the information on some representative Indian medicinal plants used against viral contagions that could be anticipated as a conceivable therapeutic strategy against the coronaviruses infection including COVID-19.

9.4.1 OCIMUM SANCTUM

Ocimum sanctum is known as Tulsi, also called holy basil for its religious and spiritual sanctity; it is a native plant to the Indian subcontinent. It is greatly recommended for its curative properties since more than 3,000 years ago. Many preclinical and clinical reports have evidenced its therapeutic efficacy as an adaptogenic, anti-carcinogenic, antibacterial, antidiabetic, anti-inflammatory, antiviral, cardioprotective, and immunity-boosting properties (Jamshidi and Cohen, 2017). Although Eugenol (1-hydroxy-2-methoxy-4-allylbenzene), the functional constituent in *O. sanctum* L. has been documented to be principally accountable for the medicinal potential, ursolic acid, euginal, limatrol, carvacrol, methyl carvicol, caryophyllene, and sitosterol are the other bioactive constituents (Pattanayak et al., 2010). *Ocimum sanctum* is being consumed for pain, cough, cold, diarrhea, fever, bronchitis, and respiratory ailments since time immemorial, which are also common symptoms related to coronaviruses-borne infections, including COVID-19 (Goothy et al., 2020). It has been documented that *Ocimum sp.* possesses tremendous antiviral properties against both DNA and RNA viruses (Chiang et al., 2005), owing to its immunomodulatory activity via a humoral and cell-mediated immune response (Mediratta et al., 2002). Studies have documented that *O. sanctum* leaf extract stimulate T & B lymphocytes, particularly Th1 subset, evidenced by enhanced IL-2 production *in vitro and in vivo* in albino Wistar rats (Goel et al., 2010). Moreover, randomized double-blind placebo-controlled trials of ethanolic extract of Tulsi leaves on healthy humans were reported to increase immune response with enhancement in the level of IFN-γ, IL-4, T-helper cells, and Natural Killer (NK) (Mondal et al., 2011). These studies further corroborate the role of *Ocimum* sp. as an antiviral agent. Recently molecular docking study on *Ocimum*-derived phytochemicals, i.e., Ursolic acid, Vicenin, and Isorientin 4'-O-glucoside 2"-O-p-hydroxybenzoagte have hinted at their role as an inhibitor against SARS-CoV-2 Mpro (Main protease), suggesting its possible use against this global contagion (Shree et al., 2020).

9.4.2 ZINGIBER OFFICINALE

Zingiber officinale, known as Ginger, is a common Indian spice and traditional medicinal plant having numerous pharmacological activities, such as antibacterial, antiviral, anti-hypertensive, antioxidant, analgesic antipyretic properties (Chrubasik et al., 2005), and also has been reported to be effective against respiratory ailments (Akoachere et al., 2002; Sharma and Chakotiya, 2020). The fresh extract of *Z. officinale* exhibits antiviral properties against the human respiratory syncytial virus by blocking its attachment and internalization. It also triggers mucosal cells to secrete IFN-β that believably counteract viral infection (Chang et al., 2013). Moreover, it exhibits immunomodulatory properties by stimulating humoral and cell-mediated immune responses (Carrasco et al., 2009). The bioactive composites of *Z. officinale* such as gingerdion, zingiberene, nevirapine, 6-gingediol, 6-gingerol, β-sitosterol, germacrene, 6-shogaol, methyl-6-shogaol, α-linalool, etc., restrain viral duplication where β-sitosterol is the most effective inhibitor of reverse transcriptase (RT) enzyme (Gautam et al., 2020). Recently major components of *Z. officinale* were evaluated against coronavirus (SARS-CoV-2) to explore the effectual inhibitors of the entry point of infection by ligand–receptor binding docking study. The virus SARS-CoV-2 initiates its infection to the human body via the interaction of its spike (S) protein with the human Angiotensin-Converting Enzyme 2 (ACE2)—a functional SARS-CoV receptor of the host cells (Xie et al., 2020b). It was revealed that phytochemicals Gingerenone and zingiberene possess an effective binding affinity for angiotensin-converting enzyme 2 (ACE-2), which is required for host cell entry and subsequent viral replication. Whereas phytochemicals shoagol, zingerone, and zingiberene showed higher binding with the extracellular domain of serine protease Transmembrane protease Serine 2 (TMPRSS2). It is pertinent to note that cascading of the attachment and subsequent entry of the SARS-CoV-2 virus is the utmost critical phase of pathogenesis facilitated by Spike protein, ACE-2, and additionally supported by TMPRSS2 (Sharma and Chakotiy, 2020). Thus, prohibiting such stages may help to plot probable drug targets against this virus. These results were also corroborated by another in silico study where it was documented that *Z. officinale* reduced viral load and shedding of SARS-CoV-2 in the nasal passage, indicating the involvement of certain components of *Z. officinale* having substantial affinity against virus spike protein and host ACE-2 receptor (Haridas et al., 2021). Moreover, Ginger also possesses strong antioxidant, anti-inflammatory, bronchodilatory effect, induces airway smooth muscle relaxation, and attenuates the airway inflammation and resistance (Mao et al., 2019), thus could prevent severe damage to the lungs, which is also applicable in COVID-19.

9.4.3 *CURCUMA LONGA*

Curcuma longa—commonly known as turmeric is an Indian spice widely used in Indian traditional medicine owing to its therapeutic properties like antiviral, lipid-lowering, antimicrobial, analgesic, antiproliferative, anti-inflammatory, and immunomodulatory activity (Dwivedi et al., 2020). Major bioactives reported from *C. longa* are curcuminoids, turmerones, and sesquiterpenoids. Curcuminoids (a mixture of curcumin, Demethoxycurcumin, and Bisdemethoxycurcumin) are responsible for the main pharmacological activities of turmeric (Thimmulappa et al., 2021), and *Curcumin* is the most abundant bioactive curcuminoid in turmeric. Curcumin acts against many viruses responsible for respiratory ailments, for example, respiratory syncytial virus, influenza A virus (Praditya et al., 2019), and SARS-CoV (Wen et al., 2007).

Numerous studies using computational modeling tools have predicted that curcumin possesses a binding affinity for ACE2 receptor of host and spike protein of virus, and thus could anticipate as a curative agent for restraining virus attachment to the host (Pandey et al., 2021; Maurya et al., 2020). Curcumin also intervenes in several respiratory diseases by modulating inflammation and oxidative stress via various mechanisms, e.g., inhibition of lipoxygenase (LOX), cyclooxygenase-2 (COX-2), and inducible nitric oxide synthase (iNOS) enzymes that mediate inflammatory processes (Menon and Sudheer, 2007). Moreover, it also blocks NF-κB activation—a transcription factor responsible for regulating inflammation via tumor necrosis factor α (TNF-α) in most ailments (Hewlings and Kalman, 2017). Curcumin also triggers NRF2—a transcription factor and chief modulator of the cell's antioxidant defences, comprising GSH-biosynthesizing enzymes and heme oxygenase-1 (Dai et al., 2018). In addition to antioxidant property (direct action), curcumin lessens reactive oxygen species (ROS) by restraining NADPH oxidase (indirect action) (Huang et al., 2015).

Interestingly, curcumin also blocks IL-6 trans-signaling and thus inhibits the advancement of inflammatory ailments (Zhao et al., 2016a). These studies postulate a strong notion that curcumin could lessen virus-induced oxidative stress, cytokine syndrome, and tissue injury; thus it could be an effectual therapeutic against COVID-19 (Thimmulappa et al., 2021).

9.4.4 *GLYCYRRHIZA GLABRA*

Glycyrrhiza glabra—commonly known as Mulethi is a dried and processed root of *G. glabra*— a widely used medicinal plant found in traditional formulas of Indian culture to treat symptoms of viral respiratory tract infections, for instance, sore throat or dry cough (Fiore et al., 2008). Various biologically active compounds have been reported from *G. glabra*, e.g., alkaloids, glycosides, flavonoids, phenolics, saponins, tannins, terpenes, and steroids (Mamedov and Egamberdieva, 2019). Glycyrrhizin (GLR)—a triterpenoid saponin from *G. glabra* inhibits SARS-associated coronavirus (SARS-CoV) by restraining its adsorption and penetration during the early steps and also its replication in Vero cells (Cinatl et al., 2003). GLR has been reported to cure respiratory ailments and ARDS by inhibiting the levels of high-mobility group proteins B1 (HMGB1) and IL-33 via modulating the release of proinflammatory cytokines during the preliminary stage of the syndrome from stimulated inflammatory cells (Fu et al., 2016). GLR also regulates anti-inflammatory and immunomodulatory activity via multiple pathways such as MAPK and Toll-like receptors signaling pathways (Zhao et al., 2016b) declining IL-6 release from macrophages, thus leading to reduced cytokine storm induction. Moreover, molecular docking studies suggested that GLR stimulates cholesterol-reliant lipid rafts disorganization—a critical event for the entry of coronavirus into cells (Bailly and Vergoten, 2020).

Recently, through *in silico* approach, it was showed that the constituents of licorice, i.e., Glyasperin A & Glycyrrhizic acid affect the replication process of SARS-CoV-2 (Sinha et al., 2020), where ACE-2 receptor was interrupted by the former and replication was repressed by the latter.

9.4.5 *HOUTTUYNIA CORDATA*

Houttuynia cordata is a perennial herb that propagates in the moist and shady regions in Asian countries. It possesses effectual antiviral action specifically against enveloped viruses of clinical significance. The water extract of *H. cordata* has been reported to exhibit a biphasic response (Lau et al., 2008). It exhibits an immunomodulatory effect in mouse splenic lymphocytes by stimulating the propagation of CD4+ and CD8+ T cells along with substantial enhancement in cytokines, IL-2 and IL-10 secretion. Thus, it may help to prevent SARS-CoV infection by boosting cell-mediated immunity. Moreover, it also suppresses SARS coronavirus 3C-like protease and RNA-dependent RNA polymerase (Lau et al., 2008).

Recently, two bioactive phytocompounds—6-Hydroxyondansetron and Quercitrin from *H. cordata* have been documented to possess potential against three main replication proteins of SARS-CoV-2, i.e., Main protease (Mpro), ADP ribose phosphatase (ADRP), and Papain-like protease (PLpro) (Das et al., 2021) via molecular docking studies.

9.4.6 *TINOSPORA CORDIFOLIA*

Tinospora cordifolia is also known by diverse names such as Guduuchi, Guluuchi, Amrita, Guduchi sattva in Ayurveda, and Giloya in folk. It is a deciduous shrub found in the tropical region of India and recognized as 'nectar of life' because of its immune-strengthening potential. Besides, it is also known for its antioxidant, antidiabetic, cytotoxic, and antihepatotoxic activity (Sharma et al., 2012). Its bioactive components—Tinocordioside, Magnoflorine, Syringin, and Cordifolioside A are well-known for its immunomodulatory action (Sharma et al., 2012). Tinocordiside *from T. cordifolia* has been reported to potentially inhibit SARS-CoV-2 Mpro (Main protease), using *a* molecular docking study (Shree et al., 2020). Recently, various secondary metabolites from *T. cordifolia*, i.e., Columbin, *N*-Maslinic acid, trans-feruloyl-tyramine-diacetate, Tinosporide, Amritoside (C, B & A), Palmatoside (G & F), and Tinocordifolin are reported as quintessential bioactives based on docking count against SARS-CoV-2 Mpro (Thakkar et al., 2021).

9.4.7 *WITHANIA SOMNIFERA*

Withania somnifera—traditionally recognized as 'Indian Ginseng' and 'Ashwagandha' is a well-known pharmaceutical plant in Indian Ayurveda and ethnic medicines to treat various human ailments (Mishra et al., 2000). *W. somnifera* possess flavonoids, alkaloids and steroidal lactones like somnine, somniferinine, somniferine, withananine, withanine, withaferin A & B, pseudo-tropine, pseudowithanine tropane, anahydrine, anaferine, choline, isopelletierine, 17-hydroxywithaferin A, dihydrowithaferin A, 27-deoxywithaferin, withanolides, withanone, withanosines, Quercetin, and Quercetin-3-O-galactosyl-rhamnosyl-glucoside (Hirayama et al., 1982). It is widely claimed to have antimicrobial, hepatoprotective, antidepressant, anti-inflammatory, antioxidant, anti-stress, anticonvulsant, cardioprotective, antitumor, antiviral, and immunomodulatory properties (Mishra et al., 2000). The plant and its derivatives have been described for its efficient antiviral action against herpes simplex type-1, hepatitis, H1N1 influenza, coxsackie virus, HIV, infectious bursitis, and coronaviruses—involving SARS-CoV and SARS-CoV-2 (Akram et al., 2018; Straughn and Kakar, 2020). A plethora of *in silico* analyses recently has depicted numerous conceivable antiviral modes of action of the bioactives from *W. somnifera* via impeding host cell protease (TMPRSS2), Viral protease (3CLpro & PLpro), RNA polymerase (RdRp), host receptor ACE-2, and viral hydrophobic envelope (E) protein (involved in the formation of envelope, replication, budding, and release of virus progeny from the host) (Shree et al. 2020; Kumar et al. 2020; Chikhale et al. 2020; Tripathi et al. 2020; Straughn and Kakar 2020; Abdullah-Alharbi 2021).

Interestingly, *W. somnifera* also inhibits and lessens numerous proinflammatory molecules, i.e., IL-2, -6, & -7, TNF-α, IFN-γ thus clinically improves the COVID-19-associated cytokine storm (Kashyap et al., 2020). Moreover, it also acts as an antioxidant agent by inhibiting the viral-induced expression of glutathione, ROS generation (Kashyap et al., 2020) against SARS-COV-2-associated oxidative stress in the lungs.

9.4.8 *ALLIUM SATIVUM*

Allium sativum, frequently known as garlic, is a valuable herb used since ancient times for culinary and medicinal purposes in India. It possesses antimicrobial, immunomodulatory, antimutagenic, anti-inflammatory, and antitumor properties as well (Donma and Donma, 2020), owing to the presence of sulfur-containing compounds. The most important biologically active sulfur constituents are Allicin, Alliin, ajoenes, vinyldithiins, and diallyl sulfide (El-Saber Batiha et al., 2020). *A. sativum* has been reported to show antiviral activity against various viruses, especially influenza and infectious bronchitis virus—a coronavirus (Mohajer-Shojai et al. 2016; El-Saber Batiha et al. 2020). Interestingly, frequent use of garlic promotes internal antioxidant defence and reduces adverse oxidative effects by various modes, e.g., increasing glutathione and cellular antioxidant enzymes, ROS modulation and its prevention by inhibiting NADPH oxidase one, and stimulating Nrf2 and heme-oxygenase one enzyme transcription (El-Saber Batiha et al., 2020). Garlic supplementation significantly stimulates CD4+ & CD8+ T cells, macrophages, dendritic, and NK cells. Moreover, garlic is also associated with other valuable immunological and hormonal effects, as evidenced by the decline in leptin, leptin receptor, peroxisome proliferator-activated receptor-gamma (PPAR-γ), and IL-6 (Donma and Donma, 2020). Thus, above-mentioned characters hint that *A. sativum* may be beneficial as a preventive measure against coronaviruses infection via boosting the immune system and repressing proinflammatory cytokines. Moreover, docking studies have explored that the synergistic interactions of various substances of the garlic essential oil exhibited an inhibitory effect on the ACE2 receptor and the PDB6LU7 protein of the SARS-Co V-2virus (Thuy et al., 2020).

9.4.9 *PHYLLANTHUS EMBLICA*

Phyllanthus emblica—commonly known as Amla is a native plant of India and has extensively been investigated for different biological activities such as anti-inflammatory, antioxidant, antiaging, antidiabetic, and hepatoprotective properties (Gaire and Subedi, 2014). Interestingly, in Ayurveda, Amla is known to be an effective immunomodulator and rejuvenator against senescence and degenerative issues by promoting longevity, enhancing digestion, reducing fever, and cough (Gaire and Subedi, 2014).

Corilagin, quercetin, ellagic acid, gallic acid, leutolin, chebulagic acid, and chebulinic acid are the representative polyphenols from *Phyllanthus* spp. that have been reported to exhibit miscellaneous pharmacological effects (Jantan et al., 2014). Amla substantially lessens the immunosuppressive effect on the proliferation of lymphocyte via restoring INFγ and IL-2 levels (Sai Ram et al., 2002). It also increases anti-inflammatory cytokines and reduces proinflammatory cytokines levels (Chatterjee et al., 2011). Moreover, there are many reports on its antiviral properties on various types of viruses like Herpes simplex virus and Hepatitis B virus (Xiang et al. 2011; Lv et al. 2014). Moreover, polyphenolics present in *P. emblica* act as a free radical scavenger (Chaphalkar et al., 2017) and may provide strong protection against CCL4-induced pulmonary fibrosis by inhibiting lipid peroxidation and increasing levels of SOD and glutathione peroxidase (Tahir et al., 2016).

Besides the above-mentioned plants, there are numerous other Indian medicinal plants recommended by AYUSH, Government of India, in the form of formulations or ingredients as such that are well-known for boosting immunity and possess antiviral and anti-inflammatory properties. Therefore, these can propose excellent leads against coronaviruses infection, especially COVID-19. A list of some of these medicinal plants with their details has been provided in Table 9.1.

TABLE 9.1

List of Some Indian Medicinal Plants Recommended by AYUSH Targeting COVID-19

S.N.	Name of the Plant/ Family	Plant Part Used	Mechanism of Action	*In silico* Studies
	Adhatoda vasica (Acanthaceae)	Leaves	Downregulation of hypoxia, inflammation, TGF-β1, enhance adaptive immunity, reduce the viral load in SARS-CoV2 infected Vero cells (Gheware et al., 2021).	Alkaloid vasicine was reported as a potential target against COVID-19 (Thangaraju et al., 2021).
	Aegle marmelos (Rutaceae)	Root, stem bark, fruits	Immunostimulatory, anti inflammatory, antioxidant (improve SOD CAT, GSH, and GPx) (Yadav et al., 2020).	Seselin—a compound from leaf extract interacts with COVID-19 main protease, SARS-CoV-2S protein and free enzyme of the SARS-CoV-2 main protease (Nivetha et al., 2021).
	Anacyclus pyrethrum (Asteraaceae)	Root	Immunomodulatory, anti-inflammatory, antimicrobial, antioxidant (Jawhari et al., 2021).	Gamma sitosterol from a. pyrethrum was reported to be antiviral against COVID-19, as it acts on the main protease (3CLpro) (Vincent et al., 2020).
	Andrographis paniculata (Acanthaceae)	Leaves	Anti-inflammatory, antibacterial, antitumor, antidiabetic, anti-HIV, antifeedant, and antiviral anti-SARS-CoV-2 activity (Jayakumar et al., 2013; Sangiamsuntorn et al., 2021).	Andrographolide—a bioactive was reported as an effective inhibitor of SARS-CoV-2 main protease (Enmozhi et al., 2021)
	Azadirachta indica (Meliaceae)	Leaves	Antioxidant, antiviral, anti-inflammatory, antimicrobial (Atawodi and Atawodi, 2009).	Nimbolin A, Nimocin, and Cycloartanols were predicted to be the inhibitor of SARS-CoV-2 M(Membrane) and E (Envelope) proteins that are crucial for virus assembly and budding. (Borkotoky and Banerjee, 2020).
	Carica papaya (Caricaceae)	Leaves, fruits	Immunomodulatory, anti inflammatory, antimicrobial, antioxidant (Dwivedi et al., 2020).	Flavonoids are reported to interact with a variety of SARS-Coronavirus-2 protein targets (Hariyono et al., 2021).
	Cassia occidentalis (Fabaceae)	Aerial part, seeds	Antibacterial, antifungal, antimalarial, anti-inflammatory, antioxidant, hepatoprotective, and immunosuppressive activity (Bhagat and Saxena, 2010).	Cassiaoccidentalins A–C from *Cassia occidentalis* reported with excellent binding against SARS-CoV-2 3C-Like Protease (Bondhon et al., 2020).
	Cocculus hirustus (Menispermaceae)	Whole plant	Immunomodulatory, anti-inflammatory, antimicrobial, antioxidant (Logesh et al., 2020)	-

(Continued)

TABLE 9.1 *(Continued)*

List of Some Indian Medicinal Plants Recommended by AYUSH Targeting COVID-19

S.N.	Name of the Plant/ Family	Plant Part Used	Mechanism of Action	*In silico* Studies
	Cordia myxa (Boraginaceae)	Fruits	Anti-inflammatory, immunomodulatory, antimicrobial antioxidant (Al-Snafi, 2016)	-
	Cynodon dactylon (Poaceae.)	Whole plants	Anti-inflammatory, antiviral, immunomodulatory, antimicrobial, antioxidant, antipyretic (Mangathayaru et al., 2009).	-
	Jatropha curcas (Euphorbiaceae)	Leaves, roots	Anti-inflammatory, antiviral, immunomodulatory, antimicrobial, antioxidant, anti-HIV (Sharma and Singh, 2012).	-
	Mollugo cerviana (Molluginaceae)	Whole plant	Antiviral against COVID-19 (Adhikari et al., 2021), anti-inflammatory, antimicrobial, antioxidant (Napagoda et al., 2020).	
	Nigella sativa (Ranunculaceae)	Seeds	Anti-inflammatory, antiviral, effective against respiratory disorders, immunomodulatory, antimicrobial, antioxidant (Kooti et al., 2016).	Caryophyllene oxide, α-bergamotene, and β-bisabolene from *Nigella sativa* was reported with excellent binding against SARS-CoV-2 (Duru et al., 2021).
	Piper nigrum (Piperaceae)	Fruits	Antioxidant, antiviral, anti-inflammatory, antimicrobial (Takooree et al., 2019), Antiviral against COVID-19 (Roshdy et al., 2020).	Piperdardiine and piperanine has displayed significant inhibitory potential against coronavirus's selected targets (Rajagopal et al., 2020).
	Solanum nigrum (Solanaceae)	Seeds	Antioxidant, antiviral, anti-inflammatory, and antipyretic (Campisi et al., 2019).	-
	Valeriana wallichii (Valerianaceae)	Roots	Antioxidant, antiviral, anti-inflammatory, antimicrobial (Sundaresan and Ilango, 2018).	-
	Vitex negundo (Verbanaceae)	Leaves	anti-inflammatory, antioxidant, insecticidal, antimicrobial (Zheng et al., 2015).	The synergistic interaction of ursolic acid, oleanolic acid, isovitexin, and 3β-acetoxyolean-12-en-27-oic acid exhibits repressive activity against papain-like protease (PLpro) of SARS CoV-2 (Mitra et al., 2021).

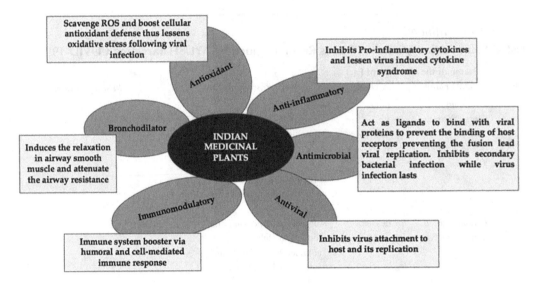

FIGURE 9.1 Schematic representation of the possible mode of action of Indian medicinal plants against coronavirus infections.

9.5 CONCLUSION AND PROSPECTS

In this review, we attempted to highlight some of the representative Indian medicinal plants that could be used against coronaviruses, with special emphasis on SARS-CoV-2. Uses of plant-based medicines possess several benefits over conventional ones, i.e., low cost, less or no side effects, and as an alternative to antibiotic-induced drug resistance. There is a broad scope of Indian medicinal plants against various acute and chronic ailments, and they have been used since traditional times. All the plants described in this review have been used for ages in India against various health issues, especially against virus-induced respiratory ailments. Although the exact mechanism of action of some of these plants is still unknown, we are of opinion that immune-boosting and modulation, anti-inflammatory, and antioxidant properties are some of the mechanisms by which these plant-based medicines act (Figure 9.1). Thus, we speculate that such properties might be of importance against SARS-CoV-2—a coronavirus. Moreover, docking studies on various such plants also corroborated the available literature so far, which hinted that these bioactive compounds from these plants may be of immense importance in developing the future drug against coronaviruses.

Nevertheless, further preclinical and clinical findings are required to validate their usage in terms of the correct dosage, toxicity levels (if any), and efficacy.

REFERENCES

Abdullah-Alharbi, R. (2021). Structure insights of SARS-CoV-2 open state envelope protein and inhibiting through active phytochemical of ayurvedic medicinal plants from *Withania somnifera*. *Saudi Journal of Biological Sciences*, *28*(6), 3594–3601.

Adhikari, B., Marasini, B.P. Rayamajhee, B. et al. (2021). Potential roles of medicinal plants for the treatment of viral diseases focusing on COVID-19: a review. *Phytotherapy Research*, *35*(3), 1298–1312.

Ahmad, S., Zahiruddin, S. Parveen, B. et al. (2021). Indian medicinal plants and formulations and their potential against COVID-19–preclinical and clinical research. *Frontiers in Pharmacology*, *11*, 1–34.

Akoachere, J.F.T.K., Ndip, R.N., Chenwi, E.. et al. (2002). Antibacterial effects of *Zingiber officinale* and *Garcinia kola* on respiratory tract pathogens. *East African Medical Journal*, *79*(11), 588–592.

Akram, M., Tahir, I.M., Shah, S.M.A. et al. (2018). Antiviral potential of medicinal plants against HIV, HSV, Influenza, Hepatitis, and Coxsackievirus: a systematic review. *Phytotherapy Research*, *32*(5), 811–822.

Al-Snafi, A.E. (2016). The pharmacological and therapeutic importance of *Cordia myxa*-A review. *IOSR Journal Of Pharmacy*, 6(6), 47–57.

Alamgeer, W.Y., Asif, H. et al. (2018). Traditional medicinal plants used for respiratory disorders in Pakistan: a review of the ethno-medicinal and pharmacological evidence. *Chinese Medicine*, 13(1), 48.

Atawodi, S.E. and Atawodi, J.C. (2009). *Azadirachta indica* (Neem): a plant of multiple biological and pharmacological activities. *Phytochemistry Reviews*, 8(3),601–620.

Bailly, C. and Vergoten, G. (2020). Glycyrrhizin: an alternative drug for the treatment of COVID-19 infection and the associated respiratory syndrome. *Pharmacology & Therapeutics*, 214, 107618.

Bermejo-Martin, J.F., Almansa, R., Menéndez, R. et al. (2020). Lymphopenic community acquired pneumonia as signature of severe COVID-19 infection. *Journal of Infection*, 80(5), e23–e24.

Bhagat, M. and Saxena, A.K. (2010). Evaluation of *Cassia occidentalis* for *in-vitro* cytotoxicity against human cancer cell lines and antibacterial activity. *Indian Journal of Pharmacology*, 42(4), 234.

Bondhon, T.A., Rana, Md. A.H., Hasan, A. et al. (2020). Evaluation of phytochemicals of *Cassia occidentalis* L. for their binding affinities to SARS-CoV-2 3C-like protease: an *in-silico* approach. *Asian Journal of Research in Infectious Diseases*, 4(4), 8–14.

Borkotoky, S. and Banerjee, M. (2020). A computational prediction of SARS-CoV-2 structural protein inhibitors from *Azadirachta indica* (Neem). *Journal of Biomolecular Structure and Dynamics*, 39(11), 4111–4121.

Campisi, A., Acquaviva, R., Raciti, G. et al. (2019). Antioxidant activities of *Solanum nigrum* L. leaf extracts determined in *in-vitro* cellular models. *Foods*, 8(2), 63.

Carrasco, F.R., Schmidt, G., Romero, A.L. et al. (2009). Immunomodulatory activity of *Zingiber officinale* Roscoe, *Salvia officinalis* L. and *Syzygium aromaticum* L. essential oils: evidence for humor- and cell-mediated responses. *Journal of Pharmacy and Pharmacology*, 61(7), 961–967.

Chan, J.F.W., Zhang, A.J., Yuan, S. et al. (2020). Simulation of the clinical and pathological manifestations of coronavirus disease 2019 (COVID-19) in a golden Syrian hamster model: implications for disease pathogenesis and transmissibility. *Clinical Infectious Diseases*, 71(9), 2428–2446.

Chang, J.S., Wang, K.C., Yeh, C.F. et al. (2013). Fresh ginger (*Zingiber officinale*) has anti-viral activity against human respiratory syncytial virus in human respiratory tract cell lines. *Journal of Ethnopharmacology*, 145(1), 146–151.

Chaphalkar, R., Apte, K.G., Talekar, Y. et al. (2017). Antioxidants of *Phyllanthus emblica* L. bark extract provide hepatoprotection against ethanol-induced hepatic damage: a comparison with silymarin. Oxidative Medicine and Cellular Longevity, *2017*, 1–10.

Chatterjee, A., Chattopadhyay, S., and Bandyopadhyay, S. K. (2011). Biphasic eEffect of *Phyllanthus emblica* L. extract on NSAID-induced ulcer: an antioxidative trail weaved with immunomodulatory effect. *Evidence-Based Complementary and Alternative Medicine, 2011*, 1–13.

Chen, G., Wu, D., Guo, W. et al. (2020a). Clinical and immunological features of severe and moderate coronavirus disease 2019. *Journal of Clinical Investigation*, 130(5), 2620–2629.

Chen, L., Li, X., Chen, M. et al. (2020b). The ACE-2 expression in human heart indicates new potential mechanism of heart injury among patients infected with SARS-CoV-2. *Cardiovascular Research*, 116(6), 1097–1100.

Chen, T., Wu, D., Chen, H., et al. (2020c). Clinical characteristics of 113 deceased patients with coronavirus disease 2019: retrospective study. *The British Medical Journal*, 368, m1091.

Chen, Y. and Li, L. (2020). SARS-CoV-2: virus dynamics and host response. *The Lancet Infectious Diseases*, 20(5), 515–516.

Chiang, L.C., Ng, L.T., Cheng, P.W. et al. (2005). Antiviral activities of extracts and selected pure constituents of *Ocimum basilicum*. *Clinical and Experimental Pharmacology and Physiology*, 32(10), 811–816.

Chikhale, R.V., Gurav, S.S. Patil, R.B. et al. (2020). Sars-Cov-2 host entry and replication inhibitors from Indian Ginseng: an *in-silico* approach. *Journal of Biomolecular Structure and Dynamics*, 39(12), 4510–4521.

Chrubasik, S., Pittler, M.H., and Roufogalis, B.D. (2005). *Zingiberis rhizoma*: a comprehensive review on the ginger effect and efficacy profiles. *Phytomedicine*, 12(9), 684–701.

Cinatl, J., Morgenstern, B., Bauer, G. et al. (2003). Glycyrrhizin, an active component of liquorice roots, and replication of SARS-associated coronavirus. *The Lancet*, 361(9374), 2045–2046.

Cock, I.E. and Van Vuuren, S.F. (2020). The traditional use of Southern African medicinal plants for the treatment of bacterial respiratory diseases: a review of the ethnobotany and scientific evaluations. *Journal of Ethnopharmacology*, 263, 113204.

Dai, J., Gu, L., Su, Y. et al. (2018). Inhibition of curcumin on influenza a virus infection and influenzal pneumonia via oxidative stress, TLR2/4, P38/JNK MAPK and NF-KB pathways. *International Immunopharmacology*, 54, 177–187.

Das, S.K., Mahanta, S., Tanti, B. et al. (2021). Identification of phytocompounds from *Houttuynia cordata* Thunb. as potential inhibitors for SARS-CoV-2 replication proteins through GC–MS/LC–MS characterization, molecular docking and molecular dynamics simulation. *Molecular Diversity, 26*(1), 365–388.

Diao, B., Wang, C., Tan, Y. et al. (2020). Reduction and functional exhaustion of T cells in patients with coronavirus disease 2019 (COVID-19). *Frontiers in Immunology, 11*, 827.

Donma, M.M. and Donma, O. (2020). The effects of *Allium sativum* on immunity within the scope of COVID-19 infection. *Medical Hypotheses, 144*, 109934.

Duru, C.E., Duru, I.A., and Adegboyega, A.E. (2021). *In-silico* identification of compounds from *Nigella sativa* seed oil as potential inhibitors of SARS-CoV-2 targets. *Bulletin of the National Research Centre, 45*(1), 57.

Dwivedi, M.K., Sonter, S., Mishra, S. et al. (2020). Antioxidant, antibacterial activity, and phytochemical characterization of *Carica papaya* flowers. *Beni-Suef University Journal of Basic and Applied Sciences, 9*(1), 23.

El-Saber Batiha, G., Magdy Beshbishy, A., Wasef, L.G., et al. (2020). Chemical constituents and pharmacological activities of Garlic (*Allium sativum* L.): a review. *Nutrients, 12*(3), 872.

Enmozhi, S.K., Raja, K., Sebastine, I., et al. (2021). Andrographolide as a potential inhibitor of SARS-CoV-2 main protease: an *in-silico* approach. *Journal of Biomolecular Structure and Dynamics, 39*(9), 3092–3098.

Fiore, C., Eisenhut, M., Krausse, R. et al. (2008). Antiviral effects of *Glycyrrhiza* species. *Phytotherapy Research, 22*(2), 141–148.

Fu, J., Lin, S., Wang, C., et al. (2016). HMGB1 regulates IL-33 expression in acute respiratory distress syndrome. *International Immunopharmacology, 38*, 267–274.

Fu, Y., Cheng, Y., and Wu, Y. (2020). Understanding SARS-CoV-2-mediated inflammatory responses: from mechanisms to potential therapeutic tools. *Virologica Sinica, 35*(3), 266–271.

Gaire, B.P. and Subedi, L. (2014). Phytochemistry, pharmacology and medicinal properties of *Phyllanthus emblica* Linn. *Chinese Journal of Integrative Medicine*, 1–8.

Gautam, S., Gautam, A., Chhetri, S., et al. (2020). Immunity against COVID-19: potential role of Ayush Kwath. *Journal of Ayurveda and Integrative Medicine, 13*(1), 100350.

Gheware, A., Dholakia, D., Kannan, S., et al. (2021). *Adhatoda vasica* attenuates inflammatory and hypoxic responses in preclinical mouse models: potential for repurposing in COVID-19-like conditions. *Respiratory Research, 22*(1), 99.

Goel, A., Singh, D.K., Kumar, S., et al. (2010). Immunomodulating property of *Ocimum sanctum* by regulating the IL-2 production and its mRNA expression using rat's splenocytes. *Asian Pacific Journal of Tropical Medicine, 3*(1), 8–12.

Goothy, S.S.K., Goothy, S., Choudhary, A., et al. (2020). Ayurveda's holistic lifestyle approach for the management of coronavirus disease (COVID-19): possible role of Tulsi. *International Journal of Research in Pharmaceutical Sciences, 11*(Special Issue 1),16–18.

Guan, W., Ni, Z., Hu, Y., et al. (2020). Clinical characteristics of coronavirus disease 2019 in China. *New England Journal of Medicine, 382*(18), 1708–1720.

Haridas, M., Sasidhar, V., Nath, P., et al. (2021). Compounds of *Citrus medica* and *Zingiber officinale* for COVID-19 inhibition: *In-silico* evidence for cues from ayurveda. *Future Journal of Pharmaceutical Sciences, 7*(1), 13.

Hariyono, P., Patramurti, C., Candrasari, D.S., et al. (2021). An integrated virtual screening of compounds from *Carica papaya* leaves against multiple protein targets of SARS-Coronavirus-2. *Results in Chemistry, 3*, 100113.

He, L., Ding, Y., Zhang, Q., et al. (2006). Expression of elevated levels of pro-inflammatory cytokines in SARS-CoV-infected ACE2+ cells in SARS patients: relation to the acute lung injury and pathogenesis of SARS. *The Journal of Pathology, 210*(3), 288.

Hewlings, S. and Kalman, D. (2017). Curcumin: a review of its effects on human health. *Foods, 6*(10), 92.

Hirayama, M., Gamoh, K., and Ikekawa, N. (1982). Stereoselective synthesis of Withafein A and 27-Deoxywithaferin A1. *Tetrahedron Letters, 23*(45), 4725–4728.

Huang, C., Wang, Y., Li, X., et al. (2020). Clinical features of patients infected with 2019 novel Coronavirus in Wuhan, China. *The Lancet, 395*(10223), 497–506.

Huang, S.L., Chen, P.Y., Wu, M.J., et al. (2015). Curcuminoids modulate the PKCδ/NADPH oxidase/reactive oxygen species signaling pathway and suppress matrix invasion during monocyte–macrophage differentiation. *Journal of Agricultural and Food Chemistry, 63*(40), 8838–8848.

Jamshidi, N. and Cohen, M.M. (2017). The clinical efficacy and safety of Tulsi in humans: a systematic review of the literature. *Evidence-Based Complementary and Alternative Medicine, 2017*, 1–13.

Jantan, I., Ilangkovan, M., Yuandani et al. (2014). Correlation between the major components of *Phyllanthus amarus* and *Phyllanthus urinaria* and their inhibitory effects on phagocytic activity of human neutrophils. *BMC Complementary and Alternative Medicine, 14*(1), 429.

Jawhari, F.Z., Moussaoui, A.E.L., Bourhia, M. et al. (2021). *Anacyclus pyrethrum* Var. *pyrethrum* (L.) and *Anacyclus pyrethrum* Var. *depressus* (Ball) Maire: correlation between total phenolic and flavonoid contents with antioxidant and antimicrobial activities of chemically characterized extracts. *Plants, 10*(1), 149.

Jayakumar, T., Hsieh, C.Y., Lee, J.J. et al. (2013). Experimental and clinical pharmacology of *Andrographis paniculata* and its major bioactive phytoconstituent andrographolide. *Evidence-Based Complementary and Alternative Medicine, 2013*,1–16.

Kashyap, V.K., Dhasmana, A., Yallapu, M.M. et al. (2020). *Withania somnifera* as a potential future drug molecule for COVID-19. *Future Drug Discovery, 2*(4), FDD50.

Kooti, W., Hasanzadeh-Noohi, Z., Sharafi-Ahvazi, N. et al. (2016). Phytochemistry, pharmacology, and therapeutic uses of black seed (*Nigella sativa*). *Chinese Journal of Natural Medicines, 14*(10), 732–745.

Kumar, V., Dhanjal, J.K., Kaul, S.C. et al. (2020). Withanone and caffeic acid phenethyl ester are predicted to interact with main protease (Mpro) of SARS-CoV-2 and inhibit its activity. *Journal of Biomolecular Structure and Dynamics, 39*(11), 3842–3854.

Lau, K.M., Lee, K.M., Koon, C.M. et al. (2008). Immunomodulatory and anti-SARS activities of *Houttuynia cordata*. *Journal of Ethnopharmacology, 118*(1), 79–85.

Li, C., He, Q., Qian, H. et al. (2021). Overview of the pathogenesis of COVID-19 (Review). *Experimental and Therapeutic Medicine, 22*(3), 1011.

Liang, W., Feng, Z., Rao, S. et al. (2020). Diarrhoea may be underestimated: a missing link in 2019 novel Coronavirus. *Gut, 69*(6), 1141–1143.

Logesh, R., Das, N., Adhikari-Devkota, A. et al. (2020). *Cocculus hirsutus* (L.) W. Theob. (Menispermaceae): a review on traditional uses, phytochemistry and pharmacological activities. *Medicines, 7*(11), 69.

Lv, J.J., Wang, Y.F., Zhang, J.M. et al. (2014). Anti-Hepatitis B virus activities and absolute configurations of sesquiterpenoid glycosides from *Phyllanthus emblica*. *Organic and Biomolecular Chemistry, 12*(43), 8764–8774.

Mamedov, N.A. and Egamberdieva, D. (2019). Phytochemical constituents and pharmacological effects of Licorice: a review. In: *Plant and Human Health,* Springer International Publishing, Cham, *Vol. 3*, 1–21.

Mangathayaru, K., Umadevi, M., and Reddy, C.U. (2009). Evaluation of the immunomodulatory and DNA protective activities of the shoots of *Cynodon dactylon*. *Journal of Ethnopharmacology, 123*(1), 181–184.

Mao, Q.Q., Xu, X.Y., Cao, S.Y. et al. (2019). Bioactive compounds and bioactivities of ginger (*Zingiber officinale* Roscoe). *Foods, 8*(6), 185.

Maurya, V.K., Kumar, S., Prasad, A.K. et al. (2020). Structure-based drug designing for potential antiviral activity of selected natural products from ayurveda against SARS-CoV-2 spike glycoprotein and its cellular receptor. *Virus Disease, 31*(2), 179–193.

Mediratta, P.K, Sharma, K.K., and Singh, S. (2002). Evaluation of immunomodulatory potential of *Ocimum sanctum* seed oil and its possible mechanism of action. *Journal of Ethnopharmacology, 80*(1), 15–20.

Menon, V.P. and Sudheer, A.R. (2007). Antioxidant and anti-inflammatory properties of Curcumin. *The Molecular Targets and Therapeutic Uses of Curcumin in Health and Disease,* Springer, Boston, MA, vol. 595, 105–125.

Mishra, A.P., Saklani, S., alehi, B. et al. (2018). *Satyrium nepalense*, a high altitude medicinal orchid of Indian Himalayan region: Chemical profile and biological activities of tuber extracts. *Cellular and Molecular Biology, 64*(8), 35–43.

Mishra, L.C., Singh, B.B., and Dagenais, S. (2000). Scientific basis for the therapeutic use of *Withania somnifera* (Ashwagandha): a review. *Alternative Medicine Review, 5*(4), 334–346.

Mitra, D., Verma, D., Mahakur, B. et al. (2021). Molecular docking and simulation studies of natural compounds of *Vitex negundo* L. against papain-like protease (PL pro) of SARS-CoV-2 (Coronavirus) to conquer the pandemic situation in the world. *Journal of Biomolecular Structure and Dynamics, 40*(12), 1–22.

Mohajer-Shojai, T., Ghalyanchi Langeroudi, A., Karimi, V. et al. (2016). The effect of *Allium sativum* (Garlic) extract on infectious bronchitis virus in specific pathogen free embryonic egg. *Avicenna Journal of Phytomedicine, 6*(4), 458–267.

Mondal, S., Varma, S., Bamola, V.D. et al. (2011). Double-blinded randomized controlled trial for immunomodulatory effects of Tulsi (*Ocimum sanctum* Linn.) leaf extract on healthy volunteers. *Journal of Ethnopharmacology, 136*(3), 452–456.

Mukherjee, P.K., Rai, S., Kumar, V. et al. (2007). Plants of Indian origin in drug discovery. *Expert Opinion on Drug Discovery, 2*(5),633–657.

Napagoda, M., Gerstmeier, J., Butschek, H. et al. (2020). The anti-inflammatory and antimicrobial potential of selected ethnomedicinal plants from Sri Lanka. *Molecules, 25*(8), 1894.

Nivetha, R., Bhuvaragavan, S., and Janarthanan, S. (2021). Inhibition of multiple SARS-CoV-2 proteins by an antiviral biomolecule, seselin from *Aegle marmelos* deciphered using molecular docking analysis. *Journal of Biomolecular Structure and Dynamics, 1*–12.

Pandey, P., Rane, J.S., Chatterjee, A., et al. (2021). Targeting SARS-CoV-2 spike protein of COVID-19 with naturally occurring phytochemicals: an in silico study for drug development. *Journal of Biomolecular Structure and Dynamics, 39*(16), 6306–6316.

Pattanayak, P., Behera, P., Das, D. et al. (2010). *Ocimum sanctum* Linn. A reservoir plant for therapeutic applications: an overview. *Pharmacognosy Reviews, 4*(7), 95.

Praditya, D., Kirchhoff, L., Brüning, J. et al. (2019). Anti-infective properties of the golden spice *Curcumin*. *Frontiers in Microbiology, 10*, 912.

Prasad, L.V. (2002). Indian system of medicine and homoeopathy. In: *Traditional Medicine in Asia*. Chaudhury, R.R. and Rafei, U.M. (eds.), World Health Organization, Regional Office for South-East Asia, New Delhi, India, 283–286.

Rajagopal, K., Byran, G., Jupudi, S. et al. (2020). Activity of phytochemical constituents of black pepper, ginger, and garlic against coronavirus (COVID-19): an *in-silico* approach. *International Journal of Health & Allied Sciences, 9*(5),S43–S50.

Roshdy, W.H., Rashed, H.A., Kandeil, A. et al. (2020). EGYVIR: an immunomodulatory herbal extract with potent antiviral activity against SARS-CoV-2. *PLOS ONE, 15*(11), e0241739.

Sangiamsuntorn, K., Suksatu, A., Pewkliang, Y. et al. (2021). Anti-SARS-CoV-2 activity of *Andrographis paniculata* extract and its major component andrographolide in human lung epithelial cells and cytotoxicity evaluation in major organ cell representatives. *Journal of Natural Products, 84*(4), 1261–1270.

Sai Ram, M., Neetu, D., Yogesh, B. et al. (2002). Cyto-protective and immunomodulating properties of Amla (*Emblica officinalis*) on lymphocytes: an *in-vitro* study. *Journal of Ethnopharmacology, 81*(1), 5–10.

Sharma, R.K. and Chakotiya, A.S. (2020). Phytoconstituents of *Zingiber officinale* targeting host viral protein interaction at entry point of SARS-CoV-2 a molecular docking study. *Defence Life Science Journal, 5*(4), 268–277.

Sharma, S.K. and Singh, H. (2012). A review on pharmacological significance of genus *Jatropha* (Euphorbiaceae). *Chinese Journal of Integrative Medicine, 18*(11), 868–880.

Sharma, U., Bala, M., Kumar, N. et al. (2012). Immunomodulatory active compounds from *Tinospora cordifolia*. *Journal of Ethnopharmacology, 141*(3), 918–926.

Shi, S., Qin, M., Shen, B. et al. (2020). Association of cardiac injury with mortality in hospitalized patients with COVID-19 in Wuhan, China. *JAMA Cardiology, 5*(7), 802–810.

Shree, P., Mishra, P., Selvaraj, C. et al. (2020). Targeting COVID-19 (SARS-CoV-2) main protease through active phytochemicals of ayurvedic medicinal plants–*Withania somnifera* (Ashwagandha), *Tinospora cordifolia* (Giloy) and *Ocimum sanctum* (Tulsi)—a Molecular docking study. *Journal of Biomolecular Structure and Dynamics, 40*(1), 190–203.

Sinha, S.K., Prasad, S.K., Islam, Md.A. et al. (2020). Identification of bioactive compounds from *Glycyrrhiza glabra* as possible inhibitor of SARS-CoV-2 spike glycoprotein and non-structural protein-15: a pharmacoinformatics study. *Journal of Biomolecular Structure and Dynamics, 39*(13), 4686–4700.

Straughn, A.R. and Kakar, S.S. (2020). Withaferin A: a potential therapeutic agent against COVID-19 infection. *Journal of Ovarian Research, 13*(1), 79.

Su, S., Wong, G., Shi, W. et al. (2016). Epidemiology, genetic recombination, and pathogenesis of coronaviruses. *Trends in Microbiology, 24*(6), 490–502.

Sundaresan, N. and Ilango, K. (2018). Review on *Valeriana* Species-*Valeriana wallichii* and *Valeriana jatamansi*. *Journal of Pharmaceutical Sciences and Research, 10*(11), 2697–2701.

Tahir, I., Khan, M.R., Shah, N.A. et al. (2016). Evaluation of phytochemicals, antioxidant activity and amelioration of pulmonary fibrosis with *Phyllanthus emblica* leaves. *BMC Complementary and Alternative Medicine, 16*(1), 406.

Takooree, H., Aumeeruddy, M.Z., Rengasamy, K.R.R. et al. (2019). A systematic review on black pepper (*Piper nigrum* L.): From folk uses to pharmacological applications. *Critical Reviews in Food Science and Nutrition, 59*(sup1), S210–S243.

Thakkar, S.S., Shelat, F., and Thakor, P. (2021). Magical bullets from an indigenous indian medicinal plant *Tinospora cordifolia*: an *in-silico* approach for the antidote of SARS-CoV-2. *Egyptian Journal of Petroleum, 30*(1), 53–66.

Thangaraju, P., Sudha, S.T.Y., Pasala, P.K. et al. (2021). The role of *Justicia adhatoda* as prophylaxis for COVID-19–analysis based on docking study. *Infectious Disorders-Drug Targets, 21*(8), e160921190441.

Thimmulappa, R.K., Mudnakudu-Nagaraju, K.K., Shivamallu, C. et al. (2021). Antiviral and immunomodulatory activity of Curcumin: a case for prophylactic therapy for COVID-19. *Heliyon, 7*(2), e06350.

Thuy, B.T.P., My, T.T.A., Hai, N.T.T. et al. (2020). Investigation into SARS-CoV-2 resistance of compounds in garlic essential oil. *ACS Omega, 5*(14), 8312–8320.

To, K.K.W., Tsang, O.T.Y., Leung, W.S. et al. (2020). Temporal profiles of viral load in posterior oropharyngeal saliva samples and serum antibody responses during infection by SARS-CoV-2: an observational cohort study. *The Lancet Infectious Diseases, 20*(5), 565–574.

Tripathi, M.K., Singh, P., Sharma, S. et al. (2020). Identification of bioactive molecule from *Withania somnifera* (Ashwagandha) as SARS-CoV-2 main protease inhibitor. *Journal of Biomolecular Structure and Dynamics, 39*(15), 5668–5681.

Vellingiri, B., Jayaramayya, K., Iyer, M. et al. (2020). COVID-19: a promising cure for the global panic. *Science of the Total Environment, 725*, 138277.

Vincent, S., Arokiyaraj, S., Saravanan, M. et al. (2020). Molecular docking studies on the anti-viral effects of compounds from *Kabasura kudineer* on SARS-CoV-2 3CLpro. *Frontiers in Molecular Biosciences, 7*, 613401.

Wang, D., Hu, B., Hu, C. et al. (2020). Clinical characteristics of 138 hospitalized patients with 2019 novel coronavirus-infected pneumonia in Wuhan, China. *JAMA, 323*(11), 1061–1069.

Wen, C.C., Kuo, Y.H., Jan, J.T. et al. (2007). Specific plant terpenoids and lignoids possess potent antiviral activities against severe acute respiratory syndrome coronavirus. *Journal of Medicinal Chemistry, 50*(17), 4087–4095.

Wyganowska-Swiatkowska, M., Nohawica, M., Grocholewicz, K. et al. (2020). Influence of herbal medicines on HMGB1 release, SARS-CoV-2 viral attachment, acute respiratory failure, and sepsis. A literature review. *International Journal of Molecular Sciences, 21*(13), 4639.

Xiang, Y., Pei, Y., Qu, C. et al. (2011). *In-vitro* anti-herpes simplex virus activity of 1,2,4,6-Tetra-O-Galloyl-β-d-Glucose from *Phyllanthus emblica* L. (Euphorbiaceae). *Phytotherapy Research, 25*(7), 975–982.

Xie, J., Tong, Z., Guan, X. et al. (2020a). Critical care crisis and some recommendations during the COVID-19 epidemic in China. *Intensive Care Medicine, 46*(5), 837–840.

Xie, Y., Karki, C.B., Du, D. et al. (2020b). Spike proteins of SARS-CoV and SARS-CoV-2 utilize different mechanisms to bind with human ACE2. *Frontiers in Molecular Biosciences, 7*, 392.

Yadav, V.K., Singh, G., Jha, R.K. et al. (2020). Visiting bael (*Aegle marmelos*) as a protective agent against COVID-19: a review. *Indian Journal of Traditional Knowledge, 19*(4), S153–S157.

Yang, X., Yu, Y., Xu, J. et al. (2020). Clinical course and outcomes of critically Ill patients with SARS-CoV-2 pneumonia in Wuhan, China: a single-centered, retrospective, observational study. *The Lancet Respiratory Medicine, 8*(5), 475–481.

Yoshikawa, T., Hill, T., Li, K. et al. (2009). Severe acute respiratory syndrome (SARS) coronavirus-induced lung epithelial cytokines exacerbate SARS pathogenesis by modulating intrinsic functions of monocyte-derived macrophages and dendritic cells. *Journal of Virology, 83*(7), 3039–3048.

Zhao, H.M., Xu, R., Huang, X.Y. et al. (2016a). Curcumin suppressed activation of dendritic cells via JAK/STAT/SOCS signal in mice with experimental colitis. *Frontiers in Pharmacology, 7*, 455.

Zhao, Y.K., Li, L., Liu, X. et al. (2016b). Explore pharmacological mechanism of Glycyrrhizin based on systems pharmacology. *Zhongguo Zhongyao Zazhi, 41*(10), 1916–1920.

Zheng, C.J., Li, H.Q., Ren, S.C. et al. (2015). Phytochemical and pharmacological profile of *Vitex negundo*. *Phytotherapy Research, 29*(5), 633–647.

Zhu, N., Zhang, D., Wang, W. et al. (2020). A novel coronavirus from patients with pneumonia in China, 2019. *New England Journal of Medicine, 382*(8), 727–733.

10 Scientific Evaluation and Biochemical Validation of Efficacy of Extracts and Phytochemicals from Zimbabwe
A Review of Prospects for Development of Bioproducts

Stanley Mukanganyama

CONTENTS

10.1 INTRODUCTION

The World Health Organization reports that more than 80% of the world's population depends on traditional medicine to cater for their primary health care (WHO, 2019). The search for new pharmacologically active agents obtained by screening microbial and plant extracts has led to the discovery of many clinically useful drugs that play a major role in the treatment of human diseases (Atanasov et al., 2015). It is important to know the biochemistry of plants to avoid issues of toxicity. Zimbabweans use traditional medicines, since they are affordable and believed to be safe than conventional drugs.

DOI: 10.1201/9781003206620-10

10.2 MICROBIAL INFECTIONS

Bacterial and fungal infections have become a major medical problem due to the rise in frequency of multi-drug resistance. Pathogenic bacteria can grow, divide, and spread in the body, leading to infectious diseases (Rahman et al., 2020). Other bacteria are opportunistic pathogens and do not cause infections in healthy humans but cause diseases when the immune system is compromised (Price et al., 2017).

10.3 PLANTS AS SOURCES OF ANTI-INFECTIVE COMPOUNDS

Phytochemical screening research serves to identify the active compounds from plant medicine and verify their efficiency against the disease-causing microbes. Determination of toxic components that may be harmful to human health in the traditional medicines can also be carried out (Mensah et al., 2019). Many significant drugs, including artemisinin, paclitaxel, and quinine are products of phytochemical research inspired by the ethno-traditional uses of their respective plants. Crude plant medicines used by most people in the developing countries are far from being safe. This is because some may contain toxic compounds and heavy metals (Vaikosen and Alade, 2017). Therefore, there is a need for studies that may lead to the discovery of new and alternative antimicrobials, which are relatively safer to humans.

10.4 EVALUATION AND BIOCHEMICAL VALIDATION

Antimicrobial susceptibility tests (AST) are a group of procedures that are used to determine which specific antibiotics a particular microorganism is sensitive to (Bonev et al., 2008). In addition to helping design therapy regimens for infections, AST are also used in drug discovery, for *in vitro* investigations of extracts and pure drugs as potential antimicrobial agents. A variety of laboratory methods can be used to evaluate the *in vitro* antibacterial activity of a compound. Some of the AST methods include diffusion methods, thin-layer chromatography bioautography, dilution methods, time-kill test, ATP bioluminescence, and flow cytofluorometric method (Reller et al., 2009). The broth microdilution assay is one of the most basic AST methods and is an example of the dilution methods. They are the most appropriate for the determination of the minimum inhibitory concentrations (MIC), as they are able to offer possible estimate of concentration of the tested antimicrobial agent (Ambaye et al., 1997). When determining the MIC endpoint of broth microdilution assays, viewing chromogenic devices is often used to discern growth in the wells for the determination of both antifungal and antibacterial activities. The 3-(4,5-dimethylthiazol-2-yl)-2,5-diphenyltetrazolium bromide (MTT) assay is used for assessing cell metabolism. The assay is done by checking for the activity of the NADPH-dependent cellular reductase enzymes, which in turn reflects the number of viable cell present. These enzymes can reduce the tetrazolium dye to insoluble formazan, which is purple (Braissant et al., 2020),

10.4.1 ANTIBACTERIAL EVALUATION

In our initial work, we evaluated the antibacterial activity of ethanolic extracts of 19 (Table 10.1) Zimbabwean plants using the agar diffusion assay and determined the MIC and the minimum bactericidal activity using ampicillin as reference.

Accumulation of rhodamine 6G in bacteria was used to determine the activity of extracts as drug efflux pump inhibitors (EPIs) (Chitemerere and Mukanganyama, 2011). At least eight extracts exhibited activity against all bacteria (Table 10.2). The MIC and the minimum bactericidal concentration (MBC) of plant extracts ranged from 0.05 to 0.5 mg/mL and 0.06 mg/mL to >1 mg/mL, respectively.

TABLE 10.1
The Scientific and Vernacular Names of Some of the Plant Extracts Used in Antibacterial Evaluation as Well as the Traditional Uses of Some of the Plants

Family	Botanical Name	Local Name Voucher	Voucher	Plant Part Tested	Major Traditional Use (Reference)
Asteraceae	*Vernonia adoensis* Sch. Bip. ex Walp.	*Musikavakadzi*	C1 E7	Leaves	Boiled and decoction drunk to cure TB. (Kisangau et al., 2007)
Anacardiaceae	*Mangifera indica* L.	*Mumango*	N 17 E4	Stems and twigs	Used for coughs and diarrhoea
Chrysobalanaceae	*Parinari curatellifolia* Planch. Ex	*Muhacha*	C6 E1, E7	Stems and twigs, Leaves	Herpes
Leguminosae	*Xeroderris stuhlmannii* (Taub.) Mendoça & E.P. Sousa	*Murumanyama*	C4 E4	Stems and twigs	Stomach ailments (Iwawela et al., 2007), mastitis, backache (Ruffo, 1991)
Proteacea	*Faurea saligna*	*Mutsatsati*	C13 E7	Leaves	
Myrtaceae	*Callistemon citrinus* Skeels		UZ2 E7	Leaves	Antibacterial, Haemorrhoid treatment. (Oyedeji et al., 2009)
Moraceae	*Ficus sycomorus* L.	*Mukuyu*	SCC1, E7	Leaves	
Rubiaceae	*Fadogia stenophylla* Welw. ex Hiern		SCC2, E7	Leaves	
Vitaceae	*Cyphostemma viscosum* (Gilg & R.E.Fr.) Desc. ex Wild & R.B. Drumm.	*Fodya yemusango*	SCC4 E7	Leaves	
Rubiaceae	*Fadogia ancylantha* Schweinf.	*Masamba emusango*	SCC7;E7		
Ranunculaceae	*Clematopsis scabiosifolia* Hutch.		SCC8;E7	Leaves	
Ranunculaceae	*Salons delagoense*	*Nhundurwa*		Leaves, fruits	Used to treat scabies in children (Chigora et al., 2007)
Myrtaceae	*Syzigium cumini* Linn. Skeels	*Muboo*	C15, E7, E 12	Leaves, fruits	Anti-inflammatory herb, treatment of dysentery (Kumar et al., 2008)
Rhamnaceae	*Zyziphus mucronata* Willd.	*Muchecheni*	C7 E7	Stems	Snake bites and stomach ache (Ruffo, 1991)
Celastraceae	*Gymnosporia senegalensis* Loes	*Chizhuzhu musosawafa*	N5 E7	Leaves	
Verbenaceae	*Lantana camara* var. aculeata (L.) Moldenke	*Mbarambati*	UZ1 E7	Leaves	Ring worm infections (Chigora et al., 2007)
Fabaceae	*Brachystegia boehmii* Taub	*Mupfuti*	N7 E7	Leaves	
Rubiaceae	*Catunaregum spinosa* Thunb	*Mutsvairachuru*	C 17	Leaves	

Source: Chitemerere and Mukanganyama (2011).

TABLE 10.2

A Summary Table for the Minimum Inhibitory Concentration (MIC) and the Minimum Bactericidal Concentration (MBC) Assays

Microorganism	G+/G–	Plant Species	MIC[a] µg/mL	MBC[b] µg/mL	MIC	amp
Escherichia coli	G–	C. citrinus	125	250	8	16
		V. adoensis	188	500		
		Parinari curatellifolia (stems)	188	500		
		P. curatellifolia	300	500		
		Mangifera indica	47	250		
Pseudomonas aeruginosa	G–	V. adoensis	188	1,000	16	125
		C. citrinus	63	250		
		P. curatellifolia (stems)	ND	>1,000		
		P. curatellifolia	188	500		
		M. indica	46	250		
Staphylococcus aureus	G+	V. adoensis	94	1,000	< 2	8
		C. citrinus	47	250		
		P. curatellifolia	375	500		
		P. curatellifolia (stems)	250	500		
		Faurea sp	ND	>1,000		
Bacillus subtilis	G+	C. citrinus	24	63	4	8
		Asteraceae family (flowers)	292	1,000		
		V. adoensis	188	188		
		Faurea sp	500	>1,000		
		M. indica	40	125		
Bacillus cereus	G+	V. adoensis	188	1,000	4	8
		C. citrinus	63	125		
		P. curatellifolia (stems)	179	500		
		Lantana camara	ND	>1,000		
		P. curatellifolia	375	>1,000		

Leaf extracts of *Mangifera indica*, *Callistemon citrinus*, and *Vernonia adoensis* were bactericidal against *Staphylococcus aureus*, *Pseudomonas aeruginosa*, *Escherichia coli*, *Bacillus cereus*, and *Bacillus subtilis*, while the other extracts from *Parinari curatellifolia*, *Lantana camara*, and *Faurea* spp were bacteriostatic. The three extracts from *M. indica*, *C. citrinus*, and *V. adoensis* were effective drug efflux inhibitors when using Rhodhamine 6G (R6G) as a probe substrate. All the subsequent antibacterial work was based on the initial screening of these plants from Zimbabwe. Evaluation of bacterial membrane integrity of two species of bacteria—*S. aureus* and *P. aeruginosa*, in the presence of ethanolic leaf extracts of two plant species—*C. citrinus* and *V. adoensis* was also carried out (Chitemerere and Mukanganyama, 2014). Bacterial efflux pump inhibition

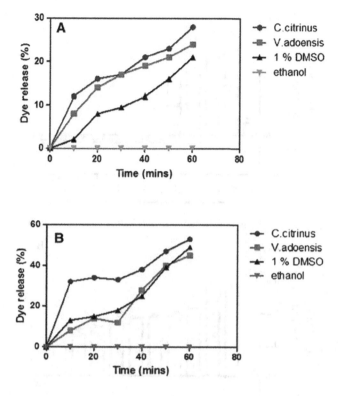

FIGURE 10.1 Measurement of 3,3′-Dipropylthiadicarbocyanine Iodide (diSC3-5) dye release overtime from *S. aureus* (A) and *P. aeruginosa*. (B) Membranes in the presence of permeabilizing agent 1% DMSO, negative control ethanol and plant extracts *C. citrinus* and *V. adoensis*. Extracts from *C. citrinus* and *V. adoensis* showed increased dye release compared to the negative control (ethanol) (Chitemerere and Mukanganyama, 2014).

using both leaf extracts was determined by monitoring the transport of Rhodamine 6G (R6G) across the cell membrane, and IC$_{50}$ values were obtained. Membrane permeabilizing properties of both the extracts were also evaluated using the membrane potential sensitive dye 3'3 dipropylthiadi-carbocyanine (diSC3–5). Extracts of *C. citrinus* and *V. adoensis* inhibited bacterial efflux pumps. Both the plant extracts showed some significant effects on permeability of the bacterial membrane (Figure 10.1).

It was suggested that these plant extracts may, therefore, provide new lead compounds for developing potential EPIs or permeabilising agents that could aid the transport of antibacterial agents into bacterial cells. Root extracts from *Cissus welwitschii* and *Triumfetta welwitschii* were screened for antibacterial activity against *E. coli* and *B. cereus* (Moyo and Mukanganyama, 2015a). It was discovered that the methanolic root extracts of *T. welwitschii* were more effective than the aqueous extracts, with an MIC and MBC values of 0.125 mg/mL and 0.5 mg/mL, respectively, against *E. coli* and *B. cereus*. Nucleic acid leakage in *B. cereus* and *E. coli,* and protein leakage in *E. coli* were observed after exposure to the extract from *T. welwitschii*. The extracts from *T. welwitschii* had greater antibacterial activity than the extracts from *C. welwitschii*. Since extracts from *T. welwitschii* showed greater antibacterial activity, they could serve as a potential source of lead compounds for that could be developed into antibacterial agents. Since we had found these preliminary results on this plant, we carried out further phytochemical studies to characterise the major compounds in this plant were carried out. (Mombeshora and Mukanganyama, 2019; Mombeshora et al., 2021). Alkaloids are plant secondary metabolites that have been shown to have potent pharmacological

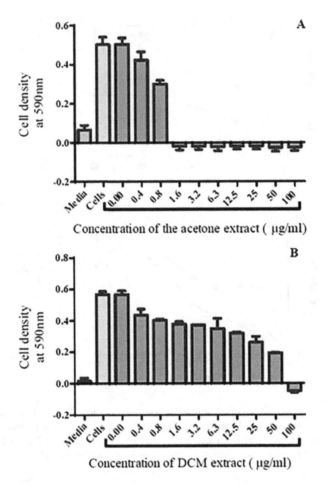

FIGURE 10.2 The effect of most potent *V. adoensis* extracts on growth of *P. aeruginosa*, where (A) is effect from DCM extract and (B) is effect from acetone extract. Concentrations of extract ranged from 100 to 0.4 µg/mL. The bacteria concentration used was 1.6×10^6 cfu/mL. Values are expressed as mean cell density at 590 nm wavelength±the standard deviation (*n*=4) (Mozirandi and Mukanganyama, 2017). As the concentration of the acetone and DCM extracts were increased, the cell density decreased giving an MIC 1.6 and 100 µg/mL, respectively.

activities (Mabhiza et al., 2016). The effect of alkaloids from *C. citrinus* and *V. adoensis* leaves on bacterial growth and efflux pump activity was evaluated on *S. aureus* and *P. aeruginosa*. The alkaloids from *C. citrinus* were the most potent against *S. aureus*, whilst the effects on *P. aeruginosa* by both the plant alkaloids were bacteriostatic. *P. aeruginosa* was the most susceptible to drug efflux pump inhibition by alkaloids from *C. citrinus*. These alkaloids, thus, may serve as potential sources of compounds that can act as lead compounds for the development of plant-based antibacterial and/ or their adjunct compounds. The antibacterial activity of *V. adoensis* extracts was also determined *in vitro* on *S. aureus* and *Pseudomonas aeruginosa* by the broth microdilution method and time-kill assays. All the extracts had an inhibitory effect on the growth of both the bacteria (Mozirandi and Mukanganyama, 2017). The acetone extract was the most potent in inhibition of growth with an MIC of 1.6 µg/mL and an MBC of 6.3 µg/mL against *P. aeruginosa* (Figure 10.2).

Extracts from *V. adoensis* showed antibacterial properties against *P. aeruginosa* and *S. aureus*, and protein leakage was proposed as a possible mechanism of action elicited by the extracts. Since the leaves of *V. adoensis* have been reported to have antimicrobial activity (Mozirandi and Mukanganyama, 2017; Chitemerere and Mukanganyama, 2011, 2014), a follow-up study was carried

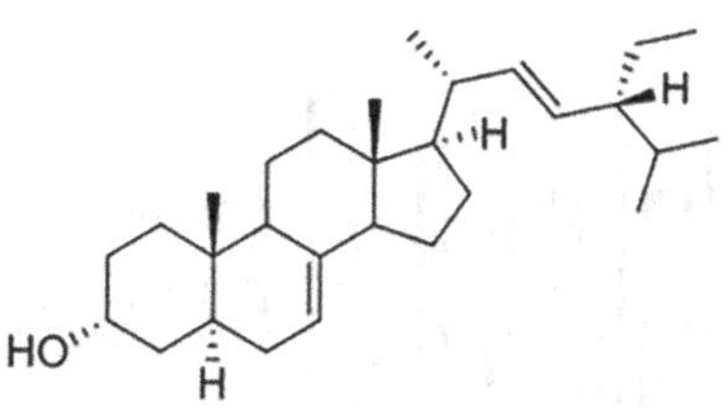

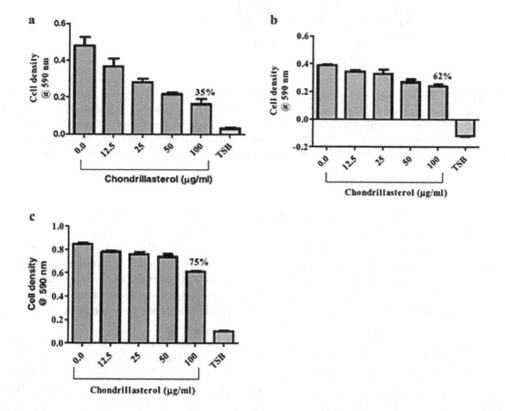

FIGURE 10.3 Structure of chondrillasterol isolated from the leaves of *V. adoensis*. Data to elucidate the structure was obtained by NMR and MS studies (Mozirandi et al., 2019).

FIGURE 10.4 The effect of chondrillasterol on growth of bacterial pathogens. The pathogens were susceptible to chondrillasterol and (a) is % remaining viable cells for *P. aeruginosa*, and (b) is % of remaining viable cells for *K. pneumoniae* while (c) is % of remaining viable cells for *S. aureus* following exposure to chondrillasterol. Concentrations of 1.6×10^6 cfu/mL of bacteria were used. Values are expressed as mean cell density at 590 nm wavelength ± the standard deviation ($n = 4$). TSB is tryptic soy broth. Chondrillasterol reduced the cell density by 35%, 63% and 75% at the highest concentration of 100 µg/mL for *P. aeruginosa*, *K. pneumoniae* and *S. aureus,* respectively (Mozirandi et al., 2019).

out to isolate the bioactive compounds from the leaf extract and evaluate their antibacterial activity on *S. aureus*, *Klebsiella pneumoniae*, and *P. aeruginosa* (Mozirandi et al., 2019). The compound isolated from *V. adoensis* was identified as chondrillasterol (Figure 10.3).

Data to elucidate the structure was obtained by NMR and MS studies. (Source: Mozirandi et al., 2019).

Chondrillasterol exhibited 25%, 38%, and 65% inhibition of growth on *S. aureus*, *K. pneumonia*, and *P. aeruginosa*, respectively (Figure 10.4).

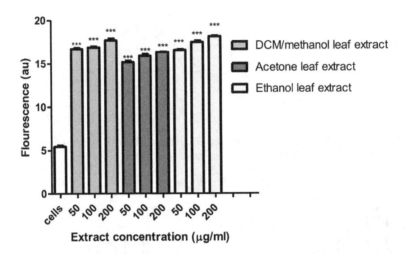

FIGURE 10.5 Fluorescence of propidium iodide bound to nucleic acids of *P. aeruginosa* cells after exposure to the acetone, ethanol and DCM/methanol leaf extracts from *Triumfetta welwitschii*. Cells with no extract were used as the control. Values are for mean±standard deviation (error bar) for *n*=3. The asterisks indicate a significant difference from the control with ***$p < 0.001$. As the concentration of the extracts was increased, there was a corresponding increase fluorescence compared to the controls (Mombeshora and Mukanganyama, 2019).

   T. welwitschii plant roots are traditionally used to treat symptoms of diarrhoea and fever, suggesting that the plant extracts possess antimicrobial (Moyo and Mukanganyama, 2015a) and immunomodulatory effects (Mombeshora and Mukanganyama, 2019). Extracts of leaves from *T. welwitschii*, were evaluated against *B. subtilis*, *S. aureus*, *P. aeruginosa*, *Streptococcus pneumoniae*, *Streptococcus pyogenes*, and *Klebsiella pneumoniae*. The results showed that the acetone, ethanol, and dichloromethane:methanol extracts had the most potent antibacterial activities against *P. aeruginosa* (ATCC 27853). All the three extracts caused membrane disruption of *P. aeruginosa*, as shown by nucleic acid leakage (Figure 10.5).

   The leaf extracts of *T. welwitschii* were shown to have antibacterial effects against *P. aeruginosa*. To further explore the antibacterial potential of these phytochemical components, the phytochemical profile of the dichloromethane:methanol leaf extract from *T. welwitschii* was investigated using ultra-performance liquid chromatography-tandem mass spectrometry (UPLC-MS/MS) (Mombeshora et al., 2021). Compounds were isolated from the extract using column chromatography and thin-layer chromatography. B1 was isolated from the fraction eluted by 90 hexane:10 ethyl acetate using column chromatography. B1 showed antibacterial activity against *P. aeruginosa* with the MIC and the MBC value of 25µg/mL. A total of 28 peaks were detected and identified using UPLC-MS/MS. The three most abundant phytochemicals identified were catechin, umbelliferone, and a luteolin derivative (Figure 10.6).

   Seven extracts from *P. curatellifolia* leaves were prepared using serial exhaustive extraction of nonpolar to polar solvents (Mawire et al., 2021). The microbroth dilution method was used to evaluate antimicrobial bioactivities of extracts. The extracts significantly inhibited growth of *Klebsiella pneumoniae* and *S. aureus* with the DCM:methanol extract showing inhibition of 75% and 90%, respectively, at 100 µg/mL concentrations.

10.4.2 ANTIFUNGAL EVALUATION

Thirty-eight Zimbabwean medicinal plant extracts were investigated for antifungal activity against *Candida albicans* and *Candida krusei* (Mangoyi and Mukanganyama, 2011). These plants were

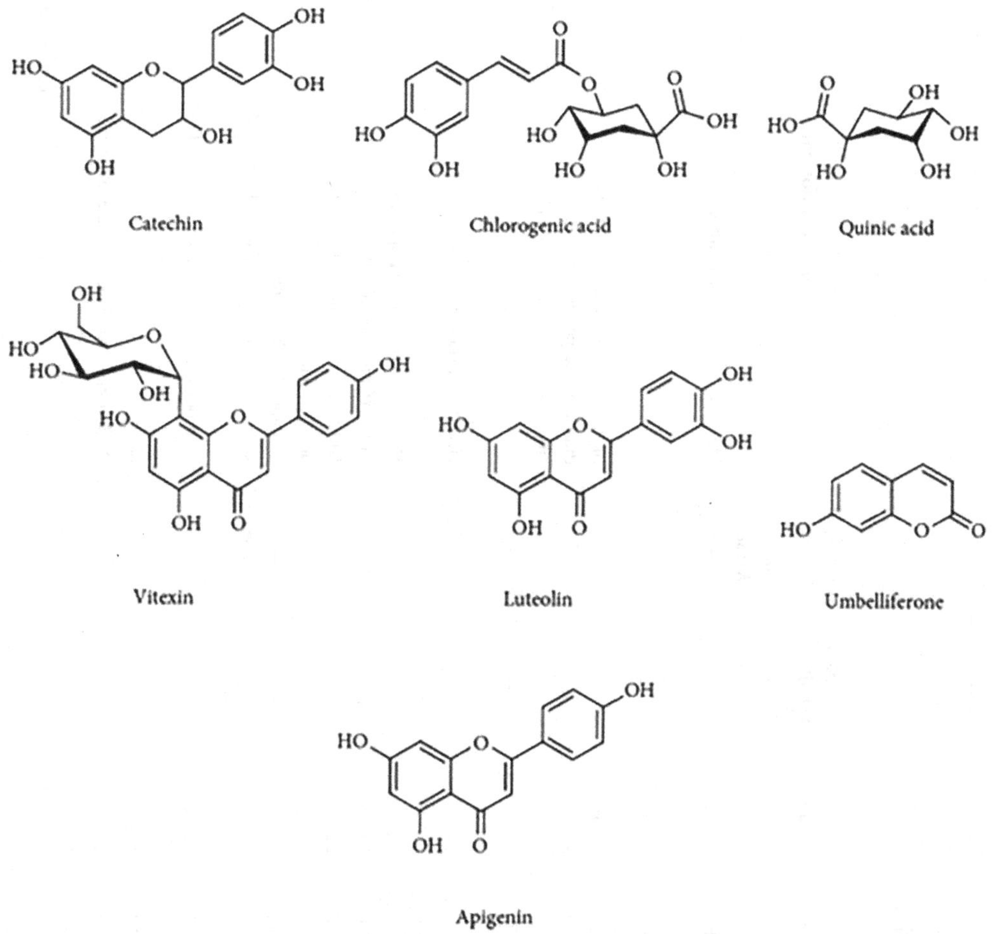

Catechin   Chlorogenic acid   Quinic acid

Vitexin   Luteolin   Umbelliferone

Apigenin

FIGURE 10.6 Chemical structures of some of the compounds identified from the DCM: methanol leaf extract of *T. welwitschii*. The phytochemical profile of the dichloromethane: methanol leaf extract from *T. welwitschii* was investigated using ultra-performance liquid chromatography-tandem mass spectrometry (UPLC-MS/MS) (Mombeshora et al., 2021).

reported to be used for the traditional treatment of various ailments, including fungal infections (Table 10.3).

Nineteen plant extracts, among the 38 investigated, showed significant antifungal activity. MIC values ranged from 0.08 to 0.63 mg/mL for both *C. albicans* and *C. krusei*. MFCs ranged from 0.31 to 2.5 mg/mL. *Combretum zeyheri* extract had the highest antifungal activity in all the cases (Table 10.4).

The antifungal activities of six *Combretum* species (*C. zeyheri*, *Combretum apiculatum*, *Combretum molle*, *Combretum kraussii*, *Combretum elaegnoides*, and *Combretum imberbe*) were investigated against *C. albicans* and *C. krusei* (Mangoyi et al., 2012). The extract from *C. zeyheri* had the highest antifungal activity with MIC values of 0.08 and 0.16 mg/mL. There was a need to isolate, identify, and characterize compounds from *C. zeyheri* leaves, which are responsible for its antifungal activity (Mangoyi et al., 2015). The compound 5-hydroxy-7,4'-dimethoxyflavone (Figure 10.7) was found to be active against *C. albicans* using broth dilution method.

TABLE 10.3

Zimbabwean Medicinal Plants Evaluated for Antifungal Activity

Family	Plant Material Family (Crude Extract) Scientific	Plant Material Family (Crude Extract) Local	Voucher	Part Used	Antifungal Activity *Candida Albicans*	Antifungal Activity *Candida Krusei*	Ethnomedicinal Information
Leguminosae	*Xeroderris stuhlmannii.* (Taub) Mendonca and E.P Sousa	*Murumanyama*	C4 E7	Leaves	+	–	Mastitis and backache (Ruffo, 1991)
Chrysobalanaceae	*Parinari curatellifolia*	*Muhacha*	C6 EI C6 E7	Roots, leaves	+ +	– –	Skin rashes, tuberculosis, chronic diarrhoea, herpes zoster, herpes simplex (Chigora et al., 2007)
Combretaceae	*Combretum molle* Engl. & Diels	*Mudziyaishe*	C9 E7	Leaves	+++	+++	–
Anarcadiaceae	*Rhus lancea* Engl	*Muchokochiana*	C11 E10 C11 E7	Flowers, leaves	++ ++	– –	Headache (Maliwichi, 2000)
Myrtaceae	*Syzygium cordatum* Hochst.ex C.krauss	*Mukute*	C12 E10 C12 E7	Flowers, leaves	+ +++	+ +++	Herpes zoster, herpes simplex, skin rashes (Chigora et al., 2007)
Proteaceae	*Faurea saligna* Harv	*Mutsatsati*	C13 E7	Leaves	+	–	Bilharzia and helminthiasis (Baerts and Lehmann)
Combretaceae	*Combretum zeyheri* Sond	*Muruka, mupembere-kono, muchenja*	N6 E7	Leaves	+++	+++	Coughs, diarrhoea, rectal prolapse, snake bites, and stomachache (Ruffo, 1991)
Fabaceae	*Cajanus cajan* (L.) Millsp.	–	N8 E7	Leaves	+++	–	Aphrodisiac, oral candidiasis (Ruffo, 1991)
Myrtaceae	*C. citrinus*	–	UZ2 E7	Leaves	++	++	–
Olacaceae	*Oxalis lutifolia* De Wild	*Gungwe, Kahungwarara*	UZ8 E7 UZ8 E1	Leaves, roots	+ +	– –	Abdominal pain (Chinemana et al., 1985)

(Continued)

TABLE 10.3 (Continued)
Zimbabwean Medicinal Plants Evaluated for Antifungal Activity

Family	Plant Material Family (Crude Extract)	Plant Material Family (Crude Extract)	Voucher	Part Used	Antifungal Activity	Antifungal Activity	Ethnomedicinal Information
Araliaceae	Cussonia natalensis Sond	Mutobvi, mufenje	UZ9 E7	leaves	++	–	–
Euphobiaceae	Croton gratissimus Burch.	Gunukira, Mufandemenge, mugugu, mubvukuta	UZ13 E7	Leaves	+	–	Respiratory disorders (Roodt, 1998)
Aloaceae	Aloe vera barbadensis Miller	Gavakava	UZ14 E7	Leaves	+	–	Wounds and inflammatory skin disorders (Gelfand et al., 1985)
Rubiaceae	Catunaregum spinosa Thunb	Murovaduri	C5 E7	Leaves	–	–	Intestinal worms, gonorrhoea, and syphilis (Ruffo, 1991)
Fabaceae	Brachystegia boehmii Taub	Mupfuti	N7 E7	Leaves	–	–	–
Euphorbiaceae	Uapaca kirkiana Muell. Arg.	Muzhanje	UZ15 E12	Fruit	–	–	–
Myrtaceae	Syzigium cumini (Linn.)m Skeels	Muboo	C15 E7	Leaves	–	–	–
Asteraceae	Bidens pilosa Linn. var.	Tsine	N1 E7	Leaves	–	–	Anti-inflammatory, anti-rheumatic (Wang et al., 2003). Wounds and relapsing fevers in children, oral candidiasis (Ruffo, 1991)
Verbenaceae	Lippia javanica (Burm.f.) Spreng	Zumbani	C3 E7	Leaves	–	–	Coughs, colds and fever; influenza, measles, malaria, and stomach ache (Chigora et al., 2007)
Malvaceae	Abelmoschus esculentus	Derere	UZ17 E12	Fruit	–	–	–
Anarcardiaceae	Mangifera indica (L)	Mumango	UZ18 E7	Leaves	–	+	Astringent, gonorrhoea, asthma, prolonged ejaculation, anthelmintic (Kadavul and Dixit, 2009)

(Continued)

TABLE 10.3 (Continued)
Zimbabwean Medicinal Plants Evaluated for Antifungal Activity

Family	Plant Material Family (Crude Extract)	Plant Material Family (Crude Extract)	Voucher	Part Used	Antifungal Activity	Antifungal Activity	Ethnomedicinal Information
Rosaceae	*Prunus cerasoides* D. Don	-	UZ5 E7	Leaves	-	-	-
Verbenaceae	*Lantana camara*	*Mbarambati*	UZ1 E7	Leaves	-	-	Ring worm infections
Rhamnaceae	*Zyziphus mucronata*	*Muchecheni*	C7 E7	Leaves	-	-	Paste of leaves treats boils, Carbuncles, and swollen glands; leaf infusion is taken against chest complaints (Roodt, 1998b) Snake bites and stomachache (Ruffo, 1991)
Clusiaceae	*Garcinia huillensis* Welw	*Mutunduru*	C10 E7	Leaves	-	-	Treatment of Cryptococcal meningitis (Chigora et al., 2007)
Celastraceae	*Gymnosporia senegalensis*	*Chizhuzhu, musosawafa*	N5 E7	Leaves	-	-	Coughs, pneumonia, and tuberculosis
Solanaceae	*Solanum mauritianum* Scop	-	UZ10 E7	Leaves	-	-	Remedy for stomach ache and diarrhoea. It is also used to treat respiratory ailments and tuberculosis (Carolus and Porter, 2004)

Source: Mangoyi and Mukanganyama (2011).

-: no inhibitory activity, +: slight inhibitory, ++: medium inhibitory activity, and +++: high inhibitory activity.

TABLE 10.4
Minimum Inhibitory Concentrations and Minimum Fungicidal Concentrations of Extracts Towards the *Candida* species

Plant Species (Crude extract)	Inhibition of Growth	*Candida albicans*	*Candida krusei*
Combretum zeyheri (leaves)	Zone of inhibition[a]	18.5±0.7 mm	18±0.2 mm
	MIC	0.08 mg/mL	0.16 mg/mL
	MFC	0.31 mg/mL	0.31 mg/mL
Combretum molle (leaves)	Zone of inhibition	17±0.1 mm	15±0.3 mm
	MIC	0.31 mg/mL	0.31 mg/mlL
	MFC	0.63 mg/mL	1.25 mg/mL
Cussonia natalensis (leaves)	Zone of inhibition	16±1.4 mm	–
	MIC	0.31 mg/mL	
	MFC	1.25 mg/mL	
Syzigium cordatum (leaves)	Zone of inhibition	15±0.1 mm	12±0.1 mm
	MIC	0.63 mg/mL	0.63 mg/mL
	MFC	1.25 mg/m	2.5 mg/mL
Syzigium cordatum (bark)	Zone of inhibition	15±0.1 mm	12±0.6 mm
	MIC	0.63 mg/mL	0.63 mg/mL
	MFC	1.25 mg/mL	2.5 mg/mL
Positive control (miconazole)	Zone of inhibition	20±0.8 mm	22.5±0.7 mm
	MIC	0.31 mg/mL	0.63 mg/mL
	MFC	0.31 mg/mL	0.63 mg/mL
Negative control (DMSO)	Zone of inhibition	6 mm	6 mm

Source: Chimponda and Mukanganyama (2011).

[a] Results are the average (±SD) of two separate antifungal susceptibility test (Each antifungal susceptibility test was followed by a disk diffusion assay done in quadruplicate.).

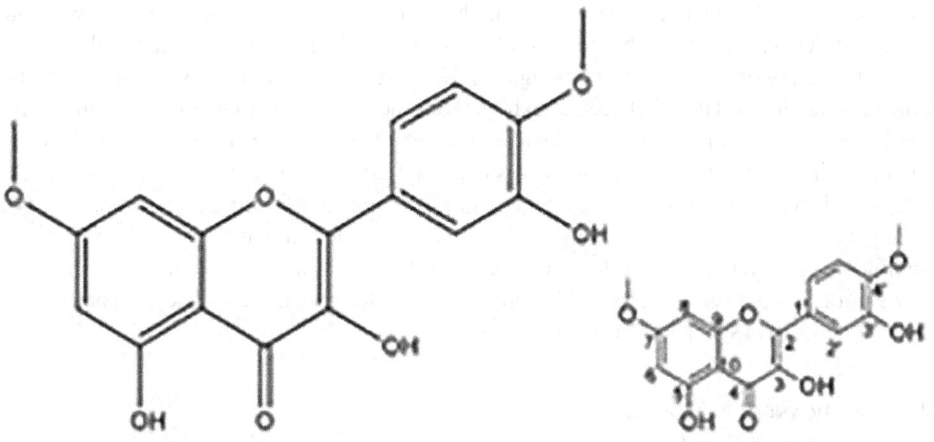

FIGURE 10.7 Flavonoids isolated from *C. zeyheri* from Norton, Mashonaland West, Zimbabwe. The flavonoids were characterized as (a) 5-hydroxy-7,4′-dimethoxyflavone and (b) 3,5,7-Trihydroxyl-3′,4′-dimethoxyflavone (Mangoyi et al., 2015).

Lampranthus francisci is an ornamental succulent plant. In Zimbabwe, the fresh sap from the leaves is used to treat fungal scalp infections (Moyo and Mukanganyama, 2015b). The activity of *L. francisci* fresh and dry acetone, ethanol, hydroethanolic, and aqueous extracts against *C. albicans* and *C. krusei* was determined. The hydroethanolic extracts were the most effective extracts against *C. albicans*. The fresh ethanol extract was the most effective extract against *C. krusei*. The dry acetone extract, dry ethanol extract, and fresh and dry aqueous extracts promoted the growth of *C. krusei*. *L. francisci* was shown to have fungicidal activity and boosted the growth of immune cells, thus, validating its use in ethnomedicine. Sachikonye and Mukanganyama (2016) investigated the antifungal activities of selected flavonoids against *C. albicans* and *C. krusei in vitro*. Epicatechin had the highest antifungal activity with MIC values of 88 and 213 µg/mL for *C. krusei* and *C. albicans*, respectively, whilst ferulic acid was shown to have inhibitory effects on drug efflux. It was concluded that epicatechin and ferulic acid might serve as templates for the development of novel antifungal agents.

10.4.3 ANTIMYCOBACTERIAL EVALUATION

Tuberculosis (TB) was thought to have been eradicated in the early second half of the past century (Chimponda and Mukanganyama, 2012) but remains an important public health problem, accounting for 8 million new cases per year. Despite improvements in chemotherapy, the treatment of TB is severely affected by the development of multi-drug resistance in Mycobacterium tuberculosis (Mtb) strains (Higuchi et al., 2008). Thirty ethanol extracts from 19 selected plants from Zimbabwe were screened against surrogate mycobacterial species for Mtb: *Mycobacterium aurum* and *Corynebacterium glutamicum* (Table 10.5) (Chimponda and Mukanganyama, 2012).

V. adoensis and *M. indica* extracts had the highest growth inhibitory activity against *M. aurum* and *C. glutamicum*, respectively (Table 10.6). The extract from *P. curatellifolia* showed drug efflux inhibitory effects on *M. aurum* and *C. glutamicum*.

Five *Combretum* plant species—*Combretum imberbe*, *C. zeyheri*, *Combretum hereroense*, *Combretum elaeagnoides*, and *Combretum platypetalum* were tested against a virulent *Mycobacterium smegmatis* and *Mycobacterium aurum* (Magwenzi et al., 2014). Only the ethanolic extract from *Combretum imberbe* was active on *M. smegmatis*, and it had an MIC of 125 µg/mL in the broth microdilution assay. However, using Sabouraud dextrose broth, *C. platypetalum* was found to have antimycobacterial effects that were not detected when using the agar-disc diffusion assay. The antimycobacterial effects of the leaf extracts from *Syzygium guineense* on *M. aurum* and *M. smegmatis* were also shown by Deeds et al. (2016). The methanol extract of *Syzygium guineense* was found to be the most potent extract that inhibited mycobacterial growth. The antimycobacterial activity of *V. adoensis* and its possible mode of action against *M. smegmatis* was also investigated (Mautsa and Mukanganyama, 2017). The ethyl acetate extract from the leaves was found to be the most effective against *M. smegmatis* with an MIC and MBC of 63 and 125 µg/mL, respectively. Significant nucleic acid and protein leakage in *M. smegmatis* was observed after exposure to the leaf extract. Alkaloid extracts from leaves of *C. zeyheri*, *C. platypetalum*, *C. molle*, and *C. apiculatum* were also assessed for antimycobacterial activity (Nyambuya et al., 2017). It was found that the extracts from *C. zeyheri* had significant antimycobacterial effects and thus could serves as a source of antimycobacterial compounds. Root extracts from *T. welwitschii* were also found to have antimycobacterial activity against selected Mycobacteria species (Marime et al., 2014).

10.4.4 ANTICANCER EVALUATION

WHO estimates that cancer is the leading cause of death from non-communicable diseases (WHO, 2020). A selected group of 14 medicinal plants was screened for antiproliferative activity against two human leukaemia cell lines—Jurkat T and Wil 2 *in vitro* (Table 10.7) (Mukanganyama et al., 2012).

TABLE 10.5
Ethnobotanical Information of Investigated Plants for Antimycobacterial from Zimbabwe

Family	Botanical Name	Local Name	Voucher	Plant Part Tested	Major Traditional Use (Reference)
Malvaceae	*Abelmoschus esculentus* Moench	*Derere*	N15E10	Fruit	Okra is a traditional food plant. Moench It is used for the treatment of bronchitis, heart diseases, and tuberculosis of the lungs (www.greenpatio, 2010).
Aloaceae	*Aloe vera barbadensis* Miller	*Gavakava*	N16E4	Leaves	Widely used for external Miller treatment of minor wounds and inflammatory skin disorders. It is used for the treatment of pneumonia, tuberculosis, and cough (Gelfand et al.,1985).
Rubiaceae	*Catunaregam spinosa* Thunb	*Mutsvairachuru Murovaduri*	C5E7 C5E4 C5E1	Leaves Bark Roots	Leaves are used for pulmonary infections, bark is an astringent. C5E1 Roots (Serada et al., 2002). Various respiratory ailments (Gelfand et al., 1985).
Leguminosae	*Rhynchosia insignis* (O.Hoffm.) R.E.Fr.	*Mukoyo*	C15E7	Leaves	Treatment of diarrhoea (Chinemana et al., 1985).
Euphobiaceae	*Croton gratissimus* Burch.	*Gunukira Mufandemenge Mugugu Mubvukuta*	UZ13E7	Leaves	Roots and bark infusions treat respiratory disorders (Roodt, 1998).
Proteaceae	*Faurea saligna* Harv	*Mutsatstsi*	C13E7	Leaves	Treatment of bilharzias helminthiasis (Baerts & Lehmann, 1989).
Clusiaceae	*Garcinia huillensis* Welw	*Mutunduru*	C10E7	Leaves	Treatment of cough, pneumonia. and tuberculosis (Gelfand et al., 1985).
Celastraceae	*Gymnosporia senegalensis* Loes	*Chishuzhu Musosawafa*	N3E7	Leaves	Treatment of coughs, pneumonia, and TB (Gelfand et al., 1985).
Verbenaceae	*Lippia javanica*	*Zumbani*	C3E7 C3E4 C3E7	Leaves Bark Roots	Leave infusion used to treat coughs, colds and fever, influenza, measles, malaria, and stomach ache (Van Wyk, 2009).
Anarcardiaceae	*Mangifera indica* (L)	*Mumango*	N17E4	Stems Twigs	Treatment of tuberculosis (Kisangau et al., 2007).
Olacaceae	*Olax obtusifolia* De Wild	*Gungwe Kahungwarara*	UZ8E7 UZ8E4	Leaves Stems	Treatment of abdominal pain (Chinemana et al., 1985).
Chrysobalanaceae	*Parinari curatellifolia* Planch, ex Benth	*Muhacha*	C6E7 C6E4	Leaves Bark	Skin rashes, tuberculosis, chronic diarrhoea, herpes zoster, herpes simplex (Chigora et al., 2007).

(Continued)

TABLE 10.5 (*Continued*)

Ethnobotanical Information of Investigated Plants for Antimycobacterial from Zimbabwe

Family	Botanical Name	Local Name	Voucher	Plant Part Tested	Major Traditional Use (Reference)
Anarcadiaceae	*Rhus longipes* Engl	*Muchokochiana* *Mudzambuya*	C11E10 C11E7	Flowers Leaves	Treatment of headache (Maliwichi, 2000).
Solanaceae	*Solanum mauritianum* Scop	–	UZ10E7	Leaves	Treatment of menorrhagia (Lewu & Afolayan, 2009).
Myrtaceae	*Syzygium guineense* Guill & Perr	*Muboo*	C8E7	Leaves	Chest pain (De Boer et al., 2004).
Combretaceae	*Terminalia sericea* Burch. ex DC.	*Mukonono*	C14E7	Leaves	Roots treat diarrhoea and stomach (Van Wyk, 2009); a hot infusion of the root bark treats pneumonia (Roodt, 1998).
Asteraceae	*V. adoensis* Sch.Bip.ex Walp	*Musikavakadzi*	C1E10 C1E7 C1E1	Flowers Leaves Roots	The leaves are used for treatment of TB (Kisangau et al., 2007).
Leguminosae	*Xerroderris stuhlmannii* (Taub) Mendonca E.P Sousa.	*Murumanyama*	C4E7 C4E4	Leaves Bark	Mastitis and backache (Ruffo, 1991).
Rhamnaceae	*Ziziphus mucronata* Willd.	*Muchecheni*	C7E7 C7E4	Leaves Stems	Paste of leaves treats boils, carbuncles, and swollen glands; leaf infusion is taken against chest complaints (Roodt, 1998).

Source: Chimponda and Mukanganyama (2012).

The five most potent medicinal plants showed the following order of potency against Wil 2 cell line: *P. curatellifolia > Aloe barbadensis > Croton gratissimus > Syzigium guineense > V. adoensis* with IG_{50}s of 93, 115, 148, 149.8, and 130 µg/mL, respectively.

T. welwitschii extracts were investigated for anticancer activity against Jurkat T cells (Moyo and Mukanganyama, 2015c). The extracts decreased cell viability in a dose-dependent and time-dependent manner and induced apoptosis in the Jurkat T cells. The antiproliferative activity of *Maerua edulis* against human leukaemic Jurkat-T cell line was also investigated by Sithole and Mukanganyama, (2017). Anticancer activity of the aqueous, acetone, hexane, and methanol extracts of *M. edulis* root were evaluated against Jurkat-T cells, using 1, 3-bis (2-chloroethyl)-1-nitrosourea (BCNU) as a reference drug. Only the methanol extract significantly inhibited the growth of Jurkat-T cells, and the effects were found to be irreversible. The leaf extracts of *Dolichos kilimandscharicus* were tested for their antiproliferative efficacy and cytotoxicity effects (Sithole et al., 2020). The methanol extract was found to have the most effect against Jurkat-T cell with an IC_{50} of 33.56µg/mL. UPLC-MS analysis of the leaf extracts led to the identification of 23 compounds from the ethanol extract, and these were suggested to be responsible for the observed effects. Rutin, quercetin, luteolin, apigenin, hispidulin, kaempferol derivatives, as well as caffeoylquinic acid were some of the compounds that were identified in the extracts (Figure 10.8).

TABLE 10.6

Minimum Inhibitory Concentrations and Minimum Bactericidal Concentrations of Extracts Towards the Mycobacterial Species

Plant Extract	[a]Zone of Inhibition (mm) at 500 (µg/disc) [b]MIC (µg/disc) [c]MBC (µg/disc) M. aurum	Plant Extract	Zone of Inhibition (mm) at 500 (µg/disc) MIC (µg/disc) MBC (µg/disc) C. glutamicum
V. adoensis leaves	[a]28±1 [b]31 [c]250	M. indica (leaves)	17±1 250 >500
Faurea saligna leaves	18±1 250 > 500	V. adoensis Leaves roots	15±1 250 >500
Syzygium guineense leaves	19±2 –	Parinari curatellifolia stems	13±1 125 >500
Xerroderris stuhlmannii leaves	18±1 31 125	Ziziphus mucronata stems	13±1 125 500
Parinari curatellifolia stems	18±1 8 63	Lippia javanica roots	12±1 32 >500
Parinari curatellifolia leaves	18±1 63 125		
Rifampicin	29±2 2 16	Rifampicin	40±1 <1 63

Source: Chimponda and Mukanganyama (2012).

[a] Results are the average (± SD) of two separate antibacterial susceptibility test (Each antibacterial susceptibility test was followed by a disk diffusion assay done in quadruplicate). The zone of inhibition being determined at a concentration of 500 µg/disk.

[b] MIC – Minimum inhibition concentration

[c] MBC – Minimum bactericidal concentration

C. zeyheri and *C. platypetalum* have been shown to have anticancer, antibacterial, antituberculosis, and antifungal effects in both *in vivo* and *in vitro* studies (Chiramba and Mukanganyama, 2016). The antiproliferative effects of compounds isolated from *C. zeyheri* and *C. platypetalum* on Jurkat T and HL-60 cancer cell lines in combination with doxorubicin and/or chlorambucil were carried out by Wende et al. (2021). It was shown that the compounds CP 404, CP 409 from *C. platypetalum* (Figure 10.9) inhibited the growth of Jurkat T cells *in vitro*. The combination of the compounds with anticancer drugs enhanced their anticancer effects.

10.4.5 ANTIBIOFILM EVALUATION

Biofilms are huge communities of microbes that are attached to a surface and play an important role in persistence of bacterial infections (Muhammad et al., 2020). Biofilms have also been implicated in hospital-acquired infections (Haque et al., 2018). Biofilm formation process includes five stages

TABLE 10.7

Plants That Were Used in Anticancer Study, Their Ethnobotanical Uses in Zimbabwe and Other Countries

Family	Plant Name and Authority	Vernacular Name	Voucher Number	Traditional Medicinal Uses of Plants
Fabaceae	*Cajanus cajan* (Druce)		N8E7	Stomach ailments (Iwawela et al., 2007).
Myrtaceae	*C. citrinus* (Curtis Skeels)		UZ2E7	Haemorrhoid treatment (Oyedeji et al., 2009).
Combretaceae	*Terminalia pruniodes* (Lawson)	*Mudziyashe*	N6E7	Diarrhoea (Ruffo et al., 1991).
Asphodelaceae	*Aloe barbadensis* (Mill.)	*Gavakava*	N11E7	Sap is used to treat skin rushes and the leaves are prepared
Combretaceae	*Combretum apiculatum* (L.)	*Muruka*	C9E7	Coughs, diarrhoea, snake bites stomach ache (Ruffo et al., 1991)
Araliaceae	*Cussonia natalensis* (Sond)	*Mutobvi*	UZ9E7	Diarrhoea (Ruffo et al., 1991).
Euphorbiaceae	*Croton gratissimus* (Burch)	*Gunukira*	UZ13E7	Malaria, rabies, gonorrhoea, wounds, Ascariasis, internal worms (Iwawela et al., 2007).
Euphorbiaceae	*Euphorbia tiraculli* (L.)		N10E7	Removal of benign moles using the latex (Iwawela et al., 2007). Rabies treatment.
Chrysobalanaceae	*Parinari curatellifolia* (Planch ex Benth)	*Muhacha*	C6E7	Facilitates conception in women (Chigora et al., 2007).
Myrtaceae	*Syzigium guineense* (Will D.C)	*Mukute*	C12E7	Tuberculosis, fevers (Chigora et al., 2007).
Anacardiaceae	*Rhus lancea* (Barkely)	*Muchokochiana*	C11E7	Stomach ailments, fevers (Chokunonga et al., 2004).
Fabaceae	*Xeroderris stuhlmanni* (Mend)	*Murumanyama*	C4E7	Stomach ailments (Iwawela et al., 2007).
Asteraceae	*V. adoensis* (Bip ex Walp)	*Musikavakadzi*	C1E7	Induce birth or carry out abortions (Iwawela et al., 2007).

Source: Mukanganyama et al. (2012).

et al., 2017). The effects of extracts from the leaves of *C. zeyheri* were investigated in *C. albicans* and *C. krusei*. The methanol extract, hexane extract, DCM extract, and DCM–methanol extract showed potent inhibition of biofilm formation in *C. albicans*, whilst in *C. krusei* only the water extract and ethanol extract showed inhibition of biofilm formation (Mtisi et al., 2018). The effect of chondrillasterol on biofilms formed by *P. aeruginosa* was investigated by Mozirandi et al., (2019), and it inhibited biofilm formation. Chondrillasterol is thus a useful template for the development of new antimicrobial agents with both antibacterial and antibiofilm activity. The antibiofilm effect of the leaf extract of *T. welwitschii* was investigated in *P. aeruginosa* (Mombeshora et al., 2021). It was found that the leaf extract and a compound B1 had antibiofilm activity. Chipenzi et al., (2020) investigated the effects of tormentic acid and leaf extracts isolated from *Callistemon viminalis*. Production of extracellular polymeric DNA and polysaccharides from biofilms were also determined. Tormentic acid and the extracts caused a significant decrease in the biofilm extracellular polysaccharide content of *S. pyogenes*, caused detachment of biofilms, and decreased the release

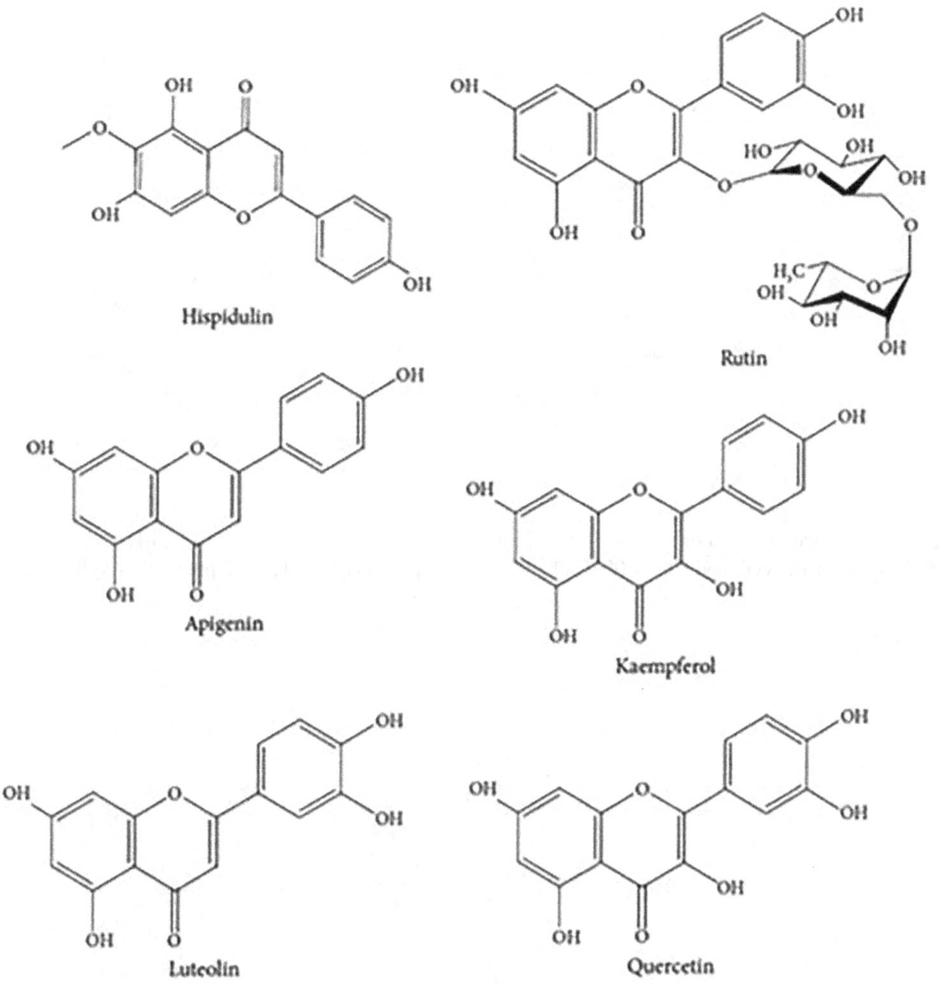

FIGURE 10.8 Chemical structures of compounds identified from UPLC–MS analysis of crude leaf ethanol extract from *D. kilimandscharicus*. The compounds are probably responsible for the aforementioned therapeutic benefits (Sithole et al., 2020).

that include adhesion, attachment, micro-colony formation, maturation, and dispersal as shown in Figure 10.10.

Bacterial biofilms undergo five stages that can be described as a lifecycle. The first stage is initial attachment; here planktonic bacterial cells initiate to colonize a surface by secreting extracellular polymeric substances (EPS), shown in green. As bacterial population increases, cells secrete more EPS, which irreversibly adheres them to each other and to the surface. The colony matures and then disperses. Source: (Stoodey et al., 2002).

Agents that inhibit biofilm formation in *M. tuberculosis* have the potential to reduce the disease treatment period and improve the quality of tuberculosis chemotherapy (Bhunu et al., 2017). *P. curatellifolia* leaf extracts have been used to treat symptoms like tuberculosis in ethnomedicinal practices (Mawire et al., 2021). The effect of the leaf extracts of *P. curatellifolia* on *M. smegmatis* growth and biofilm formation was investigated (Bhunu et al., 2017). The ethanol extract, dichloromethane extract, and water extract effectively inhibited biofilm formation in *M. smegmatis* (Bhunu

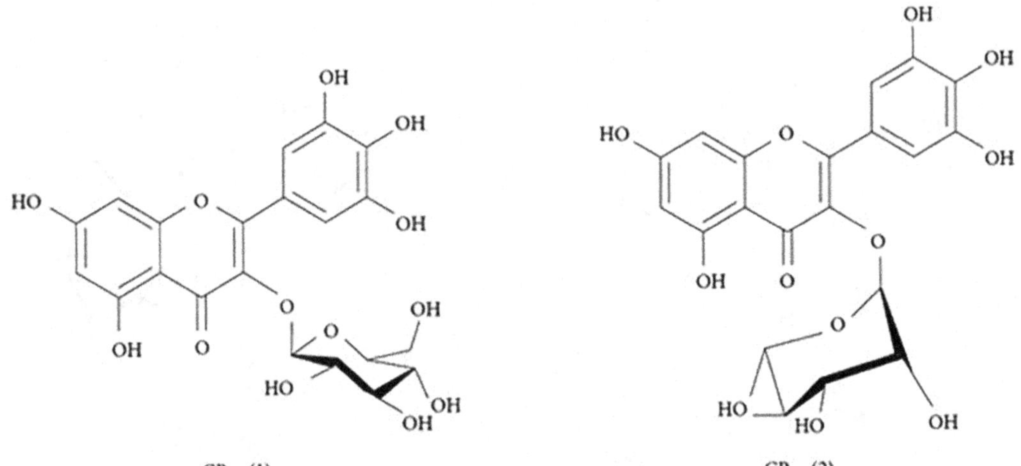

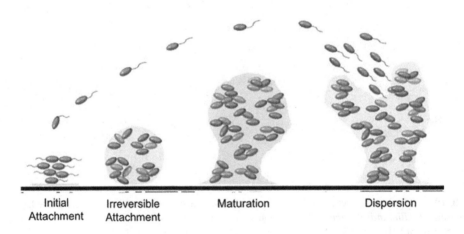

FIGURE 10.9   Structures of compounds isolated from *C. platypetalum*. CP 404 (1): 3-O-(β-D-glucopyranosyl)-3′,4′,5′,5′,7-pentahydroxyflavone; CP 409: 3-O-(β-L-rhamnopyranosyl)-3′,4′,5′,5,7-pentahydroxyflavone.

| Initial Attachment | Irreversible Attachment | Maturation | Dispersion |

FIGURE 10.10   The biofilm lifecycle. Bacterial biofilms undergo five stages that can be described as a life-cycle. The first stage is initial attachment, 1. Here planktonic bacterial cells initiate to colonize a surface by secreting extracellular polymeric substances (EPS). In 2 and 3 there is micro colony formation. In 4, there is biofilm dispersal and in 5, we have bacterial attachment on a new surface. As bacterial population increases, cells secrete more EPS which irreversibly adheres them to each other and to the surface. The colony matures and then disperses. (Self, 2022. BIA laboratory, Department of Biotechnology and Biochemistry, University of Zimbabwe.)

of extracellular DNA and capsular polysaccharides from biofilms of *P. aeruginosa* and *S. aureus*. Thus, the plant extracts and/isolated compounds have potential for antibiofilm activities in various microbes.

10.4.6   ANTIOXIDANT EVALUATIONS, ANTI-INFLAMMATORY, AND ANTIDIABETIC EVALUATIONS

Inflammation is mediated by activated inflammatory cells to protect the body against injury in the form of a wound or an infection (Chirisa and Mukanganyama, 2016). Oxidative stress resulting from accumulation of reactive oxygen species has been associated with disease (Boora et al., 2014). The search for natural antioxidants of plant origin is necessitated by the side effects associated with synthetic antioxidants currently available. In developing countries, problems are associated with the means of managing diabetes using antidiabetic drugs due to availability and affordability. A variety of plants are being studied for their antidiabetic potential (Makopa et al., 2020). The antioxidant activity of *C. zeyheri*, *C. platypetalum*, and *P. curatellifolia* extracts were investigated by determining nitrite radical scavenging ability (Boora et al., 2014). The aqueous and ethanolic leaf extracts of *C. zeyheri*, *C. platypetalum*, and *P. curatellifolia* extracts exhibited nitrite radical scavenging activity and were deemed potential sources of natural antioxidants. The anti-inflammatory and antioxidant activities of Zimbabwean medicinal plant extracts were investigated by determining the effects on COX-1 and COX-2 activity, the erythrocyte membrane stabilization, and albumin denaturation inhibition assays (Chirisa and Mukanganyama, 2016). *C. zeyheri*, *C. molle*, and *P. curatellifolia* inhibited COX-2, and extracts from *P. curatellifolia* showed antioxidant activity using the DPPH and the TMAMQ free radical scavengers. The antidiabetic potential of the leaf extracts from *Persea americana* was determined against mammalian α-glucosidase *in vitro* (Makopa et al., 2020). The methanolic leaf extract showed potent inhibitory activity on α-glucosidase enzyme in a time-dependent and dose-dependent manner. Increased expression of haematopoietic prostaglandin D_2 synthase (H-PGDS) is responsible for allergic reactions promoting the inflammatory processes (Lee et al., 2020). Extracts from *C. molle* (Moyo et al., 2014) and *P. curatellifolia* (Chimponda and Mukanganyama, 2015) were shown to inhibit H-PGDS. These extracts may be potential sources of lead compounds for development of anti-inflammatory and anti-allergic compounds.

10.4.7 TOXICOLOGICAL EVALUATIONS

Medicinal plants are perceived to be non-toxic by the public (Mapfunde et al., 2016). Toxicity studies, however, have indicated that they can cause numerous side effects; therefore, evaluation of safety is required. The effects of constituents of *C. zeyheri* on mammalian cells were investigated, and it was found that the alkaloids, saponins, and ethanol extracts were non-toxic towards mouse peritoneal cells and Jurkat T cells (Mapfunde et al., 2016). Concentrations of less than 5 mg/mL were not haemolytic to sheep erythrocytes. *P. curatellifolia*, among its other important pharmacological activities, has been shown to have significant antiproliferative activity on cancer cell lines (Mukanganyama et al., 2012). Toxicity studies were carried out to determine the safety profile of *P. curatellifolia* (Kundishora et al., 2020). It was found that *P. curatellifolia* leaf extract was not toxic to both erythrocytes and immune cells, but the water extract was found to have immunostimulatory effects (Figures 10.11 and 10.12 respectively).

Similarly, extracts from *T. welwitschii* at 100 µg/mL and *V. adoensis* (5 mg/mL) were found not to have haemolytic effects on sheep erythrocytes (Mozirandi et al., 2017; Mombeshora et al., 2019).

10.5 DEVELOPMENT OF HERBAL BIOPRODUCTS

There are many kinds of herbal formulations such as topical formulations and systemic formulations (Musthaba et al., 2010). They are believed to have less side effects than conventional formulations. The development of bioproducts from plant natural products would be an opportunity to create some employment for people to establish agroforests so that they can reap the benefits of planting and conserving the medicinal plants. Topical formulations are used to treat diseases relating to the skin, and they are used externally only (Hagen and Baker, 2017). Skin ailments are seen in many people; therefore, there is a need to develop herbal topical formulations that are believed to contain less side effects and are more affordable. Topical formulations moisturise the skin, keeping it hydrated, protects the skin from direct sunlight and other harmful environmental conditions, and provide antimicrobials to fight against infections (Verma et al., 2016).

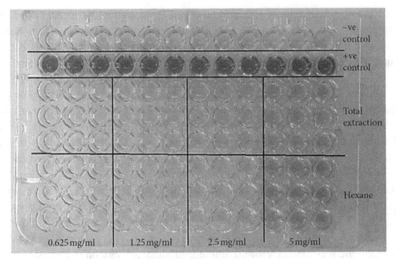

FIGURE 10.11 A 96-well plate for haemolysis of sheep erythrocytes exposed to *P. curatellifolia* crude (total) and hexane extract. Negative control (–ve control) was a measure of spontaneous haemolysis and contained centrifuged erythrocyte suspension in PBS from which an aliquot of the supernatant was withdrawn and mixed with Drabkin's reagent. Positive control (+ve control) contained uncentrifuged erythrocytes in PBS from which an aliquot was withdrawn and mixed with Drabkin's reagent to give 100% haemolysis (Kundishora et al., 2020).

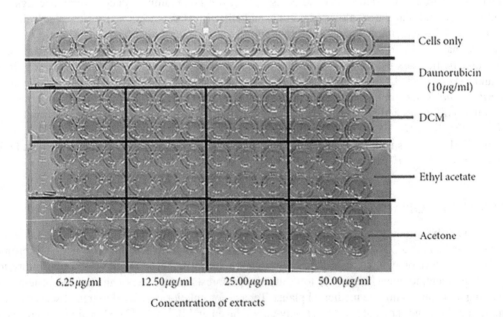

FIGURE 10.12 Image showing a 96-well plate for the MTT assay using mouse peritoneal cells, exposed to different solvent extracts from *P. curatellifolia* leaves (black dots in the wells indicate purple-coloured peritoneal cells that have sedimented at the bottom of the well). Cells only in row were the negative control, while the positive control was daunorubicin 10 μg/mL, a standard anticancer drug (Kundishora et al., 2020).

FIGURE 10.13 Antibacterial cream developed from *C. molle* extract. Post formulation test results showed that the topical cream was still active against *S. aureus*. (Self, 2021. BIA laboratory, Department of Biotechnology and Biochemistry, University of Zimbabwe.)

10.5.1 ANTIBACTERIAL HERBAL BIOPRODUCTS

Since the plant extracts from Zimbabwe were shown to have antibacterial activities, topical creams were prepared. The antibacterial cream was prepared based on the Hydrophylic–Lipophlic balance (HLB) method, as published by Chifamba et al. (2013). The method helps with the calculation of the amount of each emulsifying agent that should be added to the formulation so that the water and oil phases do not separate. When preparing the antibacterial cream, the extract was added to the water phase ingredients, since it was soluble in water. Four formulations were prepared that had 1%, 2%, 5%, and 10% of the extract by weight, respectively. Examples of an antibacterial cream developed from *C. molle* extracts are shown in Figure 10.13.

10.5.2 ANTIFUNGAL BIOPRODUCTS

Invasive candidiasis is ranked as a fatal infection that mainly affects critically ill patients as well as immunocompromised patients (Friedman and Schwartz, 2019). Topical antifungal agents exhibit various modes of action and a few adverse effects. This makes the topical route the most suitable for treatment of skin infections (Kaur et al., 2021). Adequate knowledge of the skin barrier structure and drug permeation properties are vital for rational progress in the development of topical formulations (Simões et al., 2018). After determining the potency of the antifungal activities of several plant extracts, the extracts were used as the active ingredients in the topical antifungal cream formulation (Moin et al., 2020). An antifungal cream made from *L. francisci* was developed (Figure 10.14).

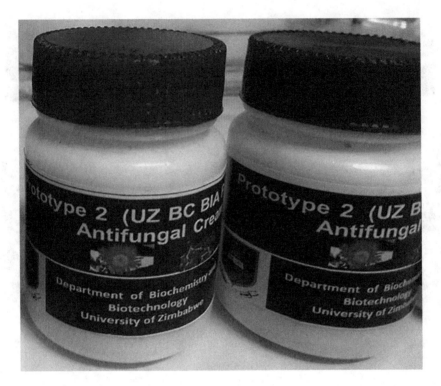

FIGURE 10.14   Antifungal cream developed from *Lampranthus francisci* extracts.

10.6   CONCLUSION AND PROSPECTS

The transition of a natural compound from a screening hit through a drug lead to a marketed drug is associated with increasingly challenging demands for compound amount, which often cannot be met by re-isolation from the respective plant sources (Atanasov et al., 2015). Of the reckoned 250,000–500,000 existing plant species, only a tiny proportion has been scientifically researched for bioactivities. To tap into the vast capacities of antimicrobial and anticancer properties of herbal plants, our research group has focussed primarily on the use of extracts to produce herbal products. The sustainable harvesting of plant parts for the development of phytomedicines (Mangoyi et al., 2014) will go a long way in conserving the sources of these plants from the rural areas of Zimbabwe, where the plants are mainly found. The data provided in these reviewed studies have shown that it is possible to scientifically evaluate the efficacy of natural plant products, validate their uses in ethnomedicines, as well as provide commercial outputs that can be used for the wellness of local populations. Further work is now in progress regarding the commercialisation of these indigenous heritable products from Zimbabwe through the established Innovation and Research Centres at tertiary institutions in Zimbabwe.

ACKNOWLEDGEMENTS

This work was supported by the Swedish International Development Agency (SIDA) through the International Science Programmes (ISP IPICS:ZIM01, Uppsala University, Uppsala, Sweden), Centre for Emerging and Neglected Diseases (CEND), University of California, Berkeley, and the University of Zimbabwe. The International Foundation in Sciences (IFS F/3413-01-03F, Stockholm, Sweden) supported the study under the title: *Screening Natural Plant Products from Selected*

Plants from Zimbabwe As a Source of Anti-infective Compounds for Phytomedicines Development. Funding bodies played no role in the design of the study; collection, analysis, and interpretation of data; and in writing the manuscript. The author also acknowledges the contribution of various students from 2010 to 2022 from the Biomolecular Interactions Analyses Laboratory, Department of Biotechnology and Biochemistry, University of Zimbabwe. The author acknowledges the assistance of Mr. Christopher Chapano—a taxonomist, with the National Herbarium and Botanical Gardens, Harare, Zimbabwe, in the authentication of the plant sample identity. The author would like to thank Mr. Noel Mukanganyama—a herbalist from Centenary, Zimbabwe, who showed them the *C. welwitschii* plant and provided information on ethnomedicinal uses.

REFERENCES

Ambaye, A., Kohner, P.C., Wollan, K.L., Roberts, L.D., and Cockerill, F.R. (1997). Comparison of agar dilution, broth microdilution, disk diffusion, E-Test, and BACTEC radiometric methods for antimicrobial susceptibility testing of clinical isolates of the nocardia asteroides complex. *Journal of Clinical Microbiology, 35*(4), 847–852.

Atanasov, A., Waltenberge, G.B., Pferschy-Wenzig, E.-V., Linderd, T., Wawroscha, C., Uhrine, P., et al. (2015). Discovery and resupply of pharmacologically active plant derived natural products: a review. *Biotechnology Advances, 33*, 1582–1614. http://doi.org/10.1016/j.biotechadv.2015.08.001.

Bhunu, B., Mautsa, R., and Mukanganyama, S. (2017). Inhibition of biofilm formation in *Mycobacterium smegmatis* by *Parinari curatellifolia* leaf extracts. *BMC Complementary and Alternative Medicine, 17*(1), 285. http://doi.org/10.1186/s12906-017-1801-5.

Bonev, B., Hooper, J., and Parisot, J. (2008). Principles of assessing bacterial susceptibility to antibiotics using the agar diffusion method. *Journal of Antimicrobial Chemotherapy, 61*, 1295–1301. http://doi.org/10.1093/jac/dkn090.

Boora, F., Chirisa, F., and Mukanganyama, S. (2014). Evaluation of nitrite radical scavenging properties of selected Zimbabwean plant extracts and implications for anti-oxidative uses in food preservation. *Journal of Food Processing, 2014, 1–7*, Article ID 918018. http://doi.org/10.1155/2014/918018.

Braissant, O., Astasov-Frauenhoffer, M., Waltimoand, T., and Bonkat, G. (2020). A review of methods to determine viability, vitality, and metabolic rates in microbiology. *Frontiers in Microbiology, 11*, 547458. https://doi.org/10.3389/fmicb.2020.

Chifamba, J., Manyarara, T., and Derera, P. (2013). Investigation of the suspending properties of Dicerocaryum zanguebarium and Adansonia digitata mucilage as structured vehicles. International Journal of Science and Research, 4(4), 1696–1700.

Chimponda, T. and Mukanganyama, S. (2010). Antimycobacterial activities of selected medicinal plants from Zimbabwe against *Mycobacterium aurum and Corynebacterium glutamicum. Tropical Biomedicine, 27*(3), 595–610.

Chimponda, T. and Mukanganyama, S. (2015). Evaluation of selected Zimbabwean plant extracts as inhibitors of hematopoietic prostaglandin D2 synthase. *Journal of Herbs, Spices and Medicinal Plants, 21*, 243–258. http://doi.org/10.1080/10496475.2014.954073.

Chipenzi, T., Baloyi, G., Mudondo, T., Sithole, S., Chi, G.F., and Mukanganyama, S. (2020). An evaluation of the antibacterial properties of tormentic acid congener and extracts from *Çallistemon viminalis* on selected ESKAPE pathogens and effects on biofilm formation. *Advances in Pharmacological and Pharmaceutical Sciences, 2020*, 14, Article ID 8848606. https://doi.org/10.1155/2020/8848606.

Chiramba, O and Mukanganyama, S. (2016). Cytotoxic effects of combretum platypetalum welw. ex M.A. lawson subsp. oatesii (rolfe) exell (combretaceae) leaf extracts on jurkat T-cells and reversal of effects by reduced glutathione. *Journal of Biologically Active Products from Nature, 6*, 250–265. http://doi.org/10.1080/22311866.2016.1232626.

Chirisa, E., and Mukanganyama, S. (2016). Evaluation of *in vitro* anti-inflammatory and antioxidant activity of selected Zimbabwean plant extracts. *Journal of Herbs, Spices & Medicinal Plants, 22*(2), 157–172. http://doi.org/10.1080/10496475.2015.1134745.

Chitemerere, T. and Mukanganyama, S. (2011). *In-vitro* antibacterial activity of selected medicinal plants from Zimbabwe. *The African Journal of Plant Science and Biotechnology, 5*, 1–7.

Chitemerere, T.A. and Mukanganyama, S. (2014). Evaluation of cell membrane integrity as a potential antimicrobial target for plant products. *BMC Complementary and Alternative Medicine, 14*, 278. http://doi.org/10.1186/1472-6882-14-278.

Friedman, D.Z.P. and Schwartz I.S. (2019). Emerging fungal infections: new patients, new patterns, and new pathogens. *Journal of Fungi*, *5*, 67. http://doi.org/10.3390/jof5030067.

Hagen, M. and Baker, M. (2017). Skin penetration and tissue permeation after topical administration of diclofenac. *Current Medical Research and Opinion*, *33*(9), 1623–1634, http://doi.org/10.1080/03007995.2017.1352497.

Haque, M., Sartelli, M., McKimm, J., and Bakar M.A. (2018). Health care-associated infections – an overview. *Infection and Drug Resistance*, *11*, 2321–2333. Published online 15 Nov 2018 http://doi.org/10.2147/IDR.S177247.eCollection 2018. https://www.who.int/news-room/fact-sheets/detail/cancer. (accessed 29 July 2021).

Kaur, N., Bains, A., Kaushik, R., Dhull, S.B., Melinda, F., and Chawla, P. (2021). A review on antifungal efficiency of plant extracts entrenched polysaccharide-based nanohydrogels. *Nutrients*, *13*, 2055. https://doi.org/10.3390/ nu13062055.

Kundishora, A., Sithole, S., and Mukanganyama, S. (2020). Determination of the cytotoxic effect of different leaf extracts from *Parinari curatellifolia* (Chrysobalanaceae). *Journal of Toxicology*, *2020*, 11, Article ID 8831545. https://doi.org/10.1155/2020/8831545.

Lee, K., Lee, S.H., and Kim, T.H., (2020). The biology of prostaglandins and their role as a target for allergic airway disease therapy. *International Journal of Molecular Medicine*, *21*, 1851. http://doi.org/10.3390/ ijms21051851.

Mabhiza, D., Chitemerere, T., and Mukanganyama, S. (2016). Antibacterial properties of alkaloid extracts from *Callistemon citrinus* and *Vernonia adoensis* against *Staphylococcus aureus* and *Pseudomonas aeruginosa*. *International Journal Medicinal Chemistry*, *2016*, 7, Article ID 6304163. http://doi.org/10.1155/2016/6304163.

Magwenzi, R., Nyakunu, C., and Mukanganyama, S. (2014). The effect of selected combretum species from Zimbabwe on the growth and drug efflux systems of *Mycobacterium aurum* and *Mycobacterium smegmatis*. *Journal of Microbial and Biochemical Technology*, *2014*, S3–003. http://doi.org/10.4172/1948-5948. S3-003.

Makopa, M., Mangiza, B., Banda, B., Mozirandi, W., Mombeshora, M., and Mukanganyama, S. (2020). Antibacterial, antifungal, and antidiabetic effects of leaf extracts from *Persea americana* Mill. (Lauraceae). *Biochemistry Research International*, 2020, 10, Article ID 8884300. https://doi.org/10.1155/2020/8884300.

Mangoyi, R. and Mukanganyama, S. (2011). *In vitro* antifungal activities of selected medicinal plants from Zimbabwe against *Candida albicans* and *Candida krusei*. *The African Journal of Plant Science and Biotechnology*, *5*, 8–14.

Mangoyi, R., Chitemerere, T., Chimponda, T., Chirisa, E., and Mukanganyama, S. (2014). Multiple anti-infective properties of selected plant species from Zimbabwe, Chapter 14. In: Novel Plant Bioresources: Applications in Food, Medicine and Cosmetics, 1st edition, Brijesh, K.T., Tomas, N., and Nicolas, M.H., Gurib-Fakim, A. (eds.), John Wiley & Sons, Ltd, Hobokon, NJ, pp. 179–190.

Mangoyi, R., Mafukidze, W., Marobela, K., and Mukanganyama, S. (2012). Antifungal activities and preliminary phytochemical investigation of combretum species from Zimbabwe. *Journal of Microbial and Biochemical Technology*, *4*, 37–44.

Mangoyi, R., Midiwo, J., and Mukanganyama, S. (2015). Isolation and characterization of an antifungal compound 5-hydroxy-7, 4'-dimethoxyflavone from *Combretum zeyheri*. *BMC Complementary and Alternative Medicine*, *15*(1), 405. http://doi.org/10.1186/s12906-015-0934-7.

Mapfunde, S., Sithole, S., and Mukanganyama, S. (2016). *In vitro* toxicity determination of antifungal constituents from *Combretum zeyheri*. *BMC Complementary and Alternative Medicine*, *16*, 162. http://doi.org/10.1186/s12906-016-1150-9.

Marime, L., Chimponda, T., Chirisa, E., and Mukanganyama, S. (2014). Antimycobacterial effects of *Triumfetta oletschia* extracts on *Mycobacterium aurum* and *Mycobacterium smegmatis*. *Journal of Antimicrobials Photon*, *129*, 319–332.

Mautsa, R. and Mukanganyama, S. (2017). *Vernonia adoensis* leaf extracts cause cellular membrane disruption and nucleic acid leakage in *Mycobacterium smegmatis*. *Journal of Biologically Active Products from Nature*, *7*, 2, 140–156. http://doi.org/10.1080/22311866.2017.1324321.

Mawire, P., Mozirandi, W., Heydenreich, M., Chi, G.F., and Mukanganyama S. (2021). Isolation and antimicrobial activities of phytochemicals from *Parinari curatellifolia* (Chrysobalanaceae). *Advances in Pharmacological and Pharmaceutical Sciences*, *2021*, 18, Article ID 8842629.. https://doi.org/10.1155/2021/8842629.

Mensah, M.L.K., Komlaga, G., Forkuo, A.D., Firempong, A.K., Anning, C., and Dickson, A.R. (2019). Toxicity and Safety Implications of Herbal Medicines Used in Africa. *Herbal Medicine*, 63. http://doi.org/10.5772/intechopen.72437.

Moin, S., Farooqi, J., Jabeen, K., Laiq, S., and Zafar, A. (2020). Screening for triazole resistance in clinically significant Aspergillus species; report from Pakistan. *Antimicrobial Resistance and Infection Control, 9* ,62.https://doi.org/10.1186/s13756-020-00731-8.

Mombeshora, M. and Mukanganyama, S. (2019). Antibacterial activities, proposed mode of action and cytotoxicity of leaf extracts from *Triumfetta oletschia* against *Pseudomonas aeruginosa. BMC Complementary and Alternative Medicine, 19,* 315.

Mombeshora, M., Chi, G. F., and Mukanganyama, S. (2021). Antibiofilm activity of extract and a compound isolated from *Triumfetta oletschia* against *Pseudomonas aeruginosa. Biochemistry Research International, 2021,* 13, Article ID 9946183. https://doi.org/10.1155/2021/9946183.

Moyo, B. and Mukanganyama S. (2015b). The anticandidial and toxicity properties of *Lampranthus franscici. Journal of Mycology, 2015,* 15, Article ID 898202. http://doi.org/10.1155/2015/898202.

Moyo, B. and Mukanganyama, S. (2015a). Antibacterial effects of *Cissus oletschia* and *Triumfetta oletschia* extracts against *Escherichia coli and Bacillus cereus. International Journal of Bacteriology, 2015,* Article ID 162028.

Moyo, B. and Mukanganyama, S. (2015c). Anti-proliferative activity of T. *oletschia* extract on Jurkat T cells *in vitro. Biomed Research International, 2015,* 10, Article ID 817624. http://doi.org/10.1155/2015/817624.

Moyo, R., Chimponda, T., and Mukanganyama, S. (2014). Inhibition of hematopoietic prostaglandin D2 Synthase (H-PGDS) by an alkaloid extract from *Combretum ole. BMC Complementary and Alternative Medicine, 14,* 221. http://doi.org/10.1186/1472-6882-14-221.

Mozirandi, W. and Mukanganyama, S. (2017). Antibacterial activity and mode of action of *Vernonia adoensis* (Asteraceae) extracts against *S. aureus* and *Pseudomonas aeruginosa. Journal of Biologically Active Products from Nature, 7*(5), 341–357. http://doi.org/10.1080/22311866.2017.1378922.

Mozirandi, W., Tagwireyi, D., and Mukanganyama, S. (2019). Evaluation of antimicrobial activity of chondrillasterol isolated from *Vernonia adoensis* (Asteraceae). *BMC Complementary and Alternative Medicine, 19,* 249. https://doi.org/10.1186/s12906-019-2657-7.

Mtisi, B., Sithole, S., Mombeshora M., and Mukanganyama, S. (2018). Inhibition of biofilm formation in *Candida albicans* and *Candida krusei* by *Combretum zeyheri* leaf extracts. *Journal of Bacteriology and Mycology, 5*(3),1068.

Muhammad, M.H., Idris, A.L., Fan, X., Guo, Y., Yu, Y., Jin, X., et al. (2020). Beyond risk: bacterial biofilms and their regulating approaches. *Frontiers in Microbiology.* https://doi.org/10.3389/fmicb.2020.00928.

Mukanganyama, S., Dumbura, S.C., and Mampuru, L. (2012). Anti-Proliferative effects of plant extracts from Zimbabwean medicinal plants against human leukaemic cell lines. *The African Journal of Plant Science and Biotechnology, 6,* 14–20.

Mushayavanhu, D., Chimponda, T. and Mukanganyama, S. (2016). *Syzygium guineense* (Myrtaceae.Willd.) leaf extracts inhibit drug efflux and enhance protein leakage in *Mycobacterium aurum* and *Mycobacterium smegmatis. Journal of Biologically Active Products from Nature, 6,* 5–6, 352–364. http://doi.org/10.108 0/22311866.2016.1268067.

Musthaba, M., Baboota, S., MD Athar, T., Y Thajudeen, K., Ahmed, S., and Ali, J. (2010). Patented herbal formulations and their therapeutic applications. *Recent Patents on Drug Delivery & Formulation, 4*(3), 231–244. http://doi.org/10.2174/187221110793237538. PMID: 20626333.

Nyambuya, T., Mautsa, R., and Mukanganyama, S. (2017). Alkaloid extracts from *Combretum zeyheri* inhibit the growth of *Mycobacterium smegmatis. BMC Complementary and Alternative Medicine, 17,* 124. http://doi.org/10.1186/s12906-017-1636-0.

Price, L.B., Hungate, B.A., Koch, B.J., Davis, G.S., Liu, C.M. (2017). Colonizing opportunistic pathogens (COPs): the beasts in all of us. *PloS Pathogens, 13* (8), e1006369. https://doi.org/10.1371/journal. Ppat.1006369.

Rahman, T., Sobur, A., Islam, S., Ievy, S., Hossain, J., El Zowalaty, M.E., Rahman T., et al. (2020). Zoonotic diseases: aetiology, impact, and control. *Microorganisms, 8,* 1405. http://doi.org/10.3390/ microorganisms8091405.

Reller, L.B., Weinstein, M., Jorgensen, J.H., and Ferraro, M.J. (2009). Antimicrobial susceptibility testing: a review of general principles and contemporary practices. *Clinical Infectious Diseases, 49,* 1749–1755. https://doi.org/10.1086/647952.

Sachikonye, M. and Mukanganyama, S. (2016). Antifungal and drug efflux inhibitory activity of selected flavonoids against *Candida albicans and Candida krusei. Journal of Biologically Active Products from Nature, 6*(3), 223–236. http://doi.org/10.1080/22311866.2016.1231078. http://doi.org/10.1080/223118 66.2016.1231078.

Simões, D., Miguel, S.P., Ribeiro, M.P., Coutinho, P., Mendonça, A.G., and Correia, I.J. (2018). Recent advances on antimicrobial wound dressing: a review. *European Journal of Biopharmaceutics, 127,* 130–141. http:// doi.org/10.1016/j.ejpb.2018.02.022. Epub 2018 Feb 17.

Sithole, S., and Mukanganyama, S. (2017). Evaluation of the antiproliferative activity of *Maerua edulis* (Capparaceae) on jurkat-T cells. *Journal of Biologically Active Products from Nature*, 7(3), 214–227. http://doi.org/10.1080/22311866.2017.1342561.

Sithole, S., Mushonga, P., Nhamo, L.N., Chi, G.F., and Mukanganyama, S. (2020). Phytochemical fingerprinting and activity of extracts from the leaves of *Dolichos kilimandscharicus* (Fabaceae) on jurkat-T Cells. *BioMed Research International*, 9. http://doi.org/10.1155/2020/1263702.

Stoodley, P., Sauer, K., Davies, D.G., and Costerton, J.W. (2002). Biofilms as complex differentiated communities. *Annual Review of Microbiology*, 56, 187–209. http://doi.org/10.1146/annurev.micro.56.012302.160705.

Vaikosen, E.N. and Alade, G.O. (2017). Determination of heavy metals in medicinal plants from the wild and cultivated garden in Wilberforce Island, Niger Delta region, Nigeria. *Journal of Pharmacy & Pharmacognosy Research*, 5, 129–143, ISSN 0719-4250 http://jppres.com/jppres.

Verma, A., Jain, A., Hurkat, P., and Jain, S.K. (2016). Transfollicular drug delivery: current perspectives. *Research and Reports in Transdermal Drug Delivery*, 5, 1–17.

Wende, M., Sithole, S., Chi, G.F., Stevens, M.Y., and Mukanganyama, S. (2021). The Effects of Combining Cancer Drugs with Compounds Isolated from *Combretum zeyheri* Sond. and *Combretum platypetalum* Welw. ex MA Lawson (Combretaceae) on the Viability of Jurkat T Cells and HL-60 Cells. *BioMed Research International*, *2021*, 10. Article ID 6049728. https://doi.org/10.1155/2021/6049728.

WHO. Global Report on Traditional and Complementary Medicine (2019). World Health Organization, Geneva . Licence: CC BY-NC-SA 3.0 IGO.

11 Sustainable Agricultural Practices of Industrially Utilized Tropical Medicinal Plants

Nikita Patel, Swetal Patel, Sherif Babatunde Adeyemi,
Abdullahi Alanamu Abdul Rahaman,
Kehinde Stephen Olorunmaiye and Ramar Krishnamurthy

CONTENTS

DOI: 10.1201/9781003206620-11

11.1 INTRODUCTION

Plants are perceived as important natural resources, and they have been a good source of indigenous medicine throughout the globe. According to a WHO survey, nearly 80% of the world populace, including tribal and rural people, uses herbs as a part of their general healthcare routine, because of their poor economic condition and cultural acceptability (Pal and Shukla, 2003; Hussain et al., 2018).

The annual global medicinal plant trade is around 62 billion USD, which is anticipated to rise by 500 billion USD by 2050. However, due to increased commercial trade and human needs, the wild species are generally overexploited because of a lack of knowledge in their cultivation practices and agrotechnology (Schippmann et al., 2002). Cultivating medicinal plants is critical for preserving and conserving exotic plant species to meet market demand on a global scale. Medicinal plants are used either whole or in combination, and roughly 25–30% of medicinal plant derivatives are used in conventional treatment (Ramakrishnappa, 2002). Due to their varied applications, plants and their phytoconstituents create new medications with improved therapeutic potentials (Mohan et al., 2021).

India and China are the world's largest producers of herbs and their purified products. India ranks second to meet the market need for medicinal plants' crude and finished raw materials (Vasisht et al., 2016). Sustainable cultivation and harvesting practices may help to avoid overexploitation of medicinal plants found in the wild. As long as plants are cultivated sustainably using biotechnological approaches, phenotypic and genotypic diversity can be maintained while yield and quality increase (Canter et al., 2005) Distribution, sustainable agrotechnology, growth tactics, active components, proximate elements, and mineral composition of industrially employed tropical medicinal plants are presented in Tables 11.1 and 11.2.

11.2 INDUSTRIALLY IMPORTANT TROPICAL MEDICINAL PLANTS

11.2.1 *Acorus calamus* Linn (Sweet Flag)

Acorus calamus Linn. (Sweet flag) is a flowering plant species with psychoactive properties. It is indigenous to Himalayan tracts of India and grows easily in the wild without much consideration (Singh et al., 2011). It requires marshy and swampy habitats for proper growth and yield (Kasture and Krishnamurthy, 2016). *A. calamus* leaves are 0.5–2 cm wide with undulated or curved margins (Figure 11.1a). Rhizomes are branched, aromatic, and 2–2.5 cm thick. *A. calamus* rarely bears flowers, but they are cylindrical and covered with multitude of round spikes if flowers are present. They form spadix, which reaches 4.9–8.9 cm during expansion. Fruits are generally small and berry-like in shape, which may vary according to their ecotypes (Chandra and Prasad, 2017).

TABLE 11.1

Proximate Constituents of Industrially Utilized Tropical Medicinal Plants

Medicinal Plant	Ash Value (%)	Moisture Content (%)	Carbohydrate Content (%)	Crude Protein (%)	Crude Fat (%)	Crude Fibre (%)	References
Acorus calamus (Rhizome)	17.3	68.2	37.26	15.62	0.00057	6.5	Manju et al. (2014)
Adhatoda zeylanica (Leaves)	10.57	58.2	16.4	15.6	1.6	6.4	Javed et al. (2010); Kumar and Sinha (2013); Chanu and Sarangthem (2014); Tandel et al. (2018)
Andrographis paniculata (Leaves)	16.13	28.1	36.3	1.5	2.0	15.7	Alobi et al. (2015)
Asparagus racemosus (Roots)	6.2	9.59	58.6	5.6	< 1	18.5	Tripathi et al. (2015); Karunarathne et al. (2020)
Baliospermum montanum	–	–	–	–	–	–	–
Bacopa/ Centella (Leaf)	1.9/16.4	88.4/84.6	5.9/6.7	2.1/2.4	0.6/0.2	1.05/1.06	Devendra et al. (2018); Hashim (2011); Seneviratne et al. (2015); Chandrika et al. (2015)
Canavalia gladiata (Seeds)	3.20	8.27	54.94	22.97	2.38	8.24	Ibeto et al. (2019)
Cassia angustifolia	10.65	8.81	65.76	10.55	4.22	10.61	Hussain et al. (2009)
Coleus forskalaei (Roots)	7.19–8.16	8.03–10.70	6.50–12.68	2.73–7.65	–	–	Srivastava et al. (2002)
Convolvulus pluricaulis	18.77	7.38	–	–	–	–	Sethiya et al. (2010)
Costus speciosus (Leaves)	1.98	89.01	5.4	1.73	0.29	–	Nadeeshani et al. (2018)
Uraria/ Pseudarthria (Leaves)	10.4/6.95	10.55/1095	–	–	–	–	Nitesh et al. (2012); Jayanthy et al. (2013)
Dioscorea sp. (Tuber)	0.62–1.02	76.43–82.76	11.86–19.48	1.57–3.85	0.13–0.14	0.63–1.28	Chadikun et al. (2019)
Eclipta alba	15.91	8.73	–	–	–	–	Regupathi and Chitra (2015)
Glycyrhizza glabra	10.3	5.32	24.6	24.1	10.7	–	Tuyen et al. (2010); Lim (2016)
Guggul	6.42–11.9	9.23–12.19	–	–	–	–	Thakur et al. (2018)
Gymnema sylvestre (Leaves)	2.04	79.69	7.17	10.29	0.50	0.29	Perera and Pavitha (2017)
Moringa oleifera (Leaves)	8.05–10.38	7.55–8.65	47.55–56.25	22.99–29.36	4.03–9.51	6.00–9.60	Sultana (2020)

(Continued)

TABLE 11.1 (*Continued*)

Proximate Constituents of Industrially Utilized Tropical Medicinal Plants

Medicinal Plant	Ash Value (%)	Moisture Content (%)	Carbohydrate Content (%)	Crude Protein (%)	Crude Fat (%)	Crude Fibre (%)	References
Mucuna pruriens (Seeds)	3.8	10.1	12.65	9.83	7.95	5.4	Rane et al. (2019)
Chlorophytum/ Cucurligo (Roots)	2.5/7.66	6.5/8.46	6/35.78	8.05/5.89	8.33/0.0043	5/-	Nalawade and Rajaram (2017); Panda et al. (2010); Raaman et al. (2009)
Ocimum sanctum (Leaves)	10.07	80.23	1.63	0.17	3.18	4.72	Patel et al. (2019)
Piper longum	8.3	15.70	-	-	-	-	Sharma et al. (2019)
Plantago ovata	2.7	12.5	48.6	17.4	6.7	24.6	Romero-Baranzini et al. (2006)
Plumbago zeylanica	2.05	10.71	5.52	4.223	4.127	-	Kanungo et al. (2012); Singh et al. (2018)
Rauvolfia serpentina	7.20	10.60	-	-	-	-	Rungsung et al. (2014)
Rubia cordifolia	4.32	95.5	93.82	3.06	2.58	-	Soni et al. (2010)
Sida acuta	6.33	9.03	55.30	19.13	0.67	9.50	Rami et al. (2014)
Stevia rebaudiana (Leaves)	7.41	5.37	61.9	11.4	3.73	15.5	Gasmalla et al. (2014); Abou- *Arab* et al. (2010)
Tinospora cordifolia (Stem)	7.96	34.39	69.80	7.74	1.99	56.42	Hussain et al. (2009); Rahal et al. (2014)
Withania somnifera (Roots)	6.93	15.68	20.24	6.26	6.89	44	Hameed and Hussain (2015)

11.2.1.1 Propagation

Generally, *A. calamus* is propagated via rhizomes/roots. Additionally, it can be grown with seeds. The field is thoroughly ploughed before the onset of rain. Farmyard manure (FYM) is applied at the rate of 15 mt/ha along with nitrogen and phosphorous as a basal requirement (Kasture and Krishnamurthy, 2016). Flowering and fruiting happen in July. After 6–8 months of proper plant maturation, the plants are harvested. A higher yield was observed at planting space of 30×30 cm, which yielded 13,150 kg/ha rhizome (Tiwari et al., 2012; Kasture and Krishnamurthy, 2016) The Indian domestic consumption of *A. calamus* root/rhizome is estimated at around 500–1,000 MT dry rhizome annually (Imam et al., 2013).

11.2.1.2 Chemical Constituents and Utilization

A. calamus or Vacha is rich in bioactive constituents such as alkaloids, phenols, flavonoids, saponins, tannins, mucilage, and bitter principles (Imam et al., 2013; Muchtaromah et al., 2017; Babar et al., 2020). Different solvent extracts of rhizome reported the presence of benzene, 1,2-dimethoxy-4-(2-propenyl), shyobunone, α-asarone, β-asarone, 7-tetracycloundecanol, 4,4,11,11-tetramethyl, 4a,

TABLE 11.2

Mineral Composition of Industrially Utilized Tropical Medicinal Plants

Medicinal Plant	Calcium mg/kg	Magnesium mg/kg	Iron mg/kg	Zinc mg/kg	Potassium mg/kg	References
Acorus calamus (Rhizome)	1585.66	644	89.37	18.3	1507	Manju et al. (2014); FAO (2017)
Adhatoda zeylanica (Leaf)	1,010	23	530	1.79	560	Javed et al. (2010); Kumar and Sinha (2013); Chanu and Sarangthem (2014)
Andrographis paniculata (Leaves)	106.3	124.3	0.466	0.266	125.6	Alobi et al. (2015)
Asparagus racemosus (Root)	3366.5	2082.5	417.0	14.65	Trace amount	Krishnamoorthy and Chidambaram (2019)
Bacopa/Centella (Leaf)	2,020/1740	3,675/870	328.5/52	78/32	-/3450	Hashim (2011); Ranovona et al. (2019); Devendra et al. (2018) Muszyńska et al. (2018)
Canavalia gladiata (Seeds)	2,320	119.8	56.9	23.9	298.5	Ibeto et al. (2019)
Cassia angustifolia	-	-	209.6	50	-	Hussain et al. (2009)
Coleus forskalaei (Roots)	-	-	-	27–67	-	Srivastava et al. (2002)
Convolvulus pluricaulis	-	531.57	2,460	64.35	9.99	Sethiya et al. (2010)
Costus speciosus (Leaves)	3,730	610	55.1	4.03	5.31	Nadeeshani et al. (2018)
Uraria/ Pseudarthria (Leaves)	1,810/N.A.	2,100/-	-	-	6,700/-	Hari et al. (2014)
Dioscorea sp.	170	210	5.2	6.21	8,160	Chandrasekara et al. (2016); Padhan and Panda (2020); Padhan et al. (2018)
Eclipta alba	1.0	-	8.95	1.04	35.8	Hussain and Khan (2010); Choudhury et al. (2017)
Glycyrhizza glabra	1,470	1,200	200	44	1,750	Ercisli et al. (2008)
Guggul	-	-	-	-	-	-
Gymnema sylvestre (Leaves)	1,580	604.8	19.26	24.02	-	Dey and Kazi (2015)
Moringa oleifera (Leaves)	16,700	5,400	160	33.3	1,236	Gopalakrishnan et al. (2016); Kumssa et al. (2017)
Mucuna pruriens (Seeds)	5,250	1,630	950	210	130	Ezeokonkwo and Sunday (2015)
Chlorophytum/ Cucurligo (Roots)	34.5/595	203/591.7	251.8/45.6	48.72/15.7	0.065/46.2	Nalawade and Rajaram (2017); Raaman et al. (2009)

(Continued)

TABLE 11.2 (*Continued*)
Mineral Composition of Industrially Utilized Tropical Medicinal Plants

Medicinal Plant	Calcium mg/kg	Magnesium mg/kg	Iron mg/kg	Zinc mg/kg	Potassium mg/kg	References
Ocimum sanctum (Leaves)	35.8	14.9	863.0	39.33	20.4	Patel et al. (2019)
Piper longum	3,444	1353.9	291.3	14.6	13,520	Bhat et al. (2010)
Plantago ovata	1,600	63.5	21.7	99.4	1,000	Bukhsh et al. (2007)
Plumbago zeylanica	-	-	-	-	-	-
Rauvolfia serpentina	3.2	1	37.8	53.8	0.4	Harisaranraj et al. (2009)
Rubia cordifolia	-	-	-	-	-	-
Sida acuta	144.28	1221.1	-	325.12	-	Nwankpa et al. (2015); Rami et al. (2014)
Stevia rebaudiana (Leaves)	1,770	3,260	5,890	1,260	2,115	Abou-Arab et al. (2010)
Tinospora cordifolia (Stem)	102.23	-	26.05	7.34	8,450	Rahal et al. (2014); Modi et al. (2020)
Withania somnifera (Roots)	1.835	11.15	7.586	0.099	7.595	Hameed and Hussain (2015)

FIGURE 11.1   (a) *Acorus calamus* cultivation field and harvesting; (b) *Adhatoda zeylanica* cultivation field; (c) *Andrographis paniculata* nursery seedling and cultivation field; (d) *Asparagus racemosus* cultivation field and tuber harvest; (e) *Asparagus racemosus* dried tuber; (f) *Baliospermum montanum* cultivation field intercropping with *C. asiatica*; (g) *C. asiatica* under shade net cultivation; (h) *Bacopa monnierri* cultivation field; (i) *C. gladiata* plant with pod development stage; (j) *Cassia angustifolia* cultivation field; (k) *Coleus forskohlii* cultivation field; and (l) *Convolvulus pluricaulis* plant at harvesting stage.

7-methano-4Ah-naphth[1,8a-b] oxirene, octahydro-4,4,8,8-tetramethyl, *n*-hexadecanoic acid, and 9,12-octadecadienoic acid (Pokhrel et al., 2020). Traditionally, root/rhizome, stem, and leaves are used in the treatment of diarrhoea, indigestion, kidney or liver infection, eczema, arthritis, neuralgia, sinusitis, asthma, fever, bronchitis, numbness, cough, inflammation, loss of hair, and other disorders (Kasture et al., 2016; Khwairakpam et al., 2018). *α*Asarone and *β*-asarone are the main components present in *A. calamus* followed by other bioactive, which are attributed to the medicinal property of this plant (Imam et al., 2013; Chatterjee et al., 2021).

11.2.2 *Adhatoda zeylanica* Medic (Ardusi)

Adhatoda zeylanica Medic. (syn *Justicia adhatoda* Linn.) is a gregarious species of plant that is commonly known as Ardusi in Gujarati (Figure 11.1b). It is distributed in the sub-Himalayan tracts of India and is also grown in different ecological zones of India. *A. zeylanica* is a small herbaceous shrub with 6–8 cm long and 3–4 cm broad leaves (Singh and Kushwaha, 2005; Mehta, 2016). The plant has been used in the Indian System of Medicine for ages because of its various medicinal properties. Seeds of *Adhatoda* are non-endospermic, rugose, and sub-orbicular (Mehta and Bajaj, 2018). The colour of seeds varies from light brown to dark brown. Leaves are generally bitter, and fruit capsules are globular in shape (Gantait and Panigrahi, 2018)

11.2.2.1 Propagation

Adhatoda can be easily propagated through tender stem cuttings. Propagation via seeds is limited because of the poor germination rate and seed viability (Damor et al., 2019). Before propagation of *Adhatoda* on ridges and mounds through stem cutting, the field must be ploughed and levelled properly. Stem cuttings with three to four nodes are ideal for planting. The seed germination phase starts from mid-July, and flowering starts from mid-September to February. It starts blooming in December, where it sets fruit in late December, which lasts till April (Madhuri et al., 2014; Mehta, 2016). Almost all the plant parts possess medicinal value, and hence leaves, stems, and roots must be correctly harvested. Harvesting is generally carried out from December to January, but the entire plant is harvested after 2 years to get a higher yield of roots. Around 10–11 t of material is harvested from a 1 ha area (Tandel et al., 2018).

11.2.2.2 Chemical Constituents and Utilization

A. zeylanica has a variety of bioactive compounds that have a variety of health advantages. Alkaloids, phenols, flavonoids, tannins, and a variety of other phytoconstituents are found in it. Alkaloids such as vasicine and vasicinone are abundant in the leaves, branches, and roots. Apart from quinazoline alkaloids, *Adhatoda* contains significant amounts of vasicine, sterols, and glycosides (Kirtikar and Basu, 1994; Khalekuzzaman et al., 2008). Vasicine being the major alkaloid is present at the rate of 1.3–1.4%. Flowers are rich in quercetin and kaempferol along with 2-4 dihydroxy chalcone-4-glucoside (Bhartiya and Gupta, 1982; Rawat et al., 1994). In terms of pharmacological property, this plant's parts possess antibacterial, antioxidant, anti-inflammatory, anti-tussive, abortifacient, wound healing, and antimicrobial properties. Leaves and twigs are also used as green manure (Jahan and Siddiqui, 2012; Ahmed et al., 2013; Singh and Singh, 2018).

11.2.3 *Andrographis paniculata* (Burm. f.) Wall. Ex. Nees (Kalmegh)

Andrographis paniculata (Burm. f.) (Kalmegh) is a herbaceous plant native to Asia's tropical and subtropical regions, Tibet, and few other countries such as China, Thailand, Myanmar, and Indonesia (Jayakumar et al., 2013). It is a perennial herb that grows to a height of 50 cm–1 m (Figure 11.1c). The stem is quadrangular with multiple branches, and the leaves are arranged in opposite directions. The flowers are in a raceme with short petioles. They bear 0.5 cm–2.5 cm fruit capsules with multiple seeds, usually yellowish-brown (Shahjahan et al., 2013; Saffie et al., 2019).

11.2.3.1 Propagation

A. paniculata can be grown/propagated via stem cuttings or seeds on varied soils with less fertility. It generally requires a tropical climate for proper growth, development, and yield. Vegetative propagation is beneficial in some instances, but it is generally propagated through seeds by direct seeding or transplanting with proper nursery techniques for large-scale production (Samantaray et al., 2001). This plant grows luxuriantly during the onset of rain and starts flowering during September, which lasts up to December. Fruiting lasts up to December until temperature differences arise in the northern plains (Chauhan and Krishnamurthy, 2020). After proper maturation, the plant is appropriately uprooted, yielding 2.5 t/ha—a reasonable profit to growers and consumers.

11.2.3.2 Chemical Constituents and Utilization

A. paniculata (Kalmegh) is considered "Kings of Bitter," as it is exceedingly bitter because of the presence of numerous phytoconstituents in it (Shahjahan et al., 2013). *A. paniculata* is used in curing and treating diseases since time immemorial. The whole plant, i.e., stem, leaves, and inflorescence/flowers are used in traditional and current drug delivery systems (Farooqi and Sreeram, 2010). Phytochemical analysis revealed that this plant is abundant in flavonoids, phenols, diterpenoid lactones, and tannins, all of which contribute to the plant's pharmacological capabilities as an antioxidant, anti-inflammatory, anti-diabetic, and antibacterial. Andrographolide is the primary ingredient of this plant and has been demonstrated in numerous studies to have potent anti-cancer properties (Mishra et al., 2007; Aminah et al., 2021).

11.2.4 *Asparagus racemosus* Willd (Shatavari)

Asparagus racemosus (Shatavari) is native to the tropics and subtropics of India. Around 350 species of *Asparagus* are available in India, from which 20 to 22 species are used in the Indian system of medicine (Krishnamurthy et al., 2004). The plant is generally tall and woody prickly perennial climber (1–2 m in length) bearing tuberous roots with spinescent spine-like leaves (Figure 11.1d and e). Flowers are white and tiny, bearing subglobose berries as fruits (Alok et al., 2013).

11.2.4.1 Propagation

Asparagus is propagated via seeds and root crowns. This plant usually grows in various soils, but deep ploughing is required, followed by harrowing after certain days. Broad ridges are required for plantation. Seeds are sown during the onset of rain, and the roots are harvested in winter after proper maturation (Krishnamurthy et al., 2004).

11.2.4.2 Chemical Constituents and Utilization

Asparagus racemosus is rich in essential bioactive constituents such as shatavarin I–IV (saponins), racemosol (alkaloids), phenols, flavonoids, and vitamins (Singh et al., 2018). Almost all the parts of this plant, such as roots, stem, leaves, and flowers, are used to treat several diseases like throat infection, tuberculosis, dyspepsia, male genital dysfunction, and female reproductive system-related problems. Apart from treating various ailments, this plant possesses antidepressant, antioxidant, diuretic, anti-HIV, anti-plasmodial, and immunostimulant properties (Singla et al., 2014).

11.2.5 *Baliospermum montanum* (Willd.) Muell. Arg. (Danti)

Baliospermum montanum (Willd.) Muell.-Arg. (Danti) from Euphorbiaceae family is monoecious undershrub with multiple shoots arising from the base. This plant is generally 1.8–3.5 m in height with lanceolate leaves in the upper part and three to five lobed ovate leaves at the base (Figure 11.1f). Danti is distributed in subtropical and tropical regions of India, Malaya, Nepal, and Burma (Joshi et al., 2017). This plant bears unisexual flowers that start blooming between January and February, and forms two to three lobed capsules as fruits with oily seeds (Das, 2008).

11.2.5.1 Propagation

Both seeds and shoot cuttings can be used to propagate Danti. Cuttings have demonstrated a sprouting rate of 60–70%, compared to seeds with a germination rate of approximately 50%. Generally, vegetative propagation by shoot cuttings is preferred through raised beds or by planting the cuttings into polybags. Before propagating this plant, the land is prepared by proper levelling, harrowing, and ploughing to get a good yield. After proper maturity of fruit capsules, seeds are harvested, and the whole plant is individually dug out because every part of this plant is used to prepare herbal formulations (Das, 2008; Bijekar et al., 2014).

11.2.5.2 Chemical Constituents and Utilization

Almost every part of *B. montanum* has been used medicinally due to its numerous and essential phytoconstituents. Roots of this plant possess baliospermin, montanin, 12-deoxy-5-hydroxyphorbol 13-myristate, 12-deoxyphorbol 13-palmitate, and 12-deoxy-16-hydroxyphorbol 13-palmitate as major bioactive constituents that are diterpene hydrocarbon in nature (Mali and Wadekar, 2008). The leaves are exceptionally high in flavonoids, phenols, hexacosamol, 8-sitosterol, and 8-D-glucoside, all of which contribute to the plant's therapeutic properties (Mukherjee et al., 1980; Pasqua et al., 2003). Danti is known to have anti-inflammatory, anti-cancer, antibacterial, anthelmintic, immunomodulatory, hepatoprotective, and free radical scavenging properties from a pharmacological perspective (Wadekar et al., 2008a; Johnson et al., 2010; Lalitha and Gayathiri, 2013)

11.2.6 Brahmi (*Centella asiatica* L. Urban & *Bacopa monnieri* [L.] Pennel)

Brahmi—an essential medicinal plant with a wide range of therapeutic potential—has been utilized mainly by traditional medicine practitioners for ages. Mandukaparni (*Centella asiatica*) is a perennial herb with rooted and long nodes and internodes (Krishnamurthy et al., 2006), whereas Nirbrahmi (*Bacopa monnieri*) is a creeping herbaceous plant mostly found in marshy and wet areas. Both the plants are used to prepare various herbal formulations in the Indian system of medicine (Aparna et al., 2015). *C. asiatica* possesses reniform crenate leaves with the elongated petiole (Figure 11.1g), and *Bacopa* possesses simple, succulent, and oppositely arranged oblanceolate leaves, on the other hand (Figure 11.1h) (Patel et al., 2020; Lal and Baraik, 2019). Flowers are usually white or white-to-purple tinged in *Centella* and *Bacopa*, respectively. They possess bracteoles with short pedicels during flowering in *Bacopa,* whereas *Centella* bears flowers as fascicled umbels (Krishnamurthy et al., 2006; Belwal et al., 2019).

11.2.6.1 Propagation

Bacopa and *Centella* can be propagated readily via stem cuttings with rootlets and internodes, as seed germination studies have not shown encouraging results. *Centella* can be planted between February and March, while *Bacopa* can be planted between May and July because it grows sumptuously during rainy seasons (Mathur et al., 2002; Patel et al., 2020). *Centella* is harvested after 90 days of growth when the plants have reached full maturity and are usually conducted in the summer to dry harvested leaves properly. *Bacopa* can be harvested after 70–90 days, with September and October being the optimal months for collecting the entire plant and manually separating the succulent leaves for a higher yield of chemical components (Krishnamurthy et al., 2006; Singh et al., 2020).

11.2.6.2 Chemical Constituents and Utilization

The major chemical constituents present in *Bacopa* and *Centella* are Bacoside and Centelloid, which are generally pentacyclic triterpenoid in nature. However, there are many bioactive constituents with various therapeutic potentials. Other bioactive constituents include asiaticoside, madecassoside, asiatic acid, centellin, and centellicin in *Centella* and 12 analogues of Bacoside in *Bacopa,* including two common flavonoids apigenin and luteolin with 3-phenylethanoid glycoside and 4-cucurbitacin (Murthy et al., 2006; Yu et al., 2007). *C. asiatica* and *B. monnieri* have been used ethnomedicinally for their neuroprotective, anti-cancer, anti-ulcer, anti-depressant, anti-diabetic, memory boosting, antioxidant, antibacterial, anti-inflammatory, and analgesic properties. The entire plant is used to manufacture various formulations due to the presence of bioactive components (Prakash et al., 2017; Bala et al., 2018).

11.2.7 *Canavalia gladiata* (Jacq.) DC. (Sword Bean)

Canavalia gladiata (Family: Fabaceae) is an underutilized plant with important medicinal and pharmacological properties. It is indigenous to tropical and subtropical regions of Asian and African countries. They have also been domesticated in West Indies and Australia (Moteetee, 2016). Morphologically, *C. gladiata* is a perennial climber with about 4–10 m in height (Figure 11.1i). They possess trifoliate leaves with large pubescent acuminate leaflets (Purseglove, 1968; Indriani et al., 2019). There are two varieties of sword beans, i.e., early and late variety with white and pink flowers about 3–4 cm long. Seed pods are large and 3.5–5 cm broad (Das et al., 2016).

11.2.7.1 Propagation

C. gladiata is propagated via seeds. Before sowing the seeds, the land is prepared by proper ploughing and by applying 5 MT farmyard manure throughout the land. Both red and white seed varieties of sword beans have varied growth and agronomic traits. With a proper irrigation system, these beans can be cultivated in all seasons, including Kharif and rabi. The maturation period is of 110–120 days. However, tender pods are harvested after 70–75 days after sowing (Ekanayake, 2000; Xianzong, 2017)

11.2.7.2 Chemical Constituents and Utilization

Being an underutilized plant *C. gladiata* is rich in amino acids, minerals, protein, carbohydrates, and lectins (Kulkarni et al., 2016). It possesses antioxidant, antimicrobial, anti-inflammatory, haematopoietic, hepatoprotective, and anti-angiogenic activities (Kumar & Reddy, 2014). Monogalloyl hexoside, gallic acid, digallic acid, ellagic acid, and trigalloyl hexoside are the major phenolic compounds that contribute to various biological activities exhibited by the plant (Gan et al., 2016). Sword beans are generally used for culinary purposes as stews, soup, or fermented food products (Bosch, 2004). Roasted beans are also used as a substitute for coffee beans in Central America and as ornamental plants in some other regions (Ekanayake, 2000; Seena and Sreedhar, 2005). It was also reported that these plants were also used as a cover crop or forage crop (Bosch, 2004).

11.2.8 *Cassia angustifolia* Vahl. (Senna)

Cassia angustifolia (Senna), Family-Cesalpinaceae is an anthropoid drug used basically for laxative properties. Although about 25–26 species of *Cassia* have been documented to have potential laxative properties across the world, *C. angustifolia* and *C. acutifolia* (Alexandrian Senna) are by far the most extensively utilised species in various pharmacopoeias due to their bulk availability and laxative property. This plant is indigenous to Sudan, but it is widely distributed in India in different ecological regions, including Gujarat, Rajasthan, Tamil Nadu, Karnataka, and Andhra Pradesh (Ramchander et al., 2017; Kumar and Jnanesha, 2017). *C. angustifolia* is a perennial undershrub of about 1–2 m tall (Figure 11.1j). Leaves are primarily large and pinnately compound. Flowers are zygomorphic with 12–13 mm long pedicels and 6–9 mm wide yellow–green sepals. They bear dehiscent hairy pods with multiple seeds (Săvulescu et al., 2018).

11.2.8.1 Propagation

Cassia angustifolia is primarily propagated by seeds after proper ploughing and land preparation. Generally, seeds are sown/seeded between September and November, and 10 t/ha farmyard manure is applied to facilitate the basal nutrient requirement for proper growth and yield. (Kumar and Jnanesha, 2017). Mostly, well-drained soil is essential, as this plant is sensitive to water-logged conditions (Sastry et al., 2007). Flowers begin to bloom 50–60 days after sowing, and harvesting is done in batches. After 90 days, the first batch is harvested, and the second batch is harvested after 30 days. This is done to obtain the optimum yield of sennoside.

11.2.8.2 Chemical Constituents and Utilization

C. angustifolia, a traditional medicinal herb, contains phenols, flavonoids, tannins, alkaloids, anthraquinones, and saponins. Sennoside A and Sennoside B are the main bioactive components of *Cassia*, along with quercimeritrin, scutellarein, and rutin (Ahmed et al., 2016). Both pods and leaves have sennoside A and B with minor other sennoside compounds. Additionally, carbohydrates such as mannose, sucrose, fructose, and glucose are included (Agarwal and Bajpai, 2010). *C. angustifolia* is pungent and traditionally used to relieve constipation. Apart from its laxative effect, this plant offers various biological activities, including anti-cancer, anti-parasitic, free radical scavenging, and anti-microbial activity (Ahmed et al., 2016; Jnanesha et al., 2018). According to the traditional medicinal system, both pods and leaves are included in a variety of Ayurveda remedies (Laghari et al., 2011).

11.2.9 *Coleus forskohlii* Briq

Coleus forskohlii Briq, belonging to the family Lamiaceae, is an essential medicinal plant with varied pharmacological and biological activities. It is a perennial herb native to India and is also found in Nepal, Sri Lanka, Thailand, and Burma (Lokesh et al., 2018). *C. forskohlii* grows about 1–2 ft in height, and flowers are like those of other family members. Roots are thick, succulent, and fasciculate (Figure 11.1k). While the entire plant is enriched with bioactive compounds, roots are usually employed for pharmacological and therapeutic purposes (Srivastava et al., 2002).

11.2.9.1 Propagation

Coleus can be easily propagated by the terminal or rooted stem cuttings and seeds (Patel, 2016). Cuttings are generally planted between June and July, and are harvested after 4–5 months of planting for matured tubers (Bhattacharjee et al., 2020).

11.2.9.2 Chemical Constituents and Utilization

The tuberous succulent roots of *Coleus* possess forskolin (diterpene in nature), which has beneficial properties. Apart from this, *Coleus* is also rich coleoside, α-cedrol, 4β, 7 β, 11-enantioeudesmantriol, α-amyrin, myrianthic acid, uvaol, betulic acid, coleonolic acid, and euscaphic acid (Kavitha et al., 2010). Being aromatic, the essential oil from the roots, stem, and leaves also includes 3-decanone, bornyl acetate, caryophyllene oxide, limonene, sesquiterpene hydrocarbon, β-sesquiphellandrene, γ-eudesmol, β-phellandrene, α-pinene, α-copaene, sabinene, βcaryophyllene, and α-humulene (Paul et al., 2013). *C. forskohlii* is therapeutic to treat heart disease, glaucoma, asthma, obesity, microbial infections, and thrombosis (Kavitha et al., 2010; Ganash and Qanash, 2018).

11.2.10 *Convolvulus pluricaulis* Choisy (Shankhpushpi)

Convolvulus pluricaulis is a nootropic herb used in the treatment of multiple ailments. This herb is native to the northwest regions of India, and is also spread throughout tropical Africa and Sri Lanka (Austin, 2008; Nisar et al., 2012). It is perennial, with branches about 30 cm long and elliptical leaves arranged in an alternate position (Figure 11.1l). According to the traditional medicinal system, this herb is reported as brain/nervine tonic and is used in different dosage forms to boost memory (Ganie et al., 2015; Sethiya et al., 2015).

11.2.10.1 Propagation

C. pluricaulis is propagated through seeds. Propagules are raised during June–July. Flowers are bluish, white, or purple, whereas, fruits are globose. The plant is harvested after proper maturity between September and October (Agarwa et al., 2014).

11.2.10.2 Chemical constituents and Utilization

Carbohydrates abound in *C. pluricaulis* (Shankhpushpi), including glucose, maltose, and rhamnose. They include a high concentration of secondary metabolites such as phenolics, flavonoids, glycosides, triterpenoids, kaempferol, taraxerone, delphinidine, -sitosterol, N-hexacosanol, taraxerol, hydroxycinnamic acid, and steroids. According to the Ayurvedic medicine system, the entire plant is used in the treatment of various diseases. *C. pluricaulis* has several natural biostimulants and shankhpuspi, which help with wear and tear mechanisms and weight loss (Bhowmik et al., 2012; Saroya and Singh, 2018). Ethnomedicinal importance and utilization of *C. pluricaulis* include treatment or curing of anaemia, anorexia, anxiety, asthma, hypertension, bronchitis, blood disorders, burning/heating sensation, dementia, diabetes, epilepsy, leukoderma, cough, menorrhagia, vertigo, syphilis, ulcer, urinary infections, and pyrexia (Balkrishna et al., 2020).

11.2.11 *Costus speciosus* (Koen ex. Retz.) Sm. (Insulin Plant)

Costus speciosus (Family: Costaceae) is a perennial plant distributed in tropical and subtropical regions of Africa, Asia, and America. Both *C. speciosus* and *C. pictus* are known as Insulin plants (Naik et al., 2017). It is succulent, growing to a height of around 2–3 m, with lanceolate leaves and dense creeping roots at the base (Figure 11.2a) (Dubey et al., 2010; Choudhury et al., 2012). *C.*

FIGURE 11.2 (a) *Costus speciosus* cultivation field; (b) *Desmodium gangeticum* cultivation field; (c) *Desmodium gangeticum* harvested plants; (d) *Pseudarthria viscida* cultivation field; (e) *Uraria picta* plant; (f) *Dioscorea* cultivation field; (g) *Eclipta alba* plant; (h) *Glychrhizza glabra* plant; (i) *Commiphora wightii* plant and gum resin harvest; (j) *Gymnema sylvestre* cultivation field and leaf harvest(dry); (k) *Moringa oleifera* cultivation field; and (l) *Mucuna pruriens* cultivation field.

speciosus is used by tribal people to control diabetes, and it also possesses significant phytopharmacological effects. (Khan and Ramu, 2018).

11.2.11.1 Propagation

Due to low seed viability and germination rate, *Costus* is propagated vegetatively by stem cuttings or
culms (Rani et al., 2012). This plant thrives in moist, humid environments. Between July and August,
flowers begin to bloom, producing fruits with a white fleshy aril. The ripening of fruit is followed by
leaf shedding and the plant's eventual death during November, while the rhizome remains dormant
from December to March, with new stems growing in April (Benny, 2004; Rajesh et al., 2009).

11.2.11.2 Chemical Constituents and Utilization

C. speciosus is used for food, medicine, and ornamental purposes. It has diverse phytochemicals
and bioactive constituents, including phenols, flavonoids, steroids, alkaloids, glycosides, saponins,
and tannins. It also possesses diosgenin, costusosides, eremanthin, dioscin, prosapogenins A, prosapogenins B, β-sitosterol, β-carotene, β-D-glucoside, aliphatic hydroxyl ketones, triterpenes, starch
mucilage, fatty acid, abscisic acid, α-amyrinsterate, β-amyrin, lupeol palmitates, and corticosteroid
from the leaves and rhizomes (Maji et al., 2020). Ethnomedicinally, it is used to treat hyperlipidemia, headache, inflammation, cancer, asthma, bronchitis, anaemia, jaundice, and pneumonia
(Pawar and Pawar, 2014; Al-Attas et al., 2015).

11.2.12 *DESMODIUM GANGETICUM* (L.) DC (SHALPARNI)

Desmodium gangeticum (L.) DC is an undershrub native to the sub-Himalayan zones of India. It is also
distributed throughout Africa, China. Malay, Burma, and Ceylon. It grows to a height of approximately
1.0–2.0 m (Figure 11.2b and c) and has alternate unifoliate broad leaves (Vishwakarma et al., 2009).
Roots are a crucial component of Dashmoolarishta. *D. gangeticum* is also a nerve tonic, which is why it
is used to treat typhoid, chronic fever, and inflammation (Kirtikar and Basu, 1918; Pathak et al., 2005).

11.2.12.1 Propagation

D. gangeticum is easily propagated from seeds and does not require pretreatment for germination.
It thrives on loamy and clayey soils. Between March and April, nursery propagules are raised.
Flowering and fruiting occur twice a year in May/June and September/October (Das, 2008).

11.2.12.2 Chemical Constituents and Utilization

D. gangeticum is used in the preparation of various ayurvedic formulations. Studies have revealed
desmodin, gangetin, and hordenine as the major bioactive constituent of *D. gangeticum*. Apart from
this, *Pseudarthria viscida* (Figure 11.2d) and *Uraria picta* (Figure 11.2e) are also used in the preparation of Dashmoolarishta. Both the allied species are part of *Desmodium* or Dashmmol drug plant
(Naik and Krishnamurthy, 2018). Desmodin, gangetin, and hordenine have been shown to exhibit a
variety of biological activities. Ethnomedicinally, Desmodium is used to alleviate oxidative stress,
i.e., scavenging free radicals, and manage skin diseases, boils, blisters, whooping cough, fever, and
dysentery (Singh et al., 2015).

11.2.13 *DIOSCOREA* SPP.

Dioscorea spp. (Dioscoreaceae Family) is a highly beneficial tuberous plant. It is often referred to as
Yam. Its rhizome, roots, and tubers have been used medicinally since prehistoric times. *Dioscorea*
spp. occurs throughout Asia, Africa, and Latin America (Viruel et al., 2016). *Dioscorea* spp. is a
climber (Figure 11.2f), which sprouts from tuber/rhizome (Govaerts et al., 2007). There are 600 species in the Genus *Dioscorea*, from which ten species are being cultivated and domesticated. It includes

white yam, yellow yam, water yam, air potato, bitter yam, Chinese yam, and lesser yam. They are consumed as a local staple food with important medicinal properties (Behera et al., 2009).

11.2.13.1 Propagation

Dioscorea spp. can be propagated in various ways, including tubers, minisets, and vine cuttings approximately 8 cm long, as seed stock propagation is arduous, complicated, and time consuming. Seeds are typically light and flat, making them disperse easily (Kumar et al., 2017). It bears dioecious flowers, and fruits are in small capsules. Usually, *Dioscorea* spp. is planted carefully on ridges, mounds, holes, and flat surfaces with proper mulching to maintain soil moisture. Harvesting is carried out after 8–9 months of planting, indicated by the dieback of leaves and vines (Andres et al., 2017).

11.2.13.2 Chemical Constituents and Utilization

Dioscorea spp. contains a variety of secondary metabolites, with saponins being the most abundant. Apart from saponins, other metabolites *viz.*, phenolics, quinones, clerodane diterpenes, cyanidins, and diarylheptanoids from *Dioscorea* spp. have been identified and quantified (Ou-Yang et al., 2018). Compounds like diosgenin and more than 100 steroidal saponins are also reported from *Dioscorea* spp. (Jesus et al., 2016). Despite being a major food source, *Dioscorea* spp. has been associated with various biological actions, including cytotoxic, neuroprotective, immunomodulatory, anti-allergic, anti-inflammatory, and neuroprotective. It also possesses anti-proliferative, contraceptive, and anti-microbial properties (Salehi et al., 2019)

11.2.14 *Eclipta alba* (L.) Hassk (Bhringraj)

Eclipta alba (L.) Hassk (Asteraceae Family) is known in India and Bangladesh as Bhringraj (also known as Keshraj, which means Ruler of hair) due to its extensive ethnomedicinal properties. It is distributed in India, Bangladesh, China, Thailand, and Brazil. It is an annual herb with 3–5 cm long blackish-green leaves that prefers moist conditions (Figure 11.2g). Typically, the stem is pubescent, flat, erect, or extensively branched, and of a darkish green colour (Soni and Soni, 2017). This plant is well-known in the Indian system of medicine for treating skin disorders, hair greying, hepatoprotective, and respiratory diseases (Mithun et al., 2011).

11.2.14.1 Propagation

E. alba can be propagated by seeds and stem cuttings. Propagules can be raised between February and March but they are transplanted between April and August upon the climatic conditions. It bears white compressed achene and narrowly winged flowers, whereas fruiting occurs between October and November (Pareek and Kumar, 2015; Soni and Soni, 2017).

11.2.14.2 Chemical Constituents and Utilization

Eclipta is an appetite stimulator and mild bowel regulator. It has varied bioactive constituents such as alkaloids, flavonoids, glycosides, coumestans majorly, and de-methyl wedelolactone and wedelolactone. The entire plant is known to have important ethnomedicinal and pharmacological properties in treating cough, asthma, senility, headache, and problems associated with hair fall and greying of hair. Apart from this, they are known to possess biological activity such as anti-cancer, analgesic, antioxidant, anti-myotoxic, antiviral, anti-hepatotoxic, spasmogenic, anti-inflammatory, hypotensive, and ovicidal properties (Mithun et al., 2011; Balakrishnan et al., 2018)

11.2.15 *Glychrhizza glabra* Linn (Jethimadh)

Glychrhizza glabra Linn (Fabaceae Family) is commonly known as Jethimadh. It is a perennial plant of about 2–2.5 m in height with multipotent pharmacological properties. It is distributed throughout

India, including Punjab, Jammu Kashmir, and sub-Himalayan tract. It has 8–15 cm long pinnate leaves with 10–17 leaflets (Figure 11.2h). Rhizome/roots are responsible for treating stomach ulcers, digestion problems, fever, epilepsy, and rheumatism (Nassiri-asl et al., 2012; Lohar et al., 2020).

11.2.15.1 Propagation

Cuttings are frequently used to propagate *G. glabra*, since seeds have a very low germination rate. Propagules are mostly raised during the spring season by adequate field preparation and the use of FYM as a basal nutrient requirement. Flowers are purple, which bear glabrous compressed pods with reniform seeds (Kriker, 2013). Harvesting takes place between November and December, as this plant matures during the winter season (Khaitov et al., 2021).

11.2.15.2 Chemical Constituents and Utilization

G. glabra possesses substantial bioactive constituents and is well-known in Siddha, Unani, and Ayurveda for treating various ailments. The roots/rhizomes of the plant possess ethnomedicinal property. Apart from this, it is also used as a flavouring agent. Several studies have reported the secondary metabolites present in *G. glabra,* which include glycyrrhizin, triterpenoids, glycosides, saponins, quercetin derivatives, flavones, liquiritic acid, liquoric acid, licuraside, liquiritoside with glycyrrhetinic acid, and glycyrrhizin being the major phytochemical constituent (Yang et al., 2015; Rizzato et al., 2017). This plant is therapeutically known to be anti-tussive, anti-tumour, antimicrobial, antiviral, anti-diabetic, anti-coagulant, anti-cancer, and possesses immunomodulatory properties (Batiha et al., 2020).

11.2.16 Guggul

Commiphora wightii (Arnott) Bhandari and *Boswellia serrata* (Roxb) belonging to the Burseraceae family are called Guggal. These plants are indigenous to different ecological zones of Asia and Africa. The gum resins (Figure 11.2i) extracted from both the allied species are used in Ayurveda for ages (Thosar and Yende, 2009). Morphologically, *C. wightii* is 2–3 m tall with trifoliate leaves, and while *B. serrata* possesses lanceolate–ovate leaflets. The gum resin extracted from these allied species can alleviate tumours, obesity, sores, obesity, ulcers, and other ailments (Siddiqui, 2011; Reddy et al., 2012).

11.2.16.1 Propagation

Leafless stem cuttings are usually preferred for raising propagules of guggul, as poor germination is observed when raised through seeds. They are also propagated through air layering and plant tissue culture techniques (Jitendra et al., 2009). Flowers are sessile in *C. wightii* and appear throughout the year, whereas small and white flowers are observed in *B. serata*. These plants prefer sandy-to-loamy soil for their growth and development. It requires at least 6–7 years for the plant to get a good yield of gum resin and is usually harvested between November and February (Jain and Nadgauda, 2013; Soumya et al., 2019).

11.2.16.2 Chemical Constituents and Utilization

Guggul is rich in Guggulusterone, i.e., M, E, and Z, steroids, resin, volatile oils, M-dehydoguggulsterone, Guggulsterol(I–V), lignans, triterpenoids, flavones, and ferrulates. It also contains monoterpenes, diterpenes, and boswellic acid (Hazra et al, 2018). Traditionally, guggul is used in treating gout, arthritis, rheumatoid, body ache, weight loss, and inflammation. The gum obtained from this plant is also utilised to treat endometritis, bronchitis, dyspepsia, bone fractures, and thrombosis (Rout et al., 2012). Certain pharmacological activities of guggul include cardioprotective, anti-hyperglycemic, thyroid-stimulating, fibrinolytic, hypolipidemic, cytotoxic, and antimicrobial activities (Bhardwaj et al., 2019).

11.2.17 *Gymnema sylvestre* R. Br. (Gudmar)

Gymnema sylvestre R.Br. (Asclepiadeace Family) is indigenous to India, Africa, Japan, Malaysia, Srilanka. Australia, Indonesia, and Vietnam. This plant is believed to neutralize body sugar, and hence it is commonly known as gurmar/gudmar. *G. sylvestre* is a perennial plant with elliptical or ovate leaves (Figure 11.2j) that grows up to 500–600 m tall (Saneja et al., 2010). The stem is usually hairy and brown. The characteristic feature of this plant is its bitter and astringent taste, which can mask the taste of sweetness for few hours (Kanetkar et al., 2007; Krishnamurthy et al., 2015).

11.2.17.1 Propagation

Nodal cuttings and seeds can both be used to propagate *G. sylvestre*. Flowers are zygomorphic and start blooming between April and November, whereas fruiting occurs between December and March (Pandey, 2012). Field preparation is carried out by applying farmyard manure before commencing the cultivation of *G. sylvestre*, and harvesting is generally carried out at 12th month after flowering to get high productivity of leaves with gymnemic acid. (Krishnamurthy et al., 2015).

11.2.17.2 Chemical Constituents and Utilization

G. sylvestre is a good source of bioactive constituents with ethnomedicinal and modern applications. Leaves contain various triterpene saponins, alkaloids, anthraquinones, flavonoids, and glycosides (Khramov et al., 2008). Gymnemic acid (I–VII) and Gymnema saponins are the major constituents of this plant and are well-known for treating diabetes (Tiwari et al., 2014). Additionally, it is used to treat snakebites, malaria, rheumatoid arthritis, obesity, and hyperlipideamia, among other conditions. Some of the biological activities exhibited by *G. sylvestre* include antioxidant, antibacterial, anti-arthritic, anti-inflammatory, anti-cancer, hepatoprotective, and immunomodulatory action (Yaseen and Shahid, 2020).

11.2.18 *Moringa oleifera* Lam. (Drumstick Tree)

Moringa oleifera Lam. (Moringaceae Family) is a fast-growing nutritionally rich tree species with various medicinal, biological, pharmacological, and nutraceutical properties. The entire plant is used to heal numerous ailments, hence garnering it a moniker "miracle tree" (Patel and Krishnamurthy, 2021). It is indigenous to tropical and subtropical regions of India and Africa. However, it is also grown in different ecological zones of Brazil, Zimbabwe, Bangladesh, and Pakistan. *Moringa* can easily withstand mild frost and drought conditions, reaching up to 12 m in height (Figure 11.2k). Leaves are compound, bipinnate, and composite, bearing three to five opposite leaflets (Abubakar et al., 2011; Gadzirayi et al., 2013).

11.2.18.1 Propagation

Stem cuttings and seeds are used in growing *Moringa*, and it is carried out with the onset of rain. However, with proper irrigation facilities, it can be planted at any time of the year. Flowers start blooming during the winter season, and pods harvesting is carried out after 160–180 days of sowing. Leaf material is primarily harvested after 90 days of sowing (Gadzirayi et al., 2013; Sekhar et al., 2017).

11.2.18.2 Chemical Constituents and Utilization

Moringa is known to have diverse bioactive constituents, such as glucosinolates (Glucomoringin and Glucosoonjnain), alkaloids, phenols, flavonoids, tannins, saponins with essential amino acids, protein, and minerals (Patel and Krishnamurthy, 2021). The entire plant, including the gum resin, has biological and therapeutic properties. Biological activities include antioxidant, anti-cancer, anti-tumour, anti-hypertensive, antimicrobial, anti-trypanosomal, anti-leishmanial, antiviral, anti-fertility, anti-inflammatory, and anti-depressant activity (Kaur et al., 2015; Abd-Rani et al., 2018).

Apart from the biological and pharmacological activities, it is enriched with nutraceutical properties that can help combat malnutrition and anaemia (Falowo et al., 2018).

11.2.19 *Mucuna pruriens* L. (DC). (Velvet beans)

Mucuna pruriens L. (DC) (Fabaceae Family) is an underutilized legume with anti-Parkinson properties. It is an annual climber that can reach a height of 50 ft and has trifoliate broad leaves (Figure 11.2l). This species is indigenous to India and Africa. The active ingredient L-Dopa (5–6%) is used to treat neurodegenerative disorders. It produces an abundance of green foliage and is utilised as a green cover or animal food (Krishnamurthy et al., 2003; Natarajan et al., 2012).

11.2.19.1 Propagation

Seeds are used as a starting material or propagating material in the case of *M. pruriens* by applying 10 MT/ha of farmyard manure as a part of field preparation. Flowers are zygomorphic to actinomorphic and bear pods with itching trichomes (Krishnamuthy et al., 1996; Krishnamurthy et al., 2003). This plant matures after 7–8 months of germination, and pods are harvested when they turn brown (Krishnamurthy et al., 2005).

11.2.19.2 Chemical Constituents and Utilization

M. pruriens, an underutilized legume, is an excellent source of bioactive compounds such as phenols, flavonoids, tannins, and alkaloids. Additionally, it contains both albumin and globulin protein fractions and carbohydrates, calcium, phosphorus, and amino acids. It also contains significant oleic, linoleic, palmitic, and other minerals (Siddhuraju et al., 1996; Patel et al., 2020). Pharmacologically, it has antioxidant, anti-tumour, aphrodisiac, anti-diabetic, antibacterial, analgesic, antivenin, and anti-inflammatory properties (Kumar et al., 2016; Pathania et al., 2020). Seeds are generally utilized for medicinal purposes because of the presence of L-Dopa, whereas leaves cover crops and feed non-gastric non-ruminant animals (Peniche-Gonzalez et al., 2018).

11.2.20 Musli—Safed Musli and Kali Musli

Chlorophytum borivilianum Santapau and Fernandes (Liliaceae Family), and *Cucurligo orchoides* Gaertn (Amarylladaceae) are commonly known as Safed musli and Kali musli, respectively. They are under endangered species because of overexploitation and collection of these plants from the wild (Chauhan et al., 2010; Singh et al., 2012). It is reported that this plant is distributed throughout India and Africa. *C. borivilianum* has lanceolate leaves (Figure 11.3a), whereas *C. orchoides* has simple 15–45 cm long leaves (Figure 11.3b) attached at the base. Roots are elongated, and hence both Safed (White roots) and Kali musli (Black roots) are differentiated based on the colour of their roots (Singh et al., 2006; Khanam et al., 2013).

11.2.20.1 Propagation

Roots stocks and seeds are used as propagating material in terms of musli. Rootstock is mainly preferred because of the poor germination rate of seeds (Kothari and Singh, 2003; Mehta and Singh, 2014). *C. borivilianum* requires optimum soil and temperature for proper growth, and are unable to bear higher temperatures. Therefore, they are planted from July to October, and they are harvested between March and April after proper maturity (Juju et al., 2017; Khanam et al., 2013). However, only a few discs and roots are removed to facilitate the next planting/cropping season. *C. orchoides* is planted between February and March and is harvested after 7 months to get high-yielding rhizomes (Tewari and Singh, 2019).

11.2.20.2 Chemical Constituents and Utilization

Fleshy roots of musli (Safed and Kali musli) are a good source of phytochemicals and are used in the traditional system of medicine for ages. Green leaves are utilized by the tribal belt of Gujarat, Chhattisgarh, and Madhya Pradesh for culinary purpose (Vijaya and Chavan, 2009). *C. borivilianum* and *C. orchoides* abound with over 110 phytocompounds, including tannins, phenols, flavonoids, cucurligoside, glycosides, saponins, sapogenins, mucilage, fructans, and furostane and spirostane glycosides (Kaushik, 2005; Jena et al., 2021). They are aphrodisiacs and are used in the treatment of male sexual dysfunction. Apart from this, they have antioxidant, hepatoprotective, oestrogenic, and immunomodulatory activities (Wang et al., 2021).

11.2.21 *Ocimum sanctum* Linn. (Tulsi)

Ocimum sanctum (Lamiaceae family) is known as the "Queen of herbs" and "Holy Basil" because of its potent medicinal and therapeutic potential (Mahajan et al., 2016). It is distributed and grown across the globe. This plant is a subshrub, erect, multi-branched, about 35–60 cm tall, with green-to-purple leaves (Figure 11.3c). The stem is hairy with aromatic leaves and toothed petiole (Pattanayak et al., 2010). Holy basil is used to treat headaches, malaria, and coughs and bolsters the immune system (Joshi et al., 2017).

11.2.21.1 Propagation

O. sanctum can be grown from seeds and nodal cuttings. It thrives well in moist conditions, and soil preparation is accomplished by spreading 15 t of farmyard manure/ha during April and June. Harvesting is carried out after 3 months of planting and subsequently after 60–65 days regularly to get a good yield of essential oil (Bhattacharjee et al., 2020).

FIGURE 11.3 (a) *Chlorophytum borivilianum* root harvest; (b) *Curculigo orchioides* cultivation field and tuber harvest; (c) *Ocimum sanctum* plant; (d) *Piper longum* plant; (e) *Piper chaba* plant; (f) *Plantago ovata* cultivation field; (g) *Plumbago zeylanica* cultivation field; (h) *Rauwolfia serpentina* cultivation field; (i) *Sida cordifolia* plant; (j) *Stevia rebaudiana* cultivation field; (k) *Tinospora cordifolia* cultivation field and stem harvest; and (l) *Withania sonmifera* cultivation field and dry root harvest.

11.2.21.2 Chemical Constituents and Utilization

O. sanctum have important plant secondary metabolites such as phenolics, flavonoids, triterpenoids, neolignans, fatty acids, β-sitosterol, mucilage, minerals, and various proximate constituents (Patel et al., 2019). Pharmacologically, it has antioxidant, anti-cancer, radiation protection, anti-stress, anti-diabetic, anti-inflammatory, mosquitocidal, and antimicrobial property. *O. sanctum* is used as a whole or combined with other herbs to treat ulcers, vomiting, skin diseases, and other ailments (Bano et al., 2017; Singh and Chaudhuri, 2018).

11.2.22 *Piper longum* L. (Pippali)

Piper longum L. (Piperaceae family), well-known as long pepper, is a perennial climber found in India, America, and Sri Lanka. It has cordate leaves with woody roots (Figure 11.3d) and was the most valued plant during the Roman era. It is used for analgesic and carminative purposes (Zaveri et al., 2010). *Piper chaba* Hunter is also an allied species of the Piperaceae family known to substitute for *Piper longum*. Both the species have various biological properties, and *P. chaba* is a creeper plant that spreads through the ground morphologically (Figure 11.3e). It may also grow around trees and have ovate leaves found mostly in West Bengal, India (Haque et al., 2018).

11.2.22.1 Propagation

Stem or vine cuttings are used as propagating material. The seedlings are raised by transplanting them in well-drained soil. Field operations include proper ploughing, harrowing, and weeding at proper intervals. Male flowers are larger and slender when compared to female flowers and bear small ovoid berries as fruits. Their planting season varies from June to July and September to October, and the fruits are harvested when they turn brown (Rameshkumar et al., 2011; Islam et al., 2020).

11.2.22.2 Chemical Constituents and Utilization

Piper longum is reported to possess various biological properties, including analgesic, anti-inflammatory, anti-fertility, antimicrobial, antioxidant, anti-diabetic, anti-arthritis, anti-helminthic, hepatoprotective, anti-asthamic, anti-cancer, nootropic, and anti-angiogenic properties (Yadav et al., 2020). It is because of the presence of a substantial amount of phytoconstituents such as piperine (alkaloid in nature), pipernonaline, piperettine, methyl piperine, asarinine, pellitorine, piperlonguminine, piperlongumine, retrofractamide A, pergumidiene, and a complex mixture of essential oils along with vitamins A, E, caryophyllene, p-cymene, pentadecane, thujone, terpinolene, zingiberene, and dihydrocarveol (Gani et al., 2019; Sharma et al., 2020).

11.2.23 *Plantago ovata* Forsk—(Isabgul)

Plantago ovata (Plantaginaceae family) is cultivated and distributed throughout temperate and tropical regions of India, primarily in Gujarat, Haryana, and Punjab. Being annual, *P. ovata* is stemless, with narrow and linear leaves arranged in an alternate position (Figure 11.3f). Seeds are ovoid and concave at one side, with membranous cover as seed coat. It is widely used to cure constipation, dysentery, and diarrhoea (Sarfraz et al., 2017).

11.2.23.1 Propagation

Seeds serve as a starting/propagating material in *P. ovata*. However, recent tissue culture techniques are also involved in providing quality material for planting. *P. ovata* requires cool and dry conditions for proper growth and yield (Ghadheri et al., 2012). Field preparation plays an essential role in getting a good yield of this plant, and it requires proper ploughing, harrowing, weeding, and removal of clods. Seeds are sown between October and November, and harvest is done between February and March when the spikes turn brown (Bannayan et al., 2008; Kumar et al., 2016).

11.2.23.2 Chemical Constituents and Utilization

Seeds and husk are the most widely used part of *P. ovata* because of their essential phytocompounds. It accounts for the availability of mucilaginous fibre, polysaccharides, phenols, iridoids, cumatines, tannins, sterols, and alkaloids to treat various ailments. Recent research indicates that *P. ovata* may have various pharmacological applications in treating disease conditions, including hyperlipidemia, hypercholesterolemia, piles, rheumatoid arthritis, and constipation. (Marjhan and Kalam, 2018). The husk is also used to prepare Orodispersible tablets (ODT) with improved rheological properties (Abbas et al., 2021).

11.2.24 *PLUMBAGO ZEYLANICA* LINN.—(CHITRAK)

Plumbago zeylanica (Plumbaginaceae—Family) is a perennial plant distributed throughout India, and is mostly cultivated in the southern part of India for its roots and milky white juice obtained from the leaves. Being an undershrub, it grows 1–2 m tall with simple and opposite leaves (Figure 11.3g) (Chauhan, 2014). Roots are striated and light in colour with excellent medicinal properties. Along with *P. zeylanica*, the roots of *P. rosea* are also utilized in China and India (Chaudhari and Chaudhari, 2015).

11.2.24.1 Propagation

Stem cuttings with few nodes and seeds are primarily used as propagating material. However, seeds are not recommended for cultivation because of their erratic seed viability and poor growth rate. The flowers of *P. zeylanica* are whitish and red in *P. rosea* (Patel et al., 2016). They start blooming between September and November, and bear fruits between January and February. Harvesting of roots is preferred after attaining maturity of 12 months to the date of sowing. Due to their sensitivity to water-logged conditions, field preparation is critical to obtaining a high production of roots, and deep ploughing in conjunction with disc ploughing is essential to sustain that output (Chaplot et al., 2006; Jalalpure, 2011).

11.2.24.2 Chemical Constituents and Utilization

P. zeylanica possesses biodynamic properties because of the presence of various phytoconstituents in it. Phenols, flavonoids, tannins, glycosides, and sterols are present in *P. zeylanica*. Plumbagin—a stirring yellowish compound present abundantly in roots of this plant is napthoquinone in nature (Roy and Bharadvaja, 2017; Aleem, 2020). Pharmacologically, it has anti-diabetic, anti-cancer, antiinflamatory, larvicidal, and wound-healing properties. Several studies have revealed the efficacy of green synthesized nanoparticles using *P. zeylanica* on pathogenic microbial strains (Patel and Krishnamurthy, 2015; Velammal et al., 2016).

11.2.25 *RAUWOLFIA SERPENTINA* BENTH. EX KURZ.—(SARPAGANDHA)

Rauwolfia serpentina (Apocynaceae—Family) is a perennial undershrub bush with lanceolate–ovate leaves (Figure 11.3h). It is distributed around tropical regions of India, Burma, and Sri Lanka. Roots comprise of tuberous taproot system (Isha et al., 2021). The entire plant is therapeutically important since the ancient times, and, specifically, they are used as an antidote for insect stings, snake venom, and chomps (Bunkar, 2017).

11.2.25.1 Propagation

R. serpentina can be cultivated through seeds, stem cuttings, root cuttings, and recently through *in vitro* plant tissue culture techniques. They rely on neutral-to-acidic soil for proper growth and development. This plant is usually harvested after a year of transplanting to get a high yield of roots (Biradar et al., 2016; Subandi and Dikayani, 2018).

11.2.25.2   Chemical Constituents and utilization

R. serpentina has been explored and utilized due to its prevailing therapeutic properties and phytocompounds being majorly alkaloids including indole alkaloid—resperine, ajamalicine, ajmaline, indocine, resperiline, serpentinine, serpentine, and various other isolated compounds (Rukachaisirikul et al., 2017). Apart from this, tannins, phenols, and flavonoids have also been reported. Roots are the most widely used part of this plant, because the association of larger moieties of alkaloids contribute to the clinical properties (Kumari et al., 2013; Kalam and Majeed, 2020).

11.2.26   *RUBIA CORDIFOLIA* L.—(MANGISTHA)

Rubia cordifolia (Rubiaceae—Family) is a perennial climbing plant with its stem attached to a woody base. Roots are flexuose with reddish bark. It is cultivated in India, Malaysia, Japan, Africa, and Indonesia, among other places (Khare, 2004). It is well-known for the red dye extracted from the roots—a highly effective blood purifier. Leaves are ovate, arranged in whorls of four, and bears terminal flowers with globose/didymous fruits (Deshkar et al., 2008; Nyeem et al., 2018).

11.2.26.1   Propagation

R. cordifolia is propagated through nodal root cuttings and seeds. However, seeds are preferred for large-scale cultivation to balance the cost-effective ratio. Field operation includes adequate ploughing, harrowing, planting, and weeding soil to maintain its texture and porosity. Farmyard manure is applied as a basal nutrient source, and harvesting is carried out 2 years after planting to get a good yield of roots (Verma et al., 2016; Ali et al., 2020).

11.2.26.2   Chemical Constituents and Utilization

R. cordifolia, commonly known as Manjistha, is a well-known herb proven to be effective against tuberculosis, cancer, contusion, metrorrhagia, and rheumatism (Zhao et al., 2011). Napthohydroquinones and hydroquinones are the major phytochemical constituents, followed by terpenoids, saponins, tannins, and quinones. The root dye is used as a food colourant. Pharmacologically, this medicinal herb has been reported to have anti-inflammatory, hepatoprotective, anti-diabetic, cardioprotective, anti-arthritic, wound healing, nephroprotective, antimicrobial, and anti-allergic properties (Kumari et al., 2021; Srinivasulu et al., 2021).

11.2.27   *SIDA CORDIFOLIA* L.—(BALA)

Sida cordifolia L. (Malvaceae—Family) is an undershrub with densely stellate, hairy, and ovate leaves with stout stems and roots (Figure 11.3i). It is distributed throughout Asia, Africa, and the southern part of America (Sankar et al., 2012; Galal et al., 2015). However, another allied species *Sida acuta* is used as an adulterant in ayurvedic preparations because of similar chemical constituents in *S. cordifolia*. Moreover, *S. acuta* is also an undershrub with lanceolate leaves, and it is distributed in tropical regions of India. The allied species are known as Bala or Nagabala in various ancient scriptures of medicine (Mohideen et al., 2002; Tcheghebe et al., 2017).

11.2.27.1   Propagation

S. cordifolia can be easily propagated via stem cuttings and seeds. Before the sowing process, the land is prepared by continuous ploughing, and removing pebbles and weeds to get a fine tilth soil (Krizevski et al., 2009). Early flowering is observed when it is grown through stem cutting. The field is usually enriched with minerals by applying FYM and green manure at regular intervals, and manual harvesting is carried out after attaining the maturity of 1 year (Pramanick et al., 2015).

11.2.27.2   Chemical Constituents and Utilization

Sida spp. is rich in flavonoids, alkaloids, and phyto-steroids. It possesses analgesic, anti-inflammatory, depressive, anti-fertility, antibacterial, anti-cancer, and nephroprotective characteristics pharmacologically. It has been reported that seeds possess alkaloid, mucin, resin, and fatty acids, whereas the aerial parts and roots are rich in palmitic acids, β-sitosterol, stearic acid, asparagine, choline, betaine, and resin, respectively (Khurana et al., 2016; Rodrigues et al., 2020)

11.2.28   *Stevia rebaudiana* Bert.

Stevia rebaudiana Bert. (Asteraceae—Family) is a perennial plant smaller in size and can adapt to varied climatic conditions. It can grow up to 3–80 cm in height with sessile and lanceolate–oblanceolate trichotomous leaves on upright woody stems (Figure 11.3j) (Ramesh et al., 2006). This plant is well distributed in India, China. Brazil, Thailand, and Japan. The leaves contain an abundant natural sweetener, which is mostly used for diabetic patients (Pandey, 2018).

11.2.28.1   Propagation

Stevia rebaudiana can be cultivated/propagated through seeds, stem cuttings, and plant tissue culture techniques (Singh et al., 2017; Pandey, 2018). It bears tubular flowers with an achene single-seeded fruit. Manual harvesting is carried out 4–5 months after planting and subsequently after every 3 months until 3 years (Rayaguru and Khan, 2008).

11.2.28.2   Chemical Constituents and Utilization

Stevioside, steviolbioside, and rebaudioside (A—E) are the major chemical constituents in *S. rebaudiana* followed by glycosides, tannins, alkaloids, and flavonoids. Additionally, it was discovered that the leaves are high in niacin, riboflavin, thiamine, carotene, and dulcoside, contributing to the plant's herbomineral and therapeutic properties (Goyal et al., 2010; Ahmad et al., 2020).

11.2.29   *Tinospora cordifolia* (Willd.) Miers—(Galo/Giloy)

Tinospora cordifolia (Menispermaceae—Family) is an immunomodulatory plant with immune-boosting properties. It is distributed in the tropics and subtropics of India. Based on their morphological and chemical constituent similarity, there are three main *Tinospora* species *viz., T. crispa, T. malabarica,* and *T. cordifolia*. They are closely related and frequently employed as adulterants in many ayurvedic formulations and preparations. Leaves are simple-cordate and alternate on a fibrous stem with greyish brown to cream bark having rosette lenticels (Figure 11.3k) (Modi et al., 2020; Yates et al., 2021).

11.2.29.1   Propagation

T. cordifolia can be cultivated through stem cuttings and seeds. For large-scale cultivation, vegetative propagation is most widely preferred (Patel et al., 2013; Kumawat et al., 2019). *T. cordifolia* usually prefers light and well-drained sandy–loamy soil. As a basal nutrient requirement, farmyard manure is applied during land preparation to maintain soil nutrient density. Flowers are yellow, which produce drupelets fruits with tiny seeds. Harvesting is carried out when the plant matures during the autumn season (Kumar and Jnanesha, 2017; Kumar and Jnaneha, 2019).

11.2.29.2   Chemical Constituents and Utilization

Pharmacologically, this herbal plant is reported to have antioxidant, antibacterial, anti-diabetic, anti-cancer, anti-stress, and hepatoprotective properties. Chemically, *T. cordifolia* contains steroids, lignans, terpenes, and alkaloids, the main base of its pharmacological properties. Apart from this, Giloy is also helpful in treating fever, dysentery, skin disorders, and urinary tract infections (UTI)

(Patel et al., 2020). The stem is the most popular component, since it contains numerous chemical elements such as diterpene glucoside and amritoside A–D (Sharma et al., 2019).

11.2.30 *WITHANIA SONMIFERA* (L.) DUNAL,—(ASHWAGANDHA)

Withania sonmifera (L.) Dunal (Solanaceae—Family) is an evergreen perennial plant widely distributed in India and across the world's tropical regions. It is widely used as indigenous medicine for ages. Leaves are simple, entire, and arranged alternately on the stem. The entire plant is covered with trichomes and has fleshy brownish-white roots (Figure 11.3l). The roots, followed by stems and leaves, are used for therapeutic purposes (Gupta and Rana, 2007; Kothari et al., 2015).

11.2.30.1 Propagation

Withania somnifera (Ashwagandha) is cultivated through seeds and stem cuttings. To ensure feasibility and homogeneity in large-scale production, *in vitro* tissue culture techniques have been used to cultivate this plant (Patel and Krishnamurthy, 2013). Field operation includes ploughing, harrowing, and disking to remove clods and granulate the land to a fine tilth soil. Usually, farmyard manure and organic fertilizers are applied to maintain the nutrient density of soil. Harvesting is carried out after 150–180 days of planting/sowing (Jat et al., 2021).

11.2.30.2 Chemical Constituents and Utilization

Roots of *Withania* are used in the Indian traditional system of medicine for the treatment of various diseases. Pharmacologically, it has antioxidant, haematopoietic, and immune-boosting properties (Patel and Krishnamurthy, 2013). They are prescribed for bronchitis, female disorders, and hiccups. Phytochemically, roots have alkaloids, steroids, amino acids, volatile oil, and withaniol, whereas the leaves have withanolide and withaferin A. Fruits, bark, and stems are known to have amino acids and certain minerals, respectively (Uddin et al., 2012; Kherde et al., 2020).

11.3 CONCLUSION AND PROSPECTS

Cultivation of highly demanded and largely consumed medicinal plants used in industry are important in order to maintain consistency in quality and efficacy, and therefore commercialization process of medicinal plants have the potential to replace crops as a sustainable source of employment and livelihood improvement in both developed and developing countries due to the high and growing interest in these products. Technological advances used during the propagation process will meet the high standards of the pharmaceutical industries for high-quality medicinal plant resources. Additionally, issues including incorrect identification, contamination, pest infestation, harvesting, limited supply, and processing techniques for plant materials are ought to be solved

ACKNOWLEDGEMENTS

We would like to acknowledge Dr. R. Nagaraja Reddy(Scientist, DMAPR-Anand) and Former Zandu Employee Dr. Vinayak R. Naik for their help and support. We are also thankful to the management of Uka Tarsadia University, Bardoli, Gujarat, India for providing necessary research facilities to conduct the study.

REFERENCES

Abbas, S., Sherazi, M., Amjad Khan, Alyami, H.S, Muhammad Latif, Zia-Ur-Rahman Qureshi, and Muhammad Hassham Hassan Bin Asad. (2021) Investigation of *Plantago ovata* husk as pharmaceutical excipient for solid dosage form (Orodispersible tablets). *BioMed Research International, 2021*, 1–10. https://doi.org/10.1155/2021/5538075.

Abou-Arab, A. Esmat, A. Azza Abou-Arab, and Ferial Abu-Salem, M. (2010) Physico-chemical assessment of natural sweeteners steviosides produced from *Stevia rebaudiana* Bertoni plant. *African Journal of Food Science, 4*(5), 269–281.

Agarwal, P., Sharma, B., Fatima, A.,and Jain, S.K. (2014) An update on Ayurvedic herb *Convolvulus pluricaulis* Choisy. *Asian Pacific journal of tropical biomedicine, 4*(3), 245–252.

Agarwal, V. and Bajpai, M. (2010) Pharmacognostical and biological studies on Senna & its products: an overview. *International Journal of Pharma and Bio Sciences, 1*(2), PS107

Ahmad, J., Khan, I., Blundell, R., Azzopardi, J., and Mahomoodally, M.F. (2020) Stevia rebaudiana Bertoni: an updated review of its health benefits, industrial applications and safety. *Trends in Food Science & Technology, 100*, 177–189.

Ahmed, S., Gul, S., Gul, H., and Bangash, M.H. (2013) Dual inhibitory activities of *Adhatoda vasica* against cyclooxygenase and lipoxygenase. *International Journal of Endorsing Health Science Research, 1*(1), 14–17.

Ahmed, S.I., Hayat, M.Q., Tahir, M., Mansoor, Q., Ismail, M., Keck, K.,and Bates, R.B. (2016) Pharmacologically active flavonoids from the anticancer, antioxidant and antimicrobial extracts of *Cassia angustifolia* Vahl. *BMC Complementary and Alternative Medicine, 16*(1), 1–9.

Al-Attas, A.A., El-Shaer, N.S., Mohamed, G.A., Ibrahim, S.R., and Esmat, A. (2015) Anti-inflammatory sesquiterpenes from *Costus speciosus* rhizomes. *Journal of Ethnopharmacology, 176*, 365–374.

Aleem, M. (2020) Anti-inflammatory and anti-microbial potential of *Plumbago zeylanica* L.: a review. *Journal of Drug Delivery and Therapeutics, 10*(5), 229–235.

Ali, A., Aslam, M.,and Chaudhary, S. (2020) "A review on pharmacognostic and therapeutic uses of *Rubia cordifolia. Journal of Drug Delivery and Therapeutics, 10*(6), 195–202.

Alobi, N.O., Eja, M.E., Okoi, A.I., Uno, U.A.,and Asuqo, B.G. (2015) Comparative evaluation of the nutrient composition of *Andrographis paniculata* and *Gongronema latifolium. New York Science Journal, 8*(12), 16–20.

Alok, S., Jain, S.K., Verma, A., Kumar, M., Mahor, A.,and Sabharwal, M. (2013) Plant profile, phytochemistry and pharmacology of *Asparagus racemosus* (Shatavari): a review. *Asian Pacific Journal of Tropical Disease, 3*(3), 242–251.

Aminah, N.S., Tun, K.N.W. Kristanti, A.N. Aung, H.T. Takaya, Y. and Choudhary M.I. (2021) Chemical constituents and their biological activities from Taunggyi (Shan state) medicinal plants. *Heliyon 7*(2), e06173.

Andres, C., AdeOluwa, O., and Bhullar, G.S. (2017) Yam (Dioscorea spp.)-A rich staple crop neglected by research. *Encyclopedia of Applied Plant Sciences*, vol. 2, Academic Press, 435–441.

Aparna, V., Mallya, S.V., Srikanth, P., and Sunil Kumar, K.N. (2015) Comparative pharmacognosy of two medhya dravyas, Brahmi (*Bacopa monnieri* Linn.) and Mandukaparni (*Centella asiatica* Linn.). *The Journal of Phyto Pharmacology, 4*(1), 1–5.

Austin, D.F. *Evolvulus alsinoides* (Convolvulaceae): an American herb in the old world. *Journal of Ethnopharmacology, 117*(2), 185–198.

Babar, P.S., Deshmukh, A.V., Salunkhe, S.S.,and Chavan, J.J. (2020) Micropropagation, polyphenol content and biological properties of Sweet Flag (*Acorus calamus)*: a potent medicinal and aromatic herb. *Vegetos, 33*(2), 1–8.

Bala, S., Jothi Priya, A., and Gayatri Devi, R. (2018) Physiological and pharmacological effects of Bacopa monnieri. *Drug Invention Today, 10*(11), 2179–2182.

Balakrishnan, P., Sekar, G.K.,Ramalingam, P.S., Nagarasan, S., Murugesan, V.,and Shanmugam K. (2018) Distinctive pharmacological activities of *Eclipta alba* and it's Coumestan Wedelolactone. *Indo American Journal of Pharmaceutical Sciences, 5*(4), 2996–3002.

Balkrishna, A., Thakur, P.,and Varshney, A. (2020) Phytochemical profile, pharmacological attributes and medicinal properties of *convolvulus prostratus*–A cognitive enhancer herb for the management of neurodegenerative etiologies. *Frontiers in Pharmacology* 11, 171.

Bannayan, M., Nadjafi, F., Azizi, M., Tabrizi, L., and Rastgoo, M. (2008) Yield and seed quality of *Plantago ovata* and *Nigella sativa* under different irrigation treatments. *Industrial Crops and Products, 27*(1), 11–16.

Bano, N., Ahmed, A., Tanveer, M., Khan, G.M., and Ansari, M.T. (2017) Pharmacological evaluation of *Ocimum sanctum. Journal of Bioequivalence and Bioavailablity, 9*(3), 387–392.

Batiha, G.E.S, Magdy Beshbishy, A., El-Mleeh, A., Abdel-Daim, M.M., and Prasad Devkota, H. (2020) Traditional uses, bioactive chemical constituents, and pharmacological and toxicological activities of *Glycyrrhiza glabra* L.(Fabaceae). *Biomolecules 10*(3), 352.

Behera, K.K., Sahoo, S. and Prusti, A. (2009) Relative agronomic performance of different Dioscorea species found in different parts of Orissa. *Nature and Science 7*(3), 25–35.

Belwal, T., Andola H.C., Atanassova, M.S., Joshi, B., Suyal, R., Thakur, S., Bisht, A., Jantwal, A., Bhatt, I.D., and Rawal. R.S. (2019) Gotu Kola (*Centella asiatica*). *Nonvitamin and Nonmineral Nutritional supplements*, Academic Press, USA, 265–275.

Benny, M. (2004) Insulin plant in gardens. *Natural Product Radiance, 3,* 349–350.

Bhardwaj, M. and Alia, A., (2019) *Commiphora wightii* (Arn.) Bhandari. review of its botany, medicinal uses, pharmacological activities and phytochemistry. *Journal of Drug Delivery and Therapeutics, 9*(4-s), 613–621.

Bhartiya, H. P. and Gupta, P. C. (1982) A chalcone glycoside from the flowers of *Adhatoda vasica*. *Phytochemistry, 21*(1), 247.

Bhat, R.Kiran, K. Arun, A.B.,and Karim, A. A. Determination of mineral composition and heavy metal content of some nutraceutically valued plant products. *Food Analytical Methods, 3*(3), 181–187.

Bhattacharjee, T., Sen, S., Chakraborty, R., Maurya, P.K.,and Chattopadhyay, A. (2020) Cultivation of medicinal plants: special reference to important medicinal plants of India. *Herbal Medicine in India*, Springer, Singapore, 101–115.

Bhowmik, D., Sampath Kumar, K.P., Paswan, S., Srivatava, S., Yadav, A., and Dutta, A. Traditional indian herbs *Convolvulus pluricaulis* and its medicinal importance. *Journal of Pharmacognosy and Phytochemistry, 1*(1), 44–51.

Bijekar, S.R., Gayatri, M.C., and Rajanna, L. (2014) Phytochemical profile of *Baliospermum montanum* (Wild.) Muell. Arg. *International Journal of Innovative Science and Research, 12*, 37–42.

Biradar, N., Hazar, I., and Chandy, V. (2016) Current insight to the uses of *Rauwolfia*: a review. *Research and Reviews: A Journal of Pharmacognosy, 3*(3), 1–4.

Bosch, C.H. (2004) Canavalia gladiata (Jacq.) DC. [Internet] Record from PROTA4U. In: Grubben, G.J.H., and Denton, O.A. (eds.), PROTA (Plant Resources of Tropical Africa/Ressources végétales de l'Afrique tropicale). Wageningen, Netherlands.

Bukhsh, E., Malik, S.A., and Ahmad, S.S. (2007) Estimation of nutritional value and trace elements content of *Carthamus oxyacantha, Eruca sativa,* and *Plantago ovata. Pakistan Journal of Botany, 39*(4), 1181.

Bunkar, A.R. (2017) Therapeutic uses of *Rauwolfia serpentina. International Journal of Advance Science Research, 2*(2), 23–26.

Canter, P.H., Thomas, H., and Ernst, E. (2005) Bringing medicinal plants into cultivation: opportunities and challenges for biotechnology. *TRENDS in Biotechnology, 23*(4), 180–185.

Chadikun, P., Sakya, A.T., Cahyani, V.R., and Sri Budiastuti, M.T. (2019) Physicochemical characterization of *Dioscorea* spp. in Manokwari Regency, West Papua. *Advances in Engineering Research, 194*, 24–29.

Chandra, D. and Prasad, K. (2017) Phytochemicals of *Acorus calamus* (Sweet flag). *Journal of Medicinal Plants Studies, 5*(5), 277–281.

Chandrasekara, A. and Josheph Kumar, T. (2016) Roots and tuber crops as functional foods: a review on phytochemical constituents and their potential health benefits. *International Journal of Food science 2016*, 1–15.

Chandrika, U.G. and Prasad Kumara, P.A. (2015) Gotu Kola (Centella asiatica): nutritional properties and plausible health benefits. *Advances in Food and Nutrition Research, 76*, 125–157.

Chanu, W.S. and Sarangthem, K. (2014) Phytochemical constituents of *Justicia adhatoda* linn. found in Manipur. *Indian Journal of Plant Science, 3*(2), 2319–3824.

Chatterjee, M. and Mukherjee, A. (2021) Evaluation of bioactivity of green nanoparticles synthesized from traditionally used medicinal plants: a review. *Evidence Based Validation of Traditional Medicines*, Springer, Singapore, 799–815.

Chaudhari, S.S. and Chaudhari, G.S. (2015) A review on plumbago zeylanica linn.-A divine medicinal plant. *International Journal of Pharmaceutical Sciences Review and Research, 30*(2), 119–127.

Chauhan J. and Krishnamurthy, R. (2020) Water hyacinth based organic farming as a sustainable method for cultivation of *Andrographis Paniculata*: a medicinal plant of great prominence. *International Journal of Creative Research Thoughts,8*(8), 1837–1849

Chauhan, M. A review on morphology, phytochemistry and pharmacological activities of medicinal herb *Plumbago zeylanica* Linn. *Journal of Pharmacognosy and Phytochemistry, 3*(2), 95–118

Chauhan, N.S., Sharma, V., Thakur, M., and Dixit, V.K. (2010) *Curculigo orchioides:* the black gold with numerous health benefits. *Journal of Chinese Integrative Medicine, 8*(7), 613–623.

Choudhury, B.H., Baruah, A.M., and Das, P. (2017) Minerals and arsenic composition of twenty five indigenous leafy vegetables of Jorhat district of Assam state India. *Asian Journal of Chemistry, 29*(10), 2138–2142

Choudhury, N., Chandra, K.J., and Ansarul, H. (2012) "Effect of *Costus speciosus* Koen on reproductive organs of female albino mice. *International Research Journal of Pharmacy 3*(4), 200–202.

Damor, P., Kumar, H., Chinapolaiah, A., Damors, R., Manjesh, Gn., Thondaiman, V. (2019) Vegetative propagation of *Adhatoda vasica* a medicinal plant: effect of indole-3-butyric acid (IBA) on stem cuttings. *Journal of Pharmacognosy and Phytochemistry, 8,* 1176–1180.

Das, A. (2008) *Agro-Techniques of Selected Medicinal Plants.* National Medicinal Plants Board, New Delhi, India, 139–143.

Das, A., Subhash Babu, G.S., Yadav, M.A., Singh, A.R., Baishya, L.K., Rajkhowa, D.J.,and Ngachan S.V. (2016) Status and strategies for pulses production for food and nutritional security in north-eastern region of India. *Indian Journal of Agronomy, 61,* 43–57.

Deshkar, N., Tilloo, S., and Pande, V. A comprehensive review of *Rubia cordifolia* Linn. *Pharmacognosy Reviews, 2*(3), 124–134.

Devendra, P., Patel, S.S., Birwal, P., Basu, S., Deshmukh, G., and Datir, R. Brahmi (*Bacopa monnieri*) as functional food ingredient in food processing industry. *Journal of Pharmacognosy and Phytochemistry, 7*(3), 189–194.

Dey, P. and Kazi, L.K. (2015) Quantitative estimation of some essential minerals of *Gymnema sylvestre* as a potential herb in counteracting complications of diabetes. *International Journal of Research Studies in Biosciences, 3*(1), 71–74.

Dubey, S., Vijendra, K.V., Amit, K.S., Amit, K.J., and Tiwari, A. (2010) Evaluation of diuretic activity of aqueous and alcoholic rhizomes extracts of *Costus speciosus* Linn in wister Albino rats. *International Journal of Research in Ayurveda and Pharmacy (IJRAP), 1*(2), 648–652.

Ekanayake, S., Jansz, E. R., and Nair, B.M. (2000) Literature review of an underutilized legume: *Canavalia gladiata* L. *Plant foods for Human Nutrition, 55*(4), 305–321.

Ercisli, S., Coruh, I., Gormez, A., Sengul, M., and Bilena, S. "Total phenolics, mineral contents, antioxidant and antibacterial activities of *Glycyrrhiza glabra* L. roots grown wild in Turkey. *Italian Journal of Food Science, 20*(1).

Ezeokonkwo Mercy, A. and Okafor Sunday, N. (2015) Proximate composition and mineral analysis of *Mucuna utilis* (Velvet Bean). *IOSR Journal of Applied Chemistry,8*(10), 42–45.

Falowo, A.B.,. Mukumbo, F.E., Idamokoro, E.M., Lorenzo, J.M., Afolayan, A.J., and Muchenje, V. (2018) Multi-functional application of *Moringa oleifera* Lam. in nutrition and animal food products: a review. *Food Research International, 106,* 317–334.

FAO (2017) *Food Composition Database for Biodiversity.* Version 4.0- BioFoodComp4.0. Rome, Italy.

Farooqi, A.A. and Sreeramu, B.S. (2010) Cultivation of Medicinal and Aromatic Crops, universine Press Private ltd, 1–8

Gadzirayi, C.T., Mudyiwa, S.M., Mupangwa, J.F., and Gotosa, J. (2013) Cultivation practices and utilisation of *Moringa oleifera* provenances by small holder farmers: case of Zimbabwe. *Asian Journal of Agricultural Extension, Economics & Sociology, 2*(2), 152–162.

Galal, A., Raman, V., and Khan, I.A. (2015) *Sida cordifolia,* a traditional herb in modern perspective–a review. *Current Traditional Medicine, 1*(1), 5–17.

Gan, R.-Y, Lui, W.-Y., and Corke, H. (2016) Sword bean (*Canavalia gladiata*) as a source of antioxidant phenolics. *International Journal of Food Science & Technology, 51*(1), 156–162.

Ganash, M. and Qanash, S. (2018) Phenolic acids and biological activities of *Coleus forskohlii* and *Plectranthus barbatus* as traditional medicinal plants. *International Journal of Pharmacology, 14*(6), 856–865.

Ganie, S.H., Ali, Z., Das, S., Srivastava, P.S.,and Sharma, M.P. (2015) Identification of Shankhpushpi by morphological, chemical and molecular markers. *European Journal of Biotechnology and Bioscience, 3*(2), 01–09.

Gantait, S. and Panigrahi, J. (2018) In vitro biotechnological advancements in Malabar nut (*Adhatoda vasica* Nees): Achievements, status and prospects. *Journal of Genetic Engineering and Biotechnology, 16*(2), 545–552.

Gasmalla, M.A.A., Yang, R., and Hua, X. (2014) *Stevia rebaudiana* Bertoni: an alternative sugar replacer and its application in food industry. *Food Engineering Reviews, 6*(4), 150–162.

Ghadheri, F.F., Alimagham, S.M., Kameli, A.M., and Jamali, M. (2012) Isabgol (Plantago Ovata Forsk) seed Germination and Emergence as Affected by Environmental Factors and Planting, *International Journal of Plant Production, 6*(2) 185–194.

Gopalakrishnan, L., Doriya, K.,and Kumar, D.S. (2016) *Moringa oleifera*: A review on nutritive importance and its medicinal application. *Food Science and Human Wellness, 5*(2), 49–56.

Govaerts, R., Wilkin, P.,and Saunders, R.M.K.(2007) *World Checklist of Dioscorales: Yams and Their Allies.* Royal Botanic Gardens Kew.

Goyal, S.K., Samsher, and Goyal, R.K. (2010) *Stevia* (*Stevia rebaudiana*) a bio-sweetener: a review. *International Journal of Food Sciences and Nutrition*, *61*(1), 1–10.

Hameed, I., and Hussain, F. (2015) Proximate and elemental analysis of five selected medicinal plants of family Solanaceae. *Pakistan Journal of Pharmaceutical Sciences*, *28*(4), 1203–1215

Haque, M.E., Roy, A.C., and Rani, M. (2018) Review on Phytochemical and Pharmacological Investigation of *Piper chaba* Hunter. *International Journal of Scientific & Engineering Research*, *9*(3), 937–941

Harisaranraj, R., Suresh, K. and Saravanababu, S. (2009) Evaluation of the chemical composition *Rauwolfia serpentina* and *Ephedra vulgaris*. *Advances in Biological Research*, *3*(5–6), 174–178.

Hashim, P. (2011) *Centella asiatica* in food and beverage applications and its potential antioxidant and neuroprotective effect. *International Food Research Journal*, *18*(4), 12–15.

Hazra, A.K., Sur, T.K., Chakraborty, B., and Seal, T. (2018) HPLC analysis of phenolic acids and antioxidant activity of some classical ayurvedic Guggulu formulations. *International Journal of Research of Ayurveda & Pharmacy*, *9*(1), 112–117.

Hussain, I and Hamayun, K. (2010) Investigation of heavy metals content in medicinal plant, *Eclipta alba* L. *Journal of the Chemical Society of Pakistan*, *32*(1), 28–33.

Hussain, J., Abdul Latif K., Rehman, N., Zainullah, K.F., Hussain, S.T.,and Shinwari, Z.K. (2009) Proximate and nutrient investigations of selected medicinal plants species of Pakistan. *Pakistan Journal of Nutrition*, *8*(5), 620–624.

Hussain, J., Abdul, L.K., Najeeb, R., Muhammad, H., Zabta, K.S., Wasi, U., and In-Jung, L. (2009) Assessment of herbal products and their composite medicinal plants through proximate and micronutrients analysis. *Journal of Medicinal Plants Research*, *3*(12), 1072–1077.

Hussain, W., Badshah, L., Ullah, M., Ali, M., Ali, A., and Hussain, F. (2018) Quantitative study of medicinal plants used by the communities residing in Koh-e-Safaid Range, Northern Pakistani-Afghan borders. *Journal of Ethnobiology and Ethnomedicine*, *14*(1), 1–18.

Ibeto, A.U., Isiaka, A.A., and Temitayo, M.B. (2019) Comparative study of the chemical composition of Jack Bean (*Canavalia ensiformis*) and Sword Bean (*Canavalia gladiata*) seeds. *Journal of Chemical Research*, *1*(2), 246–255.

Imam, H., Riaz, Z., Azhar, M., Sofi, G., and Hussain, A. () Sweet flag (*Acorus calamus* Linn.): an incredible medicinal herb. *International Journal of Green Pharmacy (IJGP)*, *7*(4), 288–96.

Imam, H., Riaz, Z., Azhar, M., Sofi, G., and Hussain, A. (2013) Sweet flag. *An overview International Journal of Green Pharmacy*, *106*(67), 65.

Indriani, N.P., Mustafa, H.K., Ayuningsih, B., and Rochana, A. (2019) Production and nitrogen, phosphorus and calcium absorption of sword bean leaf (*Canavalia gladiata)* in application of rock phosphate and VAM inoculation. *Legume Research-An International Journal*, *42*(2), 238–242.

Isha, K., Madhusudan, S., Walia, B.,and Chaudhary, G. (2021) *Rauwolfia serpentine* (Sarpgandha): a review based upon its phytochemistry and ayurvedic uses. *International Journal of Current Research*, *13*(03), 16727–16734.

Islam, M.T., Hasan, J., Snigdha, H.M.S.H., Ali, E.S., Sharifi-Rad, J., Martorell, M., and Mubarak, M.S. (2020) Chemical profile, traditional uses, and biological activities of *Piper chaba* Hunter: A review. *Journal of Ethnopharmacology*, *257*, 112853.

Jahan, Y., and Siddiqui, H.H. (2012) Study of antitussive potential of *Glycyrrhiza glabra* and *Adhatoda vasica* using a cough model induced by sulphur dioxide gas in mice. *International Journal of Pharmaceutical Sciences and Research*, *3*(6), 1668.

Jain, N. and Nadgauda, R.S. (2013) *Commiphora wightii* (Arnott) Bhandari—a natural source of guggulsterone: facing a high risk of extinction in its natural habitat. *American Journal of Plant Sciences*, *4*(6A).

Jalalpure, S.S. (2011) A comprehensive review on *Plumbago zeylanica* Linn. *African Journal of Pharmacy and Pharmacology*, *5*(25), 2738–2747.

Jat, R.S., Basak, B.B., and Gajbhiye, N.A. (2021) Organic manures and biostimulants fostered soil health and increased the harvest quality of the medicinal herb ashwagandha. *Agronomy Journal*, *113*(1), 504–514.

Jayakumar, T., Hsieh, C.-Y., Lee, J.-J., and Sheu, J.-R. (2013) Experimental and clinical pharmacology of *Andrographis paniculata* and its major bioactive phytoconstituent andrographolide. *Evidence-Based Complementary and Alternative Medicine*, *2013*.

Jayanthy, A., Prakash Kumar, U., and Remashree, A.B. (2013) Effect of spacial and temporal variations in cellular characters and chemical contents in *Pseudarthria viscida* (l.) W. & Arn.–a medicinal plant. *International Journal of Pharmacognosy and Phytochemical Research*, *5*(3), 177–182.

Jena, M.K., Kancharla, S., and Kolli, P. (2021) *Curculigo Orchioides* (Kali Musli): a plant with huge medicinal properties. *Plant Cell Biotechnology And Molecular Biology*, 22(27–28), 48–56.

Jesus, M., Martins, A.P., Gallardo, E., and Silvestre, S. (2016) Diosgenin: recent highlights on pharmacology and analytical methodology. *Journal of Analytical Methods in Chemistry*, 2016, 4156293.

Jitendra, S., Kumawat, P.C., Raj, K., Manmohan, J.R., Pandey, S.B.S., and Singh, S. S. Propagation of guggal [*Commiphora wightii* (Arnott)] Bhand through cuttings. *Indian Journal of Agroforestry*, 11(2), 76–79.

Jnanesha, A.C., Kumar, A., Vanitha, T.K., and Verma, D.K. (2018) Opportunities and challenges in the cultivation of senna (*Cassia angustifolia* (Vahl.)). *International Journal of Herbal Medicine*, 6, 41–43.

Johnson, M., Wesely, E.G., Zahir Hussain, M.I., and Selvan, N. (2010) *In vivo* and *in vitro* phytochemical and antibacterial efficacy of *Baliospermum montanum* (Willd.) Muell. Arg. *Asian Pacific Journal of Tropical Medicine*, 3(11), 894–897.

Joshi, R.K., Setzer, W.N., and Da Silva, J.K. (2017) Phytoconstituents, traditional medicinal uses and bioactivities of Tulsi (*Ocimum sanctum* Linn.): a review. *American Journal of Essential Oil and Natural Products*, 5(1), 18–21.

Joshi, V.K., Joshi, A., and Dhiman, K.S. (2017) The ayurvedic pharmacopoeia of India, development and perspectives. *Journal of Ethnopharmacology*, 197, 32–38.

Kalam, M.A., and Majeed, S. (2020) Asrawl-*Rauwolfia Serpentina* (L) Benth. Ex. Kurz, an effective drug of Unani system of medicine for neurological and cardiological disorders: a review. *World Journal of Pharmacy and Pharmaceutical Sciences*, 9(3), 915–925.

Kanetkar, P., Singhal, R., and Kamat, M. (2007) *Gymnema sylvestre*: a memoir. *Journal of Clinical Biochemistry and Nutrition*, 41(2), 77–81

Kanungo, S., Nahak, G., Sahoo, S.L., and Sahu, R.K. (2012) Antioxidant activity and phytochemical evaluation of *Plumbago zeylanica* Linn. *in vivo* and *in vitro*. *International Journal of Pharmacy and Pharmaceutical Sciences*, 4(Suppl 4), 522–26.

Karunarathne, Y.A.U.D., Amarasinghe, A.P.G., Weerasooriya, T.R., Samarasinghe, U.K.A.,and Arawwawala, L.D.A.M. (2020) *Asparagus racemosus* (Willd) of Indian origin: in terms of physico-chemical, phytochemical and nutritional profiles. *Scholars International Journal of Traditional and Complementary Medicine*, 3(7), 140–143.

Kasture, A. and Krishnamurthy, R. (2016) Evaluation of the chemical variability and phytochemical analysis among Indian germplasm of *Acorus calamus* Linn using GC-MS and FTIR. *European Journal of Medicinal Plants*, 1–11.

Kasture, A., Krishnamurthy, R., and Rajkumar, K. (2016) Genetic variation in the endangered Indian sweet flag (*Acorus calamus* L.) estimated using ISSR and RAPD markers. *Journal of Applied Research on Medicinal and Aromatic Plants*, 3(3), 112–119.

Kaur, G., Invally, M., Sanzagiri, R., and Buttar, H.S. (2015) Evaluation of the antidepressant activity of *Moringa oleifera* alone and in combination with fluoxetine. *Journal of Ayurveda and integrative medicine*, 6(4), 273.

Kaushik, N. (2005) Saponins of *Chlorophytum* species. *Phytochemistry Reviews*, 4(2–3), 191–196.

Kavitha, C.,Rajamani, K., and Vadivel, E. (2010) *Coleus forskohlii* a comprehensive review on morphology, phytochemistry and pharmacological aspects. *Journal of Medicinal Plants Research*, 4(4), 278–285.

Khaitov, B., Urmonova, M., Karimov, A., Sulaymanov, B., Allanov, K., Israilov, I., and Sottorov, Licorice (*Glycyrrhiza glabra*)—growth and phytochemical compound secretion in degraded lands under drought stress. *Sustainability*, 13(5), 2923.

Khalekuzzaman, M., Rahman, M.S., Rashid, M.H., and Hossain, M.S. (2008) High frequency *in vitro* propagation of *Adhatoda vasica* Nees through shoot tip and nodal explants culture. *Journal of Bio-Science*, 16, 35–39.

Khan, F. and Ramu, A. (2018) Spectral characterization of bioactive compounds from *Costus pictus* and *Costus speciosus*. *International Journal of ChemTech Research*,11(9), 353–363.

Khanam, Z., Singh, O., Singh, R., and Bhat, I.U.H. (2013) Safed musli (*Chlorophytum borivilianum*): a review of its botany, ethnopharmacology and phytochemistry. *Journal of Ethnopharmacology*, 150(2), 421–441.

Khare, C.P. (2004) *Encyclopedia of Indian Medicinal Plants: Rational Western Therapy, Ayurvedic and Other Traditional Usage, Botany*, Springer.

Kherde, S.D., Parmar, K.M., Tawar, M.G., Prasad, S.K., and Itankar, P.R. (2020) Study on impact of different climatic zones on physicochemical and phytochemical profile of *Withania somnifera* (L.) Dunal. *Indian Journal of Traditional Knowledge*, 19(3), 486–493.

Khramov, V.A., Spasov, A.A., and Samokhina, M.P. (2008) Chemical composition of dry extracts of *Gymnema sylvestre* leaves. *Pharmaceutical Chemistry Journal*, 42(1), 29–31.

Khurana, N., Sharma, N., Patil, S., and Gajbhiye, A. (2016) Phyto-pharmacological properties of *Sida cordifolia*: a review of folklore use and pharmacological activities. *Asian Journal of Pharmaceutical and Clinical Research*, 2, 52–58.

Khwairakpam, A.D., Damayenti, Y.D., Deka, A., Monisha, J., Roy, N.K., Padmavathi, G., and Kunnumakkara, A.B. (2018) *Acorus calamus*: a bio-reserve of medicinal values. *Journal of Basic and Clinical Physiology and Pharmacology*, 29(2), 107–122.

Kirtikar, K.R. and Basu, B.D. (1918) *Indian Medicinal Plants*, PuSudhindra Nath Basu, M.B. Panini Office, Bhuwanéswari Asrama, Bahadurganj.

Kirtikar, K.R. and Basu, B.D. (1994) *Indian Medicinal Plants* (second edition), vol. II, Lalit Mohan Basu, Allahabad, 870–872.

Kothari, S. and Singh, K. (2003) Production techniques for the cultivation of safed musli (*Chlorophytum borivilianum*). *The Journal of Horticultural Science and Biotechnology*, 78(2), 261–264.

Kriker, S. (2013) Effect of climate on some morphological and chemical characteristics of the plant *Glycyrrhiza glabra* L. in two arid regions of southern Algeria. *Egyptian Academic Journal of Biological Sciences, H. Botany*, 4(2), 1–9.

Krishnamoorthy, C and Chidambaram, R. (2019) Study on phytochemicals, functional groups and minerals in *Asparagus racemosus* root. *Asian Journal of Chemistry*, 31(7), 1546–1548.

Krishnamurthy, R., Chandorkar, M.S., Kalzunkar, B.G., Palsuledesai, M.R., Pathak, J.M., and Rajendra, G. (2004) Agronomic manipulations for improving the productivity of Shatavari (*Asparagus racemosus*) under irrigated and rainfed conditions. *Journal of Medicinal and Aromatic Plant Sciences*, 26, 740–745.

Krishnamurthy, R., Chandorkar, M.S., Kalzunkar, E.G., Pathak, J.M., and Rajendra G. (2006) Studies on agronomic practices for growing *Centella asiatica* (L.) urban in high rainfall localities under open and partial shade of mango orchards. *Indian Journal of Horticulture*, 63(1), 76–80.

Krishnamurthy, R., Chandorkar, M.S., Palsuledesai, M.R., Kalzunkar, B.G., Pathak, J.M., and Gupta, R. (2005) Genetic variability and correlation studies in yield associated traits of *M. pruriens* (L.) DC. germplasm. *Journal of Non- Timber Forest Products*, 12(3), 135–145

Krishnamurthy, R., Chandorkar, M.S., Pathak, J.M., Adedayo, A.D., and Gupta, R.B. (2015) Selection of elite lines from accessions of *Gymnema sylvestre* (Gudmar) based on characterization of foliage and gymnemic acid yield. *International Journal of Medicinal Plants*, 108, 596–605

Krishnamurthy, R., Suvagia, P., Pathak, J.M., Parikh, K.M., and Bhikne, N.S. (1996) Selection of Mucuna lines for higher L-Dopa and improved yield through conventional breeding. *Proceedings of the Conference on Breeding Research on Medicinal Plants*, 2, 231–234.

Krizevski, R. and Lewinsohn, E. (2009) *Germination of Bala (Sida cordifolia L., Malvaceae), An Ayurvedic Plant*. IV International Symposium on Breeding Research on Medicinal and Aromatic Plants-ISBMAP2009 860, Ljubljana, Slovenia, 219–222.

Kulkarni, S.S., Lanjewar, R.B., and Gadegone, S.M. (2016) A review on levodopa and beta-sitosterol and its pharmacological actions in *Bauhinia racemosa, Canavalia gladiata, Vigna vexillata* medicinal plants. *Journal of Medicinal Plants*, 4(4), 259–264.

Kumar, A. and Jnanesha, A.C. (2017) Cultivation, utilization and role of medicinal plants in tradition medicine in deccan eco-climate. *International Journal on Agricultural Sciences*, 8(1), 98–103.

Kumar, A. and Jnanesha, A.C. (2017) Enhancing the income of the farmer by cultivating Senna in low rainfall area. *Popular kheti*, 5(01), 14–17.

Kumar, A. and Jnanesha, C. (2019) Cultivation and medicinal properties of Giloy [*Tinospora cordifolia* (Thunb.) Miers]. novel aspects in nanotechnology and applications and future challenges. *Medicinal and Aromatic Plants*, 17, 57–58.

Kumar, A., Gupta, C., Nair, D.T., and Salunke, D.M. (2016) MP-4 contributes to snake venom neutralization by *Mucuna pruriens* seeds through an indirect antibody-mediated mechanism. *Journal of Biological Chemistry*, 291(21), 11373–11384.

Kumar, J.B., Basak, B., Jat, R.S., and Reddy, N. (2016) Medicinal and aromatic plants making healthy and wealthy. *Indian Horticulture*, ICAR-Directorate of Medicinal & Aromatic Plants Research, Anand,Gujarat, vol. 61, no. 6.

Kumar, M., Dandapat, S., and Sinha, M. (2013) Determination of Nutritive Value and Mineral Elements of Five-Leaf Chaste Tree (*Vitex negundo* L.) and Malabar Nut (*Adhatoda vasica* Nees). *Academic Journal of Plant Sciences*, 6(3), 103–108.

Kumar, P. and Narsimha Reddy, Y. (2014) Protective effect of *Canavalia gladiata* (sword bean) fruit extracts and its flavanoidal contents, against azathioprine-induced toxicity in hepatocytes of albino rats. *Toxicological & Environmental Chemistry*, 96(3), 474–481.

Kumar, S., Das, G., Shin, H.-S., and Patra, J.K. (2017) *Dioscorea* spp.(a wild edible tuber): a study on its ethnopharmacological potential and traditional use by the local people of Similipal Biosphere Reserve, India. *Frontiers in Pharmacology, 8*, 52.

Kumari, R., Rathi, B., Rani, A., and Bhatnagar, S. (2013) *Rauvolfia serpentina* L. Benth. ex Kurz.: phytochemical, pharmacological and therapeutic aspects. *International Journal of Pharmaceutical Sciences Review and Research, 23*(2), 348–355.

Kumawat, N.K., Kishore, D., Vaishya, J.K., Balakrishnan, P., and Nesari, T.M. (2019) *Tinospora cordifolia*: a wonderful miracle herb of 21st century of India. *Medicinal Plants-International Journal of Phytomedicines and Related Industries, 11*(2), 117–122.

Kumssa, D.B., Joy, E.J.M., Young, S.D., Odee, D.W., Louise Ander, E., and Broadley, M.R. (2017) Variation in the mineral element concentration of *Moringa oleifera* Lam. and *M. stenopetala* (Bak. f.) Cuf.: role in human nutrition. *PloS One, 12*(4), e0175503.

Laghari, A.Q., Memon, S., Nelofar, A., and Laghari, A.H. (2011) Extraction, identification and antioxidative properties of the flavonoid-rich fractions from leaves and flowers of *Cassia angustifolia*. *American Journal of Analytical Chemistry, 2*(08), 871.

Lal, S. and Baraik, B. (2019) Phytochemical and pharmacological profile of *Bacopa monnieri*-an ethnomedicinal plant. *International Journal of Pharmaceutical Science & Research, 10*(3), 1001–1013.

Lalitha, P. and Gayathiri, P. (2013) *In vitro* anti-inflammatory and phytochemical properties of crude ethyl acetate extract of *Baliospermum montanum* leaf (Muell–Arg). *African Journal of Biotechnology, 12*(39), 5743–5748.

Lim, T.K. *Glycyrrhiza glabra*. In: *Edible Medicinal and Non-Medicinal Plants*. Springer, Dordrecht, 354–457.

Lohar, A.V., Wankhade, A.M., Faisal, M. and Jagtap, A. (2020) A review on *Glycyrrhiza glabra* linn (liquorice)-an excellent medicinal plant. *European Journal of Biomedical, 7*(7), 330–334.

Lokesh, B., Deepa, R., and Divya, K. (2018) Medicinal Coleus (*Coleus forskohlii* Briq): a phytochemical crop of commercial significance. *Journal of Pharmacognosy and Phytochemistry, 7*, 2856–2864.

Madhuri, T., Krishnamurthy, R., Pathak, J.M., Chandorkar, M.S., and Tandel, D.H. (2014) Effect of foliar application of hormones and nutrients on yield of Ardusi (*Adhatoda zeylanica*). *Trends in Biosciences, 7*(13), 1416–1417.

Mahajan, R., Khinda, P.K., Gill, A.S., Kaur, J., Saravanan, S.P., Shewale, A., Taneja, M., and Joshi, V. (2016) Comparison of efficacy of 0.2% chlorhexidine gluconate and herbal mouthrinses on dental plaque: an *in vitro* comparative study. *European Journal of Medicinal Plants, 13*(2), 1–11.

Maji, P., Dhar, D.G., Misra, P., and Dhar, P. (2020) *Costus speciosus* (Koen ex. Retz.) Sm.: current status and future industrial prospects. *Industrial Crops and Products, 152*, 112571.

Mali, R.G. and Wadekar, R.R. (2008) *Baliospermum montanum* (Danti): ethnobotany, phytochemistry and pharmacology-a review. *International Journal of Green Pharmacy (IJGP), 2*(4), 194–199.

Marjhan, H.N. and Kalam, M.A. (2018) Review on pharmacological properties of ispaghula (*Plantago ovata* Forsk.) husk and seed. *Hamdard University Bangladesh, 4*(2), 23.

Mathur, S., Gupta, M.M. Ram, M., Sharma, S. and Kumar, S. (2002) Herb yield and bacoside-a content of field-grown *Bacopa monnieri* accessions. *Journal of Herbs, Spices & Medicinal Plants, 9*(1), 11–18.

Mehta, J. (2016) Phenology of *Adhatoda vasica* a multifarious useful medicinal plant. *International Journal of Applied Research, 2*(7), 791–794.

Mehta, J. and Bajaj, A. (2018) SEM studies of *Adhatoda vasica*: an endangered traditional medicinal plant of multifarious uses. *International Journal of Recent Trends in Science and Technology, 27*(1), 278–282.

Mehta, J. and Nama, K.S. (2014) A review on ethanomedicines of *Curculigo orchioides* Gaertn (Kali Musli): black gold. *International Journal of Pharmacy & Biomedical Research, 1*(1), 12–6.

Mithun, N. M., Shashidhara, S., and Vivek Kumar, R. (2011) *Eclipta alba* (L.) a review on its phytochemical and pharmacological profile. *Pharmacologyonline, 1*(1), 345–357.

Modi, B., Kumari Shah, K., Shrestha, J., Shrestha, P., Basnet, A., Tiwari, I., and Prasad Aryal, S. (2020) Morphology, biological activity, chemical composition, and medicinal value of *Tinospora Cordifolia* (willd.) miers. *Advanced Journal of Chemistry-Section B*, 36–54.

Mohan, E., Suriya, S., Shanmugam, S., and Rajendran, K. (2021) Qualitative phytochemical screening of selected medicinal plants. *Journal of Drug Delivery and Therapeutics, 11*(2), 141–144.

Mohideen, S., Sasikala, E., and Gopal, V. (2002) Pharmacognostic studies on *Sida acuta* burm. f. *Ancient Science of Life, 22*(1), 57.

Moteetee, A.N. (2016) *Canavalia* (phaseoleae, fabaceae) species in South Africa: naturalised and indigenous. *South African Journal of Botany, 103*, 6–16.

Muchtaromah, B., Ahmad, M. Koestanti, E., Yuni Ma'rifatul, A., and Velayati Labone, A. (2017) Phytochemicals, antioxidant and antifungal properties of *Acorus calamus, Curcuma mangga,* and *Allium sativum. KnE Life Sciences, 3*(6), 93–104.

Murthy, P.B.S., Raju, V.R., Ramakrisana, T., Chakravarthy, M.S., Kumar, K.V., Kannababu, S., and Subbaraju, G.V. (2006) Estimation of twelve bacopa saponins in *Bacopa monnieri* extracts and formulations by high-performance liquid chromatography. *Chemical and Pharmaceutical Bulletin, 54*(6), 907–911.

Muszyńska, B., Lazur, J., Szewczyk, A., and Opoka, W. (2018) Evaluation of nutritional and medicinal properties of *Bacopa monnieri* biomass and preparations. *Acta poloniae Pharmaceutica-Drug Research, 75*(6), 1353–1361.

Nadeeshani, H., Wimalasiri, K.M.S., Samarasinghe, G., Silva, R., and Madhujith, T. (2018) Evaluation of the nutritional value of selected leafy vegetables grown in Sri Lanka. *Tropical Agricultural Research, 29*(3), 255–267.

Naik, A., N., Ramar, K., and Janardan, P. (2017) Comparative physicochemical and phytochemical evaluation for insulin plant *Costus pictus* D.Don accessions. *International Journal of Applied Ayurved Research,* 420–426.

Naik, K. and Krishnamurthy, R. (2018) Anti-inflammatory activity of methanolic extracts of *Pseudarthria viscida* and *Uraria picta* against carrageenan induced paw edema in albino rat. *Open Access Journal of Medicinal and Aromatic Plants, 9*(1), 1.

Nalawade, A. and Gurav, R. (2017) Proximate analysis and elemental composition of seventeen species of *Chlorophytum* from the Western Ghats. *International Research Journal of Pharmacy, 8*(7), 131–136.

Nassiri-asl, M. and Hosseinzadeh, H. (2012) Licorice (*Glycyrrhiza* species) In: Singh R.J. (ed.), *Genetic Resources, Chromosome Engineering, and Crop Improvement*: Medicinal Plants. *6*(6), 935–958.

Natarajan, K., Narayanan, N., and Ravichandran, N. (2012) Review on *"Mucuna"*—the wonder plant. *International Journal of Pharmaceutical Sciences Review and Research, 17*(1), 86–93.

Nisar, S.S., Khan, A.A., Maaz, M., and Shiffa, M. (2012) Review of sankhahauli (*Convolvulus pluricaulis* choisy) from traditional medicine to modern science. *International Journal of Institute Pharmaceutical Life Sciences, 2,* 94–101.

Nitesh, G., Manish, G., Avninder, S.M., Sandeep, G., and Bansa, G. (2012) Pharmacognostic, phytochemical studies and antianxiety activity of *Uraria picta* leaves. *Journal of Pharmaceutical Research and Opinion, 1*(1), 6–9.

Nwankpa, P., Chukwuemeka, O.G., Uloneme, G.C., Etteh, C.C., Ugwuezumba, P., and Nwosu, D. (2015) Phyto-nutrient composition and antioxidative potential of ethanolic leaf extract of *Sida acuta* in wistar albino rats. *African Journal of Biotechnology, 14*(49), 3264–3269.

Nyeem, M.A.B. and Mannan, M.A. (2018) *Rubia cordifolia*-phytochemical and pharmacological evaluation of indigenous medicinal plant: a review. *International Journal of Physiology Nutrition and Physical Education, 3*(1), 766–771.

Ou-Yang, S.H., Jiang, T., Zhu, L., and Yi, T. (2018) *Dioscorea nipponica* Makino: a systematic review on its ethnobotany, phytochemical and pharmacological profiles. *Chemistry Central Journal, 12*(1), 1–18.

Padhan, B. and Panda, D. (2020) Potential of neglected and underutilized yams (*Dioscorea* spp.) for improving nutritional security and health benefits. *Frontiers in Pharmacology, 11,* 1–13.

Padhan, B., Biswas, M., Dhal, N.K., and Panda, D. (2018) Evaluation of mineral bioavailability and heavy metal content in indigenous food plant wild yams (*Dioscorea* spp.) from Koraput, India. *Journal of Food Science and Technology, 55*(11), 4681–4686.

Pal, S.K. and Shukla, Y. (2003) Herbal medicine: current status and the future. *Asian Pacific Journal of Cancer Prevention, 4*(4), 281–288.

Panda, S.K., Das, D., and Tripathy, N.K. (2010) Pharmacognostical studies on root tubers of *Chlorophytum borivilianum* Santapau and Fernandes. *International Journal of Pharmaceutical and Phytochemical Research, 2*(4), 13–7.

Pandey, A.K. (2012) Cultivation technique of an important medicinal plant *Gymnema sylvestre* R Br (Gurmar). *Academic Journal of Plant Sciences, 5*(3), 84–89.

Pandey, S. (2018) Morphology, chemical composition and therapeutic potential of *Stevia rebaudiana. Indo American Journal of Pharmaceutical Sciences, 5*(4), 2260–2266.

Pareek, A. and Ashwani, K. (2015) Bhringaraj (*Eclipta prostrata* L.) for mental improving ability. *World Journal of Pharmaceutical Sciences, 3*(8), 1569–1571.

Pasqua, G., Avato, P., Monacelli, B., Santamaria, A.R., and Argentieri, M.P. (2003) Metabolites in cell suspension cultures, calli, and *in vitro* regenerated organs of *Hypericum perforatum* cv. Topas. *Plant Science, 165*(5), 977–982.

Patel N., Swati, P., and Krishnamurthy, R. (2013) Indian *Tinospora* species: natural immunomodulators and therapeutic agents. *International Journal of Pharmaceutical Biological and Chemical Sciences, 2*(2), 1–9.

Patel, D.K. (2016) Vegetative propagation of *Coleus forskohlii* (Wild) briq using their stem cutting for ex-situ conservation in herbal garden. *Medicinal Aromatic Plants (Los Angel), 5*(261), 2167–0412.

Patel, H. and Krishnamurthy, R. (2015) Antimicrobial efficiency of biologically synthesized nanoparticles using root extract of *Plumbago zeylanica* as bio-fertilizer application. *International Journal of Bioassays, 4*(11), 4473–4475.

Patel, N. and Krishnamurthy, R. (2021) *In-vitro* phytochemical screening and bioactivity of *Moringa oleifera* accessions. *Bioscience Biotechnology Research Communication, 14*(1), 335–339.

Patel, N., Patel, N., Patel, S., Ingalhalli, R., Garuba, T., Ahmed, A., Oyeyinka, S., and Krishnamurthy, R. (2019) Morphology, growth variability and chemical composition of Indian and Nigerian accession of *Ocimum* Species grown in India. *Carpathian Journal of Food Science & Technology, 11*(4), 72–80.

Patel, N., Patel, S., Patel, N., and Krishnamurthy, R. (2020). Indian *Tinospora* species: as antibacterial agent and phytochemical evaluation. *Current Strategies in Biotechnology and Bioresource Technology*, Book Publisher International, vol. 2, 125–135.

Patel, N., Rana, M., Rahaman, A., and Krishnamurthy, R. (2020) Medicinal plant *Centella asiatica* (mandukaparni): from farm to pharma. *Trends in Pharmaceutical Research and Developement, 4*(1), 10–22.

Patel, P. and Krishnamurthy, R. (2013) Feasibility studies on *in vitro* mass-scale propagation of Indian Ashwagandha (*Withania somnifera*) cultivars for commercial purpose. *Journal of Pharmacognosy and Phytochemistry, 2*(2), 168–174.

Patel, S., Patel, N., AbdulRahaman A.A., and Krishnamurthy, R. (2020) Medicinal plant *Mucuna pruriens*: from farm to pharma.*Trends in Pharmaceutical Research and Development, 5*(1), 1–18.

Pathak, J.M., Krishnamurthy, R., Chandorkar, M.S., Gulkari, V., and Gupta, R. (2005) Identification of high yielding genotypes of Dashmool Shalparni (*Desmodium gangeticum*) drug plant and its cultivation under high density planting. *Indian Journal of Horticulture, 62*(4), 378–384.

Pattanayak, P., Behera, P., Das, D., and Panda, S.K. (2010) *Ocimum sanctum* Linn. a reservoir plant for therapeutic applications: an overview. *Pharmacognosy Reviews, 4*(7), 95.

Paul, M., Radha, A., Kumar, D. (2013) On the high value medicinal plant, *Coleus forskohlii* Briq. *Hygeia Journal for Drugs and Medicines, 5*(1), 69–78.

Pawar, V.A. and Pawar, P.R. (2014) *Costus speciosus*: an important medicinal plant. *International Journal of Science and Research, 3*(7), 28–33.

Peniche-Gonzalez, I.N., Sarmiento-Franco, L.A., and Santos-Ricalde, R.H. (2018) Utilization of *Mucuna pruriens* whole pods to feed lactating hair ewes. *Tropical Animal Health and Production, 50*(7), 1455–1461.

Perera, N. and Pavitha, P. (2017) Development of a sauce using *Gymnema sylvestre* leaves. *Sri Lanka Journal of Food and Agriculture, 3*(1), 29–36.

Pokhrel, S. and Chaulagain, K. (2020) Phytoconstituents and biological analysis of *Acorus calamus* rhizome of Sindhupalchowk district, Nepal. *BIBECHANA, 17*, 104–109.

Prakash, V., Jaiswal, N., and Srivastava, M.R.I.N.A.L. (2017) A review on medicinal properties of *Centella asiatica*. *Asian Journal of Pharmaceutical and Clinical Research, 10*(10), 69.

Pramanick, D.D., Maiti, G.G., and Srivastava, A. (2015) Micro-morphological study of 'BALA'plant (*Sida cordifolia* L., Malvaceae) with special reference to its propagation technique. *Journal of Medicinal Plant Studies, 3*(4), 127–131.

Purseglove, J.W. (1968) *Canavalia gladiata* (Jacq.) DC. *Tropical Crops: Dicotyledons, 245*, 333–719.

Raaman, N., Selvarajan, S., Balakrishnan, D., and Balamurugan, G. (2009) Preliminary phytochemical screening, antimicrobial activity and nutritional analysis of methanol extract of *Curculigo orchioides* Gaerten rhizomes. *Journal of Pharmacy Research, 2*(7), 1201–1202.

Rahal, A., Prakash, A., Verma, A.K., Kumar, V., and Roy, D. (2014) Proximate and elemental analyses of *Tinospora cordifolia* stem. *Pakistan Journal of Biological Sciences: PJBS, 17*(5), 744–747.

Ramakrishnappa, K. (2003) Impact of cultivation and gathering of medicinal plants on biodiversity: case studies from India. *Biodiversity and the Ecosystem Approach in Agriculture, Forestry and Fisheries. Proceedings*, Rome, Italy, 12–13, 168–195.

Ramchander, .J. P., and Middha, A. (2017) Recent advances on Senna as a laxative: a comprehensive review. *Journal of Pharmacognosy Phytochemistry, 6*(2), 349–353.

Ramesh, K., Singh, V., and Megeji, N.W. (2006) Cultivation of stevia [*Stevia rebaudiana* (Bert.) Bertoni]: a comprehensive review. *Advances in Agronomy, 89*, 137–177.

Rameshkumar, K.B., Aravind, A.A., and Mathew, P.J. (2011) Comparative phytochemical evaluation and antioxidant assay of *Piper longum* L. and *Piper chaba* hunter used in Indian traditional systems of medicine. *Journal of Herbs, Spices & Medicinal Plants, 17*(4), 351–360.

Rami, M.M., Oyekanmi, A.M., and Adegoke, B.M. (2014) Proximate, phytochemical and micronutrient composition of *Sida acuta*. *Journal of Applied Chemistry, 7*(2), 93–98.

Rane, M., Suryawanshi, S., Patil, R., Aware, C., Jadhav, R., Gaikwad, S., Singh, P., et al. (2019) Exploring the proximate composition, antioxidant, anti-Parkinson's and anti-inflammatory potential of two neglected and underutilized *Mucuna* species from India. *South African Journal of Botany, 124* 304–310.

Rani, A., Zahirah, N., Husain, K.,and Kumolosasi, E. (2018) *Moringa* genus: a review of phytochemistry and pharmacology. *Frontiers in Pharmacology, 9*, 108.

Ranovona, Z., Mertz, C., Dhuique-Mayer, C., Servent, A., Dornier, M., Danthu, P., and Ralison, C. (2019) The nutrient content of two folia morphotypes of *Centella asiatica* (L) grown in Madagascar. *African Journal of Food, Agriculture, Nutrition and Development, 19*(3), 14654–14673.

Rawat, M.S., Pant, S., and Badoni, Y.S. (1994) Negi. Biochemical investigation of some wild fruits of Garhwal Himalayas. *Progressive Horticulture, 26*, 1–2.

Rayaguru, K. and Khan, M.K. (2008) Post-harvest management of *Stevia* leaves: a review. *Journal of Food Science and Technology,. 45*, 391–397.

Reddy, C.S., Meena, S.L., Krishna, P.H., Charan, P.D., and Sharma, K.C. (2012) Conservation threat assessment of *Commiphora wightii* (Arn.) Bhandari—an economically important species. *Taiwania, 57*(3), 288–293.

Regupathi, T. and Chitra, K. (2015) Physicochemical analysis of medicinal herbs, *Eclipta alba* (L.) Hassk and *Lippia nodiflora* (Linn.). *International Journal of Pharmaceutical and Phytopharmacological Research, 4*(4), 249–251.

Rizzato, G., Scalabrin, E., Radaelli, M., Capodaglio, G., and Piccolo, O. (2017) A new exploration of licorice metabolome. *Food chemistry, 221*, 959–968.

Rodrigues, F.C., and de Oliveira, A.F.M. (2020) The genus *Sida* L.(Malvaceae): an update of its ethnomedicinal use, pharmacology and phytochemistry. *South African Journal of Botany, 132*, 432–462.

Romero-Baranzini, A.L., Rodriguez, O.G., Yanez-Farias, G.A., Barron-Hoyos, J.M., and Rayas-Duarte, P. (2006) Chemical, physicochemical, and nutritional evaluation of *Plantago* (*Plantago ovata* Forsk). *Cereal Chemistry, 83*(4), 358–362.

Rout, O.P., Acharya, R., and Mishra, S.K. (2012) Oleogum resin guggulu: a review of the medicinal evidence for its therapeutic properties. *International Journal of Research in Ayurveda and Pharmacy, 3*(1), 15–21.

Roy, A. and Bharadvaja, N. (2017) A review on pharmaceutically important medical plant: *Plumbago zeylanica*. *Journal of Ayurvedic and Herbal Medicine, 3*(4), 225–228.

Rukachaisirikul, T., Chokchaisiri, S., Suebsakwong, P., Suksamrarn, A., and Tocharus, C. (2017) A new ajmaline-type alkaloid from the roots of *Rauvolfia* serpentina. *Natural Product Communications, 12*(4), 1934578X1701200408.

Rungsung, W., Dutta, S., Mondal, D.N., Ratha, K.K., and Hazra, J. (2014) Pharmacognostical profiling on the root of *Rauwolfia serpentina*. *International Journal of Pharmacognosy and Phytochemical Research, 6*(3), 612–616.

Saffie, N., Ariff, F.F.M., Bahari, S.N.S., Abdullah, M.Z., and Taini, M.M. (2019) Germination, propagation and selection of *Andrographis paniculata* for future reeding programme. *International Journal of Agriculture, Forestry and Plantation, 8*, 168–174.

Salehi, B., Sener, B., Kilic, M., Sharifi-Rad, J., Naz, R., Yousaf, Z., Mudau, F.N., et al. (2019) *Dioscorea* plants: a genus rich in vital nutra-pharmaceuticals-a review. *Iranian Journal of Pharmaceutical Research: IJPR, 18*,(suppl1), 68.

Samantaray, S., Rout, G.R., and Das, P. (2001) Heavy metal and nutrient concentration in soil and plant growing on metalliferous chromite minespoil. *Environmental Technology, 22*(10), 1147–1154.

Saneja, A., Sharma, C., Aneja, K.R., and Pahwa, R. (2010) *Gymnema sylvestre* (Gurmar): a review. *Der Pharmacia Lettre, 2*(1), 275–284.

Sankar, R.V., Ravichandran, P., and Ravikumar, K. (2012) *Plant Resources of Tiruvannamalai District, Tamil Nadu, India*. Bishen Singh Mahendra Pal Singh, Dehradun, India, 343.

Sarfraz, R.M., Khan, H., Maheen, S., Afzal, S., Akram, M.R., Mahmood, A., and Afzal, K. (2017) *Plantago ovata*: a comprehensive review on cultivation, biochemical, pharmaceutical and pharmacological aspects. *Acta Poloniae Pharmaceutica, 74*(3), 739–746.

Saroya, A.S. and Singh, J. (2018) Phytopharmacology of Indian nootropic *Convolvulus pluricaulis*. In *Pharmacotherapeutic Potential of Natural Products in Neurological Disorders*, Springer, Singapore, 141–144.

Sastry, K.P., Rao, B.R.R., Rajput D.K., Singh C.P., and Singh. K. (2007) Cultivation and processing of *Cassia angustifolia* Vahl. in Tamil Nadu. *Advances in Medicinal Plants* (1st edition), Universities Press (India) Private Limited, Hyderabad, India, 123–129.

Săvulescu, E., Georgescu, M.I., Popa, V., and Luchian, V. (2018) Morphological and anatomical properties of the *Senna alexandrina* Mill.(*Cassia angustifolia* Vahl.). sciendo. *Agriculture for Life for Agriculture Conference Proceedings*, 1(1), 305–310.

Saxena, H.O., Soni, A., Mohammad, N., and Choubey, S.K. (2014) Phytochemical screening and elemental analysis in different plant parts of *Uraria picta* Desv.: a *Dashmul* species. *Journal of Chemical and Pharmaceutical Research*, 6(5), 756–760

Schippmann, U., Leaman, D.J., and Cunningham, A.B. (2002) Impact of cultivation and gathering of medicinal plants on biodiversity: global trends and issues. *Biodiversity and the Ecosystem Approach in Agriculture, Forestry and Fisheries*, FAO, 142–167.

Seena, S., Sridhar, K.R., and Bhagya, B. (2005) Biochemical and biological evaluation of an unconventional legume, *Canavalia maritima* of coastal sand dunes of India. *Tropical and Subtropical Agroecosystems*, 5(1), 1–14.

Sekhar, C., Venkatesan, N., Vidhyavathi, A., and Murugananthi, M. (2017) Post-harvest processing of *Moringa* and socio-economic appraisal of *Moringa* orchards in Tamil Nadu. *International Journal of Horticulture*, 7, 275–287.

Seneviratne Vajira P.S., Hapuarachchi S.D., and Perera P.K. (2015) Standardization and quality control of *Centella asiatica* Linn. (Gotukola) dried powder and capsules. *World Journal of Pharmacy and Pharmaceutical Sciences*, 5(1), 42–50.

Sethiya, N.K. and Mishra, S. (2015) Simultaneous HPTLC analysis of ursolic acid, betulinic acid, stigmasterol and lupeol for the identification of four medicinal plants commonly available in the Indian market as Shankhpushpi. *Journal of Chromatographic Science*, 53(5), 816–823.

Sethiya, N.K., Trivedi, A., Patel, M.B., and Mishra, S.H. (2010) Comparative pharmacognostical investigation on four ethanobotanicals traditionally used as Shankhpushpi in India. *Journal of Advanced Pharmaceutical Technology & Research*, 1(4), 388–395.

Shahjahan, M., Solaiman, A.H.M., Sultana, N., and Kabir, K. (2013) Effect of organic fertilizers and spacing on growth and yield of Kalmegh (*Andrographis paniculata* Nees). *International Journal of Agriculture and Crop Sciences*, 6(11), 769.

Sharma, P., Dwivedee, B.P., Bisht, D., Dash, A.K., and Kumar, D. (2019) The chemical constituents and diverse pharmacological importance of *Tinospora cordifolia*. *Heliyon*, 5(9), e02437.

Sharma, P., Kaushik, R., Khan, A.D., Malik, M.T., and Fagna, B. (2019) Pharmacognostical, physicochemical and phytochemical studies of *Piper longum* Linn. fruits. *International Journal of Pharmaceutical Education and Research*, 1(1), 19–24.

Sharma, R., Kumari, N., Ashawat, M.S., and Verma, C.P.S. (2020) Standardization and phytochemical screening analysis for herbal extracts: zingiber officinalis, rosc., *Curcuma longa* Linn., *Cinnamonum zeylanicum* Nees., *Piper longum*, Linn., *Boerhaavia diffussa* Linn. *Asian Journal of Pharmacy and Technology*, 10(3), 127–133.

Siddhuraju, P., Vijayakumari, K., and Janardhanan, K. (1996) Chemical composition and protein quality of the little-known legume, velvet bean (*Mucuna pruriens* (L.) DC.). *Journal of Agricultural and Food Chemistry*, 44(9), l2636–2641.

Siddiqui, M.Z. (2011) *Boswellia serrata*, a potential anti-inflammatory agent: an overview. *Indian Journal of Pharmaceutical Sciences*, 73(3), 255.

Singh, A.K., Srivastava, A., Kumar, V., and Singh, K. (2018) Phytochemicals, medicinal and food applications of shatavari (*Asparagus racemosus*): an updated review. *The Natural Products Journal*, 8(1), 32–44.

Singh, D. and Chaudhuri, P.K. (2018) A review on phytochemical and pharmacological properties of holy basil (*Ocimum sanctum* L.). *Industrial Crops and Products*, 118, 367–382.

Singh, D., Pokhriyal, B., Joshi, Y., and Kadam, V. (2012) Phytopharmacological aspects of *Chlorophytum borivilianum* (Safed Musli): a review. *International Journal of Research in Pharmacy and Chemistry*, 2(3), 853–859.

Singh, J., Jain, S.P., and Khanuja, S.P.S.(2006) *Curculigo orchioides* Gaertn.(Kali Musali): an endangered medicinal plant of commercial value. *Natural Product Radiance*, 5(5), 369–372.

Singh, K.P. and Kushwaha, C.P. (2005) Emerging paradigms of tree phenology in dry tropics. *Current Science*, 89(6), 964–975.

Singh, M., Saharan, V., Dayma, J., Rajpurohit, D., Sen, Y., and Sharma, A. (2017) *In vitro* propagation of *Stevia rebaudiana* (Bertoni): an overview. *International Journal of Current Microbiology and Applied Sciences*, 6(7), 1010–1022.

Singh, R. and Singh, M. (2018) *Adhatoda vasica* (Vasaka): a medicinal boon for mankind. *G-Journal of Environmental Science and Technology, 6*(2), 5–7.

Singh, R., Sharma, P.K., and Malviya, R. (2011) Pharmacological properties and ayurvedic value of Indian buch plant (*Acorus calamus*): a short review. *Advances in Biological Research, 5*(3), 145–154.

Singh, S. and Mallick, M.A. (2020) Agronomic impact of vermicompost, FYM and mixed compost on the growth of two medicinal plants-*Bacopa monnieri* L. and *Centella asiatica* L. *Plant Archives, 20*(1), 497–500.

Singh, S., Priyadarshi, A., Chaudhary, S.P., Singh, B., and Sharma, P. (2018) Pharmacognostical and phyto-chemical evaluation of chitraka (*Plumbago zeylanica* Linn.). *The Pharma Innovation Journal, 7*(6), 281–285.

Singla, R. and Jaitak, V. (2014) Shatavari (*Asparagus racemosus* wild): a review on its cultivation, mor-phology, phytochemistry and pharmacological importance. *International Journal of Pharmacy & Life Sciences, 5*(3), 742–757.

Soman, M., Nair, R., Vinod, M., Kesavamongalam, S.P., and Govindapillai, A.N. (2014) Nutritional and anti nutritional status of *Acorus calamus* L. Rhizome. *Annals. Food Science and Technology, 15*, 51–59.

Soni, H.K., Shah, B.R., Bhatt, S.B., and Sheth, N.R. (2010)Standardization of single herb capsules of man-jistha, kokam and punarnava with assessment of their nutritional value. *Journal of Pharmacy Research, 3*(8), 1899–1902.

Soni, K.K. and Soni, S. (2017) *Eclipta alba* (L.) an ethnomedicinal herb plant, traditionally use in ayurveda. *Journal of Horticulture, 4*(3), 1–4.

Soumya, K.V., Shackleton, C.M., and Setty, S.R. (2019) Harvesting and local knowledge of a cultural non-timber forest product (NTFP): gum-resin from *Boswellia serrata* Roxb. in three protected areas of the western ghats, India. *Forests, 10*(10), 1–19.

Srinivasulu, P., Pavan Kumar, P., Aruna Kumar, C., Venkateswara Rao, J., Vidyadhara, S., and Ramesh Babu, J. (2021) Evaluation of the antibacterial activity of isolated anthraquinones from the root of *Rubia cor-difolia* Linn. *Toxicology International, 27*(3&4), 193–201.

Srivastava, S.K., Chaubey, M., Khatoon, S., Rawat, A.K.S., and Mehrotra, S. (2002) Pharmacognostic evalu-ation of *Coleus forskohlii*. *Pharmaceutical Biology, 40*(2), 129–134.

Srivastava, S.K., Chaubey, M., Khatoon, S., Rawat, A.K.S., and Mehrotra, S. (2002) Pharmacognostic evalu-ation of *Coleus forskohlii*. *Pharmaceutical Biology, 40*(2), 129–134.

Subandi, M. and Firmansyah Dikayani, E. (2018) Production of reserpine of *Rauwolfia serpentina* (L) kurz ex benth through *in vitro* culture enriched with plant growth regulators of NAA and kinetin. *International Journal of Engineering & Technology, 7*(2), 274–278.

Sultan, J.I., Rahim, I., Javaid, A., Bilal, M.Q., Akhtar, P., and Ali, S. (2010) Chemical composition, mineral profile, palatability and *in vitro* digestibility of shrubs. *Pakistan Journal of Botany, 42*(4), 2453–2459

Sultana, S. (2020) Nutritional and functional properties of *Moringa oleifera*. *Metabolism Open, 8*, 100061.

Tandel, M.H., Animasaun, D.A., and Krishnamurthy, R. (2018) Growth and phytochemical composition of *Adhatoda zeylanica* in response to foliar application of growth hormones and urea. *Journal of Soil Science and Plant Nutrition, 18*(3), 881–892.

Tcheghebe, O.T., Seukep, A.J., and Tatong, F.N. (2017) Ethnomedicinal uses, phytochemical and pharmacologi-cal profiles, and toxicity of *Sida acuta* Burm. f.: a review article. *The Pharma Innovation, 6*(Part A), 1.

Tewari, S.K. and Singh, S. (2019) Agro-technologies for cultivation of medicinal plants. *Medicinal Plants, 22*, 22–151.

Thakur, K.S., Patil, P., and Gawhankar, M. (2018) Qualitative evaluation and impact of Vishesh Shodhana process on Guggul (*Commiphora mukul*). *International Journal of Pharmaceutical Sciences and Research, 9*(10), 4243–4247.

Thosar, S.L. and Yende, M.R. (2009) Cultivation and conservation of guggulu (*Commiphora mukul*). *Ancient Science of Life, 29*(1), 22.

Tiwari, P.B. Mishra, N., and Sangwan, N.S. (2014) Phytochemical and pharmacological properties of *Gymnema sylvestre*: an important medicinal plant. *BioMed Research International, 2014*, 1–18.

Tiwari, R.K.S., Das, K., Pandey, D., Tiwari, R.B., and Dubey, J. (2012). Rhizome yield of sweet flag (*Acorus calamus L.*) as influenced by planting season, harvest time, and spacing. *International Journal of Agronomy, 2012* 1–8.

Tripathi, Y.C., Tiwari, S., Anjum, N., and Tewari, D. (2015) Phytochemical, antioxidant and antimicrobial screening of roots of *Asparagus racemosus* Willd. *World Journal of Pharmaceutical Research, 4*(4), 709–722.

Tuyen, C. K., Nguyen, M.H., and Roach, P.D. (2010) Effects of spray drying conditions on the physicochemical and antioxidant properties of the Gac (*Momordica cochinchinensis*) fruit aril powder. *Journal of Food Engineering, 98*(3), 385–392.

Uddin, Q., Samiulla, L., Singh, V.K., and Jamil, S.S. (2012) Phytochemical and pharmacological profile of *Withania somnifera* Dunal: a review. *Journal of Applied Pharmaceutical Science, 2*(1), 170–175.

Vasisht, K., Sharma, N., and Karan, M. (2016) Current perspective in the international trade of medicinal plants material: an update. *Current Pharmaceutical Design, 22*(27), 4288–4336.

Velammal, S., Priya, T., Devi, A., and Amaladhas, T.P. (2016) Antioxidant, antimicrobial and cytotoxic activities of silver and gold nanoparticles synthesized using *Plumbago zeylanica* bark. *Journal of Nanostructure in Chemistry, 6*(3), 247–260.

Verma, A., Kumar, B., Alam, P., Singh, V., and Gupta, S.K. (2016) Rubia cordifolia-a review on pharmaconosy and phytochemistry. *International Journal of Pharmaceutical Sciences and Research, 7*(7), 2720–2731.

Vijaya, K. and Chavan, P. (2009) *Chlorophytum borivilianum* (Safed musli): a review. *Pharmacognosy Reviews, 3*(5), 154.

Viruel, J., Segarra-Moragues, J.G., Raz, L., Forest, F., Wilkin, P., Sanmartín, I., and Catalán, P. (2016) Late cretaceous–early eocene origin of yams (*Dioscorea, Dioscoreaceae*) in the laurasian palaearctic and their subsequent oligocene–miocene diversification" *Journal of Biogeography, 43*(4), 750–762.

Vishwakarma, U.R., Gurav, A.M., and Sharma, P.C. (2009) *In vitro* propagation of *Desmodium gangeticum* (L.) DC. from cotyledonary nodal explants. *Pharmacognosy Magazine, 5*(18), 145.

Wadekar, R.R., Agrawal, S.V., Tewari, K.M., Shinde, R.D., Mate, S., and Patil, K. (2008) Effect of *Baliospermum montanum* root extract on phagocytosis by human neutrophils. *International Journal of Green Pharmacy (IJGP,) 2*(1), 68–71.

Wang, Y., Li, J., and Li, N. (2021) Phytochemistry and pharmacological activity of plants of genus *Curculigo*: an updated review since 2013. *Molecules, 26*(11), 3396.

Xianzong, X.I.A., Ruoxi, Y.I.N., Wei, H.E., Holobowicz, R., and Gorna, B. (2017) Seed yield and quality of sword bean (*Canavalia gladiata* (Jacq.) DC.) produced in poland. *Notulae Botanicae Horti Agrobotanici Cluj-Napoca, 45*(2), 561–568.

Yadav, V., Krishnan, A., and Vohora, D. (2020) A systematic review on *Piper longum* L.: bridging traditional knowledge and pharmacological evidence for future translational research. *Journal of ethnopharmacology, 247,* 112–255.

Yang, R., Wang, L., Yuan, B., and Liu, Y. (2015) The pharmacological activities of licorice. *Planta Medica, 81*(18), 1654–1669.

Yaseen, G. and Shahid, S. (2020) Comprehensive review on phytopharmacological potential of *Gymnema sylvestre. Asian Journal of Pharmacy and Technology, 10*(3), 217.

Yates, C.R., Bruno, E.J., and Yates, M. E. (2021) *Tinospora Cordifolia*: a review of its immunomodulatory properties. *Journal of Dietary Supplements, 19*(2), 1–15.

Zaveri, M., Khandhar, A., Patel, S., and Patel, A. (2010) Chemistry and pharmacology of *Piper longum* L. *International Journal of Pharmaceutical Sciences Review and Research, 5*(1), 67–76.

Zhao, S.M., Kuang, B., Fan, J.T., Yan, H., Xu, W.Y., and Tan, N.H. (2011) Antitumor cyclic hexapeptides from rubia plants: history, chemistry, and mechanism (2005–2011). *CHIMIA International Journal for Chemistry, 65*(12), 952–956.

12 Prospects for Sustainable Cultivation of Medicinal and Aromatic Plants in India

Smitha Gingade Rajachandra Rao and Rohini Mony Radhika

CONTENTS

DOI: 10.1201/9781003206620-12

12.1 INTRODUCTION

Since the ancient time Medicinal and Aromatic Plants (MAPs) are not only a source of affordable herbal products but also source of revenue. The demand for herbal medicines, wellness products, nutraceuticals, pharmaceuticals, fragrances, cosmetics, essential oils, and aroma compounds is steadily increasing in the domestic and global markets. It has been estimated that around 90% MAPs are a collection from natural habitats of which 70% are from destructive harvesting (Ved and Goraya, 2017). According to Kalaichelvi and Swaminathan (2009) of the 360,000 plant species reported in the world, 40% plants are cultivated in India owing to its 16 agro-climatic zones. Around 7,195 plants are used in Indian System of Medicine (ISM) of which, 4,720, 1,773, 1,122, and 751 plant species are used in folk medicine, Ayurveda, Siddha, and Unani, respectively (Table 12.1).

Limited commercial cultivation has forced industries and traders to rely on the wild collections to meet the growing demands. As a result, many plants have become endangered, vulnerable, or threatened. With declining supplies from wild sources and increasing worldwide demand, intensifying the cultivation of MAPs becomes an important approach (Rao et al., 2004). However, the scope for horizontal expansion of cultivable area is not viable. But it is feasible to integrate MAPs in the existing cropping systems along with the predominant food/fodder/pulses/commercial/horticultural crops of the region.

In developing countries, where traditional healthcare systems prevail, most people depend on medicinal plants to meet their primary health needs. Many of the modern-day drugs contain plant extracts or active ingredients as an integral part. Industrialized nations also use medicinal plants as ingredients in many pharmaceuticals. Now in the present time, about 24%–27% of drugs are estimated to be derived from the plant sources (Singh, 2022). Across the globe, many cosmetics and household products contain plants of therapeutic or medicinal value. Medicinal plants are hardy and robust, and can be grown successfully in various agro-climatic conditions and wide range of soils. Medicinal herbs have wider adaptability to the adverse climatic conditions. Some crops are best suited to the mixed and relay cropping. Medicinal crops do not require high land management of pest and diseases. It can be cultivated as mixed crop or as sole crop.

12.2 DEMAND AND SUPPLY OF MAPs IN INDIA

The continuous and rapid growth of the herbal-based industry globally shows the success of adaptation of the ISM in India as well as worldwide. The review of demand and supply status of MAPs in India has pointed towards continuous opacity in the herbal trade. Research and development in MAPs have led to new frontiers in dietary supplements, culinary herbs, flavour and fragrances,

TABLE 12.1
Medicinal Plant Wealth in India

World	:	3,60,000 species
India	:	40% plants grown in 16 agro-climatic zones
Ayurveda, Siddha, Unani	:	7,195 plants
Folk medicines	:	4,720 plants
Ayurveda	:	1,773 plants
Siddha	:	1,122 plants
Unani	:	751 plants

Source: Kalaichelvi and Swaminathan (2009).

aroma therapy, and natural dyes, etc. Shift in interest on use of MAPs in fields allied to medicine has opened newer markets for MAPs. With the ever-increasing interest in MAPs throughout the world, there is corresponding growth in their trade. The National Medicinal Plants Board (NMPB), New Delhi under Ministry of AYUSH, Government of India, is the Nodal agency making concerted efforts to understand the diversity of medicinal plants in trade and estimate their annual consumption. To get a holistic picture of the diversity of herbal raw drug entities in trade, the NMPB commissioned a nation-wide survey to evaluate demand and supply of medicinal plants in India during 2016–2017 (Goraya and Ved, 2017). This study reported that 960 medicinal plant species are in commercial use based on survey of domestic herbal industry, raw drug mandis and rural households, and analysis of foreign trade data. The study estimated the total trade of herbal raw drugs for the year 2005–2006 at 319,500 MT, including exports of 56,500 MT and 86,000 MT consumed by the rural households. Again, the study was repeated in the year 2014–2015 by NMPB through Indian Council of Forestry Research and Education in strategic partnership with FRLHT. They reported an inventory of 1,178 species of medicinal plants used in domestic herbal industry, rural household, present in wild and cultivation, botanicals in commercial use, and botanicals in foreign trade. The demand for herbal products in India was 512,000 MT during the year 2014–2015. This is 62% higher in volume than what has been reported during 2005–2006. This is due to a rapid increase in the export volume of botanicals from 56,500 MT in 2005–2006 to 134,500 MT in 2014–2015 i.e., more than double with an increase of 138% (Goraya and Ved, 2017).

Supply of medicinal plants is met mainly through three sources: from wild, cultivation, or import. Majority of the species are still sourced from the wild, whereas 12 major medicinal plant species (isabgol, henna, senna, mentha, tulsi, ashwagandha, ghritkumari, pippali, and pippal mool, bach, artemisia, vetiver, and kuth) are being cultivated in an area of about 202,000 hectares across the country (Goraya and Ved, 2017). About 65,000 MT of herbal raw drugs is sourced by imports of which more than 31,000 MT is on account of imports of 'Gum Arabic' alone, whereas the other major imported species include long pepper, Garcinia, and asafoetida, with a collective import volume of just about 3,600 MT (Goraya and Ved, 2017).

12.2.1 MARKET POTENTIAL AND TRADE OF MAPs: INDIAN SCENARIO

India is origin to about 17,000 species of higher plants out of which 7,500 species are known for medicinal uses. Growth of the herbal-based healthcare and wellness sector across the world is showing an exponential increase steadily. India being the native of many medicinal crops possesses the most diverse cultural traditions in the use of medicinal plants and capable of earning high foreign exchange by its export. The international trade of MAPs is dominated by only few countries. India and China are the global leaders in the production and availability of MAPs, whereas main consumers are Republic of Korea and Japan. Hong Kong, the USA, and Germany stand out as important trade centres. India is a big exporter of herbs and herbal extracts to countries such as United States of America, China, Germany, and Vietnam. In India, isabgol (*Plantago ovata*), chakoda/Powad Beej (*Senna tora*), and sonpaa (*Senna alexandrina*) were recorded as the top three exported botanical drugs with export volumes of >32,000 MT, >28,000 MT, and >13,000 MT, respectively. With respect to consumption of botanical drugs by the domestic herbal industry, *Aloe vera* stands first with an estimated consumption of 15,700 MT (dry wt.), followed by amla (14,200 MT), isabgol (13,700 MT), and harad (6,000 MT) with a consumption of >5,000 MT. India's MAP trade caters to the needs of 8,600 domestic herbal industrial units, thousands of practitioners of ISM who use medicinal herbs for preparing herbal formulations, and millions of households that use herbal raw drugs on day-to-day basis for their healthcare. Reports based on demand and supply analysis shows that 242 species of herbal raw drugs are in high commercial demand i.e., >100 MT. Further analysis of these 242 species revealed that supply source of 15 species (6%) is from import; 54 species (22%) are obtained from cultivation; 59 species (25%) are wild collected from landscapes outside forests; and 114 species (47%) are collected from forests (36 species from the Himalayan temperate forests and 78 species from tropical forests). Thus, these medicinal plant species are obtained from different sources under different local names, transported through various raw drug markets before the material finally reaches the herbal industry and the retail shops across different states in the country. Medicinal plant trade in India is mainly through the following entities:

- Trade through Conventional Herbal Raw Drug Mandis.
- Trade through Krishi Upaj Mandis.
- Trade through Specialized Herbal Mandis.
- Trade through Cooperatives/Federations/Corporations.
- Trade under Buy-back Agreements.

12.3   NEED FOR SUSTAINABLE CULTIVATION OF MAPs

Despite India's long history in the collection, use, and trade of medicinal plants, its share in the international export is only 2.5%. The main reasons behind this shortfall are the lack of regulatory framework in trade, lack of good agricultural and quality control practices, lack of organic cultivation techniques, and lack of processing technologies and well-organized marketing of MAPs.

Increasing commercial demand, unmonitored trade, habitat destruction, unregulated, and unscrupulous destructive harvesting have exerted growing pressure on plant populations in the wild, because of which of many of the naturally growing medicinal plants have moved towards extinction. The International Union for the Conservation of Nature and Natural Resources (IUCN) estimates that at least 10,000 species of MAPs are under threatened status worldwide (IUCN, 2020). Conservation of such threatened species is of utmost importance to retain their survival and use. Sustainable cultivation of these species becomes a necessity for meeting the consistent demand for their use. Export of potential herbs like Sarpagandha has been banned by the Indian government due to its threatened status. Similarly, many of the indigenous crops capable of earning substantial export value can be conserved through sustainable cultivation. In the present scenario, cultivation of MAPs has been accepted by many farmers not only as an additional source of income but also as a strategy for crop diversification. Majority of the medicinal plants are hardy in nature and capable of growing in harsh and extreme climatic conditions, thus they provide viable opportunity for cultivation in extreme conditions where other food crops or horticultural crops cannot be grown. For example, Isabgol in Rajasthan, Senna in Tamil Nadu, and Atis in high-altitude areas form a remunerative substitute for the traditional crops being grown by the progressive farmers. Moreover, the herbal-based pharmaceutical industries also promote the procurement of cultivated raw material rather than the wild collected, as they can assure and certify the quality of the material in required quantities. Companies like Dabur India Ltd., Natural Remedies, Himalaya Drug Company, Sami Labs, Patanjali, Sri Baidyanath, Emami, and Ipca Laboratories have created farmer clusters in different parts of the country for cultivation of medicinal plants according to their priority. These initiatives by the domestic herbal industries have increased the area under medicinal plant cultivation. The cultivation is also promoted by the herbal units that are engaged in making 'extracts', units making very specific formulations requiring large volume of limited number of species with consistent quality, and the firms engaged in export of herbal raw drugs. The NMPB is playing a pivotal role in promoting the cultivation of MAPs through its various schemes and programmes by providing financial support, organizing training programmes, assisting in farmer cluster formation, providing sources of quality planting material, and guiding with respect to marketing opportunities. NMPB Board supports the cultivation of 116 species with three sets of subsidy regimes for three different sets of species (NMPB website), which is 20% for 59 listed species, 50% for 38 listed species, and 75% for 19 listed species. Hence, the cultivation and utilization of our indigenous plant resources in a sustainable way is the most judicious way of conserving them for the benefit of mankind and nature.

12.4 PROPAGATION AND SUPPLY OF QUALITY PLANTING MATERIAL OF MAPs

Cultivation of MAPs and eventual returns is mainly reliant on the quality of planting material used (QPM). Quality assurance is the focus issue right from the selection of planting material to the finished product. Major factors limiting QPM production involve less attention on genetic improvement, limited knowledge on breeding systems, and lack of information on seed production, storage, and germination protocols. Greater emphasis on documentation of available information pertaining to agrotechnology, *in situ* and *ex situ* conservation programmes, and efficient methods of propagation of RET plants can help in conservation, propagation, and availability of QPM to the users. The challenges in this regard are many, as the MAPs are a large inherently complex group having varied growing habits, habitats, and breeding systems. Developing the minimum seed standards for seed certification, Good Agricultural Practices (GAP) and Distinctness, Uniformity and Stability (DUS) descriptors for different MAPs, and certification of agencies producing QPM is the need of the hour. Increased availability of QPM enhances reliability, traceability, safety, and acceptance of these commodities.

From National Ayush Mission, it is proposed to establish seed centres with research wing to stock and supply certified germplasm of priority medicinal plant species for cultivation. Research, production, and supply of quality seeds and planting material through research organisations, state agriculture universities, forest departments, NGOs, and corporate sector needs to be strengthened to meet the requirement of quality planting material of MAPs.

Though most MAPs species are regenerated sexually in nature, the multiplication rate is slow and thus insufficient to meet the commercial requirements. Hence, vegetative propagation methods like cuttings, grafting, budding, layering, runners, suckers, tissue culture, etc., can be used to supply uniform, high yield, and quality-producing propagules. In some important MAPs, such as *Centella asiatica, Bacopa monnieri, Mentha sps., Aloe barbadensis, Cissus quandrangularis, Gymnema sylvester* etc., crops raised through vegetative means is better to obtain higher yield compared to seeds. Yet, for some other spp. like *Artemisia annua, Artemisia palens, Withania somnifera, Ocimum spp., Psoralea corylipholia, Sasuurea lappa,* Velvet bean etc., propagation through seeds is the better option.

12.5 CULTIVATION OF MAPs AS SOLE CROPS

Many MAPs can be profitably grown as sole crops in different agro-climatic areas. The following are examples of success stories of the cultivation of MAPs as sole crops.

12.5.1 Cultivation of Ashwagandha (*Withania somnifera*) in Madhya Pradesh

Ashwagandha is a high-value crop adapted to drier parts of the country and valued for its roots. The largest cultivation clusters of ashwagandha are located in Neemuch district of Madhya Pradesh. This is a 6-month crop grown purely as a sole crop under direct sunlight. Sowing is done in July and becomes ready for harvest by December–January. On full maturity, the plants are uprooted and roots are spread for drying, threshed, cleaned, and then chopped into pieces. A fully mature crop yields an average of 4.4 MT of fresh roots that on drying to come to about 1.5 MT/ha. In addition to the cultivation of 'ashwagandha' in Neemuch and surrounding areas of Madhya Pradesh and Rajasthan, extensive areas have been brought under 'ashwagandha' cultivation in Guntakul and Kurnool area of Andhra Pradesh. Similarly, its cultivation has also been initiated over about 100 ha in Gadag, Hospet, and Bellary region of north Karnataka.

12.5.2 Cultivation of Pippali/Hippali (*Piper longum*) in Andhra Pradesh

Piper longum is valued for its dried spikes and roots (piplamool), which is an important raw drug in the ISM. Visakhapatnam district of Andhra dominates in the production of *P. longum* covering around 4 –80 ha area. It is a shade-loving crop of 2-year duration, and its cultivation is purely organic based only on cow and buffalo dung. Spikes are harvested at full maturity, when they turn to greenish-black in colour and then dried. Roots are harvested after 2 years, dried and sold as piplamool. About 1,500 kg of dried roots is produced from 1 acre area, and the average sale rate of dried roots is Rs. 100 per kg.

12.5.3   CULTIVATION OF *MENTHA* SPP. IN HILLS OF UTTAR PRADESH

Mentha has been cultivated for 20 years in over 50 ha area by 150 farmers in Hakkabad Khinjana and surrounding villages in Barabanki district of Uttar Pradesh. Planting is done in the month of February, first harvest is done in 60 days, and subsequent harvest is done at 50–60-days' interval. The harvested shoots are air-dried for 10–12 hours and subjected to distillation to extract 'mentha' oil. The total average annual yield of oil from both the cuts is about 200 l/ha.

12.5.4   CULTIVATION OF ROSEMARY IN KARNATAKA

Rosemary, an emerging highly demanded aromatic herb valued for its essential oil by the pharmaceutical, cosmetic, and food flavouring industry, has been brought under viable cultivation in Gadag, Bellary, and Chamrajnagar districts of Karnataka. At present, it is cultivated in about 10 ha of area. It is propagated through rooted softwood stem cuttings of 6 inch long and are planted in the field at 60×60 cm spacing. Whole herb is harvested during October and March, spread in the open for seven days, with frequent turning over to ensure uniform drying. An average yield of 3 MT per ha is expected from a well-maintained field, and the current sale price was reported to be Rs. 100 per kg of dried leaves.

12.5.5   CULTIVATION OF SENNA AND ISABGOL IN GUJARAT AND RAJASTHAN AREA

Senna and Isabgol are important export-oriented medicinal crops having laxative and purgative properties. They are dry region crops well adapted to the states of Rajasthan and Gujarat. Traded Isabgol is entirely sourced from cultivation in Rajasthan, Gujarat, and Madhya Pradesh. In Rajasthan, Isabgol is cultivated in about 2,400 ha area. Crop is propagated by seeds at 4 kg/ha. Sowing is usually done in November–December and harvested in 110–130 days (March–April). The harvested plants are air-dried and threshed with tractor/bullocks. A good crop yields 800–900 kg of seeds/ha.

'Sonamukhi' is the species of Senna that is cultivated in Rajasthan. Senna cultivation is purely organic, as the crop comes up very easily without using any fertilizers or pesticides. The processed 'senna' is largely exported to Japan, Germany, USA, and China. An estimated area of about 5,000 hectares is under 'senna' cultivation in the districts of Jalore, Jodhpur, Pali, and Barmer. A hectare of well cultivated 'senna' yields about 1.5 MT of 'farmer-grade' dried matter, which contains about 30% leaf rachis/branchlets, which fetches a price of Rs. 10–12 per kg. 'Tinnevelley senna' is widely grown as sole crop in Tirunelveli, Virudhunagar, and Thoothukudi districts of Tamil Nadu. Further, farmers from drought-prone areas of Paramakudi, Mudukulathur, Kamudhi, and Kadaladi districts are cultivating senna instead of paddy.

12.5.6   CULTIVATION OF GLORY LILLY (*GLORIOSA SUPERBA*) IN TAMIL NADU

In Tamil Nadu, Glory lilly is one of the major cash crops for more than 20 years. It is cultivated in around 3,336 ha majorly in Oddanchatam and Palani regions of Dindigul district, Dharmapuram, Mulanur, Vellakovil region of Tirupur district, and Aravakurichi region of Karur district. The seeds are exported to other countries for the extraction of principle medicinal compounds.

12.6 INTEGRATION OF MAPs IN DIFFERENT FARMING/CROPPING SYSTEM

A significant constraint in commercial cultivation of MAPs is availability of cultivable land. In view of the increasing demand for food grains production with increasing population and urbanization, there is a limited scope for bringing the lands under sole cultivation of MAPs. Several challenges like changes in demand and consumption patterns, degradation of natural resources, limited culti-vable land, water availability, climate change, biotic, and abiotic stresses etc., are emerging recently (Lele et al., 2010). Hence, adoption of integrated farming approach is need of the hour for economic upliftment of the farmers.

While diversifying crops, the need to maintain current levels of cropped area under food and related crops need not be overemphasized. Thus, it is essential to incorporate MAPs in existing cropping systems in the form of intercrops, mixed crops, multistorey crops, crop rotations, under crops etc. It is a viable practice to utilize the available resources effectively for additional remunera-tion. This leads to sustainable production of MAPs, conservation of endangered species, and also helpful to meet out huge industrial demand.

These systems not only improve the production but also improve economics, equity, and oppor-tunities for knowledge-based rural enterprise. MAPs-based cropping systems in North India have shown that farmers gain significantly higher profits from their lands (Solanki, 2018). These systems have not only influenced the economics but also paved way for agro-based enterprise in the region. Incorporating MAPs in existing farming systems improves crop production efficiency and associated enhancement of quality of life through knowledge-based enterprises.

12.6.1 NEED FOR CROPPING/FARMING SYSTEMS IN INDIA

- Climate change scenario.
- Proliferation of pest and diseases in main crops.
- Land holding size: 90% small and marginal >1.12 ha.
- Dependence on one crop.
- Avoid risks due to failure of one crop.
- Increased and regular productivity/income.
- Improvement of soil fertility.
- Water availability.
- Cultivable land availability for medicinal crops.
- Long pre-bearing stage in perennial crops.
- Congenial microclimate.
- Value added crops—medicinal crops.

The following points should be considered while selecting crops for intercropping system.

- Select the crops with different growth habits, root growth, duration, and families.
- Tall-growing crops with short-growing crops.
- Bushy crops with erect-growing crops.
- Fast growing crops with slow-growing crops.
- Deep-rooted crops with shallow-rooted crops.
- Short duration crops with long-duration crops.
- Legume crops with non-legume crops.
- Crops should have least allelopathic effect.
- Crops selected should be of different families to avoid pests and diseases.

12.6.2 INTERCROPPING

To derive the maximum benefit of soil moisture, nutrients, and other inputs, medicinal crops can be successfully grown as intercrops. Intercropping of MAPs with crops of the same category or with food and vegetable crops, and fruit trees will give extra income without affecting the growth and yield of the main crop. Co-cultivation of Kalmegh [*Andrographis paniculata* (Burm. F.) Wall ex. Nees] with food crops (maize and pearl millet), pulse crop (pigeon pea), and vegetable crop (okra) showed that intercropping of kalmegh with pigeon pea in equal ratio leads to the maximum productivity, andrographolide content, and economic returns (Verma et al., 2021).

12.6.3 CULTIVATION OF MAPs WITH PLANTATION CROPS

Among the different perennial crops, palms (coconut, arecanut, and oil palm) offer very good scope for intercropping with MAPs, as they are widely spaced crops.

12.6.3.1 Cultivation of MAPs with Coconut Crop

Coconut is an important cash crop of south India and very widely spaced crop (6–9 m), wherein, many MAPs can be effectively intercropped during early as well as later stage of crop growth period. In Karnataka, different MAPs are grown in the interspaces of coconut garden. Some of the crops like lemon grass (*Cymbopogon* spp.) are adapted well for the coconut shade and yielded better than the sole crop. But yield of most of the intercrops (tulsi, kalmegh, medicinal coleus, makoi, garden rue, arrow root, roselle, kacholam, lepidium, cowhage, ambertte, citronella, and vetiver) were reduced due to the shade effect and yield reduction was the highest in case of coleus (Basavaraju and Nanjappa, 2011). Long pepper, mandukparni, brahmi, patchouli etc., are crops that prefer shade for their growth and development, and are suitable for combined cultivation in coconut garden to fetch additional profits. Some of the crops suitable for intercropping under coconut are as follows:

Shrubs/herbs: Kacholam, kachchura (*Kaempferia galanga*), arrow root (*Maranta arundinacea*), Aadusoge, vaasa (*Adhatoda beddomei*), karigantige (*Nilgiranthus ciliates*), shatavar (*Asparagus racemosus*), nagadanthi (*Baliospermum montanum*), long pepper (*P. longum*), Java long pepper (*Piper chaba*), medicinal coleus (*Coleus aromaticus*), ghritakumari (*Aloe vera*) etc.

Aromatic grasses: Vetiver (*Vetiveria zizanioids*), lemongrass (*Cymbogon flexuosus*), citronella (*Cymbopogon winterianus*), etc.

Trees: Patanga chekke (*Caesalpinia sappan*), bilva *patre* (*Aegle marmelos*), *aane* mungu (*Oroxylum indicum*), sriparni (*Gmelina arborea*), *Stereospermum sauveolens*, ashoka *(Saraca asoca)*, agnimandha (*Premna serratifolia*).

12.6.3.2 Medicinal Plants in Arecanut Plantation

Even though majority of the crops perform better under coconut, but it will not be same in case of arecanut due to closer spacing and high level of shade. In a study conducted at Sirsi, Uttara Kannada district of Karnataka during 2004–2005 in a 5-year-old arecanut plantation, it was found that only *Stevia rebaudiana*, *A. paniculata*, and *Catharanthus roseus* could be grown profitably under the arecanut compared to *Aloe vera*, *Alpinia galangal*, *Coleus forskohlii*, and *Ocimum sanctum*, as their biomass yield reduced drastically under arecanut. *A. racemosus* and *N. ciliates* were found to be promising, as their kernel-equivalent yield was the highest throughout the study period. But, as far as the net return is considered, *Cymbopogon flexuous* and *A. racemosus* were the best option for an arecanut grower to utilize the area under arecanut (Channabasappa et al., 2008). Even *B. monnieri*, *Ocimum basilicum,* shatavari, vetiver, long pepper, Java long pepper, brahmi, *Nilagirianthus ciliatus*, periwinkle, aloe, aswagandha, senna, safed musli could be considered, as they also reported to give high returns per rupee investment and better system productivity. List of MAPs that can be profitably grown as intercrop at optimum spacing under arecanut plantation is given in Table 12.2.

TABLE 12.2
Cultivation Details of Some Important MAPs as Intercrops in Arecanut Plantation

Crop	Family	Spacing (cm)	Planting Material	Habit/Duration
Vetiveria zizanoides (Vetiver)	Poaceae	45 × 30	Root slips	15 months
A. racemosus (Asparagus)	Liliaceae	60 × 60	Roots	18 months
P. longum Linn. (Long pepper)	Piperaceae	60 × 60	Rooted cuttings	Perennial
B. monnieri (Brahmi)	Scrophulariaceae	20 × 10	Rooted cuttings	Perennial
Nilgirianthus ciliatus	Acanthaceae	60 × 60	Cuttings	15 months
Catharanthus roseus (L.) G. Don.(Periwinkle)	Apocynaceae	30 × 20	Seed	12 months
Aloe vera Linn (Aloe)	Liliaceae	60 × 45	Suckers	Perennial
Cymbopogon flexuous (Lemon grass)	Poaceae	45 × 45	Root slips,	Perennial
Ocimum basilicum (Basil)	Lamiaceae	45 × 30	Seed	3 months
Artemisia pallens) (Davana)	Asteraceae	30 × 15	Rooted cuttings	Perennial

12.6.3.3   Intercropping of Medicinal Plants in Oil Palm-Based Cropping System

Like coconut plantation, oil palm inter-space can also be effectively utilized for intercropping. The lesser galangal (*Alpinia calcarata* Rosc.) is highly adaptable to various oil palm canopy shade conditions. The plant showed better growth and yield under the shade of oil palms of age group ranging from 5 to 15 years. Substantial additional income could be obtained by growing lesser galangal as an intercrop in oil palm plantations (Jessykutty and Jayachandran, 2009).

12.6.3.4   Intercropping of Medicinal Plants in Black Pepper

Black pepper (*Piper nigrum*) is a widely spaced spice crop and considerable interspaces remain unutilized during the early years of planting, which is often subjected to soil erosion when it is devoid of vegetation or heavy weed growth, as a result more labour is required for weeding. These interspaces could be effectively utilized for growing MAPs, namely *A. beddomei* (chittadalodakam), *Desmodium gangeticum* (orila), *Pseudarthria viscida* (moovila), *Plumbago rosea* (chethikoduveli), *Niligirianthus ciliatus* (karimkurunji), *A. calcarata* (*chittaratha*), *Crysopogon zizanioides* (ramacham), and *A. racemosus* (*shathavari*) (Thankamani et al., 2012).

12.6.4   INTERCROPPING OF AROMATIC PLANTS

Lemongrass can be suitably fitted in the intercropping system involving pulses and short-duration aromatic crops like Indian basil and chamomile. Chamomile performs well as an intercrop with lemongrass, as it is a winter season crop and the vegetative growth of grass remains slow during winter months. Such practice reduces the weed problem and helps in the proper utilization of inter-row spaces between trees like *Eucalyptus*, poplar, and casuarinas under agro-forestry systems. Among the different tree species and lemongrass combinations, the highest net economic return has been obtained under poplar plus lemongrass system (Singh et al., 2006).

12.6.5   MAPs with Field Crops

The arable land is shrinking with increasing population, and there is a need to develop sustainable and area-specific cropping models to accommodate MAPs along with field crops without affecting the food grain production.

Senna (*Cassia angustifolia*) can be effectively intercropped with chickpea than safflower, linseed, mustard, wheat to obtain better senna herb yield along with the maximum net return (Rathod et al., 2010). Patchouli (*Pogostemon patchouli*) was intercropped with legumes (black gram and soybean) and vegetables (French bean and okra) at CSIR-CIMAP, Bengaluru to utilise the space during the initial growth stages of patchouli. It was found that growing patchouli with French bean is much remunerative over growing patchouli alone when gross returns were considered (Singh, 2008). Intercropping system for menthol mint was studied with other annuals (cow pea, okra, radish, chillies, and sunflower). Among various intercrops tried, radish gave good returns with the highest mint oil equivalent yield (Singh et al., 1998b). Similarly, Citronella (*Cymbopogon winterianus*)-based intercropping system in semi-arid tropics was standardized. Similarly in another study, growing lentil—a legume as intercrop was highly beneficial, as this combination over yielded the sole citronella even in terms of yield of citronella oil. Further, growing lentil decreased the nitrogen requirement and resulted in the highest oil yield at optimum nitrogen level, besides giving better economic returns (Ram et al., 2000).

12.6.6   Medicinal Plants in Multi-Story Cropping System

In this system, crops of different heights are grown together on the same piece of land, utilizing land, water, and space most efficiently and economically. Under the shade of traditionally grown multipurpose trees that produce flowers, fruit, nuts, timber, etc., coffee, tea, and cocoa can be intercropped. Tall and perennial medicinal trees like amla (*Emblica officinalis*), ashok (*Saraca indica*), bael (*A. marmelos*), custard apple (*Anona squamosa*), drumstick (*Moringa oleifera*), *Eucalyptus globules*, sandalwood (*Santalum album*), and soap nut tree (*Sapindus mukorossi*) can be intercropped with annual crops during the initial years until the tree canopy covers the ground. These intercrops manage to provide some income to farmers when the main trees have not started yielding. *Eucalyptus citriodora* can be effectively grown along with coffee, lemongrass, and palmarosa. As its growth is somewhat slow in the first 4 years, intercropping with pineapple, yam, and vegetables are proved to be profitable (KAU, 2014).

12.6.7   Crop Rotation

Crop rotation is growing two or more crops in a sequence in the same land within a specified time period (usually 1 calendar year). This system allows utilization of land throughout or most part of the year, helps in conservation of the soil fertility, irrigation water and reduces pest incidence (Ram and Kumar, 1996; Kumar et al., 2002). In the Indo-Gangetic plains, the rice-wheat crop rotation with corn mint (*Mentha arvensis*) has been effectively practiced in a large area.

12.6.8   Sequential Cropping

Growing two or more crops in a year in a sequence on the same field is referred to as sequential cropping. The succeeding crop is planted after the preceding crop has been harvested. The aromatic crops, palmarosa (45 cm × 30 cm spacing), lemongrass, Java citronella, Jamrosa, and geranium were grown throughout the year with 60 cm × 30 cm spacing in one series (continuous cropping) of treatments, while in the other series (sequential cropping) agricultural crops sorghum, red gram (pigeon pea), maize (60 cm × 20 cm spacing), pearl millet, sunflower (with 45 cm × 45 cm spacing), and castor (90 cm × 30 cm spacing) were grown in *kharif* (rainy season), followed by geranium in *rabi* (winter season). Sequential cropping of maize sunflower and sorghum with rose-scented geranium as a rabi crop was more productive and profitable (Rs. 99,270/ha–Rs. 106,925/ha) as compared to continuous cropping of aromatic crops (Rao et al., 2000).

12.6.9 ALLEY CROPPING

Alley cropping is growing different agricultural/horticultural crops in the lane between widely spaced rows of woody plants. The tree-crop competition was less during the initial stages resulting in at par yield, later on yield tends to decrease in subsequent harvest (Wolz, 2018).

12.6.10 MAPs AND AGROFORESTRY

It is a land management system where agricultural crops are grown with tree crops and livestock simultaneously or sequentially on the same piece of land. Aromatic crops like *Mentha* and *Cymbopogon* spp. can be economically intercropped with *Populus deltoides* or *Eucalyptus spp.* up to initial 5 years, and can also be grown on field bunds and soil conservation bunds in croplands (Fonzen and Oberholzer, 1984). Scented geranium with *Eucalyptus citriodora* (lemon- or citron-scented gum), aromatic crops with fuel and timber trees such as *Leucana leucocephala*, *Casuarina* sp., and *Grevillea robusta* can be intercropped economically. Vetiver can be planted on contour strips, on farm bunds, or as a live hedge barrier, which can control soil/water erosion. Hardy medicinal trees like amla, ashoka, arjuna etc., can be grown in community lands in villages and degraded lands (Chadhar, 2001). Similarly, in West Bengal, *Costus speciosus* is recommended as a major under-storey crop in social forestry programmes (Konar and Kushari, 1989). Tree growth in agroforestry system was found to be improved possibly due to the inputs given to the intercrops (Rao et al., 2004). Further, these intercropping systems gave 2–4 times greater economic returns than sole cropping.

12.6.11 SILVI-MEDICINAL SYSTEM

Silvi-medicinal system is the integration of trees and medicinal plants to produce a wide range of products like food, fodder, fruit, fibre, pulp, and medicinal plants. In this system, shade-tolerant medicinal plants would be integrated as lower strata species in multi-strata system. It would be cultivated in a short cycle in the existing stands of the plantation crops. Tall and perennial medicinal trees are planted at wide spacing in this system as shade providers and boundary markers. These trees allow 30%–50% of light to reach the ground, which is ideal for growing some MAPs. The interspaces in between the trees are utilized for growing agriculture or medicinal crops. Trees such as *A. marmelos, A. squamosa, Azadirachta indica, E. officinalis, Jatropha curcas Moringa, Prunus africana, Santalum album, S. indica, Sapindus mukorossi, Terminalia chebula, T. Arjuna* etc., can be grown as intercrop with annual crops during initial years until the tree canopy covers the ground. However, the time up to which intercropping can be done depends upon the spacing and nature of the trees grown.

12.7 MAPs FOR DRY LANDS

Dry land agriculture is characterized by uncertain monsoon resulting in low crop yield. Despite irrigation facilities available in the near future, 60%–70% of the land remains dry land, and it needs to be exploited. In India, out of the net sown area of 143 M ha, the dry land accounts for 93.13 M ha (68.4%). This dry land agriculture accounts for 40% population, 60% of cattle heads, and contributing 44% to the total food grain production in India. Coarse cereals, pulses, oil seeds, and cotton are the principal dry land crops. Traditional crops are no more economical to the dry land farmers. Medicinal plants have higher demand in the market and are found to be highly remunerative than traditional dry land crops. MAPs suitable for dry land farming are ashwagandha,guduchi, gudmar, kalmegh, makoi, sarpagandha, senna, shatavarai, tulsi (Narashima Reddy, 2006).

12.8   MAPs FOR WASTELANDS

Specific crop should be evaluated for each category of the wasteland. Medicinal plant cultivation may seem highly attractive, but it has certain limitations that should not be overlooked while venturing into its cultivation in wasteland. Some MAPs are suitable for wasteland that does not have much fertility; thereby unutilized land can be brought into cultivation. Such plants are senna (*C. angustifolia*), periwinkle (*V. rosea*), medicinal coleus (*Coleus forskholii*), keelanelli (*Phyllanthus niruri*), shatavari (*A. racemosus*), medicinal solanum (*Solanum khasianum*), aswagantha (*W. somnifera*), aloe (*Aloe vera*), gloriosa (*G. superba*), poonaikali (*Mucuna pruriens*), and jatropa (*Jatropa carcasana*).

12.9   ORGANIC NUTRIENT MANAGEMENT IN MEDICINAL AND AROMATIC CROPS

The NMPB, New Delhi and the Food and Agriculture Organization (FAO) insist that MAPs should preferably be cultivated organically. Today's emerging healthcare system relies largely on organically grown MAP products compared to their conventional counterparts (Aishwath and Tarafdar, 2008). For example, organically grown Psyllium (*P. ovata*) husk fetched six times higher price than the conventional produce. The price of organically grown senna leaves was two times higher, and pods rated 40% more than the conventional produce. Therefore, it is essential that India has to shift from chemical or conventional farming to organic farming in the medicinal and aromatic sector. Good number of evidence exists on the response of important medicinal and aromatic crops to mineral fertilizers. Brahmi, sarpagandha, long pepper, guggal, kalmegh, lemon grass, palmorosa, citronella, plants exhibited visual changes in colour and morphology of plants. Organic farming relies mainly on the natural breakdown of organic matter, using techniques like green manure, composting, crop rotation, concentrated manures, bio-fertilizers, and liquid organic manures (*jeevamruta, beejamrutha, panchagavya etc.,*). All the above soil inputs are the major sources of organic matter, which adds humus in the soil on decomposition.

Smitha et al. (2010b, 2015) and Rao et al. (2010) revealed that long pepper (*P. longum* L.) is an organic manure loving crop and application of 40 t Farm yard manure+2 t vermicompost+2 t neem cake+10 kg/ha bio-fertilizers helped in realizing better growth, yield, and quality parameters. Cultivation of Safed musli with the application of 10–15 t/ha FYM for improved yield, macro and micronutrients uptake, as well as soil properties was well documented (Chandra et al., 2003). The improved productivity of ambrette, brahmi, king of bitters, long pepper with the application of farm yard manure is well documented (Rao et al., 2010). The application of distillery effluent at 100 cm^3/ha encouraged the growth of amla buddings and improved properties of the soil (Singh and Gaur, 2004). Similarly, in aromatic crops like palmarosa, davana, jamarosa, menthol mint and South American marigold, the application of farmyard manure enhanced the herb and oil yield (Rao et al., 2001; Patra et al. 2000). Application of 15 t of FYM/ha in palmarosa increased the mean herb and oil yield at 8% and 10%, respectively (Maheshwari et al., 1991). Singh et al. (1998a) suggested mixing of compost with degraded soil low in organic matter in 1:2 proportions before sowing in *A. paniculata, Abelmoschus moschatus, Plumbago zeylanica, Psoralea corylifolia, P. ovata, Ruta graveolens, W. somnifera, Solanum xanthocarpum,* and *Hyoscyamus niger.* Galangal (*K. galangal* L.) intercropped with coconut had higher content of N in plants applied with FYM (24 t/ha & 32 t/ha and vermicompost (21 t/ha & 28 t/ha), respectively (Maheshwarappa et al., 2000). Velvet bean applied with vermicompost at 5 t/ha gave 30.4% more yield as compared to control (Rao and Rajput, 2005). Saha et al. (2005) reported that vermicompost and vermin wash application for *Aloe vera* crop has increased content of gel, moisture, gel ash, and aloin. Application of vermicompost (5 t/ha) in patchouli proved effective, realizing higher herbage yield and essential oil content (Singh, 2011; Devendra Kumar et al., 2014).

Application of different leguminous crop residues in sacred basil (*Ocimum sanctum*) significantly improved the soil's physicochemical and biological properties compared with control (Smitha et al., 2019). Similar results were also reported in Black Night Shade (*Solanum nigrum*), where the plants were supplied with 75% through fertilizers + 25% through poultry manure recorded better growth and yield parameters (Smitha et al., 2010a). Combined application of FYM with bio-fertilizers in *Indigofera tinctoria* improved the herbage yield (Sindhu et al., 2016). Salman (2006) in *Ocimum basilicum* and Umesha et al. (2011) in *Solanum nigrum*, Chand et al. (2012) in *Mentha arvensis* also reported similar results. In *Brahmi* (*B. monnieri*), Smitha et al. (2021) reported the application of 100% FYM (N equivalent basis) + Arka microbial consortium at 12.5 kg/ha/year resulted in the production of the highest cumulative dry herbage yield of 2 years (16 t/ha).

12.10 MAPs CULTIVATION IN PROBLEMATIC SOILS

The soils that are unfavourable for the cultivation of crops because of one or more unfavourable soil properties/characteristics like soluble salts, soil reaction, ESP, water logging, aeration, etc. adversely affect the optimum soil productivity are called problematic soils. These types of soils are often barren with less production potential for the majority of the traditional agricultural/horticultural crops. To meet the demand, it is imperative to utilize the hitherto lying ecological and economical unproductive barren lands for growing MAPs. The return from the crop production in the first year in these lands may be low when compared with normal soils, but it gets improved due to continuous cropping, and higher yield could be attained in the forthcoming years. In India, about 10 M ha land have been affected by salinity and alkalinity problems in different parts of the country, comprising Uttar Pradesh, Gujarat, West Bengal, Rajasthan, Punjab, and Haryana.

Studies have shown that some MAPs can be effectively grown on problematic soils with better economic returns compared to other traditional agricultural crops. Some of the aromatic grasses that have the potential for hyper accumulation of salt helps in reclaiming the saline soil. Palmarosa, lemon grass, and vetiver grown for 2 years on sodic soils having pH 10.6, 9.8, and 10.5, respectively, was found to reduce the pH to 9.4, 8.95, and 9.50, respectively. The continuous growth of palmarosa may reduce the sodicity and improve the physicochemical properties of sodic soils (Subash Chand et al., 2014). Like other aromatic grasses vetiver can reduce the sodicity, improve the physicochemical properties of sodic soils, and also has a potential to reduce soil and water erosion due to its well-distributed fibrous root system (Xia and Shu, 2003). Studies have also shown that isabgol (*P. ovata*) can also be grown profitably in sodic soils. Amla and Bael/wood apple (*A. marmelos*) withstand saline and alkali conditions, and can be grown on salt-affected soils.

12.11 SOILLESS CULTURE/TERRACE CULTIVATION OF MAPs

'Health security' through growing medicinal herbs in home gardens is gaining popularity, nowadays, since it has become more of a necessity than merely a hobby. With people moving towards a safer and organic way of living, the role of medicinal plants in day-to-day life has increased. Increasing urbanization, industrialization, and pressure on land resources have squeezed majority of the population into limited space in their apartments or houses, paving the need for terrace cultivation. Terrace gardens also make good use of frequently unused and wasted space. Besides the aesthetic benefit, terrace gardens also provide food, reduces temperature, maintain humidity, architectural enhancement, habitats for birds, and pollinators, recreational chances, and ecological benefits. Fresh herbs are often easy to grow in a confined space, may it be in pots on the terrace or even in a window box inside a kitchen. Many of these herbs can be eaten fresh, used as salads, juice, *chutney*, *tambuli*, decoction, curries, and are also used in our primary health care needs. These can be grown in soilless media (cocopeat, vermiculite, rockwool, pumice, wood residues, barks etc.,) in terrace garden to achieve health and nutritional security. Herbs have to be grown as per their requirements for temperature, light, nutrients, and water to get optimum yield and quality. Some plants require full sunlight for optimum growth and yield e.g., tulsi, aloe, ashwagandha, guduchi, coleus, stevia, rosemary, thyme etc. The rule of the thumb is that the herbs should get the minimum of 4–7 hours of direct sunlight per day. On the other hand, herbs like brahmi, mandukaparni, long pepper, mints etc., require partial shade for better growth and yield. It can be accommodated in shaded places, or provision for shade can be made using 50% shade net (preferably green shade net). These can be grown organically without the use of pesticides and can be picked and used in fresh form as and when required.

12.12   HYDROPONICS/AEROPONICS PRODUCTION OF MAPs

Awareness regarding the use, cultivation, and utilization of MAPs has amplified tremendously in the recent past. The rapid climate change has brought unpredictability in weather parameters resulting in changes in the active principle contents of some MAPs. Though protected cultivation through hydroponic and aeroponic offer a solution for sustainable cultivation of MAPs, it is still at infancy. High-valued MAPs can be profitably produced using hydroponic/aeroponic technology round the year by controlling the growing conditions and mineral nutrition to stimulate higher phytochemical production. Aeroponic cultivation can be taken up in those MAPs where root is the economic part. This technique can provide high-quality roots that are free from pesticides and soil-borne diseases. Recently, many MAP species are cultivated in different hydroponic/aeroponic system with controlled temperature, humidity, and irrigation intervals with enhanced yield and active principle content (Giurgiu et al., 2017).

12.13   CONCLUSION AND PROSPECTS

MAPs are globally valuable sources of herbal products, flavours, fragrances, pharmaceuticals, and nutraceuticals, the demand for which is increasing continuously. Increasing demand is putting immense pressure on the available natural resources, leading to its extinction. Thus, to fulfil the increasing domestic and global demand as well as to conserve the biodiversity, sustainable cultivation of MAPs should be initiated across the country. *In situ* and *ex situ* conservation, GAP, organic farming, integration of MAPs in different farming systems, etc., should be adequately taken into account for the sustainable cultivation and utilization of MAPs to meet the regular supply of raw drugs apart from conserving biodiversity. India being provided with varied agro-climatic conditions can support the cultivation of one or the other medicinal plants in every zone. Medicinal crops can be cultivated as a viable and remunerative substitute for any traditional crop of a particular region. The trailing factors in promoting medicinal plant cultivation include lack of market support, price support, and dominance of traders. India owes the potential to become the world leader in medicinal plant trade if future strategies are directed towards the elimination of these constraints. Future prospects should involve location-specific systematic research for elite cultivars, steady supply of high-yielding quality planting material, developing good cultivation practices, organic farming, cropping systems, and a dependable price and market support for the betterment of farming community.

ACKNOWLEDGEMENTS

The authors are thankful to the Director of ICAR–Indian Institute of Horticultural Research for the support.

REFERENCES

Aishwath, O.P. and Tarafdar, J.C. (2008). Organic farming in medicinal and aromatic plants. *Organic Agriculture*, 157–185.

Basavaraju, T.B. and Nanjappa, H.V. (2011). Yield, quality and economics of medicinal and aromatic crops as intercrops in coconut garden. *Mysore Journal of Agricultural Science*, 45(1), 74–82.

Chadhar, S.K. (2001). Six years of extension and research activities in Social Forestry Division Jhabua (M.P.). *Vaniki Sandesh*, 25(4), 6–10.

Chandra, R., Kumar, D., and Jha, B.K. (2003). Growth and yield of Safed musli (*Chlorophytum borivilianum*) as influenced by spacing. *Indian Journal of Agricultural Science*, 73(3), 153–55.

Channabasappa, K.S., Kumar, P., and Madiwalar, S.L. (2008). Influence of arecanut yield on economic yield and quality parameters of medicinal plants. *Karnataka Journal of Agricultural Sciences*, 20(4), 880–888.

Devendra, K., Antony, J.R., Sameer, D., Arvind, B., Harsh, P., and Siddharth, M. (2014). Impact of spacing and fertilizer levels on maize yield (*Zea mays* L.) under teak based agroforestry system. *New Agriculturist*, 25(1), 83–86.

Fonzen, P. F. and Oberholzer, E. (1984). Use of multipurpose tree in hill farming in Western Nepal. *Agroforestry Systems*, 2(3), 187–197.

Giurgiu, R.M., Morar, G., Dumitras, A., Vlasceanu, G., Dune, A., and Schroeder, F.G. (2017). A study of the cultivation of medicinal plants in hydroponic and aeroponic technologies in a protected environment. *Acta Horticulturae*, 1170.

Goraya, G.S. and Ved, D.K. (2017). *Medicinal Plants in India: An Assessment of their Demand and Supply*. National Medicinal Plants Board, Ministry of AYUSH, Government of India, New Delhi and Indian Council of Forestry Research, Dehradun.

IUCN. (2020). The IUCN Red List of Threatened Species. https://www.iucnredlist.org.

Jessykutty, P.C. and Jayachandran, B.K. (2009). Growth and yield of lesser galangal (*Alpinia calcarata Rosc.*) in an oil palm based cropping system. *Journal of Medicinal and Aromatic Plant Sciences*, 31(1), 1–4.

Kalaichelvi, K. and Swaminathan, A. (2009). Alternate land use through cultivation of medicinal and aromatic plants - a review. *Agricultural Reviews*, 30(3), 176–183.

KAU. (2014). Intercropping in fruit crops. www.keralaagriculture.gov.in/htmle/package/mediplants.pdf.

Konar, J. and Kushari, D.P. (1989). Effect of leaf leachate of four species on sprouting behaviour of rhizomes, seedling growth and diosgenin content of *Costus speciosus*. *Bulletin of the Terrey Botanical Club, 116*, 339–343.

Kothari, S.K., Ram, P., and Singh, K. (1987). Studies on intercropping autumn planted sugarcane with bergamot, pepper and spear mints in *tarai* areas of Uttar Pradesh. *Indian Journal of Agronomy, 32*(4), 406–410.

Lele, S., Peter, W., Dan, B., Reinmar, S., and Kamaljit, B. (2010). Beyond exclusion: alternative approaches to biodiversity conservation in the developing tropics. *Current Opinion in Environmental Sustainability, 2*, 94–100.

Maheshwarappa, H.P., Nanjappa, H.V., Hegde, M.R., and Biddappa, C.C. (2000). Nutrient content and uptake by galangal (*Kaempferia galanga* L.) as influenced by agronomic practices as intercrop in coconut (*Cocos nucifera* L.) garden. *Journal of Spices and Aromatic crops, 9*, 65–68.

Maheshwari, S.K., Joshi, R.C., Gangadar, S.K., Chouhan, G.S., and Trivedi, K.C. (1991). Effect of farm yard manure and zinc on rainfed palmarosa oil grass. *Indian Perfumer, 35*(4), 226–229.

Narashima Reddy, K. (2006). Agri & Herbal Vision, NMPB, https://nmpb.nic.in/content/prioritised-list-medicinal-plants-cultivation.

Patra, D.D., Anwar, M., and Chand. S. (2000). Integrated nutrient management and waste recycling for restoring soil fertility and productivity in Japanese mint and mustard sequence in Uttar Pradesh. *India Agriculture, Ecosystem and Environment, 80*(3), 267–275.

Ram, M. and Kumar, S. (1996). The production and economic potential of cropping sequences with medicinal and aromatic crops in a subtropical environments. *Journal of Herbs Spices and Medicinal Plants, 4*(2), 23–29.

Ram, P., Birendra, K., Kothari, S.K., and Rajput, D.K. (2000). Productivity of Java citronella based inter-cropping systems as affected by fertility levels under *Tarai* region of Uttar Pradesh. *Journal of Medicinal and Aromatic Plant Sciences, 22*(1B), 494–498.

Rao, B.R. and Rajput, D.K. (2005). Organic farming: Medicinal and Aromatic Plants. In *National Seminar on Organic Farming - Current Scenario and Future Thrust*, ANGR Agricultural University, Hyderabad, India, 41–51.

Rao, B.R., Singh, R., Kaul, K., and Bhattacharya, A.K. (2000). Productivity and profitability of sequential cropping of agricultural crops with aromatic crops and continuous cropping of aromatic crops. *Journal of Medicinal and Aromatic Plant Sciences*, 22(1), 214–217.

Rao, G.G.E., Krishna Reddy, G.S., Vasundhara, M., Nuthan, D., Reddy, K.M., Ganiger, P.C., and Jagadeesha, N. (2010). Integrated nutrient management (INM) in long pepper (*Piper longum* L.). *The Asian Journal of Horticulture*, 5(2), 359–363.

Rao, M.R., Palada, M.C., and Becker, B.N. (2004). Medicinal and aromatic plants in agroforestry systems. *Agroforestry Systems*, 61, 107–122.

Rathod, P.S., Biradar, D.P., and Patil, V.C. (2010). Effect of different rabi intercrops on growth and productivity of senna (*Cassia angustifolia*) in northern dry zone of Karnataka. *Journal of Medicinal and Aromatic Plant Science*, 32(4), 187–192.

Saha, R., Palit, S., Ghosh, B.C., and Mitra, B.N. (2005). Performance of *Aloe vera* as influenced by organic and inorganic sources of fertilizers supplemented through fertigation. In *III WOCMAP Congress on Medicinal and Aromatic Plant*, Chiangmai, Thailand.

Salman, A.S. (2006). Effect of biofertilization on *Ocimum basilicum*, L. plant. *Egypt Journal of Agricultural Research*, 79(2), 587–606.

Sindhu, P.V., Kanakamany, M.T., and Beena, C. (2016). Effect of organic manures and biofertilisers on herbage yield, quality and soil nutrient balance in *Indigofera tinctoria* cultivation. *Journal of Tropical Agriculture*, 54(1), 16–20.

Singh, A., Singh, M., and Singh, K. (1998a). Productivity and economic viability of a palmarosa–pigeonpea intercropping system in the subtropical climate of north India. *The Journal of Agricultural Science*, 130(2), 149–154.

Singh, A., Singh, M., Singh, K., and Tajuddin, K. (1998b). Intercropping menthol mint (*Mentha arvensis* L.) for higher returns. *Journal of Medicinal and Aromatic Plant Sciences*, 20 757–758.

Singh, A.K. and Gaur, G.S. (2004). Effect of distillery effluents on soil properties and growth of anola buddings in alkaline soils. In *Organic Farming in Horticulture*. Pathak, R.K., Kishan, R.K.R., Khan, R.M., and Ram, R.A. (eds.), CISH, Lucknow, pp. 205–208.

Singh, M. (2011). Effect of organic and inorganic fertilizers on growth, yield and nutrient uptake of patchouli [*Pogostemon cablin* (Blanco) Benth.] in a semi-arid tropical climate. *Journal of Spices and Aromatic Crops*, 20(1), 48–51.

Singh, M., and Rao, R.S.G. (2008). Influence of sources and doses of N and K on herbage, oil yield and nutrient uptake of patchouli [*Pogostemon cablin* (Blanco) Benth.] in semi-arid tropics. *Industrial Crops and Products*, 29(1), 229–234.

Singh, S., Singh, A., Pravesh, R., Arya, S.J.K., Singh, M., and Khanuja, S.P.S. (2006). Cultivation and distillation technologies of lemongrass. *Journal of Medicinal and Aromatic Plant Science*, 28, 87–90.

Smitha G.R., Basak, B.B., Thondaiman, V., and Saha, A. (2019). Nutrient management through organics, biofertilizers and crop residues improves growth, yield and quality of sacred basil (*Ocimum sanctum* Linn). *Industrial Crops and Products*, 128, 599–606.

Smitha, G.R., Chandre Gowda, M., Sreeramu, B.S., Umesha, K., and Mallikarjuna Gowda, A.P. (2010a). Influence of integrated nutrient management on growth, yield and quality of *makoi (Solanum nigrum)*. *Indian Journal of Horticulture*, 67(Special Issue), 395–398.

Smitha, G.R., Kalaivanan, D., and Sujatha, S. (2021). Influence of shade and organic nutrition on growth, yield and quality of memory enhancing herb, Brahmi (*Bacopa monnieri* L.). *Medicinal Plants*, 13(1), 72–90.

Smitha, G.R., Umesha K., and Sreeramu, B.S. (2015). Soil and plant nutrient status as influenced by organic farming in Long pepper. *Open Access Journal of Medicinal and Aromatic Plants*, 6(1), 21–28.

Smitha, G.R., Umesha, K., Sreeramu, B.S., and Waman, A.A. (2010b). Influence of organic manures and biofertilizers on growth, yield and quality of long pepper (*Piper longum* L.). *South Indian Horticulture*, 58, 139–143.

Solanki, V.K. (2018). Economic and land equivalent ratio performance of herbal medicinal crops under three-tier agroforestry system. *International Journal of Current Microbiology and Applied Science*, 7(1), 2458–2463.

Subash Chand, Sanjay Kumar and Gautam, P.B.S. (2014). Suitability and profitability of medicinal and aromatic plants on different sodic soils. *Crop Research*, 48(1–3), 92–94.

Thankamani, C.K., Kandiannan, K., and Hamza, S. (2012). Intercropping medicinal plants in black pepper. *Indian Journal of Horticulture*, 69(1), 133–135.

Ved, D.K. and Goraya, G.S., (2008). Demand and supply of medicinal plants in India. *Medplant-ENVIS Newsletter on Medicinal Plants*, 1(1), 2–4.

Verma, K., Singh, K.P., Singh, A., Lothe, N.B., Singh, S., et al. (2021). Co-cultivation of a medicinal plant kalmegh [*Andrographis paniculata* (Burm. F.) Wall ex. Nees] with food crops for enhancing field productivity and resource use efficiency. *Industrial Crops and Products*, *159*, 113076.

Wolz, K.J. and DeLcuia, E.H. (2018). Alley cropping: global patterns of species composition and function. *Agriculture, Ecosystems & Environment*, 252. www.life.illinois.edu/delucia/publications/ Wolz Alley Cropping Review.pdf. 61–68.

Xia, H.P. and Shu, W.S. (2001). Resistance to and uptake of heavy metals by *Vetiveria zizanioides* and *Paspalum notatum* form lead/zinc mine tailings. *Acta Ecologica Sinica*, *21*(7), 1121–1129.

13 The Role of Indigenous Knowledge Systems in Sustainable Utilisation and Conservation of Medicinal Plants

Wilfred Otang-Mbeng, Dumisane Thomas Muloche,
Elizabeth Kola and Peter Tshepiso Ndhlovu

CONTENTS

13.1 INTRODUCTION

Many rural communities use indigenous knowledge (IK) for decision-making related to human and animal health, education, food security, and natural resources management. Similarly, many indigenous plants are sources of food, medicine, energy, and shelter (Rankoana, 2016). The utilisation of indigenous plants by local communities is not random and haphazard, rather, local communities have utilised their indigenous knowledge systems (IKS) to foster sustainable utilisation and conservation of indigenous plant resources. Only a small proportion of this knowledge has been documented, yet it remains a valuable repository that provides us with information on how numerous local communities have interacted with the flora and fauna of their changing environment (Tanyanyiwa and Chikwanha, 2011).

DOI: 10.1201/9781003206620-13

In a study by the United Nations Food and Agriculture Organization, it was projected that the world forest resources, inclusive of medicinal plants, were declining at a staggering rate of about 114,000 km^2 per year (Tanyanyiwa and Chikwanha, 2011). Awuah-Nyamekye (2013) argued that biodiversity degradations are more of social matters, and thus a condensed method with only the Eurocentric approach cannot wholly provide the sustainable utilisation and conservation of medicinal plants or avert the degradation of biodiversity, in general. Some investigators have opined that the conservation of plant resources with religious and cultural methods are more sustainable than approaches that are based on regulation or legislation (Berkes et al., 2000). Certainly, conservation strategies that are developed to restrain biodiversity decline, including unsustainable utilisation of medicinal plants, will not function effectively if they are void of indigenous approaches that reflect the culture of the people. Hence, many conservation agencies have ignited interest in the contribution of IK in the protection of the earth's biological diversity (Tanyanyiwa and Chikwanha, 2011). Therefore, changing peoples' behavioural patterns is a crucial component in the implementation and success of every biodiversity policy or conservation strategy. Soini and Dessein (2016) confirmed IK's role as the sole behavioural guide for rural communities—IK is culturally enshrined and comprised of behavioural corrective norms capable of changing local peoples' perspectives towards biodiversity resources. In this line of thinking, this chapter considers the indigenous cultures and strategies that have shaped the sustainable utilisation and conservation of medicinal plants. Indigenous practices important for the sustainable utilisation and conservation of medicinal plants by rural communities are enshrined in myths, taboos, values, folklore, traditional beliefs, rituals, and traditional institutions (Bélair et al., 2010; Richmond et al., 2013).

According to Ajani et al. (2013), IKS are very helpful towards the sustainable utilisation and conservation of medicinal plants because of their participatory, cost-effective, and sustainable values. The participatory nature of IKS emanates from the fact that IKS promotes the spirit of socialism or communalism and thus advocates for the highest level of community participation in conservation projects. Secondly, hiring conservation experts who are paid on an hourly basis in biodiversity projects is expensive, whereas, with less or no monetary reward, IK can be easily obtained from IK holders or local people (G'Nece, 2012). Because of these immense values of indigenous knowledge, international agencies responsible for biodiversity conservation have advised national conservation partners to critically consider IK during the planning and formulation of biodiversity conservation policies and strategies. For example, Article 8 of Chapter 10 of the United Nations Convention on Biological Diversity stipulates that each contracting party should preserve, respect, and maintain knowledge of local and indigenous communities embodying traditional and local lifestyles that are relevant for the sustainable use and conservation of biological diversity (G'Nece, 2012). Thus, the current review assessed the extent of IKS in the sustainable utilisation and conservation of medicinal plants. This literature is expected to identify and explore existing knowledge on IK conservation of medicinal plants and will serve as an important reference for future research in the field.

13.2 METHODS AND MATERIALS

This study followed a systematic approach, as described by Moher (2015) and Shamseer et al. (2015), called the Preferred Reporting Items for Systematic Reviews and Meta-Analyses (PRISMA). A literature search was conducted from June 2021 to October 2021. The information on indigenous ways of conserving medicinal plants and the role of IKS on the conservation status of medicinal plants were retrieved from different online databases such as Google Scholar, PubMed, Science Direct, Scopus, and JSTOR. In addition, theses, dissertations, and books were retrieved from the library of the University of Mpumalanga (UMP), South Africa. Keywords and terminologies such as Africa, ethnobotany, ethnomedicine, ethnopharmacology, IKS, sustainability, conservation, medicinal plants, myths, taboos, and cosmological beliefs were used to search for relevant articles. Furthermore, the bibliographies from the retrieved articles were searched and saved on the Endnote reference manager (Version X9). For each study, the following information was collected:

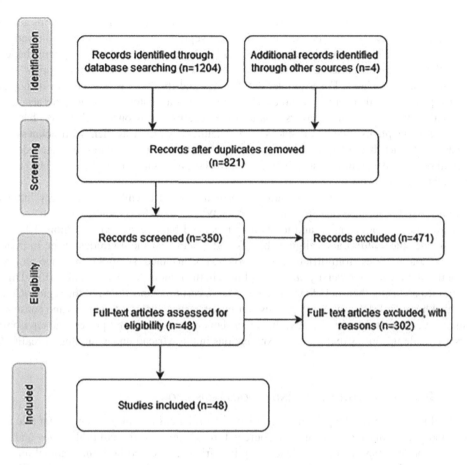

FIGURE 13.1 Preferred Reporting Items for Systematic Reviews and Meta-Analyses flow diagram for the exclusion and inclusion of articles for the current review.

Latin name of plant used, plant parts, traditional conservation methods, and conditions or therapeutic uses of plants. Articles that were excluded were review articles, those solely concerned with modern medicines or those that were not of subject matter. A total of 1,204 articles were obtained based on the keywords used for the search. Additional data was collected from four sources located at University of Mpumalanga (UMP) library. Following the removal of duplicates, the remaining seven were screened with respect to the eligibility criteria, and those publication on western methods of conservation excluded from the selection because they did not meet the criteria and the scope of the study. The study accessed the full texts of the resultant 350 publications—the research papers (systematic or literature) exclusively dealing with current western conservation methods. Thereafter, 302 articles were removed due to a lack of sufficient information regarding the scope of this review, while 48 articles were included to complete this study (Figure 13.1).

13.3 RESULTS AND DISCUSSION

13.3.1 Cosmological Beliefs in the Sustainable Utilisation of Medicinal Plants

Many African communities believe that the universe or cosmos is full of spirits, and this belief in connection with the distinct spirits in the universe or cosmos is said to be the 'cosmological' belief' (Adom, 2014). For example, the Asante's in Ghana and the Xhosas in the Eastern Cape,

South Africa believe that these spirits are organised in hierarchical order in the cosmos according to the potency of their power. For example, God (the supreme deity) is the head, followed by the ancestors, lesser spirits, and witchcraft (Adom, 2014). This cosmological belief system shapes and controls the behavioural patterns of local peoples and, therefore, impacts local peoples' interactions with medicinal plants. For example, the Asantes and the Xhosas believe that if they abuse medicinal plants and other forest resources, God and/or their ancestors will hold them accountable. Many locals believe that their ancestors monitor their negative behaviours (such as over-harvesting) towards medicinal plants and show displeasure to defaulters by afflicting them with sicknesses, sudden death, and other distressing problems (Adom, 2014). This belief regulates their attitude toward medicinal plants and encourages sustainable practices because many locals do not want to incur the punishment of their ancestors.

The use of medicinal plants as a source of primary health is embedded in every indigenous community (Abdillahi and Van Staden, 2012; Van Wyk et al., 2013). South Africa has a diverse population and rich biodiversity, and it has long-practiced biodiversity conservation. The conservation includes the cultural and religious beliefs and taboos that the divinities reside in them and protect the communities from different misfortunes. For instance, Limpopo province proclaimed some of the significant biodiversity and rich cultures in the region (Magwede et al., 2019). Different communities are using their belief systems to conserve certain plants found in the region (Constant and Tshisikhawe, 2018). Local or indigenous communities have different elements and conservation techniques that are attached to African belief systems that preserve and protect the environment. Myths are made-up stories that serve as explanations in the African understanding of reality (Jaja, 2014).

13.3.2 TRADITIONAL INSTITUTIONS/INDIGENOUS LEGISLATION

Traditional leadership are the primary custodians of rural areas and plays a significant role in protecting and preserving the indigenous resources through igniting the cultural values and norms within the communities. Traditional leadership in Africa pre-existed both the colonial and apartheid systems and was the only known system of governance among indigenous people (Koenane, 2017). Even today, despite the rapid urbanisation, traditional institutions still have an impact on local communities (Dzramedo et al., 2013), particularly in most rural areas, communities still value the custodianship of traditional leadership to protect their natural resources, such as medicinal plants and animals (Chibememe et al., 2014). Rankoana (2016) articulated that traditional leadership extends their authority duties to monitor the compliance of different rules placed for harvesting indigenous plants and the over-exploitation of the plants through different mechanisms, which vary from community to community. In South Africa, the responsibility for maintaining medicinal plants, both for their essential value as well as to ensure they continue to supply communities with food, medicine, and fuel, is inadequately implemented in many rural communities. Although South African laws recognizes both customary law and conventional law, due to the different community settings, the use and protection of medicinal plants vary from community to community through the cultural norms and values instilled within that community.

In most rural areas, the harvest of useful medicinal plants in large quantities is prohibited as it is done through the department of environmental affairs that is mandated to conserve biodiversity. However, ethics amongst these local communities play a crucial role in the sustainable harvesting of medicinal plants, with several laws and rules known as taboos to regulate their people's relationship with the environment, this includes use and protection of medicinal plants. Laws and rules that include regulating and exploitation are enforced by rites for the felling of big trees and the killing of certain animals to ensure the protection of these organisms and also to make the ecosystem secure.

For example, certain trees such as *Boscia albitrunca* (Burch.) Gilg & Benedict and *Ziziphus mucronata* Wild are not to be felled for fuelwood (Schmidt, 2022). Using traditional institutions, under the guidance of the headmen, they enforce community involvement in land preparation for farming to ensure the containment of fire hazards. Furthermore, the most useful plant species are protected by traditional leadership through regulating medicinal plant closures and open forests. This allows the environment to enable the regeneration of species in ecosystems through diversification of use, avoiding over-exploitation of the medicinal plant in that particular community. Most of these practices observed in South Africa are similar in several African countries (Asiedu-Amoako et al., 2016). Traditional leadership claims the ownership of specific plant and animal species and preventing other members of the community from using them. As a result of their royal position, these animals are safeguarded throughout the chieftaincy. Despite their importance to the community and the ecosystem, most of these tree species are thought to be extinct or endangered in their indigenous habitat.

13.3.3 SPIRITUAL BELIEFS AND WORLDVIEWS/TRADITIONAL BELIEFS

The use of medicinal plants as a source of primary health is embedded in every indigenous community (Abdillahi and Van Staden, 2012; Van Wyk et al., 2013). The use of traditional medicinal plants is often associated with tangible culture and the environment that need to be preserved and conserved for future generations. According to Rankoana (2016), the use of indigenous medicinal plant knowledge is beneficial not only for cultural conversation and biodiversity but also for primary healthcare. South Africa has a diverse population and rich biodiversity, and it has long-practiced biodiversity conservation. The conservation includes the cultural and religious beliefs and taboos that the divinities reside in them and protect the communities from different misfortunes. For instance, Limpopo province proclaimed some of the significant biodiversity and rich cultures in the region. Different communities are using their belief systems to conserve certain plants found in the region (Constant and Tshisikhawe, 2018). Local or indigenous communities have different elements and conservation techniques that are attached to African belief systems that preserve and protect the environment. Myths are made-up stories that serve as explanations in the African understanding of reality (Jaja, 2014).

13.3.4 MYTHS AND TABOO OF MEDICINAL PLANTS

According to Adu-Gyamfi (2011) 'taboo' means 'forbidden', thus, taboos are the dos and don'ts in a society, and a breach could lead to repercussions from the custodians of the society (the traditional authorities) or from the deities and the ancestors who gave out such restrictions. For example, Asante locals who abused taboos related to the felling down of medicinal plants with powerful ancestral spirits believed to be in-dwelt in them were afflicted with infertility, seizures, lack of appetite, hallucinations, and untimely death (Eshun et al., 2012). Similarly, Boaten (1998) states that there were taboos against the felling of highly economic and medicinal plants like *Butyrospermum parkii* (G.Don) Kotschy, *Elaeis guineensis* Jacq., and *Khaya ivorensii* A. Chev. Such plants could only be cut down when granted a permit by the traditional authorities and accompanied by profuse ritualistic performances.

Myths and taboos have been used to conserve various indigenous resources for sustainability (Kangalawe et al., 2014). The use of myths and taboos is based on prior knowledge of the importance of specific genetic resources. Specifically, traditional leadership and their institutions use myths and taboos to protect medicinal plants. Most of these taboos are meant to intimidate the community to haphazardly harvest these species. These taboos create fear and respect for certain places and species among communities. According to Palgrave (1983), *Z. mucronata* is considered lightning-resistant; anyone sheltering under it during a storm is regarded to be safe. It is also thought that a drought will most likely follow if this tree is cut down after the first rains. While, *Gardenia volkensii* K. Schum is often left growing in villages, as it is believed to protect against lightning (Van Wyk et al., 1997).

Brackenridgea zanguebarica Oliv is a plant of great medicinal and magical importance (Magwede et al., 2019). According to Mabogo (1990) and Netshiungani and Van Wyk (1980), this plant has been used against undesired witchdoctors, to ward off lightning as well as evil spirits, and to cure a host of different diseases; it is believed that young men are barred from having children if they touch the plants, and young women are discouraged from touching the plants because they will suffer non-stop menstruation. The plant is protected through folklore and myths that it is only harvested by elderly women only facing the sun, and these are serious tails that one would not be able to adjust to. These measures serve to keep the plant from being overharvested, and they eventually become alternative norms and taboos rather than formal legal frameworks. According to Mutshinyalo and Siebert (2010), this is one of the conservation strategies that is applied by local inhabitants to mitigate over-exploitation.

13.4 CONSERVATION OF MEDICINAL PLANTS USING IKS

13.4.1 Harvesting Practices

13.4.1.1 Restrictions on the Cutting of Green Plants

The traditional chiefs have adopted some control measures on the collection of green plants as plant sustainability. The measures include the prohibition on cutting green plants of any species in specific areas (Rankoana, 2016). In most areas, especially the villages, people are prohibited from collecting green plants due to traditional beliefs. It is also believed that ancestors will be angry and bad things will happen to the village if plants are collected from the graveyard (Rasethe et al., 2013).

Sporobolus pyramidalis P.Beauv. Aristida sp. and Aristida congesta Roem. & Schult are supposed to be collected under compliance with the traditional rules. These species are supposed to be harvested at the culms. If these grasses are collected from February to April, before the harvest of other crops, which are still in the fields, it is believed that a hailstorm will be formed, and it will destroy the crops in the fields. The conservancy of the grass species is further established by a fine set up by the chief of the community. Moreover, the collection of medicinal plants is restricted to the collection of few leaves to sustain the plant species (Rankoana, 2016) Furthermore, the cutting of Philenoptera violacea (Klotzsch) Schrire is restricted; it is believed that its cutting will result in the separation of marriage in the family (Constant and Tshisikhawe, 2018).

Harvesting of some species is restricted to a certain time and seasons of the year. Ateba et al. (2012) indicated that seasonality plays a major role when harvesting medicinal plants from the field. According to Magodielo (2018), harvesting of medicinal plants during certain seasons are prohibited according to Batswana culture. According to the Batswanas, plants are not supposed to be harvested during winter, which is from June to September. It was emphasised that the harvesting of medicinal plants should not take place during September. September is known as 'Lwetse' in Setswana languages, meaning sick; therefore, plants are prohibited to be harvested during that time. Medicinal plants are known to be sick around September. Thus, Traditional Health Practitioners (THPs) are encouraged to harvest during summer and store for the winter season (Magodielo, 2018). The THPs are forced to obey these indigenous rules to avoid the scarcity of medicinal plants. Contrarily, species such as Peltophorum africanum Sond their roots are soaked and used to cleanse the wounds, and it cannot be harvested in summer. It was reported that its harvest during summer can cause the hailstorm, heavy rains/floods (Ateba et al., 2012)

Furthermore, the timeframe is also important when harvesting medicinal plants. Medicinal plants are not supposed to be harvested during midday; it is believed that this is the time where ancestors are taking care of the plants/forest, and is the time for the plants to rest and breath (Magodielo, 2018). The preferred time for harvesting medicinal plants is in the morning because it is the time where traditional healers believe that plants are fresher and more alive.

13.4.1.2 Exclusive Harvesting of the Leaves of Certain Species

Leaves are reported to be the most effective plant parts for treating either human or livestock ailments. It is generally believed that decoction is the most effective method for treating various human or livestock ailments (Afolayan et al., 2009). Decoction is a method whereby plant parts are boiled in water to treat a certain ailments (Mkwanazi et al., 2021). Exclusive harvesting of leaves from certain plants is allowed but is restricted according to the traditional rules of the specific local community (Rasethe et al., 2013). Restrictions are mostly based around entering the sacred places. Herb-like *Lycopodiella cernua* (L.) Pic. Serm. should not be harvested, as there will be a danger after harvesting (Van Andel and Havinga, 2008). Therefore, the harvest can take place, but the headman must be told in advance about the harvest of the plants (Constant and Tshisikhawe, 2018). Exclusive harvesting of the leaves is also origin specific and time specific. According to the VhaVenda community, only tender leaves of the following plants can be harvested: *Amaranthus hybridus* Vell., *Cucurbita pepo* L., *Momordica balsamina* L., *Momordica boivinii* Baill and *Phytolacca octandra* L. However, *C. pepo* and *M. balsamina* are not supposed to be harvested by women who are menstruating, as it is believed that the plant will lose its healing powers (Constant and Tshisikhawe, 2018).

13.4.1.3 Collection of Lateral Roots from Medicinal Plant Species

Different plant parts are considered to be used for healing purposes, where roots as compared to leaves are the plant parts that are mostly used (Cheikhyoussef et al., 2011). Thus, roots are believed to carry the healing power as compared to other plants parts. During the harvesting of the lateral roots, it was reported that the dug hole must be left open and not to be closed after harvesting the roots, as this will be the reason why the patient does not get healed. It is believed that this hole will be closed as time goes by or during the activities happening around the area. It is also important not to disturb the main root system, as it will disturb some of the plants growing around the plants where roots are collected from (Magoro et al., 2010). Leaving the hole unclosed has some advantages because most THPs do not collect roots for healing from the previously dug plants, as will be indicated by the opening around the root. The THPs claim that by collecting the roots from the previously collected plants, they believe that the previous collector has taken all the healing powers. The latter indicates that roots can be collected in smaller quantities whereby sustainability will be guaranteed, whereas larger quantities will create the disappearance of the medicinal plants (Semenya et al., 2013). Plants such as *Elaeodendron transvaalense* (Burtt Davy) R.H.Archer and *Burkea africana* Hook will be affected by the large quantity collection of the roots due to their slow-growing habits (Akerele et al., 1991).

Semenya et al. (2013) further reported that Bapedi THPs from Limpopo Province in South Africa continue to collect large quantities of the plants in the fear of the scarcity of the plants. Awareness is more recommended to teach the Bapedi traditional healers on how the plants can still be collected in small quantities. This urgent awareness is supported due to the continued harvesting of the plants by uprooting the whole plant such as *Callilepis salicifolia* Oliv and *Geigeria aspera* Harv, whereby their entire root system gets removed from the ground (Magoro et al., 2010). Generally, uprooting of the whole plant leads to the disappearance of the plant species. The selection of tools to be used when harvesting medicinal plant roots is very important. The use of sophisticated sharp stools such as axes, forks, knives, and spades are more destructive as compared to the olden day's tools such as pointed wooden sticks and stone axes (Semenya et al., 2013). It is believed that these sophisticated tools might damage some of the root systems.

13.4.1.4 Seed Propagation (Sustainability)

Seed propagation is a vital method of sustainability, multiplication, and continuous food supply to the families (Rankoana, 2016). Traditional people believe in the season when propagating various seeds. According to the latter seeds are season specific. It was reported that seeds such as *Vigna subterranea* L. Verdc. are not supposed to be planted before January or/and are not supposed to be planted before *Zea mays* L. If these seeds are propagated before January, it is believed that it will reduce the chances of rainfall and cause lighting (Tshikukuvhe, 2017). The multiplication of fruits can be done through the distribution of seeds in the smallholdings for use by family members (Constant and Tshisikhawe, 2018).

13.5 THE IMPORTANCE OF PLANT CONSERVATION THROUGH IK PRACTICES HELPS STRENGTHEN CULTURAL INTEGRITY AND VALUES

It is very important to understand the IK practices before conserving the medicinal plants, because 80% of people worldwide depend on medicinal or traditional health care from medicinal plants (Kefalew et al., 2017; WHO, 2013). Due to various methods that are available concerning IK (Chirwa, 2017), these methods are, however, focusing on enlightening local livelihoods and at the same time promoting the wise use of medicinal plants (Chirwa, 2017). Information regarding the conservation of medicinal plants should be passed on to other people or generations. Kefalew et al. (2017) reported that information about the conservation of medicinal plants is only known by elders, and they pass the information to first sons, and the information will be kept secret if they do not have first sons. Conservation of medicinal plants through IK practices is very important because local people can still benefit from the knowledge that their elders passed to them, and they can reduce more travelling to get western medicines, because IKS will provide more information regarding their local medicinal plants. Information from IK deals mostly with prohibitions, whereby these prohibitions help with the increase of vegetation of medicinal plants (Ayaa et al., 2016).

13.6 CONCLUSIONS AND PROSPECTS

Biodiversity management through cultural practices helps strengthen cultural integrity and values. It is proposed that these knowledge systems should be preserved, expanded, and applied to sustain the indigenous plant resources and the livelihood of communities that depend on indigenous plants for food, medicine, fuel, and fodder. Culturally relevant species of flora and fauna as well as their habitats have been conserved using IKS. This is because these species are culturally intertwined with the customs and beliefs of the people. That is why Wilder et al. (2016) recommend to scientists and conservation planners of projects to incorporate IK systems in biodiversity conservation policies and strategies while teaming up with the local people in its planning or development. This will ensure the smooth implementation and success, rather than declining their involvement and having failures and losses of revenue (Golo and Yaro, 2013).

It is very important to understand the IK practices before conserving the medicinal plants, because majority of people worldwide depend on medicinal or traditional health care from medicinal plants, due to various methods that are available concerning IK. These methods are, however, focusing on enlightening local livelihoods and at the same time promoting the wise use of medicinal plants. Therefore, it is recommended that information regarding the conservation of medicinal plants should be passed on to other people or generations. In addition, more ethnobotanical studies should record this information for future uses.

ACKNOWLEDGEMENTS

The authors are grateful to the University of Mpumalanga and the National Research Foundation (NRF), South Africa for providing funds for this work (Grant UID 135452).

REFERENCES

Abdillahi, H.S. and Van Staden, J. (2012). South African plants and male reproductive healthcare: conception and contraception. *Journal of Ethnopharmacology, 143*, 475–480.

Adom, D. (2014). *General Knowledge in Art for Senior High Schools*. Adom Series Publications, Kumasi, Ghana.

Adu-Gyamfi, Y. (2011). Indigenous beliefs and practices in ecosystem conservation: response of the church: church and environment. *Scriptura: Journal for Contextual Hermeneutics in Southern Africa, 107*, 145–155.

Afolayan, A. and Kambizi, L.. (2009). The impact of indigenous knowledge system on the conservation of forests medicinal plants in Guruve, Zimbabwe. *Journal of Traditional Forest-Related Knowledge, 23,* 112.

Ajani, E., Mgbenka, R., and Okeke, M. (2013). Use of indigenous knowledge as a strategy for climate change adaptation among farmers in sub-Saharan Africa: implications for policy. *Asian Journal of Agricultural Extension, Economics and Sociology, 2,* 23–40.

Akerele, O., Heywood, V., and Synge, H. (1991). *Conservation of Medicinal Plants.* Cambridge University Press, UK.

Asiedu-Amoako, S., Ntiamoah, M., and Gedzi, V. (2016). Environmental degradation: a challenge to traditional leadership at Akyem Abuakwa in the Eastern Region of Ghana. *American Journal of Indigenous Studies, 1,* A1–A13.

Ateba, C.N., Kaya, H.O., Pitso, F.S., and Ferim, V. (2012). Batswana indigenous knowledge of medicinal and food plant uses for sustainable community livelihood. In: Smit, J.A. and Masoga, M.A. (eds.), *African Indigenous Knowledge Systems and Sustainable Development: Challenges and Prospects.* People's Publishers, Durban, South Africa, 68–86.

Awuah-Nyamekye, S. (2013). *Managing the Environmental Crisis in Ghana: The Role of African Traditional Religion and Culture—A Case Study of Berekum Traditional Area,* PhD thesis, University of Leeds.

Ayaa, D.D. and Waswa, F. (2016). Role of indigenous knowledge systems in the conservation of the biophysical environment among the TESO community in Busia County-Kenya. *Journal of African Journal of Environmental Science, 10*(12), 467–475.

Bélair, C., Ichikawa, K., Wong, B., and Mulongoy, K. (2010). *Sustainable Use of Biological Diversity in Socio-Ecological Production Landscapes, Background to the 'Satoyama Initiative for the Benefit of Biodiversity and Human Well-Being.* Technical Series no, 52, Secretariat of the Convention on Biological Diversity, Montreal, 1–184.

Berkes, F., Colding, J., and Folke, C. (2000). Rediscovery of traditional ecological knowledge as adaptive management. *Ecological Applications, 10,* 1251–1262.

Boaten, B.A. (1998). Traditional conservation practices: Ghana's example. *Institute of African Studies Research Review, 14,* 42–51.

Cheikhyoussef, A., Shapi, M., Matengu, K., and Mu Ashekele, H. (2011). Ethnobotanical study of indigenous knowledge on medicinal plant use by traditional healers in Oshikoto region, Namibia. *Journal of Ethnobiology and Ethnomedicine, 7,* 10–10.

Chibememe, G., Muboko, N., Gandiwa, E., Kupika, O.L., Muposhi, V.K., and Pwiti, G. (2014). Embracing indigenous knowledge systems in the management of dryland ecosystems in the great Limpopo transfrontier conservation area: the case of Chibememe and Tshovani communities, Chiredzi, Zimbabwe. *Biodiversity, 15,* 192–202.

Chirwa, K. (2017). *An Investigation into the Use and Conservation of Medicinal Plants: Case of Makoni District, Manicaland Province,* Zimbabwe, (Ph.D Dissertation, BUSE).

Constant, N.L. and Tshisikhawe, M.P. (2018). Hierarchies of knowledge: ethnobotanical knowledge, practices and beliefs of the Vhavenda in South Africa for biodiversity conservation. *Journal of Ethnobiology and Ethnomedicine, 14,* 56.

Dzramedo, B.E., Ahiabor, R., and Gbadegbe, R. (2013). The relevance and symbolism of clothes within traditional institutions and its modern impacts on the Ghanaian culture. *Journal of Art and Design Studies, 13,* 1–14.

Eshun, J.F., Potting, J., and Leemans, R. (2012). Wood waste minimization in the timber sector of Ghana: a systems approach to reduce environmental impact. *Journal of Cleaner Production, 26,* 67–78.

G'Nece, J. (2012). *The Importance of Indigenous Knowledge and Good Governance to Ensuring Effective Public Participation in Environmental Impact Assessments.* Maryland, USA. ISTF News.

Golo, B.-W.K. and Yaro, J.A. (2013). Reclaiming stewardship in Ghana: religion and climate change. *Nature and Culture, 8,* 282–300.

Jaja, J.M. (2014). Myths in African concept of reality. *International Journal of Educational Administration and Policy Studies, 6,* 9–14.

Kangalawe, R.Y., Mlele, M., Naimani, G., Tungaraza, F.S., and Noe, C. (2014). Understanding of traditional knowledge and indigenous institutions on sustainable land management in Kilimanjaro Region, Tanzania. *Journal of Soil Science, 4,* 469–493.

Kefalew, A. and Sintayehu, S. (2017). Transference of ethnobotanical knowledge and threat & conservation status of medicinal plants in Ethiopia: anthropological and ethnobotanical perspectives. *Journal of Archaeology & Anthropology, 1,* 15–19.

Koenane, M.L.J. (2017). The role and significance of traditional leadership in the governance of modern democratic South Africa. *Africa Review, 10,* 58–71.

Mabogo, D.E.N. (1990). *The Ethnobotany of the Vhavenda*. Department of Botany, University of Pretoria, Pretoria, South Africa, 1–246.

Magodielo, M. (2018). *An ethnobotanical study of African traditional medicinal plants in the Heritage Park of the North West Province*. North-West University, South Africa.

Magoro, M.D., Masoga, M.A., and Mearns, M.A. (2010). Traditional health practitioners' practices and the sustainability of extinction-prone traditional medicinal plants. *International Journal of African Renaissance Studies - Multi-, Inter- and Transdisciplinarity, 5*, 229–241.

Magwede, K., Van Wyk, B.E., and Van Wyk, A.E. (2019). An inventory of Vhavenḓa useful plants. *South African Journal of Botany, 122*, 57–89.

Mkwanazi, M.V., Ndlela, S.Z., and Chimonyo, M. (2021). Indigenous knowledge to mitigate the challenges of ticks in goats: a systematic review. *Veterinary and Animal Science, 13*, 100190.

Moher, D., Shamseer, L., Clarke, M., Ghersi, D., Liberati, A., Petticrew, M., Shekelle, P., and Stewart, L.A. (2015). Preferred reporting items for systematic review and meta-analysis protocols (PRISMA-P) 2015 statement. *Systematic Reviews, 4*(1), 1.

Mutshinyalo, T. and Siebert, S. (2010). Myth as a biodiversity conservation strategy for the Vhavenda, South Africa. *Indilinga African Journal of Indigenous Knowledge Systems 9*, 151–171.

Netshiungani, E. and Van Wyk, A. (1980). Muthavasindi-mysterious plant from Venda. *Veld & Flora, 66*, 87.

Palgrave, K.C. (1983). *Trees of Southern Africa*. Struik Publishers, Cape Town, South Africa.

Rankoana, S.A. (2016). Sustainable use and management of indigenous plant resources: a case of Mantheding community in Limpopo Province, South Africa. *Sustainability, 8*(3), 221.

Rasethe, M.T., Semenya, S.S., Potgieter, M.J., and Maroyi, A. (2013). The utilization and management of plant resources in rural areas of the Limpopo Province, South Africa. *Journal of Ethnobiology and Ethnomedicine, 9*, 1–8.

Richmond, L., Middleton, B.R., Gilmer, R., Grossman, Z., Janis, T., Lucero, S., Morgan, T., and Walton, A. (2013). Indigenous studies speak to environmental management. *Environmental Management, 52*(5), 1041–1045.

Schmidt, C. (2022). Levels of Landscape Resilience, *Landscape Resilience*. Springer, Berlin, Heidelberg, 43–95.

Semenya, S.S., Potgieter, M.J., and Erasmus, L.J.C. (2013). Indigenous plant species used by Bapedi healers to treat sexually transmitted infections: their distribution, harvesting, conservation and threats. *South African Journal of Botany, 87*, 66–75.

Shamseer, L., Moher, D., Clarke, V., Ghersi, D., Liberati, A., Petticrew, M., Shekelle, P., Stewart, K.M. (2015). Preferred reporting items for systematic review and meta-analysis protocols (PRISMA-P) 2015: elaboration and explanation. *British Medical Journal, 350*, g 7647.

Soini, K. and Dessein, J. (2016). Culture-sustainability relation: towards a conceptual framework. *Sustainability, 8*, 167.

Tanyanyiwa, V.I. and Chikwanha, M. (2011). The role of indigenous knowledge systems in the management of forest resources in Mugabe area, Masvingo, Zimbabwe. *Journal of Sustainable Development in Africa, 13*, 132–149.

Tshikukuvhe, L.D. (2017). *Exploration of Indigeneous Medicinal Knowledge of Phonda in the Vhembe District*. Masters Dissertation, University of Venda, South Africa, 1–91.

Van Andel, T. and Havinga, R. (2008). Sustainability aspects of commercial medicinal plant harvesting in Suriname. *Forest Ecology and Management, 256*, 1540–1545.

Van Wyk, B.-E., Van Oudshoorn, B. and Gericke, N. (1997). *Medicinal plants of South Africa*. Briza Publications, Pretoria, South Africa.

Van Wyk, B.-E., Van Oudtshoorn, B., and Gericke, N. (2013). *Medicinal Plants of South Africa*, 2nd edition. Briza Publications, Pretoria, South Africa.

WHO (2013). *WHO traditional medicine strategy 2014–2023*, 1–78.

Wilder, B.T., O'meara, C., Monti, L., and Nabhan, G.P. (2016). The importance of indigenous knowledge in curbing the loss of language and biodiversity. *BioScience, 66*, 499–509.

14 Metabolomics Approach
A Tool for Understanding Alzheimer's Disease

Bilqis Abiola Lawal, Hamza Ahmed Pantami,
Umar Muhammed Badeggi and Babatunde Abdulmalik Yusuf

CONTENTS

14.1 INTRODUCTION

Alzheimer's disease (AD) is a neurodegenerative disease that is the most common cause of dementia. It mainly arises due to neuronal injury in the cerebral cortex and hippocampus area of the brain, and it is mostly characterized by the formation of intraneuronal fibrillary tangles, deposition of amyloid plaque, and loss of cholinergic neurons (Alzheimer's, 2015). The clinical manifestation results in cognitive impairment, deficiencies in learning and memory, progressive loss of reasoning and judgment, and eventually dementia. The social impact is the gross impairment of daily life activities including personality changes, impairment of speech, and eventual behavioral and psychological disturbances (Rizzi et al., 2014). Other than being a major cause of dementia and hence a disruptor of regular lifestyle of a person, AD is also considered to be a disorder leading to a significant number of deaths (Dement, 2016). The global epidemiology of AD varies, and there are reports of exponential increase in the estimation of the absolute number of cases of people who have dementia (Qui and Fratiglioni, 2018). Globally, it was discovered that AD affects more than 45 million people in 2015, and this number is estimated to triple by 2050 (Pince et al., 2015). In terms of monetary value, AD and other forms of dementia are projected to cost more than 1.1 trillion by the year 2050 (Hendrix et al., 2016).

The major risk factors of AD are strongly connected to impairments in learning and in memory. AD is also closely related with aging, which has been reported to account for up to 50% of cases due to the tremendous growth of the elderly population. Aging is a complex physiological process that unfolds gradually as we get older and is also greatly influenced by increase of neuronal cell death and eventual neurodegeneration, alongside genetic and environmental conditions (Daulatzai, 2017). Aside aging, other risk factors closely associated with AD include diet, socioeconomic status, hormonal level, and vascular factors among many others. The impacts of these risk factors contribute significantly to the AD disease progression and eventually aggravate the conditions of inflammation, oxidative stress, $A\beta$ plaques, and acetylcholine degeneration (Heemels, 2016). However, a considerable fraction of this disease can affect the younger population at the early onset.

In the etiology of AD, AD can be classified into early onset Alzheimer's Disease (EOAD) and late-onset Alzheimer's disease (LOAD), although 90%–95% of AD cases belong to the LOAD category, as it usually occurs in people over the age of 65 years. The EOAD is caused by genetic mutations in amyloid precursor protein (APP) and presenilin (PS1 and PS2) proteins. It is less common, as it only affects people between the age of 30 and 65 years (Zhang et al., 2017). On the other hand, LOAD cases are more 'sporadic' and are associated with several other mechanisms, such as hypertension, dyslipidemia, hypercholesterolemia, obesity, and diabetes that have also been reported to contribute to LOAD (Mendiola-Precoma et al., 2016). Differences in gender also have a significant impact on onset of AD, as more than 60% of AD individuals are composed of post-menopausal women (Udeh-Momoh et al., 2021).

Presently, there is no known cure nor pharmacologic treatment to stop the AD process, and the current clinically approved drug only gives symptomatic relief. The first drug to be approved for the treatment of AD was Tacrine; but this was associated with hepatoxicity and later withdrawn from use (Godyń et al., 2016). There are five drugs currently approved by FDA and in use clinically; they are Donepezil (Aricept®), Rivastigmine (Exelon®), Galantamine (Razadyne®), Memantine (Namenda®), and Namzaric ® (a combination of Donepezil and extended release Memantine) (Deardorff et al., 2016). Tacrine, Donepezil, Rivastigmine, and Galantamine are acetylcholinesterase inhibitors (AChEIs), which enhances the function of acetylcholine neurotransmission in the synapses, while Memantine is a NMDA receptor antagonist, which reduces the excitoxicity that arises from excessive binding of glutamate neurotransmitter to NMDA receptor; thereby reducing neuronal damage in AD (Santos, et al., 2016). In addition to these approved drugs, other drugs categorized as disease-modifying therapies (DMTs) for the management of AD are been developed, and some are currently undergoing the pre-clinical or clinical phase of drug development. Such drugs include ponezumab (a monoclonal antibody) (Leurent et al., 2019); resveratrol (promotes protease degradation of $A\beta$ peptide) (Ji and Zhang, 2008); and secondary metabolites isolated from medicinal plants, which include Neocoylin, (-)-gallocatechin gallate, myricetin, and quercetin, withanolide A, asiatic acid, and some alkylphenolic acids (potential Beta site APP Cleaving Enzyme-1 inhibiting drugs) (Williams et al., 2011; Coimbra et al., 2018). Curcumin is also a medicinal plant isolate, which is $A\beta$ aggregation inhibitor (Corbett et al., 2012); Manzamine A and Hymenalsidine are alkaloids that function as tau hyperphosphorylation inhibitors, while Nilvadipine is a calcium channel antagonist subjected to clinical trials (Godyn et al., 2016). Chemical structures of some of these drugs and metabolites are shown in Figure 14.1.

The main limitations associated with the use of these drugs are that they give only symptomatic relief and are unable to give no permanent cure nor change the disease progression in AD (Revi, 2020). Continuous use of these drugs is also associated with side effects that span from nausea, vomiting, diarrhea, dizziness, drowsiness, headache, anorexia, dyspepsia, asthenia to anxiety, confusion, impaired balance, arrhythmias, hypertension, muscle spasms, urinary incontinence, and hallucinations (Abeysinghe

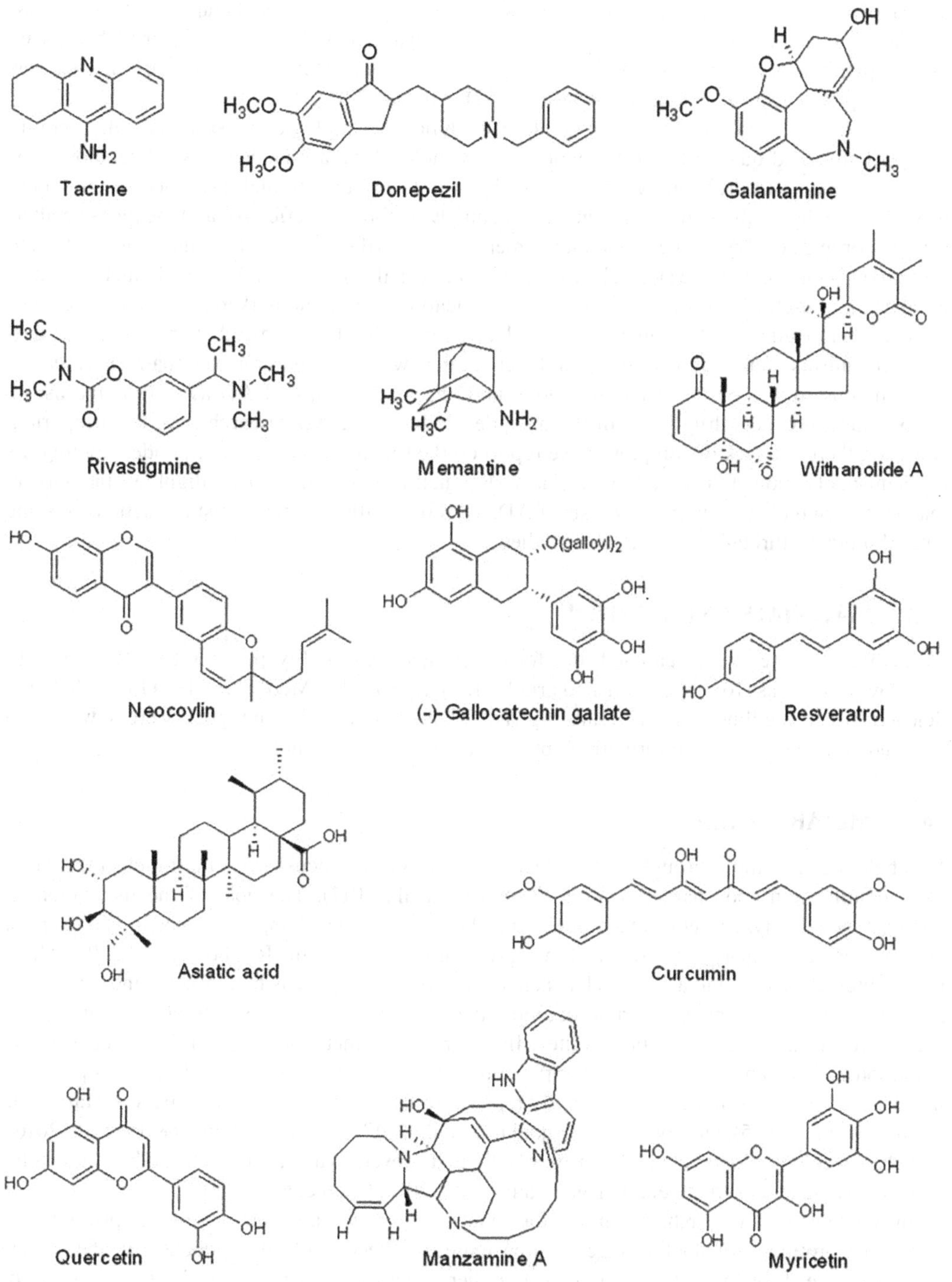

FIGURE 14.1 Some drugs and drug candidates used in pharmacotherapy of AD.

et al., 2020); Thus, the trend in on-going researches is to find therapies that would continuously manage the symptoms of AD and would only be associated with minimal side effects. Natural products have therefore served as an alternative drug candidate for AD therapy due to their multiple components, targets, and signaling pathways. However, there is limited data available of high-throughput mechanism by mechanism of action of such natural products used as drug candidate in the therapy of AD. Thus, new methods that can investigate the mechanisms of action of natural products are necessary. This has led to the exploration of metabolomics as a new approach as an effective tool for exploring the mechanisms of natural products in the therapeutic pathway of AD (Yi et al., 2017).

Metabolomics employs the use of modern analytical technologies to systematically acquire the qualitative and quantitative information of low-molecular-mass metabolites. Metabolomics is widely applicable in drug safety and toxicity by monitoring dosage and clearance in the blood, plasma, and other body fluids. Apparently, it is equally a more cost-effective and inexpensive alternative of drug discovery, as the conventional method of identifying drug leads and bringing them to the market is quite expensive (Russell et al., 2013; Weng et al., 2018). In AD research, metabolomics is particularly useful in the diagnosis of AD, identification of their biomarkers, and in the discovery of novel therapeutic targets. This has further led to the exploration of metabolomics into the anti-AD mechanisms of natural products, thereby creating new perspectives for AD treatment research.

Numerous studies on the pathophysiology and pharmacotherapy of AD as well as the use of some natural products drug leads in the drug development for AD research have been reported. However, there is a lack of comprehensive report on the role of metabolomics in understanding the mechanism of action of these natural products. Herein, this review aims to highlight few biomarkers that have been indicated in the diagnosis of AD, as well as enlist the mechanism of actions of some natural products through metabolomics studies.

14.2 MATERIALS AND METHOD

Academic databases were searched and focus was given to quality peer-reviewed papers published within the last 10 years. Published articles from Scopus, PubMed, Scifinder, Google Scholar, Science Direct, Medline, Clinical Trials.org, and Alzheimer Association reports were downloaded and studied intensively and information obtained there-in were harmonized.

14.3 METABOLOMICS

Metabolomics is a high-throughput approach that is primarily focused on small-molecule/metabolite identification and characterization (Zampieri et al., 2017). The goal of the metabolomics approach is to analyze a set of metabolites (metabolome) and the changes that occur as a result of environmental and genetic interactions in a given biological system (Rischen et al., 2019). These metabolites include amino acids, nucleic acids, vitamins, and lipids as biomarkers, and, of course, phytochemicals. It has been reported that more than 50% of drug entities currently for human use are derived from pre-existing metabolites. In other words, metabolomics can be described as a wholesome approach for addressing complex problems in drug development. The development in metabolomic findings makes it to be highly relevant in various fields, such as nutrition (Hall et al., 2008; Liu et al., 2015), environment (Lankadurai et al., 2013), chemotaxonomy (Kim et al., 2016), and pharmacology (Pontes et al., 2017), especially in discovering the most important biomarkers for a given disease and or treatment (Chagas-Paula et al., 2015; Chan et al., 2015).

In metabolomics research, the most important components are metabolite fingerprinting and metabolite profiling. Metabolite fingerprinting requires detection of all metabolites in the sample without the need for their identification. The earlier metabolite detection includes the use of chromatographic techniques such as paper and thin-layer chromatography. In recent times, some more refined techniques such as Fourier-transform infrared (FTIR), Raman, nuclear magnetic resonance (NMR), and mass spectrometry (MS) are more commonly used. NMR spectroscopy (Bingol and Brüschweiler, 2014; Nagana Gowda and Raftery, 2015) and MS (Gika et al., 2014) are usually the analytical tools often used in large-scale and in complex analyses involving millions of data variables.

Metabolomics analyses can be separated into the categories of targeted or untargeted analysis. In targeted analysis, the metabolites under investigation are usually known; MS-based metabolomics approach is usually the method for targeted analysis; on the other hand, untargeted analysis focuses on the metabolic profiling of the total complement of metabolites ('fingerprint') in a sample. NMR is commonly used in metabolomics fingerprinting studies (Jorge et al., 2015).

14.3.1 Nuclear Magnetic Resonance (NMR) Versus Mass Spectrometry (MS): The Most Common Analytical Technique in Metabolomics

Nuclear magnetic resonance (NMR) and mass spectrometry (MS) are most commonly used techniques in metabolomics profiling of samples such as plant extracts and biofluids; this is as a result of the low sample sizes that is required. NMR spectroscopy is quantitative and does not require extra steps such as separation or derivatization for sample preparation for sample preparation, however it has a low sensitivity compared with Mass Spectrometry (MS). Different ionization techniques and mass analyser technology in MS approaches can be used in order to increase the number of metabolites that can be detected (Emwas, 2015). Therefore, the high reproducibility of NMR-based techniques and the high sensitivity and selectivity of MS-based techniques accords them superiority over other analytical techniques. Comparison between the merits and demerits of these techniques is listed in Table 14.1.

14.3.2 Sample Preparations for Metabolomics Studies

14.3.2.1 Sample Preparation from Biofluids

Biofluid samples most commonly used in metabolomics are blood serum and plasma, urine, and fecal extract. However, it is important to consider several factors in choosing biofluid sample suitable for a given situation, i.e., which biofluid can reveal the most details on the biochemical system of the sample host. In addition, blood serum and plasma reflect the most endogenous metabolites (Kromke et al., 2016). Blood samples should be kept on ice for coagulation for about 20 minutes and further centrifuged at 1,500 rpm for 10 minutes at 4°C to separate the serum. Preparation of blood

---

TABLE 14.1

Comparison between the Advantages and the Disadvantages of NMR and MS Techniques in Metabolomics

Parameters	NMR Techniques	MS Techniques
Cost	• High	• Moderate
Reproducibility	• High reproducibility	• Lower reproducibility
Sensitivity	• Low sensitivity	• High sensitivity
	• Detects most organic molecules	• Detects most organic and some inorganic molecules
Resolution	• Low resolution	• High resolution
Quantitativity	• Possible with no required standards	• Possible, but with required standards
Destructivity	• Non-destructive	• Destructive
Sample preparation	• Minimal preparation required	• Moderate-to-high preparation required
	• Can be directly applied to fluids and tissues	• Extraction of metabolites is required
	• Sample recovery is possible	• GC–MS is for volatile samples; requires derivatization for non-volatile samples
		• LC–MS may form adducts
Detection	• Detects all metabolites in specific range	• Detects metabolites within broader range

Source: Adapted from Lajis et al. (2017).

plasma samples is first dispensed into tubes containing anti-coagulants such as ethylenediaminetet-raacetic acid (EDTA). The prepared biofluid samples are then frozen at temperature of–80°C and thawed with cold PBS at 4°C when ready to use. (Pinto et al., 2015).

14.3.2.2 Sample Preparation from Plants

Sample preparations in plant or microalgae for metabolomics studies usually involve collection from plant or microalgal tissue, root, fruits, leaves, and flowers (Pontes et al., 2016). The sample preparation procedure is normally carried out at low temperature, and liquid nitrogen is mostly used to quench enzymatic reactions and degradation of compounds. The sample must also be frozen at–80°C and then followed by lyophilization to remove water. The most commonly used solvent for extraction processes in plant sample preparation are methanol, perchloric acid, or methanol–chloroform/water mixed in different proportions (Wu et al., 2014); chloroform may also be the solvent of choice for non-polar compounds, while water is good for extraction sugars.

14.3.2.3 Sample Preparation from Animal Tissues

Sample preparation of animal tissues also includes extraction of the content with perchloric acid, and centrifuging to obtain the supernatant and freezing of the sample with liquid nitrogen and storing at–80°C (Ferrer et al., 2008).

14.3.3 Data Acquisition and Processing in Metabolomics

14.3.3.1 Data Acquisition Tools

Data acquisition in metabolomics employs the use of technologies such as nuclear magnetic resonance, liquid chromatography, and/or MS. This facilitates its application in identification of lead drug entity as well as disease monitoring up to its metabolism, toxicity, and even post-market surveillances (Bhogal and Balls, 2008; Puri et al., 2021). Proper identification of metabolites is usually carried out after statistical data analysis, and it is very important in data interpretation.

In the growing need to identify more novel metabolites, a number of computational-aided structural elucidation (CASE) approaches have been developed. Also, important resources employed in metabolic studies are a number of comprehensive web-based databases that assist in identifying the relevant biomarkers and drug leads. Such databases include NAPRALERT, DrugBank, the Human Metabolome Database (HMDB) ChemSpider, Lipidmaps, Massbank, MS analysis and Mestrelab, Madison Metabolomics Consortium Database (MMCD), Biological Magnetic Resonance Bank (BMRB), and Chenomx NMR Suite for NMR analysis (Romano and Joseph, 2019).

14.3.3.2 Data Processing

The first step in data analysis is to explore and find statistical methods suitable for analyzing collected data. After data acquisition and profiling, the identified metabolites are then quantified and statistically processed (Worley et al., 2013); (See schematic representation in Figure 14.2). Multivariate data analysis (MDA) is the most commonly used statistical processing method that is mostly employed to carry out an unsupervised analysis of collected data. This can be further classified into unsupervised MDA methods such as Principal Component Analysis (PCA); and Supervised MDA methods such as Partial least squares (PLS) and Partial least square discriminate analysis (PLS-DA).

14.3.3.2.1 Principal Component Analysis (PCA)

The most common method used here to analyze metabolites from different samples is the PCA. PCA is an unsupervised MDA that allows the reduction of a large number of variables, thereby producing smaller number of principal components. The produced new principal components form linear combinations with the detected metabolites, such that there are variations between the observations (Gromski et al., 2015). In metabolomics studies, PCA is recommended to be carried out first to have a rapid overview of the MDA, before the other supervised techniques such as the PLD–DA are employed (Liu, 2019).

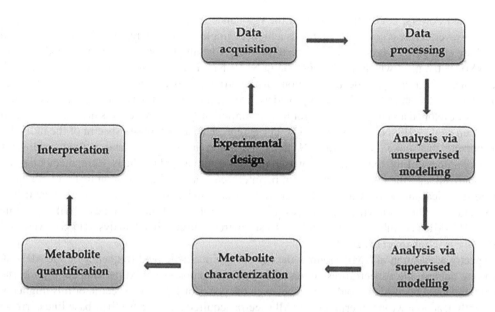

FIGURE 14.2 Schematic representation of processes involved in metabolomics studies.

14.3.3.2.2 Partial Least Squares (PLS)

PLS is a supervised MVDA method that can discriminate and separate samples data based on X variables (chemical shifts data), at the same time correlate them with one or several Y variables (external data on the same sample, e.g., bioassay results). In addition, the PLS model can also be used to predict Y of new data sets and suggests the chemical and biomarkers (Wold et al., 2001; Maulidiani et al., 2013).

14.3.3.2.3 Partial Least Square Discriminate Analysis (PLS–DA)

PLS-DA is a supervised MVDA meaning that additional special information based on the analysis conducted on the samples data are provided for discrimination. This model orders and groups samples depending on provided information about sample classes. The approach in PLS-DA method, the approach is to optimize the covariance between the independent variables X (the metabolomics data) and the corresponding dependent variables Y by finding linear subspace of the explanatory variables (Lajis et al., 2017). The main difference between the two supervised analysis (PCA and PLS-DA) is that the PLS-DA model is performed with prior data input and information obtained from this model is the correlation between the data sets.

14.3.4 METABOLITE IDENTIFICATION

Metabolite identification is one of the major tasks in metabolomics and it is usually carried out after the quantification and statistical data analysis. Proper identification of metabolites is essential for accurate biological interpretation, especially in multiple metabolomics studies. Currently, the NMR- and MS-based, either alone or in combination, are the commonly used techniques.

14.3.4.1 NMR-Based Identification

NMR spectroscopy has been one of the two most widely used analytical tools to accurately identify metabolites, actualize high-throughput analysis, and to augment new derivatization and detection methods (Emwas et al., 2019). NMR-based approach is a non-destructive technique that allows the identification of compounds through the analysis of chemical shifts or coupling constants

(Crook and Powers, 2020). It is based on the resonance frequencies of the sample in a magnetic field; however, it is less sensitive and needs a higher sample size compared with MS-based methods (Markley et al., 2017). It is a good choice of profiling for samples with relatively high abundance. 1H-NMR is the most widely used technique in NMR-based metabolomics, followed by 13C-NMR technique, which can provide information on the structural backbone for metabolite identification. 2D-NMR such as COSY (which provides spectral information between cross-peaks) is more recently developed and is a useful tool for metabolomics analysis because of its increased resolution (Bojstrup et al., 2013; Nagana et al., 2017). These represent superimposed spectra of the individual components that make up a mixture. NMR is also a non-invasive analytical method employed in metabolomic profiling, thus, extensively used in analysis of natural products. Here, metabolites can be visualized, and the relative patterns can be interpreted accordingly in combination with appropriate statistical methods. (Cardoso-Taketa et al., 2008). Some studies carried out to establish the applications of these analytical techniques include the profiling of metabolomes in wild-type mice using 1H-NMR in combination with partial least squares – discriminant analysis (PLS-DA) statistical analysis (Graham et al., 2013).

Spectra acquisition in NMR requires the use of high magnetic field frequency of 500 MHz and above to achieve high spectral resolution in analysis of complex or mixture samples. Appropriate facilities should be used to avoid overlap between the peaks of 1H-NMR spectra, or this might consequently lead to a wrong interpretation. All spectra acquired should have their base line corrected and aligned to avoid errors in the chemical shifts, and this can be done using a known solvent peak. Normalization is carried out to standardize analysis parameters, such as differences in the sample concentration and number of scans or when more than one NMR spectrometer is used in spectral recording. Finally, after NMR spectral data is processed, the data are converted into the format that can easily be executed by the statistical process of choice (Pontes et al., 2017).

14.3.4.2 MS-Based Identification

MS is an analytical technique used in separating metabolites through their ionization and resolution according to their mass-to-charge (m/z) ratios. It provides high-throughput resolution of metabolites and is commonly combined with gas chromatography (GC), liquid chromatography (LC), or capillary electrophoresis (CE). MS-based identification is higher in sensitivity and selectivity, and is highly affected by extraction procedures and matrix effect (Zhang et al., 2016).

14.3.4.2.1 Gas Chromatography–Mass Spectrometry (GC–MS)

GC–MS technique uses a fragmentation pattern in combination with retention time by use of library search. However, the data obtained is restricted to only the metabolites available in such libraries. GC–MS-based metabolomics is the analytical tool that is relatively the most inexpensive, easiest to operate, and having the highest reproducibility (Beale et al., 2018). It is, however, only suitable for volatile and semi-volatile metabolites, while chemical derivatization has to take place for the non-volatile metabolites. These derivatization procedures cannot still be applicable to heat-labile metabolites. GC–MS data processing is based on commercialized libraries such as NIST, Wiley, Replib, and many others (Vinaixa et al., 2016). GC–MS was combined with the same PLS–DA statistical analysis in a study to compare metabolic differences between AD and healthy patients; the results revealed modifications in 23 metabolites and several others (Gonzalez-Dominguez et al., 2015). Recently, GC–MS-based metabolomics has been used extensively in the evaluation of the efficacy of natural products against AD. A GC–MS-based metabolic profiling approach was performed to elucidate the metabolic phenotypes using AD cell model (Yi et al., 2017).

14.3.4.2.2 Liquid Chromatography–Mass Spectrometry (LC–MS)

LC–MS-based metabolomics is the most versatile and well-suited technique employed in metabolomics (Zhou et al., 2012). This analytical technique has wider metabolite coverage, as it is not restricted to volatile components. LC–MS is usually employed to identify and quantify very polar compounds, which include amino acids and steroids and fatty acids. It is usually performed with the aid of electron spray ionization (ESI) and atmospheric pressure chemical ionization (APCI) to easily identify the metabolite, since it does not cause fragmentation of the molecule (Yang, 2013). However, it has low reproducibility of data due to different instruments employed and the diversity of compounds available.

Hyphenated techniques with LC systems include low-resolution MS (LRMS) procedures such as quadrupole (Q) and high-resolution mass spectroscopy (HRMS) procedures such as time of flight (TOF) or Orbitrap (Chen et al., 2020). Like the GC–MS, commercialized databases are also available for LC–MS, which include Massbank and Metlin; however, they, have only been able to detect the limited number of metabolites (Horai et al., 2010).

It is a highly relevant technique in illustrating the effectiveness of natural products in AD treatment. Five ginseng species were subjected to metabolomics profiling using a LC–MS tandem technique in combination with multivariate statistical analysis techniques (PCA). The PCA of the profiling identified that these five herbs can be grouped into five different groups based on their phytochemicals (Xie et al., 2008). This result has demonstrated the reliability of LC–MS as a reliable method for rapid analysis of metabolites present in a natural product.

14.3.4.2.3 Other MS-Based Analytical Methods

Capillary electrophoresis–mass spectrometry (CE–MS) is a newly developed analytical technique relevant for metabolomics studies that relies on mass-to-charge ratio for separation of metabolites. This method requires much less amount of sample and is suitable for the separation of charged metabolites (Ramautar et al., 2009).

Two-dimensional chromatography–mass spectrometry has greater peak capacity and is suitable for analyzing complex biological samples than one-dimensional chromatography (Cook et al., 2015). It has been reported that more than a 100,000-peak capacity of metabolites was identified through RPLC and HIPIC–MS-based metabolic profiling.

Direct-infusion mass spectrometry (DI–MS) is an analytical technique that allows a high-throughput analysis and a very rapid analysis of samples. DI–MS can also be hyphenated with high-resolution MS, such as Orbitrap and Fourier transform mass spectrometry (FTMS). FTMS actually has the highest mass resolution, thereby, reaching more than 1,000,000 resolving power. FTMS is very expensive and has very high maintenance cost, hence is highly restrictive (Lisa et al., 2017). This has made DIMS very resource analytical tool and provides the most comprehensive information.

14.3.5 APPLICATION OF METABOLOMICS BIOMARKERS IN AD

Biomarkers are needed in the etiology, diagnosis pathophysiology, and treatment of AD. Currently, the diagnosis of AD is not accurate, and there is also no cure for patients with AD. However, very recently, biomarkers are employed to aid in the predictive and accurate diagnosis and treatments of AD. Lipid and sugar metabolisms are highly indicated in the metabolomic biomarkers in AD. Recently, several studies have reported aberrations in lipid metabolism such as stearoyl-CoA desaturate and sphingolipids (Hejazi et al., 2011; Trushina et al., 2013) phospholipids, which can be used to accurately identify cases of memory loss (Mapstone et al., 2014). Several reports on the aberration in the cerebral glucose metabolism is also a biomarker indicated in the pathophysiology of AD to precede loss of cognitive function. Activation of all these metabolism pathways may offer a new therapeutic strategy for the management of AD.

Biomarkers are recognized as measured characteristics that act as indicators of an underlying pathogenic process of a disease progression (McGhee et al., 2014). Here different biomarkers have been identified as related to AD, which are the amyloid β (Aβ), tau pathology, and neurodegeneration (A/T/N system). The Aβ system can be measured by PET imaging of amyloid plaques or cerebrospinal fluid (CSF) of A$\beta42$, the tau system can be measured by CSF phosphorylated tau and or tau PET imaging, while neurodegeneration can be measured by elevated levels of CSF total tau, decreased glucose metabolism detected by FDG-PET imaging, and brain atrophy by using structural MRI

(Jack et al., 2018). Metabolomics strategies have the power to compare the metabolome in biological samples under normal conditions with altered states promoted by diseases, including AD.

In addition, biochemical variation such as oxidative stress, neuro-inflammation, mitochondrial dysfunction, lipid down-regulation, and disruption of the neurotransmitter pathway is also highly significant in the AD pathogenesis. A study carried out by Ibanez et al. (2013) has proposed a substantial metabolomic profile using reversed-phase chromatography based on an ultrahigh performance LC coupled with quadrupole time-of-flight mass spectrometry (UPLC–Q–TOFMS). This platform has been able to identify taurine, creatine, xanthine, and lipid mediators among other metabolites as potential biomarkers for early AD detection (Enche Ady et al., 2017). Some identified biomarkers relevant in AD research are listed in Table 14.2.

TABLE 14.2
Some Identified Biomarkers in AD Research

Sample Type	No of Patients	Analytical Platform	Identified Biomarkers	References
Plasma	AD ($n = 15$) MCI ($n = 10$) Control ($n = 15$)	UPLC–MS–MS	Cylcarnitines, arginine, phenylalanine, creatinine, symmetric dimethylarginine, and phosphatidylcholine	Lin et al. (2019)
Serum	AD ($n = 30$) MCI ($n = 20$) Control ($n = 30$)	UPLC–QqQ–MS	Cholic acid, deoxycholic acid, and its glycine and taurine conjugated forms	Mahmoudian Dehkordi et al. (2019)
Serum and plasma	AD ($n = 188$) MCI ($n = 392$) Control ($n = 226$)	UPLC–QTOF–MS	MUFA-containing lipids PUFA-containing lipids	Barupal et al. (2019)
Plasma	AD ($n = 30$) MCI ($n = 20$) Control ($n = 30$)	UPLC–QTrap–MS	Glycochenodeoxycholic acid, glycodeoxycholic acid, and glycolithocholic acid	Marksteiner et al. (2018)
Cerebrospinal fluid (post motem)	AD ($n = 15$) Control ($n = 15$)	UPLC–QqQ–MS	Methionine sulfoxide, 3-methoxy-anthranilate, cadaverine, and guanine	Muguruma et al. (2018)
Plasma	AD ($n = 68$) Control ($n = 1,974$)	UPLC–QTrap–MS	Anthranilic acid, glutamic acid, taurine, and hypoxanthin	Chauraki et al. (2017)
Brain tissue plasma (Animal)	APP/PS1 mice ($n = 9$) Wild-type mice ($n = 9$)	UPLC–QQQ–MS	Polyamine, amino acids, and serotonin	Pan et al. (2016)
Saliva	AD ($n = 660$) MCI ($n = 83$)	FUPLC–TOF–MS	Cytidine, sphinganine 1-phosphate, and 3-dehydrocarnitine	Liang et al. (2016)
Plasma	AD ($n = 90$) MCI ($n = 77$) Control ($n = 51$)	Orbitrap–MS	Diacylglycerol levels	Wood et al. (2016)
Serum	AD ($n = 29$) MCI ($n = 18$) Control ($n = 46$)	QTrap–MS	Acetyl-L-carnitine and acyl-L-carnitine level	Cristofano et al. (2016)
Brain tissue; Cerebrospinal fluid	AD ($n = 10$) Control ($n = 10$)	UPLC–QqQ–MS	Carboxylic acid Amines	Takayama et al. (2015)
Serum	AD ($n = 23$) Control ($n = 21$)	GC–MS	Lactic acid, isocitric acid, glucose, oleic acid, adenosine, cholesterol, urea, pyroglutamate, amino acids, fatty acids, and cysteine	Gonzalez-Dominiguez., et al. (2015)

(Continued)

TABLE 14.2

Some Identified Biomarkers in AD Research

Sample Type	No of Patients	Analytical Platform	Identified Biomarkers	References
Brain tissue	AD ($n = 35$) MCI ($n = 48$) Control ($n = 40$)	UPLC–QTOF–MS	Cholesteryl esters (ChEs) including ChE 32:0, ChE 34:0, ChE 34:6, ChE 32:4, ChE 33:6, and ChE 40:4	Proitsi et al. (2015)
Serum	AD ($n = 19$) Control ($n = 17$)	UPLC–QTOF–MS ICP–MS	Phosphatidylcholines, phosphatidylethanolamines, plasmenylcholines, and plasmenylethanolamines	González-Domínguez et al. (2014)
Brain tissue	AD ($n = 1 5$) Control ($n = 15$)	1H NMR	Alanine and carnitine	Graham et al. (2014)
Brain tissue and plasma (Animal)	APP/PS1 mice (n-6) Wild-type mice ($n = 6$)	1H NMR	Ascorbate, creatine, butyric acid, aspartic acid, acetate, citrate and amino acids	Graham et al. (2013)
Plasma Cerebrospinal fluid	AD ($n = 11$) MCI ($n = 13$) Control ($n = 14$)	UPLC–QTOF–MS	Lysine, cholesterol, and sphingolipids	Trushina et al. (2013)
Brain tissue	AD ($n = 10$) Control ($n = 10$)	UPLC–TOF–MS	Spermine and spermidine	Inoue et al. (2013)
Cerebrospinal fluid	AD ($n = 75$)	UPLC–QTOF–MS	Taurine, sphingosine-1-phosphate, creatinine, and xanthine	Ibanez et al. (2012)
Plasma	AD ($n = 89$) MCI ($n = 91$) Control ($n = 46$)	UPLC–TOF–MS	Phosphatidylcholine, plasmalogens, sphingomyelins, sterols, and dihydroxybutanoic	Oresic et al. (2011)

Source: Adapted from Reveglia et al. (2021).

UPCL, ultra-performance liquid chromatography; HPLC, high-performance liquid chromatography; FUPLC, fast ultrahigh performance liquid chromatography; QTOF, quadrupole time of fight; QqQ, triple quadruple; QTrap, triple quadrupole linear ion trap; AD, patients with Alzheimer disease; MCI, patients with Mild Cognitive Impairment.

14.3.6 Applications of Metabolomics in Investigating the Anti-Alzheimer Effect of Natural Products

The journey to the development of effective drug agents with high therapeutic effects against AD has been challenging, as available therapies only give symptomatic relief. Recently, prospecting for bioactive constituents that have interesting and novel activity or mechanism of action from natural resources, such as plants, animals, and microbes, is the focus of most researchers. Natural products have, therefore, served as an alternative drug candidate for AD therapy due to their multiple components, targets, and signaling pathways. However, their complex nature makes it very difficult to validate their pharmacological mechanism of actions. The metabolomics approach has, therefore, been selected as a tool in investigating the anti-Alzheimer effect of natural products due to their effectiveness in handling multicomponent variables. The applications of LC–MS-based, GC–MS-based, and NMR-based metabolomics in investigating the effectiveness of natural products with anti-Alzheimer potentials is listed in Table 14.3.

14.4 LIMITATIONS

Metabolomics is highly applicable in the identification of bioactive natural products from plants because of their high chemical diversity and complex nature. Metabolites are usually extremely complex with a very large range of chemical diversity, thus multiple analytical techniques that

TABLE 14.3

Some Identified Metabolites from Natural Products with Anti-Alzheimer Effects

Natural Product	Botanical Source	Family	Subject	Sample	Analytical Platform	Biomarkers	References
Pinus	*Pinus species*	Pinaceae	Humans	Human melanocyte cells	HPLC–MS/MS	Flavonoids, phenolics, lignans, diterpenes, and fatty acids	Saber et al. (2021)
Centella	*Centella asiatica (L.) Urban*	Apiaceae	–	–	HPLC–QTOF	Amino acids, choline derivatives, fatty acids, flavonoids, and terpenoids	Alcazar Magana et al. (2020)
Bacopa extract	*Bacopa monnieri L.*	Plantaginaceae	AlCl$_3$-induced AD rat	Serum	UPLC–ESI/MS	Curcurbitacins, sterol glycosides, and jujubogenin glycocides	Waly et al. (2019)
Rosmarinus extract	*Rosmarinus officinalis L.*	Lamiacaeae	AlCl$_3$-induced AD rat	Serum	UPLC–ESI/MS	Phenolic diterpenes, flavonoids, lignan, phenolic acids, and dihydrochalcones	Waly et al. (2019)
Rhodiola extract	*Rhodiola crenulata*	Crassulaceae	$A\beta1$–42-induced AD mice	Hippocampus	FT–ICR–MS	Amino acids, sphingolipid, and glycerophospholipid	Zhang et al. (2019)
G-Rg1 and G-Rg2	*Panax ginseng C.A. Meyer)*	Araliaceae	APP/PS1 mice	Brain	UPLC–MS	Hypoxanthine, Sphingolipids, and some lysophosphatidylcholines (LPCs) (C 16:0. 18:0, 13:0; 15:0,18:1, and 18:3)	Li et al. (2016)
Valeriana extracts	*Valeriana amurensis P.A.Smirn. ex Kom.*	Caprifoliaceae	$A\beta$ 25–35-induced AD mice	Plasma	UPLC–MS	Amino acids, phospholipids, and lipids	Wang et al. (2016)
Kaixin-San herbal formula	*P. ginseng C. A. Mey; Polygala tenuifolia Wild; Poria cocs (Schw. Wolf); and Acorustata rinowii Schott*	-	(d-Gal)–induced and AlCl$_3$-induced AD rat	Urine; serum	UPLC–QTOF–MS	Lipids, sphingolipids, glycerophospholipid, glyoxylate, and dicarboxylate	Chu et al. (2016)
Shengmai San herbal formula	*P. ginseng C. A. Mey; Ophiopogon japonicas and Schisandra chinensis*	-	(d-Gal)–induced and AlCl$_3$-induced AD ratt	Urine	UPLC–QTOF–MS	Glyoxylate and dicarboxylate metabolism, inositol phosphate, and sphingolipids	Sun et al. (2015)
Total Ginsenoside (Triterpenoid saponins)	*Panax ginseng (C.A. Meyer)*	Araliaceae	$A\beta1$–42-induced AD mice	Plasma	UHPLC–QTOF–MS	Amino acids, sphingolipid, phospholipids, and carnitine	Gong et al. (2015)

(Continued)

TABLE 14.3 (*Continued*)

Some Identified Metabolites from Natural Products with Anti-Alzheimer Effects

Natural Product	Botanical Source	Family	Subject	Sample	Analytical Platform	Biomarkers	References
Docosahexaenoic-acid (DHA)	-		CHO-AbPP695 cells	Cell	GC–TOF–MS	Succinic acid, citric acid, malic acid, glycine, zymosterol, cholestadiene, and arachidonic acid	Bahety et al. (2014)
Green tea polyphenols	*Camellia sinensis* (L.) Kuntze	Theaceae	(d-Gal)–induced AD rats	Plasma	UPLC–MS	Amino acids, phospholipids, and lecithins	Fu et al. (2011)
Huperzine A	*Huperzia serrata* (Thunb.) Trevis	Hypericaceae	Scopolamine-induced AD rats	Brain	UPLC–TOF–MS	Amino acids	Shi and Luo (2011)
Ligustrazine	*Ligustrum chuanxiong* Hort.	Apiaceae	Scopolamine-induced AD rats	Brain	UPLC–TOF–MS	Amino acids	Shi et al. (2010)
Epimedium	*Epimedium grandiflorum* C. Morren	Berberidaceae	Scopolamine-induced AD rats	Serum	HPLC–MS	Lipids (unsaturated and saturated acid), amino acids, nucleotides, carnosine, and ergothioneine	Yan et al. (2009)
Epimedium	*Epimedium grandiflorum* C. Morren	Berberidaceae	Scopolamine-induced AD rats	Urine	NMR	Pyruvates; oxidative phosphorylation	Wu et al. (2008)

Source: Adapted from Yi et al. (2017).

UPLC–ESI/MS, ultra-performance liquid chromatography electron spray ionization mass spectrometry; FT–ICR–MS, fourier transform ion cyclotron resonance mass spectrometry; UPCL, ultra-performance liquid chromatography; HPLC, high-performance liquid chromatography; QTOF, quadrupole time of fight; (d-Gal), D-Galactose; and AlCl$_3$, Aluminum chloride.

should be sought as a single analytical platform is incapable of the comprehensive analysis required (Buckberger et al., 2018). The profiling and identification of the components of different plant extracts and fractions in combination with their corresponding biological activities using metabolomics cannot be over-emphasized. Globally, metabolomics lacks proper validation and standardization in sample collection and data processing. Also, the reproducibility of various metabolomics investigations can rarely be guaranteed, as different samples are obtained from different sources with diverse demographic distribution of age, sex, and race.

Another limitation is in the identification of biomarkers that are specific for the early diagnosis of AD and treatment targets of anti-Alzheimer drugs.

14.5 CONCLUSIONS AND PROSPECTS

Over the years, metabolomics has been recognized to be suitable for translational research and pre-clinical applications within an integrated network of biological systems. At the initial stage, application of metabolomics studies in AD research is mostly in biomarker discovery to further understand AD pathogenesis and, thus, identify plausible drug targets in its therapy. Such biomarkers identified include lipid, purine, amino acid metabolism, as well as oxidative phosphorylation.

Nowadays, Metabolomics is also widely applied in AD research by identifying natural products that have pharmacological effects, especially in animal models. Future prospects of the application of metabolomics to AD research are to validate data processing procedures to establish more accurate and reproducible metabolites. This will enhance the search for more drug leads that will ensure total treatment in patients living with AD disease.

ACKNOWLEDGEMENTS

The authors would like to acknowledge University of Ilorin, Ilorin, Nigeria, for providing the space and resources used in writing this review.

REFERENCES

Abeysinghe, A.A.D.T., Deshapriya, R.D.U.S., and Udawatte, C. (2020). Alzheimer's disease; a review of the pathophysiological basis and therapeutic interventions. *Life Sciences*, *256*, 117996.

Alcazar Magana, A., Wright, K., Vaswani, A., Caruso, M., Reed, R.L., Bailey, C.F., Nguyen, T., et al. (2020). Integration of mass spectral fingerprinting analysis with precursor ion (MS1) quantification for the characterisation of botanical extracts: application to extracts of *Centella Asiatica* (L.) urban. *Phytochemical Analysis*, *31*(6), 722–738.

Alzheimer's, Association (2015). Alzheimer's disease facts and figures. *Alzheimer's & Dementia*, *11*(3), 332–384.

Bahety, P., Tan, Y.M., Hong, Y., Zhang, L., Chan, E.C.Y., and Ee, P.L.R. (2014). Metabotyping of docosahexaenoic acid-treated Alzheimer's disease cell model. *PloS One*, *9*(2), e90123.

Barupal, D.K., Baillie, R., Fan, S., Saykin, A.J., Meikle, P.J., Arnold, M., Nho, K., et al. (2019). Sets of coregulated serum lipids are associated with Alzheimer's disease pathophysiology. *Alzheimer's & Dementia: Diagnosis, Assessment & Disease Monitoring*, *11*, 619–627.

Beale, D.J., Pinu, F.R., Kouremenos, K.A., Poojary, M.M., Narayana, V.K., Boughton, B.A., Kanojia, K., Dayalan, S., Jones, O.A.H., and Dias, D.A.(2018). Review of recent developments in GC–MS approaches to metabolomics-based research. *Metabolomics*, *14*(11), 1–31.

Bhogal, N. and Balls, M. (2008). Translation of new technologies: from basic research to drug discovery and development. *Current Drug Discovery Technologies*, *5*(3), 250–262.

Bingol, K. and Brüschweiler, R.R. (2014). Multidimensional approaches to NMR-based metabolomics. *Analytical Chemistry*, *86*(1), 47–57.

Bøjstrup, M., Petersen, B.O., Beeren, S.R., Hindsgaul, O., and Meier, S. (2013). Fast and accurate quantitation of glucans in complex mixtures by optimized heteronuclear NMR spectroscopy. *Analytical Chemistry*, *85*(18), 8802–8808.

Cardoso-Taketa, A.T., Pereda-Miranda, R., Choi, Y.H., Verpoorte, R., and Villarreal, M.L. (2008). Metabolic profiling of the Mexican anxiolytic and sedative plant *Galphimia glauca* using nuclear magnetic resonance spectroscopy and multivariate data analysis. *Planta Medica*, *74*(10), 1295–1301.

Chagas-Paula, D.A., Oliveira, T.B., Zhang, T., Edrada-Ebel, R., and Da Costa, F.B. (2015). Prediction of anti-inflammatory plants and discovery of their biomarkers by machine learninggas algorithms and metabolomic studies. *Planta Medica*, *81*(06), 450–458.

Chan, A.W., Mercier, P., Schiller, D., Bailey, R., Robbins, S., Eurich, D.T., Sawyer, M.B., and Broadhurst, D. (2016). 1H-NMR urinary metabolomic profiling for diagnosis of gastric cancer. *British Journal of Cancer*, *114*(1), 59–62.

Chu, H., Zhang, A., Han, Y., Lu, S., Kong, L., Han, J., Liu, Z., Sun, H., and Wang. X. (2016). Metabolomics approach to explore the effects of Kai-Xin-San on Alzheimer's disease using UPLC/ESI-Q-TOF mass spectrometry. *Journal of Chromatography B Analytical Technologies in the Biomedical and Life Sciences*, *1015*, 50–61.

Coimbra, J.R.M., Marques, D.F.F., Baptista, S.J., Pereira, C.M.F., Moreira, P.I., Dinis, T.C.P., Santos, A.E., and Salvador, J.A.R. (2018). Highlights in BACE1 inhibitors for Alzheimer's disease treatment. *Frontiers in Chemistry*, *6*, 178.

Cook, D.W., Rutan, S.C., Stoll, D.R., and Carr, P.W. (2015). Two dimensional assisted liquid chromatography– a chemometric approach to improve accuracy and precision of quantitation in liquid chromatography using 2D separation, dual detectors, and multivariate curve resolution. *Analytica Chimica Acta, 859*, 87–95.

Corbett, A., Smith, J., and Ballard, C. (2012). New and emerging treatments for Alzheimer's disease. *Expert Review of Neurotherapeutics, 12*(5), 535–543.

Cristofano, A., Sapere, N., La Marca, G., Angiolillo, A., Vitale, M., Corbi, G., Scapagnini, G., et al. (2016). Serum levels of acyl-carnitines along the continuum from normal to Alzheimer's dementia. *PloS one, 11*(5), e0155694.

Crook, A.A. and Powers, R. (2020). Quantitative NMR-based biomedical metabolomics: current status and applications. *Molecules, 25*(21), 5128.

Daulatzai, M.A. (2017). Cerebral hypoperfusion and glucose hypometabolism: key pathophysiological modulators promote neurodegeneration, cognitive impairment, and Alzheimer's disease. *Journal of Neuroscience Research, 95*(4), 943–972.

Deardorff, W.J. and Grossberg, G.T. (2016).A fixed-dose combination of memantine extended-release and donepezil in the treatment of moderate-to-severe Alzheimer's disease. *Drug Design, Development and Therapy, 10*, 3267.

Emwas, A.H.M. (2015). The strengths and weaknesses of NMR spectroscopy and mass spectrometry with particular focus on metabolomics research. *Metabonomics*, Humana Press, New York, 161–193.

Emwas, A.H., Luchinat, C., Turano, P., Tenori, L., Roy, R., Salek, R.M., Ryan, D., Merzaban, J.S., Kaddurah-Daouk, R. Zeri, A.C., and Gowda, G.A.N. (2015). Standardizing the experimental conditions for using urine in NMR-based metabolomic studies with a particular focus on diagnostic studies: a review. *Metabolomics, 11*(4), 872–894.

Emwas, A.H., Roy, R., McKay, R.T., Tenori, L., Saccenti, E., Gowda, G.N., Raftery, D., Alahmari, F., Jaremko, L., Jaremko, M., and Wishart, D.S. (2019). NMR spectroscopy for metabolomics research. *Metabolites, 9*(7), 123.

Ferrer, I., Martinez, A., Boluda, S., Parchi, P., and Barrachina, M. (2008). Brain banks: benefits, limitations and cautions concerning the use of post-mortem brain tissue for molecular studies. *Cell and Tissue Banking, 9*(3), 181–194.

Fu, C., Wang, T., Wang, Y., Chen, X., Jiao, J., Ma, F., Zhong, M., and Bi, K. (2011). Metabonomics study of the protective effects of green tea polyphenols on aging rats induced by d-galactose. *Journal of Pharmaceutical and Biomedical Analysis, 55*(5), 1067–1074.

Gika, H.G., Theodoridis, G.A., Plumb, R.S., and Wilson. I.D. (2014). Current practice of liquid chromatography–mass spectrometry in metabolomics and metabonomics. *Journal of Pharmaceutical and Biomedical Analysis, 87*, 12–25.

Godyń, J., Jończyk, J., Panek, D., and Malawska, B. (2016). Therapeutic strategies for Alzheimer's disease in clinical trials. *Pharmacological Reports, 68*(1), 127–138.

Gong, Y., Liu, Y., Zhou, L., Di, X., Li, W., Li, Q., and Bi, K. (2015). A UHPLC–TOF/MS method based metabonomic study of total ginsenosides effects on Alzheimer disease mouse model. *Journal of Pharmaceutical and Biomedical Analysis, 115*, 174–182.

Gonzalez-Dominguez, R., Garcia-Barrera, T., Vitorica, J., and Gómez-Ariza, J.L. (2015). Metabolomic screening of regional brain alterations in the APP/PS1 transgenic model of Alzheimer's disease by direct infusion mass spectrometry. *Journal of Pharmaceutical and Biomedical Analysis, 102*, 425–435.

Gowda, G.N. and Raftery, R. (2015). Can NMR solve some significant challenges in metabolomics? *Journal of Magnetic Resonance, 260*, 144–160.

Graham, S.F., Holscher, C., and Green, B.D. (2014). Metabolic signatures of human Alzheimer's disease (AD): 1H NMR analysis of the polar metabolome of post-mortem brain tissue. *Metabolomics, 10*(4), 744–753.

Graham, S.F., Holscher, C., McClean, P., Elliott, C.T., and Green, B.D. (2013). 1H NMR metabolomics investigation of an Alzheimer's disease (AD) mouse model pinpoints important biochemical disturbances in brain and plasma. *Metabolomics, 9*(5), 974–983.

Gromski, P.S., Muhamadali, H., Ellis, D.I., Xu, Y., Correa, E., Turner, M.L., and Goodacre, R. (2015). A tutorial review: metabolomics and partial least squares-discriminant analysis–a marriage of convenience or a shotgun wedding. *Analytica Chimica Acta, 879*, 10–23.

Hall, R.D., Brouwer, I.D., and Fitzgerald, M.A. (2008). Plant metabolomics and its potential application for human nutrition. *Physiologia Plantarum, 132*(2), 162–175.

Halouska, S. and Powers, R. (2006). Negative impact of noise on the principal component analysis of NMR data. *Journal of Magnetic Resonance, 178*(1), 88–95.

Han, X., Rozen, S., Boyle, S.H., Hellegers, C., Cheng, H., Burke, J.R., Welsh-Bohmer, K.A., Doraiswamy, P.M., and Kaddurah-Daouk, R. (2011). Metabolomics in early Alzheimer's disease: identification of altered plasma sphingolipidome using shotgun lipidomics. *PloS one, 6*(7), e21643.

Heemels, M.T. (2016). Neurodegenerative diseases. *Nature, 539*(7628), 179–180.

Hejazi, L., Wong, J.W.H., Cheng, D., Proschogo, N., Ebrahimi, D., Garner, B., and Don, A.S. (2011). Mass and relative elution time profiling: two-dimensional analysis of sphingolipids in Alzheimer's disease brains. *Biochemical Journal*, 438(1), 165–175.

Hendrix, J.A., Bateman, R.J., Brashear, H.R., Duggan, C., Carrillo, M.C., Bain, L.J., DeMattos, R., Katz, R.G., Ostrowitzki, S., Siemers, E., and Sperling, R. (2016). Challenges, solutions, and recommendations for Alzheimer's disease combination therapy. *Alzheimer's & Dementia*, 12(5), 623–630.

Horai, H., Arita, M., Kanaya, S., Nihei, Y., Ikeda, T., Suwa, K., Ojima, Y., Tanaka, K., Tanaka, S., Aoshima, K., and Oda, Y. (2010). MassBank: a public repository for sharing mass spectral data for life sciences. *Journal of Mass Spectrometry*, 45(7), 703–714.

Hu, H., Zhang, A., Han, Y., Lu, S., Kong, L., Han, J., Liu, Z., Sun, H., and Wang, X. (2016). Metabolomics approach to explore the effects of Kai-Xin-San on Alzheimer's disease using UPLC/ESI-Q-TOF mass spectrometry. *Journal of Chromatography B, Analytical Technologies in the Biomedical and Life Sciences*, 1015, 50–61.

Ibáñez, C., Simó, C., Barupal, D.K., Fiehn, O., Kivipelto, M., Cedazo-Mínguez, A., and Cifuentes, A. (2013). A new metabolomic workflow for early detection of Alzheimer's disease. *Journal of Chromatography A*, 1302, 65–71.

Ibáñez, C., Simó, C., Martín-Álvarez, P.J., Kivipelto, M., Winblad, B., Cedazo-Mínguez, A., and Cifuentes, A. (2012). Toward a predictive model of Alzheimer's disease progression using capillary electrophoresis–mass spectrometry metabolomics. *Analytical Chemistry*, 84(20), 8532–8540.

Inoue, K., Tsutsui, H., Akatsu, H., Hashizume, Y., Matsukawa, N., Yamamoto, T., and Toyo'Oka, T. (2013). Metabolic profiling of Alzheimer's disease brains. *Scientific Reports*, 3, 1–9.

Jack Jr, C.R., and Holtzman, D.M. (2013). Biomarker modeling of Alzheimer's disease. *Neuron*, 80(6), 1347–1358.

Ji, H.F. and Zhang, H.Y. (2008). Multipotent natural agents to combat Alzheimer's disease. functional spectrum and structural features 1. *Acta Pharmacologica Sinica*, 29(2), 143–151.

Jorge, T.F., Rodrigues, J.A., Caldana, C., Schmidt, R., van Dongen, J.T., Thomas-Oates, J., and António, C. (2016). Mass spectrometry-based plant metabolomics: metabolite responses to abiotic stress. *Mass Spectrometry Reviews*, 35(5), 620–649.

Kaddurah-Daouk, R. and Krishnan, K. (2009). Metabolomics: a global biochemical approach to the study of central nervous system diseases. *Neuropsychopharmacology*, 34(1), 173–186.

Kaddurah-Daouk, R., Kristal, B.S., and Weinshilboum, R.M. (2008). Metabolomics: a global biochemical approach to drug response and disease. *Annual Review of Pharmacology and Toxicology*, 48(1), 653–683.

Kim, G.H., Kim, J.E., Rhie, S.J., and Yoon, S. (2015). The role of oxidative stress in neurodegenerative diseases. *Experimental Neurobiology*, 24(4), 325.

Kim, H.K., Khan, S., Wilson, E.G., Kricun, S.D.P., Meissner, A., Goraler, S., Deelder, A.M., Choi, Y.H., and Verpoorte, R. (2010). Metabolic classification of South American Ilex species by NMR-based metabolomics. *Phytochemistry*, 71(7), 773–784.

Kim, W., Peever, T.L., Park, J.J., Park, C.M., Gang, D.R., Xian, M., Davidson, J.A., Infantino, A., Kaiser, W.J., and Chen, W. (2016). Use of metabolomics for the chemotaxonomy of legume-associated Ascochyta and allied genera. *Scientific Reports*, 6(1), 1–12.

Kromke, M., Palomino-Schätzlein, M., Mayer, H., Pfeffer, S., Pineda-Lucena, A., Luy, B., Hausberg, M., and Muhle-Goll, C. (2016). Profiling human blood serum metabolites by nuclear magnetic resonance spectroscopy: a comprehensive tool for the evaluation of hemodialysis efficiency. *Translational Research*, 171, 71–82.

Lajis, N., Maulidiani, M., Abas, F., and Ismail, I.S. (2017). Metabolomics approach in pharmacognosy, In: S. Badal and R. Delgoda (Eds.) *Pharmacognosy: Fundamentals, Applications and Strategy*, 597–616, London: Elsevier Inc..

Lankadurai, B.P., Nagato, E.G., and. Simpson, M.J (2013). Environmental metabolomics: an emerging approach to study organism responses to environmental stressors. *Environmental Reviews*, 21(3), 180–205.

Leurent, C., Goodman, J.A., Zhang, Y., He, P., Polimeni, J.R., Gurol, M.E., Lindsay, M., Frattura, L., Sohur, U.S., Viswanathan, A., and Bednar, M.M. (2019). Immunotherapy with ponezumab for probable cerebral amyloid angiopathy. *Annals of Clinical and Translational Neurology*, 6(4), 795–806.

Li, N., Liu, Y., Li, W., Zhou, L., Li, Q., Wang, X., and He, P. (2016). A UPLC/MS-based metabolomics investigation of the protective effect of ginsenosides Rg1 and Rg2 in mice with Alzheimer's disease. *Journal of Ginseng Research*, 40(1), 9–17.

Li, P., Wei, D.D., Wang, J.S., Yang, M.H., and Kong, L.Y. (2016). 1H NMR metabolomics to study the effects of diazepam on anisatin induced convulsive seizures. *Journal of Pharmaceutical and Biomedical Analysis*, 117, 184–194.

Liang, Q., Liu, H., Li, X., and Zhang, A.H. (2016). High-throughput metabolomics analysis discovers salivary biomarkers for predicting mild cognitive impairment and Alzheimer's disease. *RSC Advances, 6*(79), 75499–75504.

Lin, C.N., Huang, C.C., Huang, K.L., Lin, K.J., Yen, T.C., and Kuo. H.C. (2019). A metabolomic approach to identifying biomarkers in blood of Alzheimer's disease. *Annals of Clinical and Translational Neurology, 6*(3), 537–545.

Lísa, M., Cífková, E., Khalikova, M., Ovčačíková, M., and Holčapek, M. (2017). Lipidomic analysis of biological samples: comparison of liquid chromatography, supercritical fluid chromatography and direct infusion mass spectrometry methods. *Journal of Chromatography A, 1525*, 96–108.

Liu, H., Tayyari, F., Khoo, C., and Gu, L. (2015). A 1H NMR-based approach to investigate metabolomic differences in the plasma and urine of young women after cranberry juice or apple juice consumption. *Journal of Functional Foods, 14*, 76–86.

Liu, R. (2019). *Single Cell Metabolomics Using Mass Spectrometry: Devices, Methods and Applications*, Ph.D Dissertation in the Department of Chemistry and Biochemistry, University of Oklahoma, Norman.

MahmoudianDehkordi, S., Arnold, M., Nho, K., Ahmad, S., Jia, W., Xie, G., Louie, G., et al. (2019). Altered bile acid profile associates with cognitive impairment in Alzheimer's disease—an emerging role for gut microbiome. *Alzheimer's & Dementia, 15*(1), 76–92.

Mapstone, M., Cheema, A.K., Fiandaca, M.S., Zhong, X., Mhyre, T.R., MacArthur, L.H., Hall, W.J., Fisher, S.G., Peterson, D.R., Haley, J.M., and Nazar, M.D. (2014). Plasma phospholipids identify antecedent memory impairment in older adults. *Nature Medicine, 20*(4), 415–418.

Markley, J.L., Brüschweiler, R., Edison, A.S., Eghbalnia, H.R., Powers, R., Raftery, D., and Wishart, D.S. (2017). The future of NMR-based metabolomics. *Current Opinion in Biotechnology, 43*, 34–40.

Marksteiner, J., Blasko, I., Kemmler, G., Koal T., and Humpel, C. (2018). Bile acid quantification of 20 plasma metabolites identifies lithocholic acid as a putative biomarker in Alzheimer's disease. *Metabolomics, 14*(1), 1–10.

Maulidiani, M., Sheikh, B.Y., Mediani, A., Wei, L.S., Ismail, I.S., Abas, F., and Lajis, N.H. (2015). Differentiation of *Nigella sativa* seeds from four different origins and their bioactivity correlations based on NMR-metabolomics approach. *Phytochemistry Letters, 13*, 308–318.

McGhee, D.J.M., Ritchie, C.W., Thompson, P.A., Wright, D.E., Zajicek, J.P., and Counsell, C.E. (2014). A systematic review of biomarkers for disease progression in Alzheimer's disease. *PloS one, 9*(2), e88854.

Mendiola-Precoma, J., Berumen, L.C., Padilla, K., and Garcia-Alcocer, G., (2016). Therapies for prevention and treatment of Alzheimer's disease. *BioMed Research International*, Article ID 2589276, 1–17.

Muguruma, Y., Tsutsui, H., Noda, T., Akatsu, H., and Inoue, K. (2018). Widely targeted metabolomics of Alzheimer's disease postmortem cerebrospinal fluid based on 9-fluorenylmethyl chloroformate derivatized ultra-high performance liquid chromatography tandem mass spectrometry. *Journal of Chromatography B, Analytical Technologies in the Biomedical and Life Sciences, 1091*, 53–66.

Nagana Gowda, G.A. and Raftery, D. (2017). Whole blood metabolomics by 1H NMR spectroscopy provides a new opportunity to evaluate coenzymes and antioxidants. *Analytical Chemistry, 89*(8), 4620–4627.

Pan, X., Nasaruddin, M.B., Elliott, C.T., McGuinness, B., Passmore, A.P., Kehoe, P.G., Hölscher, C., McClean, P.L., Graham, S.F., and Green, B.D. (2016). Alzheimer's disease–like pathology has transient effects on the brain and blood metabolome. *Neurobiology of Aging, 38,* 151–163.

Pinto, J., Barros, A.S., Domingues, M.R.M., Goodfellow, B.J., Galhano, E., Pita, C., Almeida, M.C., Carreira, I.M., and Gil, A.M. (2015). Following healthy pregnancy by NMR metabolomics of plasma and correlation to urine. *Journal of Proteome Research, 14*(2), 1263–1274.

Pontes, J.G.M., Brasil, A.J.M., Cruz, G.C.F., de Souza, R.N., and Tasic, L. (2017). NMR-based metabolomics strategies: plants, animals and humans. *Analytical Methods, 9*(7), 1078–1096.

Pontes, J.G.M., Ohashi., W.Y., Brasil, A.J.M., Filgueiras, P.R., Espíndola, A.P.D., Silva, J.S., Poppi, R.J., Coletta-Filho, H.D., and Tasic, L. (2016). Metabolomics by NMR spectroscopy in plant disease diagnostic: Huanglongbing as a case study. *Chemistry Select, 1*(6), 1176–1178.

Proitsi, P., Kim, M., Whiley, L., Simmons, A., Sattlecker, M., Velayudhan, L., Lupton M.K., et al. (2017). Association of blood lipids with Alzheimer's disease: a comprehensive lipidomics analysis. *Alzheimer's & Dementia, 13*(2), 140–151.

Puri, S., Sahal, D., and Sharma, U. (2021). A conversation between hyphenated spectroscopic techniques and phytometabolites from medicinal plants. *Analytical Science Advances, 2*(11–12), 579–593.

Qiu, C. and Fratiglioni, L. (2018). Aging without dementia is achievable: current evidence from epidemiological research. *Journal of Alzheimer's Disease, 62*(3), 933–942.

Ramautar, R., Somsen, G.W., and de Jong, G.J.(2009). CE-MS in metabolomics. *Electrophoresis*, *30*(1), 276–291

Reveglia, P., Paolillo, C., Ferretti, G., De Carlo, A., Angiolillo, A., Nasso, R., Caputo, M., Matrone, C., Di Costanzo, A., and Corso, G. (2021). Challenges in LC–MS-based metabolomics for Alzheimer's disease early detection: targeted approaches versus untargeted approaches. *Metabolomics*, *17*(9), 1–18.

Revi, M. (2020). Alzheimer's disease therapeutic approaches. *Advances in Experimental Medicine and Biology*, *1195*, 105–116.

Rizzi, L., Rosset, I., and Roriz-Cruz, M. (2014). Global epidemiology of dementia: Alzheimer's and vascular types. *BioMed Research International*, Article ID 908915, 1–8.

Romano, J.D. and Tatonetti, N.P. (2019). Informatics and computational methods in natural product drug discovery: a review and perspectives. *Frontiers in Genetics*, *10*, 368.

Russell, C., Rahman, A., and Mohammed, A.R. (2013). Application of genomics, proteomics and metabolomics in drug discovery, development and clinic. *Therapeutic Delivery*, *4*(3), 395–413.

Saber, F.R., Mohsen, E., El-Hawary, S., Eltanany, B.M., Elimam, H., Sobeh, M., and Elmotayam, A.K. (2021). Chemometric-enhanced metabolic profiling of five *Pinus* species using HPLC-MS/MS spectrometry: correlation to in vitro anti-aging, anti-Alzheimer and antidiabetic activities. *Journal of Chromatography B, Analytical technologies in the biomedical and life sciences*, *1177*, 122759.

Santos, M.A., Chand, K., and Chaves, S. (2016). Recent progress in multifunctional metal chelators as potential drugs for Alzheimer's disease. *Coordination Chemistry Reviews*, *327*, 287–303.

Shi, J., Wang, Y., and Luo, G. (2011). UPLC-TOF MS-based metabonomic study on coadministration of huperzine A and ligustrazine phosphate for treatment of Alzheimer's disease. *Chromatographia*, *74*(11), 827–832.

Sun, H.Z., Wang, D.M., Wang, B., Wang, J.K., Liu, H.Y., Guan, L.L., and Liu. J.X. (2015). Metabolomics of four biofluids from dairy cows: potential biomarkers for milk production and quality. *Journal of Proteome Research*, *14*(2), 1287–1298.

Sun, H., Han, Y., Zhang, A., and Wang, X. (2015). Metabolic profiles delineate the effect of shengmai san on Alzheimer's disease in rats. In X. Wang (Ed.) *Chinmedomics*, 363–371, Boston, MA: Academic Press..

Takayama, T., Mochizuki, T., Todoroki, K., Min, J.Z., Mizuno, H., Inoue, K., Akatsu, H., Noge, I., and Toyo'oka, T. (2015). A novel approach for LC-MS/MS-based chiral metabolomics fingerprinting and chiral metabolomics extraction using a pair of enantiomers of chiral derivatization reagents. *Analytica Chimica Acta*, *898*, 73–84.

Trushina, E., Dutta, T., Persson, X.M.T., Mielke, M.M., and Petersen, R.C. (2013). Identification of altered metabolic pathways in plasma and CSF in mild cognitive impairment and Alzheimer's disease using metabolomics. *PloS One*, *8*(5), e63644.

Udeh-Momoh, C. and Watermeyer, T. (2021). Female specific risk factors for the development of Alzheimer's disease neuropathology and cognitive impairment: call for a precision medicine approach. *Ageing Research Reviews*, *71*, 01459.

Vinaixa, M., Schymanski, E.L., Neumann, S., Navarro, M., Salek, R.M., and Yanes, O. (2016). Mass spectral databases for LC/MS-and GC/MS-based metabolomics: state of the field and future prospects. *TRAC Trends in Analytical Chemistry*, *78*, 23–35.

Waly, N.E., Aborehab, N., and Helmy, M. (2019). Chemical fingerprint of *Bacopa monnieri* L. and *Rosmarinus officinalis* L. and their neuroprotective activity against Alzheimer's disease in rat model's putative modulation via cholinergic and monoaminergic pathways. *Journal of Medicinal Plant Research*, *13*(11), 252–268

Wang, G., Zhou, Y., Huang, F.J., Tang, H.D., Xu, X.H., Liu J.J., and Wang, Y.(2014). Plasma metabolite profiles of Alzheimer's disease and mild cognitive impairment. *Journal of Proteome Research*, *13*(5), 2649–2658.

Wang, X., Han, Y. Zhang, A., and Sun, H. (2015). Metabolic Profiling Provides a System for the understanding of Alzheimer's Disease in rats post-treatment with Kaixin San. In: X. Wang (Ed.) *Chinmedomics*, 347–362, Boston, MA: Academic Press.

Wang, Y., Zhao, M., Xin, Y., Liu, J., Wang, M., and Zhao, C. (2016). 1H NMR and MS based metabolomics study of the therapeutic effect of cortex fraxini on hyperuricemic rats. *Journal of Ethnopharmacology*, *185*, 272–281.

Wang, Z., Li, W., Li, J.Y., Li, L., and Li, N.J. (2016). Metabonomics study in the effect of *Valeriana amurensis* extractive for model mice with Alzheimer's disease. *Chinese Medicine Herbal*, *13*, 4–7.

Weng, H.B., Chen, H.X., and Wang, M.W. (2018). Innovation in neglected tropical disease drug discovery and development. *Infectious Diseases of Poverty*, *7*(1), 1–9.

Williams, P., Sorribas, A., and Howes, M.J.R. (2011). Natural products as a source of Alzheimer's drug leads. *Natural Product Reports*, *28*(1), 48–77.

Wold, S., Sjöström, M., and Eriksson, L. (2001). PLS-regression: a basic tool of chemometrics. *Chemometrics and Intelligent Laboratory Systems*, *58*(2), 109–130.

Wood, P.L., Locke, V.A., Herling, P., Passaro, A., Vigna, G.B., Volpato, S., Valacchi, G., Cervellati, C., and Zuliani, G. (2016). Targeted lipidomics distinguishes patient subgroups in mild cognitive impairment (MCI) and late onset Alzheimer's disease (LOAD). *BBA Clinical*, *5*, 25–28.

Worley, B., Halouska, S., and Powers, R. (2013). Utilities for quantifying separation in PCA/PLS-DA scores plots. *Analytical Biochemistry*, *433*(2), 102–104.

Wu, B., Yan, S., Lin, Z., Wang, Q., Yang, Y., Yang, G., Shen, Z., and Zhang, W. (2008). Metabonomic study on ageing: NMR-based investigation into rat urinary metabolites and the effect of the total flavone of epimedium. *Molecular BioSystems*, *4*(8), 855–861.

Wu, X., Li, N., Li, H., and Tang, H. (2014). An optimized method for NMR-based plant seed metabolomic analysis with maximized polar metabolite extraction efficiency, signal-to-noise ratio, and chemical shift consistency. *Analyst*, *139*(7), 1769–1778.

Xie, G., Plumb, R., Su, M., Xu, Z., Zhao, A., Qiu, M., Long, X., Liu, Z., and Jia, W. (2008). Ultra-performance LC/TOF MS analysis of medicinal Panax herbs for metabolomic research. *Journal of Separation Science*, *31*(6–7), 1015–1026.

Yan, S., Wu, B., Lin, Z., Jin, H., Huang, J., Yang, Y., Zhang, X., Shen, Z., and Zhang, W. (2009). Metabonomic characterization of aging and investigation on the anti-aging effects of total flavones of Epimedium. *Molecular BioSystems*, *5*(10), 1204–1213.

Yang, Q. (2013). *Fast Quantitative Analysis of Metabolic Biomarkers for Clinical Screening Using Mass Spectrometry*, Purdue University, West Lafayette, IN.

Yi, L., Liu, W., Wang, Z., Ren, D., and Peng, W. (2017). Characterizing Alzheimer's disease through metabolomics and investigating anti-Alzheimer's disease effects of natural products. *Annals of the New York Academy of Sciences*, *1398*(1), 130–141.

Zampieri, M., Sekar, K., Zamboni, N., and Sauer, U. (2017). Frontiers of high-throughput metabolomics. *Current Opinion in Chemical Biology*, *36*, 15–23.

Zhang, A., Sun, H., Yan, G., Wang, P., and Wang, X. (2016). Mass spectrometry-based metabolomics: applications to biomarker and metabolic pathway research. *Biomedical Chromatography*, *30*(1), 7–12.

Zhang, X., Jiang, X., Wang, X., Zhao, Y., Jia, L., Chen, F., Yin, R., and Han, F. (2019). A metabolomic study based on accurate mass and isotopic fine structures by dual mode combined-FT-ICR-MS to explore the effects of Rhodiola crenulata extract on Alzheimer disease in rats. *Journal of Pharmaceutical and Biomedical Analysis*, *166*, 347–356.

15 Can Phylogenetics Clear Plant Taxonomic Confusion for Conservation and Sound Bioprospecting?

Steven Mapfumo, Omolola Afolayan, Charles Petrus Laubscher, Joyce Ndlovu, Callistus Bvenura and Learnmore Kambizi

CONTENTS

15.1 INTRODUCTION

The need to thoroughly delineate species for producing accurate medicinal species inventories is of importance, considering most questions such as speciation in evolutionary biology, conservation biology, and bioprospecting depend, in part, on species inventories and our knowledge of the species. Erroneous species' boundaries or names may result in the use of the wrong plants for medicinal research. Molecular identification of plants using DNA barcodes and reconstruction of molecular phylogenies have demonstrated versatility and are exceptionally useful tools (Donoghue, 2008; Sanmartin and Ronquist, 2004; Pennington et al., 2006).

Apart from the bio-screening aspect, phylogenetics in medicinal plant use can broaden our understanding of traditional ethnobotanical knowledge. The findings that some plant lineages are more heavily used than others (Bennett and Husby, 2008), and the fact that there is a degree of agreement in those lineages between disparate cultures, (Saslis-Lagoudakis et al., 2011, Leonti et al., 2003) imply that phylogenetic relationships underlie people's selection of medicinal plants in traditional medicine and in a fashion that overcomes cultural differences. Apart from some unpublished study findings presented at conferences and other academic platforms (Bletter, 2000), such discoveries have not been examined in terms of phylogenetic framework. By superposing medicinal

DOI: 10.1201/9781003206620-15

properties on lineages with wide distributions, cross-cultural phylogenetic patterns in ethnobotany can be observed, including the agreement in usage of closely related lineages in distant cultures (Bletter, 2000).

The traditional system of medicine suffers from substitution and adulteration of medicinal herbs with closely related species, and the efficacy of the drug decreases if it is adulterated and, in some cases, can be lethal if substituted with toxic adulterants (Patel et al., 2017). There are various approaches to discovering active drug compounds from traditional medicine, collectively referred to as bioprospecting. The methods include chemical testing for active compounds, *in vitro* testing of plant extracts (Clarkson et al., 2004), and traditional ones such as organoleptic, macroscopic, microscopic, and chemical profiling methods (Techen et al., 2014). Due to the inadequacies of the above methods, medicinal plant authentication at DNA level has been introduced. Many molecular methods are increasingly getting investigated to identify medicinal plants (Cowan et al., 2006). This is because DNA authentication is more reliable (Patel et al., 2017).

Conservation of correctly identified medicinal plants is also very crucial. Cryopreservation—a method of preserving biological constructs, which has been useful in human and plant tissue preservation since the 1950s, offers opportunities for long-term conservation of medicinal plant material. Specifically, cryopreservation refers to the storage of damage-susceptible biological constructs such as organs, organelles, cells, tissues, and/or extracellular matrices under cryogenic conditions (Wen and Wang, 2010). This is normally at ultra-low temperatures of −196°C. DNA identification and verification of DNA stability of cryopreserved material play a critical role in building reliable and safe medicinal plant gene banks or libraries.

15.2   TAXONOMY AND PHYLOGENETICS

15.2.1   ADULTERATION AND MISIDENTIFICATION

The main source of livelihood for many herbalists involves trading of medicinal plants. However, unfavourable economic restraints can force herbalists to substitute rare species with inexpensive and easily accessible species, thereby camouflaging and selling the different species under the same name. As such, deliberate adulteration of coveted ingredients is often difficult to discern from cases of under- or over-differentiation or misidentification. Adulteration or misidentification occurs either intentionally or inadvertently (by misidentification). Therefore, phylogenetic techniques are used to accurately identify species and their phytochemistry at molecular level. For example, Bellakhdar et al. (1991) reported that *Colchicum autumnale* L (Bukbuka)—a plant traditionally used in the treatment of renal disorders and acute arthritis in Morocco is taken off the market because it is highly potent. Phylogenetic identification indicated some misidentification of the plant based on the vernacular name provided by the herbalist. Other cases of adulteration or misidentification comprise both samples of *l-harmal,* which instead of *harmala* L in fact was identified as *Carlina brachylepis* (Kool et al., 2012). In yet another case, Kool et al. (2012) reported that *Peganum harmala*—a plant putatively known as ʟ-harmel in the local language is actually *Carlina brachylepis* (Batt.) Meusel and A. Kästner. Furthermore, ʼaqirqarha—a relatively upmarket product whose samples have been profitably misidentified as Catananche is phylogenetically Anacyclus. Taxonomic limitations can easily be overcome by molecular phylogenetic trees that measure plant species' phylogenetic distance in medicinal flora (Faith, 1992). Hence, phylogenetic methods can subdue taxon-based cross-cultural comparisons and associated limitations, and bring about robust measures to synthesise and interpret cross-cultural patterns in medicinal floras. Phylogenetic tools have been used to show clustering of traditionally used (Saslis-Lagoudakis et al., 2011) and bioactive species (Zhu et al., 2011), and to describe cross-cultural trends in plant use (Hart and Cox, 2000). With the inclusion of molecular sources of characters and refinements to phylogenetic theory, taxonomy and evolutionary biology discipline have become more incorporated as a unit. Terms such as molecular taxonomy and phylogenetics are or should be subsumed as a single research niche, for example, systematics (Adams et al., 2009).

15.3 TAXONOMIC UNDER- AND OVER-DIFFERENTIATION

It is not unusual in some countries for a group of species to be assigned the same vernacular name. This is mainly caused by failure to separate between species that are closely related or are under taxonomic under-differentiation. A classic example is that of *'ud-mserser*, where molecular phylogenetics showed that the superior-quality samples, as differentiated by the herbalist, appeared to be related to a single genus, while those of lesser quality were also found to belong to a different genus, although they all belonged to the same family (Kool et al., 2012). Thus, over-differentiation at the taxonomic level has been widely reported and unfortunately has the potential to put the lives of herbal medicine consumers at risk (Bellakhdar et al., 1991). Such cases referred to above are currently being scientifically addressed using different phylogenetic methods that have enhanced the identification of plant species down to molecular levels. It remains, however, to be seen how accurate phylogenetic analysis can be integrated with ethnobotanical knowledge to address these setbacks completely since most of medicinal plant knowledge is derived from ethnobotany.

15.4 METHODS IN TAXONOMY

Several phylogenetic techniques have been explored in taxonomy and still more work is being done. This is due to the difference in the efficiency of these methods when they are used on different organisms, be it bacteria, fungi, plants, and even on animals. So many different results have been reported, and yet no consensus has been reached on which is the best method to use for purposes of taxonomy and species' discovery. A diverse number of different users previously had no access, meaning that the need to incorporate new sequences into a consistent universal taxonomic framework is essential and in demand. According to Macdonald et al. (2012), reference phylogenies are essential for providing a taxonomic framework for the interpretation of marker genes and metagenomic surveys, which continually displays novel species at an exceptional rate. There is evidence now that taxonomic doubts are now being addressed correctly using phylogenetic techniques.

15.5 PHYLOGENETIC TAXONOMY

Biological classifications should reflect only phylogenetic relationships, meaning that only monophyletic groups should be given formal taxonomic names (Swofford et al., 1990). Recognising the limits of the Linnaean system, Queiroz and Gauthier (1994) developed a system of 'phylogenetic taxonomy' to suggest how the rules of biological nomenclature could be reformulated with evolution as the central organising principle. The crucial advancement of phylogenetic taxonomy is the non-use of ranks as well as the definition of taxon names strictly in terms of evolutionary history. On the other hand, the possibility of different names referring to the same entity, better known as taxonomic synonymy, is better clarified by phylogenetic definitions (Queiroz and Gauthier, 1992). Although this was declared more than 20 years ago, there is evidence that this approach has not changed significantly.

According to Reynolds (2007), the field of chemosystematics was developed based on what was presumed to be shared chemistry in closely related species. However, today, instead of using chemical affinities as a basis for taxonomic classification, the distribution of chemistry is better understood by the application of phylogenetic techniques. But a combination of phytochemical and phylogenetic studies so far reveal a solid phylogenetic signal in plant chemical composition distributions that can be used as a tool in the search for novel natural products (Bay-Smidt et al., 2011; Larsen et al., 2010; Muellner et al., 2005; Wink, 2003; Wink and Mohamed, 2003).

TABLE 15.1

Different Molecular Markers That Are Recommended for Use in Plants

Molecular Marker	Sequencing Success Rate
RpoC1	The high amplification rate in the chloroplast region marks the main advantage
PsbA-trnH	Although the sequencing success for this locus was lower that of RpoC1, it was generally high for reference sequences
ITS	Once a sequence was obtained, ITS was shown to be the most useful marker, with 63.8% and 29.8% species and genus level sample identification, respectively
BLAST	To successfully identify market samples, BLAST in collaboration with critical evaluation of the absence or presence of GenBank species and species distribution data were employed
RAxML	51.3%

Source: Kool et al. (2012).

15.6   THE BARCODING OF LIFE

The Barcoding of Life initiative (Hebert et al., 2003a, b) has envisioned a standardised method to alleviate difficult species identifications and to discover cryptic species by focusing sequencing efforts on target genes, like cytochrome c oxidase subunit 1(Cox1), MatK, trnl, and others. This method has been rendered inexplicit for reliable species diagnostics in some cases, and character-based identification systems have been brought forward as preference to proceed with the Barcoding of Life Project (DeSalle et al., 2005). The CBOL Plant Working Group (2009) proposed rbcL and *matK* combined as a universal barcode for land plants but with the option to supplement it with one or two other markers such as ITS or *psbA-trnH*. Table 15.1 shows the results of an experiment that was conducted using the different markers that are specifically recommended for use in plants. These methods were used to identify samples of medicinal plants collected from a herbal market in Morocco.

The possibility to easily associate different stages of life of the same species is one of the merits of DNA barcoding of species' life stages (Blaxter, 2004). In situations where taxa are difficult to propagate in the laboratory, DNA barcoding becomes very valuable. The benefits of short DNA sequences in associating adult stages with their immature counterparts in some species have thus been reported (Barrett and Hebert, 2005; Hebert et al., 2004b). Although DNA barcoding is a limited aspect of systematics and taxonomy, the difference between classification and identification may not be very clear; thus cox1 sequencing has increasingly become a reliable tool for new species discovery (Besansky et al., 2003; Hebert et al., 2004a, b; Smith et al., 2006).

The inclusion of both organellar DNA and nDNA, for example, mtDNA markers (or plastid DNA for plants), has been proposed as an ideal molecular systematic approach, while differences between partitions can be used to augment the interpretation of the evolutionary history of the taxa under consideration (Rubinoff and Holland, 2005). The dramatic increase in interest to infer large-scale phylogenies has been due to the easy accessibility of sequence data for a wide range of organisms (Bininda-Emonds, 2000). This increased interest in phylogenomic studies that produce data sets to infer phylogeny based on countless markers has led to an increase in character sampling but for a limited number of model taxa (Dunn et al., 2008; Rokas et al., 2003; Wildman et al., 2007).

There is no doubt in molecular research that the polymerase chain reaction (PCR) has become an indispensable tool, more so, in cryopreservation efforts in museums conducting phylogenetic studies of tissues with an interest to preserve genetic diversity (Webster and Brian, 2001). The thrust of DNA barcoding is the provision of efficient methods for species' identification and thus has the potential to contribute positively to biodiversity research. DNA barcoding is thus expected to make major contributions to phylogenetics, population genetics, and taxonomic research, in general.

The routine identification of sample specimens as well as flagging unusual specimens for comprehensive taxonomic studies are some of the uses of DNA barcoding (Hajibabaei et al., 2007). For example, Kool et al. (2012) investigated the commercialised medicinal roots of some medicinal plants in the souks of Marrakech in Southern Morocco using a regional reference database as well plastid genome (*matK, psbA-trnH,* and *rpoC1*) sequence data and the nuclear genome (ITS). These are markers that have been used and are recommended specifically for plants.

15.7 INTEGRATION OF PHYLOGENETICS AND ETHNOBOTANY

15.7.1 Phylogenetic Ethnobotany

In plant-based drug discovery, phylogeny-guided techniques and time-efficient approaches are based on the theory that phytochemicals are constrained to evolutionary plant lineages (Saslis-Lagoudakis et al., 2011). The use of ethnomedicinal information in some of these approaches has been reported (Fabricant and Farnsworth, 2001). Medicinal plant reports are often used as proxies for bioactivity, which are then superimposed on phylogenetic trees, for example the work of Saslis-Lagoudakis et al. (2012). Evolutionary techniques are then used to forecast the potential bioactivity of various plant lineages based on medicinal plant use distribution across the phylogeny (Ernst et al., 2016).

Phylogenetic conservatism (Crisp et al., 2009; Prinzing et al., 2001) in herbal medicinal plants is not new. For example, Lukhoba et al. (2006) reported that the genus *Plectranthus* of the Lamiaceae family comprises 62 of 300 species with ethnomedicinal uses, the majority of which fall within the same large phylogenetic clade and, therefore, suggesting potential phylogenetic patterns in medicinal properties within the genus. This concurs with many other conclusions made by researchers who have analysed the relationship between phylogeny and the medicinal properties of plants. Regardless of the lack of quantification in these studies, the report of Forest et al. (2007) quantitatively showed a significant clumping in the phylogeny of some ethnomedicinal Cape flora of South Africa. Rønsted et al. (2008) also reported similar findings in *Narcissus* spp. with medicinal properties.

15.7.2 Phylogenetic Analysis and Bioprospecting

Instead of taking a random approach to searching for new drugs (bioprospecting), several search algorithms have been proposed to give the process a greater degree of certainty and make it more accurate. These encompass those based on ethnomedicinal information or indigenous traditional knowledge (ITK), ecological associations, evolutionary rationale, data mining, and phylogenetic rationale. But, due to the recent development of molecular phylogenetic techniques and consequent rapid development of angiosperm phylogenetic trees (APG III, The Angiosperm Phylogeny Group, 2009), the phylogenetic rationale is now the most widely accepted technique. The increase to access of information on fundamental structural and controlling elements, such as transcription factors as well as metabolites and their converting enzymes, coupled with phylogenetic association, and systems including allosteric regulation and their conservation, provides a platform for more directed bioprospecting (Muller et al., 2004).

Secondary metabolite expression of some structural types has been known to surface occasionally in various parts of the plant kingdom. Differential gene expressions or convergent evolution could be the potential sources of these discrepancies (Wink, 2003). Therefore, the metabolite constituents of plants could potentially reveal more information on bioactive patterns of plants and their interaction, instead of relying on morphological characteristics. These metabolite constituent-based categorisations have the potential to not only reveal plant phylogenetic relationships but also to harness for studying bioactive properties, exploring hidden nutritional or economic uses, and predicting medicinal properties of plants and their relationships through bioprospecting (Liu et al., 2017).

According to Lagoudakis et al. (2015), a phylogenetic framework may prove to be insightful when investigating commonalities among psychoactive plants, which are rather ubiquitous in many various cultures of the world. In the previous authors' observations, information is allocated to nodes instead of a single species at a time, thereby facilitating trait distribution studies. It has been shown in phylogenetics that phytochemical traits and medicinal uses of culturally diverse medicinal plants are shared by plants that are closely related regardless of the plants' geographic or cultural differences, meaning that these traits are not randomly distributed on the phylogeny (Lagoudakis et al., 2012; 2015). Furthermore, phylogenetic analysis has been used to reveal if patterns of cultural convergence in psychoactive plants are similar (Xavier and Molina, 2016), while some studies report that culturally important psychoactive plant phylogeny possesses a certain inclination towards targeted plant families and for specific psychoactive effects such as sedative, stimulant, and hallucinogenic effects among others (Alrashedy and Molina, 2016). According to Proches et al. (2008), the use of phylogenetic analysis can help humans to expand their diets/culinary horizons by pinpointing relatively new sources of food for propagation by revealing information on families from the existing database, which are already being exploited. But psychoactive chemicals are phylogenetically clustered, as shown by reports that indicate that psychoactive effects can only be found in groups (Alrashedy and Molina, 2016). This clearly shows the importance of using phylogenetic techniques in bioprospecting.

The phylogenetic clustering of secondary metabolites as well as of medicinal traits are widely reported (Lagoudakis et al., 2012; Lagoudakis et al., 2015; Wink, 2003, 2013; Wink et al., 2010; Xavier and Molina, 2016). The existence of similar psychoactive chemicals in lineages that are not phylogenetically related, hence suggesting convergent evolution or differential regulation of genes of similar metabolic pathways, does not take away from the fact that many traditionally used psychoactive herbs show phylogenetic conservatism in pharmacology as well as phytochemistry and may thus be explored as potential novel neurological disorder therapeutics. This, therefore, provides further evidence of the potential of psychoactive plants as springboards for the discovery of psychotherapeutic drugs (McKenna, 1995).

15.7.3 Phylogenetic Rationale

Wink (2003) suggested a phylogenetic rationale, since phylogeny forms an evolutionary link between various taxa, and thus creates a platform for common metabolic pathway discovery and empirical biomedical natural compound prospecting. Without any other additional information, the phylogenetic search algorithm (based on the taxonomic neighbourhood) can be regarded as the first step to the discovery of other means of existing secondary metabolites, novel biological molecules, and activities (Wink, 2003). The key success factor of this method relies on the assumption that there is a phylogenetic inertia for the compounds being searched. Numerous earlier workers have effectively utilised the phylogenetic avenue to discover alternative sources for existing high-value metabolites (Kingston, 2000).

The cancer treatment drug—taxol (Kingston, 2000) is a frequently mentioned example of the predictive value of the phylogenetic approach. Taxol was initially extracted from the bark of the Pacific yew, *Taxus brevifolia* Nutt. The high demand for taxol necessitates the discovery of alternative source of the compound. A different source of taxol (the precursor baccatin III) was established quickly and efficiently by seeking for the compound in *T. baccata*—a close relative of *T. brevifolia* (Kingston, 2000). This example of screening for taxol that might have taken years to discover was streamlined into a simple task by integrating the principle of genetic descent with modification, which is a phylogenetic approach.

In essence, the phylogenetic rationale could be used to bioprospect for a range of products, including searching for known chemical moieties in unknown floristic systems to unravelling newer sources of a certain bioactivity. In recent years, this has been greatly facilitated by the development of molecular phylogenetic tools and the subsequent rapid build-up of angiosperm phylogenetic trees (APG III, The Angiosperm Phylogeny Group, 2009). Phylogenetic techniques have been scientifically useful in bioprospecting, because they have enabled scientists to make more accurate molecular analysis of plant secondary metabolites, consequently leading to quicker drug discovery. The use of phylogenetic analysis in bioprospecting has received so much attention to an extent that computational methods have also been developed to make the analysis faster and more accurate. Phylogenies inference with computational methods has numerous biological and medical research applications, for example, conservation biology and drug discovery. Bioprospecting is one of the domains where these applications have been implemented.

15.8 PHYLOGENETIC TECHNIQUES IN CRYOPRESERVATION

Germplasm conservation assumes that the material is conserved under conditions ensuring genetic stability. Various contributing factors, however, associated with *in vitro* culture and conservation procedures may consequently lead to somatic variations (Srivastava et al., 2012). Hence, it is crucial to continually observe genetic stability during the period of *in vitro* conservation. Generally, the genetic fidelity of recovered plants is evaluated at the biochemical, morphological, cytological, and molecular levels (Srivastava et al., 2012). The sequence-related amplified polymorphism (SRAP) assay has thus been used to assess genetic stability and detect DNA-level variation. Martín and González-Benito (2005) observed that there existed a low variance rate of 0.01%. Similarly, a mutation frequency of 0.10% was found by randomly amplified polymorphic DNA (RAPD) assay in the chrysanthemum plants cryopreserved by encapsulation–dehydration (Martín and González-Benito, 2005). Subsequently, using **amplified fragment length polymorphism** (AFLP) to analyse the genetic stability of chrysanthemum throughout all stages of an encapsulation–dehydration cryopreservation protocol showed that most differences were affected by the preculture process of high-concentration sucrose solution, which showed that genetic variation could appear throughout the cryopreservation process, and the ultra-low temperature itself was not the only stress risk of the conservation technique. Ai et al. (2011) concluded that further monitoring of genetic stability throughout the cryopreservation process and optimising of cryopreservation protocol should be the next work. Their study showed the existence of low genetic variation in plantlets regenerated from cryopreserved shoot-tips of *Rubescens*.

As reported by Urbanova et al. (2005), phylogenetic methods so far indicate that the capability of cells to regenerate as well as differentiate into whole plants, and their viability cannot be the only factors to assess cryopreservation; however, genetic stability of the regenerated plants at DNA level can also be assessed. To evaluate DNA stability, several methods such as restriction fragment length polymorphism (RFLP), RAPD, AFLP, or the amplification of variable number tandem repeats (VNTR) sequences can be employed. But although alterations in DNA sequences are possible, these are not commonly observed after cryopreservation (Lui et al., 2017; Zhai et al., 2003). Genetic stability in recovered plants is one important aspect of cryopreservation concern. Cryopreserved materials can thus be preserved theoretically without genetic alteration for an unlimited period (Hao et al., 2002). It is essential to determine whether the materials retrieved from *in vitro* conservation are genetically identical to the material accessed.

Different phylogenetic methods and genetic markers have been used to compare genomic DNA patterns, including RFLP, RAPD fragments, simple sequence repeat (SSR) analysis, and AFLP (Kaczmarczyk et al., 2011). This confirms that phylogenetic methods have received extensive applications in testing different parameters during cryopreservation. Various studies have established the presence of genetic stability (Castillo et al., 2010) and where changes in the genome have been detected, for example in sugarcane and potato with RFLP markers, the changes could not be related to the process of cryopreservation itself (Castillo et al., 2010; Harding, 2004).

Cryopreserved tissues have been found to contain unaltered DNA fragment patterns in comparison with the control samples during the examination of embryos (Hao et al., 2002; Helliot et al., 2002; Peredo et al., 2008; Zhai et al., 2003). However, intra-clonal variation in the RAPD profiles was detected in two separate non-frozen cell lines that were pre-treated with sorbitol and dimethylsulphoxide (DMSO). However, regardless of the success of the RAPD technology in detecting genetic alternation occurrences, some reproducibility limitations in this technology have been reported, and this has led to its replacement with SSR and AFLP (Ahuja et al., 2002; Castillo et al., 2010; Urbanova et al., 2005). Some researchers have suggested the use of SSR and AFLP after cryo-procedures in future genetic fidelity studies showing mixed feelings on the use of phylogenetic methods (Salaj et al., 2007). The rapid development of DNA-based methods has also led to increased interest in plant genetic integrity assessments after *in vitro* culture tests and cryopreservation. Plant tissue culture-induced changes in genetic profiles or somaclonal variation in specimens are also well known and reported (Martín and González-Benito, 2005). Some of the genetic changes in plant, protists, and animal cells have long been linked to Dimethylsulphoxide (Me_2SO). Aronen et al. (2005) found some variations in *Abies cephalonica's* non-frozen embryonic cultures that were subjected to Me_2SO using RAPD markers, while Harding (2001) reported that RFLPs of ribosomal DNA led to polymorphism in slow-growth potato plantlets and not in cryopreserved specimens. Unfortunately, reports on genetic stability of plantlets obtained from cryopreserved apices using different techniques are scarce. AFLP polymorphic fragments have been found with a significantly diverse frequency between *Prunus* plantlets derived from frozen and non-frozen apices, using encapsulation–dehydration and slow freezing methods, although all plants that were regenerated were phenotypically true-to type (Martín and González-Benito, 2005). Such conclusions were only arrived at after the use of phylogenetic methods that allow for genetic evaluation before, during, and after cryopreservation.

A PCR-based randomly amplified DNA fingerprinting (RAF) was used to explore the genomic DNA structure in papaya plants, while the Amplified DNA Methylation Polymorphism (AMP) was used to monitor the specimens' methylation patterns. In terms of molecular analysis, the following techniques have been used to assess stability after cryopreservation; RFLP (Harding and Benson, 2000), SSR, (Harding and Benson, 2001), RAPD (Gagliardi et al., 2003), and AFLP (Turner et al., 2001). Although some minor changes have been detected in recovered plants using these techniques, the consensus among these reports indicate overall stability after cryopreservation, as outlined by Harding (2004). A phylogenetic technique termed RAF was employed to screen for changes in genomic DNA primary structure by Kaity et al. (2008). These authors reported that the main advantage of this arbitrarily primed PCR marker technique is that no prior knowledge of DNA sequence is required, allowing rapid and reproducible amplification and detection of markers from any organism.

RAPD, ISSR, and SSR analysis can be used to evaluate genetic fidelity of the cryopreserved lines of spices. The genetic fidelity of lines of spices that have been cryopreserved have previously been evaluated using ISSR, RAPD, and SSR techniques. The methods of Winirmal Babu et al. (2003, 2007) and Ravindran et al. (2004) were employed to develop these profiles. Morphological characters coupled with RAPD profiles using 24 operon primers showed genetic fidelity among randomly selected Aimpiriyan and Subhakara black pepper micropropagated plants, indicating the potential to clone commercial black pepper using micropropagation protocols (Nirmal Babu et al. 2003). *Drosophila melanogaster* specimen DNA quality has also been isolated and assessed using the quantitative PCR (QPCR) technique. The significant damage to DNA owing to cryopreservation has implications on genetic studies that employ the use of enzyme amplifications as well as phylogenetics. But, for long-term preservation of tissues, more so when used for sequencing where only a short strand is required in the PCR reaction, cryopreservation is the most reliable genomic technique that is available and will conceivably be around a long time (Webster and Brian, 2001).

15.9   CONCLUSION AND PROSPECTS

The relatedness of organisms can be questioned if only the phenotype is analysed, but, however, phylogenetic analysis has revealed such relatedness at molecular level, ushering in a much more accurate way of identifying these organisms. The existence of different languages has had a negative bearing on the taxonomy of medicinal plants in many different countries of the world. This has resulted in taxonomic under- and over-differentiation. At the same time, the existence of homonyms and synonyms has been another result of the existence of many languages, probably caused by species divergence and convergence, which has also resulted in erroneous botanical identifications. All the problems mentioned above have been and are still being resolved by the advent and use of phylogenetic techniques. These techniques have been used to analyse medicinal plants down to molecular level, and they have revolutionised the field of taxonomy ushering in a more accurate way of identifying and classifying plants. Phylogenetic techniques have also ushered in a more efficient way of discovering active drug compounds from traditional medicine, collectively referred to as bioprospecting. Using phylogenetic techniques, scientists have discovered that certain plant groups are more likely to have bioactive compounds than others, and such phylogenetic studies of plant species have helped in new drug discovery. Combining traditional knowledge of medicinal plants and phylogenetic analysis has been proposed as being able to uncover neglected species of high promise. It can be argued that the use of phylogenetic techniques in bioprospecting can enhance in terms of time taken to discover a drug and improve accuracy in drug discoveries.

Conservationists are now relying more on phylogenetic analysis to be able to correctly identify species, and craft the correct conservation approach and help in the setting up of conservation legislation, including conservation resources allocation. Conservation of biological material is only effective and dependable if it does not alter the DNA of the conserved material. Phylogenetic techniques have thus been and are still being used to test whether the DNA of cryopreserved biological material remains true-to type. It is without doubt that phylogenetic methods have proved to be indispensable in taxonomy, bioprospecting, and cryopreservation. Remarkable achievements have been made using each or a combination of the following phylogenetic methods: PCR, randomly amplified DNA fingerprinting (RAF), AMP, RFLP, RAPD, AFLP, SSR, and MSAP.

ACKNOWLEDGEMENTS

We acknowledge the National Research Foundation of South Africa: Grant No. SRUG210326591092, for supporting this work.

REFERENCES

Adams B.J, Dillman A.R, and Finlinson, C. (2009). Molecular taxonomy and phylogeny. *Root Knot Nematodes*, Chapter 5, CABI, Oxfordshire, 134.

Ahuja, S., Mandal, B.B., Dixit, S., and Srivastava, P.S. (2002). Molecular, phenotypic and biosynthetic stability in *Dioscorea floribunda* plants derived from cryopreserved shoot tips. *Plant Science*, *163*, 971–977.

Ai, P.F., Lu, L.P., and Song, J.J. (2012). Cryopreservation of *in vitro*-grown shoot-tips of *Rabdosia rubescens* by encapsulation-dehydration and evaluation of their genetic stability. *Plant Cell, Tissue and Organ Culture*, *108*, 381–387. https://doi.org/10.1007/s11240-011-0049-x.

Alrashedy, N.A. and Molina, J. (2016). The ethnobotany of psychoactive plant use: a phylogenetic perspective. Authentication of medicinal plants by DNA markers. *Peer Journal*, *4*, 2546.

Barrett, R.D.H. and Hebert, P.D.N. (2005). Identifying spiders through DNA barcodes. *Canadian Journal of Zoology*, *83*, 481–491.

Bay-Smidt, M.G.K., Jager, A.K., Krydsfeldt, K., Meerow, A.W., and Stafford, G.I. (2011). Phylogenetic selection of target species in Amaryllidaceae tribe Haemantheae for acetylcholinesterase inhibition and affinity to the serotonin reuptake transport protein. *South African Journal of Botany*, *77*, 175–183.

Bellakhdar, J., Claisse, R., Fleurentin, J., and Younos, C. (1991). Repertory of standard herbal drugs in the *Moroccan pharmacopoea*. *Journal of Ethnopharmacology*, *35*(2), 123–143.

Bennett, B.C. and Husby, C.E. (2008). Patterns of medicinal plant use: an examination of the Ecuadorian Shuar medicinal flora using contingency table and binomial analyses. *Journal of Ethnopharmacology*, *116*, 422–430.

Besansky, N.J., Severson, D.W., and Ferdig, M.T. (2003). DNA barcoding of parasites and invertebrate disease vectors: what you don't know can hurt you. *Trends Parasitol*, *19*, 545–546.

Bininda-Emonds, O.R.P., Vazquez, P.D., and Manne, L.L. (2000). The calculus of biodiversity: integrating phylogeny and conservation. *Trends Ecology and Evolution*, *15*, 92–94.

Blaxter, M.L. (2004). The promise of DNA taxonomy. *Philosophical Transactions of the Royal Society of London* B, *359*, 669–679.

Bletter, N. (2000). Cross-Cultural phylogenetic medical ethnobotany. *41st Annual Meeting of the Society for Economic Botany*, Colombia, SC.

Castillo, N.R.F., Bassil, N.V., Wada, S., and Reed, B.M. (2010). Genetic stability of cryopreserved shoot tips of *Rubus germplasm. In Vitro Cell Developmental Biological Plant, 46*, 246–256.

Clarkson, C., Maharaj, V.J., Crouch, N.R., Grace, O.M., Pillay, P., Matsabisa, G.M., Bhagwandin, G., Sith, P.J., and Folb, P.I. (2004). *In vitro* antiplasmodial activity of medicinal plants native to or naturalised in South Africa. *Journal of Ethnopharmacology, 92*(2), 177–191.

Cowan, R. S., Chase, M.W., Kress, W.J., and Savolainen. V. (2006). 300,000 species to identify: problems, progress, and prospects in DNA barcoding of land plants. *Taxon, 55*(3), 611–616.

Crisp, M.D., Arroyo, M.T.K., Cook, L.sG., Gandolfo, M.A., and Jordan, G.J. (2009). Phylogenetic biome conservatism on a global scale. *Nature, 458*, 754–756.

de Queiroz, K and Gauthier, J. (1994). Toward a phylogenetic system of biological nomenclature. *Trends in Ecology and Evolution, 9*(1), 27–31

de Queiroz, K. and Donoghue, M.J. (1990). Phylogenetic systematics or Nelson's version of cladistics? *Cladistics, 6*, 61–75.

DeSalle, R., Egan., M.G., and Siddall, M. (2005). The unholy trinity: taxonomy, species delimitation and DNA barcoding. *Philosophical Transactions of the Royal Society of London. Series B Biological Science, 360*, 1905–1916.

Donoghue, M.J. (2008). A phylogenetic perspective on the distribution of plant diversity. *Proceedings of the National Academy of Sciences, 105*, 11549–11555.

Ernst, M., Saslis-Lagoudakis, C.H., Grace, O.M., Nilsson, N., Simonsen, H.T., Horn, J.W., and Rønsted, N. (2016). Evolutionary prediction of medicinal properties in the genus *Euphorbia* L. *Scientific Reports, 6*, 30531.

Fabricant, D.S. and Farnsworth, N.R., (2001). The value of plants used in traditional medicine for drug discovery. *Environmental Health Perspectives, 109*(suppl 1), 69–75.

Faith, D.P., 1992. Conservation evaluation and phylogenetic diversity. *Biological Conservation, 61*(1), 1–10.

Faith, D.P., Magallón, S., Hendry, A.P., Conti, E., Yahara, and Donoghue, M.J. (2010). Evosystem services: an evolutionary perspective on the links between biodiversity and human well-being. *Current Opinion in Environmental Sustainability, 2*(1), 66–74.

Gagliardi, R.F., Pacheco, G.P., Carneiro, L.A., Valls, J.F.M., Vieira, M.L.C., and Mansur, E. (2003). Cryopreservation of Arachis species by vitrification of *in vitro*-grown shoot apices and genetic stability of recovered plants. *Cryo Letters, 24*, 103–110.

Gonzalez-Arnao, M.T., William, A.P., Roca, R., Escobar, R.H., and Engelmann, F. (2008). Development and large scale application of cryopreservation techniques for shoot and somatic embryo cultures of tropical crops. *Plant Cell, Tissue and Organ Culture, 92*, 1–13.

Hajibabaei, M., Singer, G.A., Hebert, P.D., and Hickey, D.A. (2007). DNA barcoding: how it complements taxonomy, molecular phylogenetics and population genetics. *Trends in Genetics, 23*(4), 167–172.

Hao, Y., You, C., and Deng. X. (2002). Effects of cryopreservation on developmental competency, cytological and molecular stability of citrus callus. *Cryo Letters, 23*, 27–35.

Harding, K. (1991). Molecular stability of the ribosomal RNA genes in *Solanum tuberosum* plants recovered from slow growth and cryopreservation. *Euphytica, 55*, 141–146.

Harding, K. (2004). Genetic integrity of cryopreserved plant cells: a review. *Cryo Letters, 25*, 3–22.

Harding, K. and Benson, E.E. (2001). The use of microsatellite analysis in *Solanum tuberosum* L. *in vitro* plantlets derived from cryopreserved germplasm. *Cryo Letters, 22*,199–208.

Harding, K. and Benson. E.E, (2000). Analysis of chloroplast DNA in *Solanum tuberosum* plantlets derived from cryopreservation. *Cryo Letters, 21*, 279–288.

Hart, K.H. and Cox. P.A. (2000). A cladistic approach to comparative ethnobotany: dye plants of the the southwestern United States. *Journal of Ethnobiology, 20*(2), 303–25.

Hebert, P.D., Cywinska, A., and Ball, S.L. (2003). Biological identifications through DNA barcodes. *Proceedings of the Royal Society of London B: Biological Sciences, 270*(1512), 313–321.

Hebert, P.D.N., Cywinska, A., Ball, S.L., and deWaard, J.R. (2004a). Biological identifications through DNA barcodes. *Proceedings of the Royal Society of London B: Biological Sciences, 270*, 313–321.

Hebert, P.D.N., Ratnasingham, S., and deWaard. J.R. (2004b). Barcoding animal life: cytochrome c-oxidase subunit 1 divergences among closely related species. *Proceedings of the Royal Society of London, 270* (suppl 1), S96–S99.

Helliot, B., Madur, D., Dirlewanger, E., and Bocaud, M.T. (2002). Evaluation of genetic stability in cryopreserved *Prunus. In Vitro Cellular Development Biology-Plant, 38*, 493–500.

Hollingsworth, P.M., Forrest, L.L., Spouge, J.L., Hajibabaei, M., Ratnasingham, S., and Ipser, J. (2009). Effect of dimethylsulfoxide on methylmethanesulfonate induced chromosomal aberrations in *Crepis capillaris* cultivated *in vitro*. *Biologia Plantarum, 35,* 137–139.

Kaczmarczyk, A., Funnekotter, B., Menon, A., Ye Phang, P., Al-Hanbali, A., Bunn, E., and Mancera, L.R. (2012). Current issues in plant cryopreservation. *Current Frontiers in Cryobiology,* ISBN: 978-953-51-0191-8, InTech. http://www.intechopen.com/books/currentfrontiers-in-cryobiology/current-issues-in-plant-cryopreservation.

Kaity, A., Ashmore, S.E., Drew, R.A., and Dulloo. E. (2008). Assessment of genetic and epigenetic changes following cryopreservation in papaya. *Plant Cell Reproduction, 27,* 1529–1539.

Kingston, D.G.I. (2000). Recent advances in the chemistry of Taxol. *Journal of Natural Products, 63,* 726–34.

Kool, A., de Boer, H.J., Krüger, A., Rydberg, A., Abbad, A., Björk, L., and Martin, G. (2012). Molecular identification of commercialized medicinal plants in Southern Morocco. *PloS One, 7*(6), 39459.

Larsen, M.M., Adsersen, A., Davis, A.P., Lledó, M.D., Jäger, A.K., and Rønsted, N. (2010). Using a phylogenetic approach to selection of target plants in drug discovery of acetylcholinesterase inhibiting alkaloids in Amaryllidaceae tribe Galantheae. *Biochemical Systematics and Ecology, 38*(5), 1026–1034.

Leonti, M., Ramirez, R.F., Sticher, O., and Heinrich. M. (2003). Medicinal flora of the Popoluca, Mexico: a botanical systematical perspective. *Economic Botany, 57,* 218–230.

Liu, K., Abdullah, A.A., Huang, M., Nishioka, T., Altaf-Ul-Amin, M.D., and Kanaya, S. (2017). Novel approach to classify plants based on metabolite-content similarity. *Biomed Research International, 2017,* 5296729. doi: 10.1155/2017/5296729.

Lukhoba, C.W., Simmonds, M.S.J., and Paton, A.J. (2006). *Plectranthus*: a review of ethnobotanical uses. *Journal of Ethnopharmacology, 103,* 1–24.

Martín, C. and. González-Benito, M.E. (2005). Survival and genetic stability of *Dendranthema grandiflora* Tzvelev shoot apices after cryopreservation by vitrification and encapsulation-dehydration. *Cryobiology, 51,* 281–289.

McDonald, D., Price, M.N., and Goodrich, J. (2012). An improved Greengenes taxonomy with explicit ranks for ecological and evolutionary analyses of bacteria and archaea. *The ISME Journal, 6*(3), 610–618. doi: 10.1038/ismej.2011.139.

Mckenna, D.J. (1995). Plant *hallucinogens*: springboards for psychotherapeutic drug discovery. *Behavioural Brain Research, 73*(1–2), 109–116.

Muellner, A.N., Samuel, R., Chase, M.W., Pannell C.M., and Greger, H. (2005). *Aglaia* (Meliaceae): an evaluation of taxonomic concepts based on DNA data and secondary metabolites. *American Journal of Botany, 92,* 534–543.

Muller, W.E.G., Batel, R., Schröder, H.C., and Muller, I.M. (2004). Traditional and modern biomedical prospecting: part I-the history and sustainable exploitation of biodiversity (sponges and invertebrates) in the Adriatic sea in Rovinj (Croatia). *Evidence Based Complement Alternative Medicine, 1*(1), 71–82.

Patel, A., Patel, R., and Jani, M. (2017). Potential use of DNA barcoding and phylogenetic analysis for the identification of medicinal plants of girnar forest. *International Journal of Botany Studies, 2*(1), 68–72.

Pennington, R.T., Richardson, J.E., and Lavin, M. (2006). Insights into the historical construction of species-rich biomes from dated plant phylogenies, neutral ecological theory and phylogenetic community structure. *New Phytologist, 172,* 605–616.

Peredo, E.L., Arroyo-García, R., Reed, B.M., and Revilla, M.A. (2008). Genetic and epigenetic stability of cryopreserved and cold-stored hops (*Humulus lupulus* L). *Cryobiology, 57,* 234–241.

Proches, S., Wilson, J.R.U., Vamosi, J.C., and Richardson, D.M. (2008). Plant diversity in the human diet: weak phylogenetic signal indicates breadth. *BioScience, 58,* 151–159.

Reynolds, T. (2007). The evolution of chemosystematics. *Phytochemistry, 68,* 2887–2895.

Rokas, A., Williams, B.L., King, N., and Carroll, S.B. (2003). Genome-scale approaches to resolving incongruence in molecular phylogenies. *Nature, 425,* 798–804.

Rønsted, N., Savolainen, V., Mølgaard, P., and Jager, A.K. (2008) Phylogenetic selection of *Narcissus* species for drug discovery. *Biochemical Systematics and Ecology, 36,* 417–422.

Rubinoff, D., and Holland, B.S. (2005). Between two extremes: mitochondrial DNA is neither the panacea nor the nemesis of phylogenetic and taxonomic inference. *Systematic Biology, 54*(6), 952–961.

Salaj, T., Panis, B., Swennen, R., and Salaj, J. (2007). Cryopreservation of embryogenic tussues of *Pinus nigra* Arn. by a slow freezing method. *Cryo Letters, 28,* 69–76.

Sanmartin, I. and Ronquist, F. (2004). Southern hemisphere biogeography inferred by event- based models: plant versus animal patterns. *Systematic Biology, 53*(2), 216–243.

Saslis-Lagoudakis, C.H., Klitgaard, B.B., Forest, F., Francis, L., and Savolainen, V. (2011). The use of phylogeny to interpret cross-cultural patterns in plant use and guide medicinal plant discovery: an example from *pterocarpus* (Leguminosae). *PLoS One, 6*(7), e22275. doi: 10.1371/journal.pone.0022275

Saslis-Lagoudakis, C.H., Savolainen, V., Elizabeth, M., Williamson, E.M., Forestc, F., Wagstaff, S.J., Baral, S.R., Watson, M.F., Pendry, C.A., and Hawkins, J.A. (2012). Phylogenies reveal predictive power of traditional medicine in bio-prospecting. *Proceedings of the National Academy of Sciences of the United States of America, 109,* 39.

Saslis-Lagoudakis, C.H., Savolainen, V., Williamson, E.M., Forest, F., Wagstaff, S.J., Baral, S.R., Watson, M.F., Pendry C.A., and Hawkins, J.A. (2012). Phylogenies reveal predictive power of traditional medicine in bioprospecting. *Proceedings of the National Academy of Sciences, 109*(39), 15835–15840.

Saslis-Lagoudakis, C.H., Williamson, E.M., Savolainen, V., and Hawkins, J.A. (2011). Cross- cultural comparison of three medicinal floras and implications for bioprospecting strategies. *Journal of Ethnopharmacology, 135,* 476–487.

Smith, M.A., Woodley, N.E., Janzen, D.H., Hallwachs, W., and Hebert, P.D.N. (2006). DNA barcodes reveal cryptic host-specificity within the presumed polyphagous members of a genus of Parasitoid flies (Diptera: Tachinidae). *Proceedings of the National Academy of Sciences of the United States of America, 103,* 3657–3662.

Srivastava, D.S., Cadotte, M.W., MacDonald, A.A.M., Marushia, R.G., and Mirotchnick, N. (2012). Phylogenetic diversity and the functioning of ecosystems. *Ecology Letters, 15*(7), 637–648.

Swofford, D.L. and Olsen. G.J. (1990). Phylogeny reconstruction. In: Hillis, D.M, Moritz, C. (eds.,) *Molecular systematics.* Sinauer Associates, Sunderland, MA, 411–501.

Techen, N., Parveen, I., Pan, Z., and Khan, I.A. (2014). DNA barcoding of medicinal plant material for identification. *Current Opinion in Biotechnology, 25,* 103–110.

The Angiosperm Phylogeny Group III, (2009). An update of the Angiosperm Phylogeny Group classification for the orders and families of flowering plants: APG III. *Botanical Journal of the Linnean Society, 161,* 105–121.

Turner, S., Krauss, S.L., Bunn, E., Senaratna, T., Dixon, K., Tan, B., and Touchell, D. (2001). Genetic fidelity and viability of *Anigozanthos viridis* following tissue culture, cold storage and cryopreservation. *Plant Science, 161,* 1099–1106.

Urbanova, M., Kosuth, J., and Cella´rova, E. (2005). Genetic and biochemical analysis of *Hypericum perforatum* L. plants regenerated after cryopreservation. *Plant Cell Report, 25*(2), 140–147. Version of cladistics. *Cladistics, 6,* 61–75.

Webster, S. and Brian, J. (2001). A quantitative study of the ability of cryopreservation to stabilize nucleic acids in *Drosophila* tissue specimens. *Drosophila Information Service, 84,* 157–165.

Wen, B. and Wang, R.L. (2010). Pretreatment incubation for culture and cryopreservation of *Sabal* embryos. *Plant Cell, Tissue and Organ Culture, 102,* 237–243.

Wildman, D.E., Uddin, M., Opazo, J.C., Liu, G., Lefort, V., Guindon, S., Gascuel, O., Grossman, L.I., Romero, R., and Goodman, M. (2007). Genomics, biogeography, and the diversification of placental mammals. *Procedures of the National Academy of Science, USA, 104,* 14395

Wink, M. (2003). Evolution of secondary metabolites from an ecological and molecular phylogenetic perspective. *Phytochemistry, 64,* 3–19.

Wink, M. and Mohamed, G.I.A. (2003). Evolution of chemical defense traits in the Leguminosae: mapping of distribution patterns of secondary metabolites on a molecular phylogeny inferred from nucleotide sequences of the rbcL gene. *Biochemical Systematics and Ecology, 31,* 897–917.

Xavier, C. and Molina. J. (2016). Phylogeny of medicinal plants depicts cultural convergence among immigrant groups in New York City. *Journal of Herbal Medicine, 6*(1), 1–11.

Zhai, Z., Wu, Y., Engelmann, F., Chen, R., and Zhao, Y. (2003). Genetic stability assessments of plantlets regenerated from cryopreserved *in vitro* cultured grape and kiwi shoot tips using RAPD. *Cryo Letters, 24,* 315–322.

Zhu, F., Qin, C., Tao, L., Liu, X., Shi, Z., Ma, X., Jia, J., Tan, Y., Cui, C., Lin, J., and Tan, C. (2011). Clustered patterns of species origins of nature-derived drugs and clues for future bioprospecting. *Proceedings of the National Academy of Sciences, 108*(31), 12943–12948.

16 Spices as Potential Human Disease Panacea

Adenike Temidayo Oladiji and Johnson Olaleye Oladele

CONTENTS

16.1 INTRODUCTION

According to International Standards Organisation (ISO) based in Geneva, spices are defined as natural plants, vegetable materials, or food products used for seasoning, flavouring, and improving/ imparting aroma in foods (Peter, 2001). They serve as ingredients in the preparation of many local and international delicacies, and are also useful in food and soup preparation in eateries and homes. Spices are typically dried seeds, barks, roots, or vegetable substance used whole or in powdered form. For instance, saffron occur as flower stigma, ginger as root, cinnamon as bark, clove as bud, and cumin as aromatic seed. They are also consumed as components of diets mixed with food items such as nuts, legumes, vegetables, fruits, cereals, milk and milk products, or as whole spices. Furthermore, spices are used as appetite stimulants and to generate visual appeal to food (Opara and

DOI: 10.1201/9781003206620-16

Chohan, 2014). They have been documented as agents that improved the nutritional quality of foods by contributing to the total nutrient intake of protein, carbohydrates, vitamins, fats, protein, and minerals. Spices are not only used as food supplements to flavour and/or add colour to foods, but they are also employed in traditional medicine as herbs in the management and treatment of diverse diseases in different regions of the world (Srinivasan, 2005). Spices are applied as herbs primarily in their isolated form from their extracts.

Spices contain several phytochemicals that have proven health benefits (Kaefer and Milner, 2011) and are sources of natural antioxidant such as flavonoids, phytoestrogens, volatile oils, phenolic diterpenes, terpenoids, phenolic acids, and carotenoids (Kennedy and Wightman, 2011). Spice phytochemicals such as piperine (black pepper), cuminaldehyde (cumin), linalool (coriander), zingerone (zinger), capsaicin (red chillies), eugenol (cloves), and curcumin (turmeric) have been reported to prevent lipid peroxidation (Oboh and Rocha, 2007; Oladele et al., 2020a). Furthermore, apigenin, fisetin, and curcumin have been shown to inhibit coronavirus-2 in an *in-silico* study and have potential as drug candidates for the treatment of COVID-19 due to their antioxidant and anti-inflammatory properties (Ademiluyi et al., 2020; Oladele et al., 2020b, c). Polyphenols from turmeric (*Curcuma longa* L.) and ginger (*Zingiber officinale*) have also demonstrated radical scavenging and antioxidant properties (Sclbert and Williamson, 2000). Spice antioxidants have gained relevance among manufactures, scientists, and consumers, because they are natural. There is an increase in the awareness among the consumer of danger posed by intake of synthetic antioxidants because of their high instability and high volatility at high temperatures. Thus, attention has been moved towards the use of natural antioxidants in food processes (Oboh and Rocha, 2007).

Some scientific studies have reported that spices displayed bioavailability enhancement ability, hypolipidaemic property, anti-inflammatory ability, antimicrobial activity, anti-carcinogenic potential, carminative attribute, digestive stimulant activity, neuroprotective effects, and are important source of natural antioxidant (Srinivasan, 2005). Epidemiological, basic, and clinical findings have also reported that nutrients from spices reduced the risk and burdens of diseases in human health (Iyer et al., 2009; Oladele et al., 2020d).

In Nigeria, the most populous country in Africa, various types of indigenous spices are used to add desired flavours and properties to the African/Nigerian foods. Common spices such as clove, chili pepper, turmeric, black pepper, sweet basil, and ginger are frequently part of regular African household meals and are used as Galenical in traditional African medicine due to their putative health benefits. For example, 'Pepper soup' is noted for its spicy, aromatic, pungency, and sensory flavour due to its constituents such as black pepper, chili peppers, ginger, alligator pepper, clove, garlic, Bastered melegueta, Ethiopian pepper, and other spices.

The upsurge in request for natural, safer, and cheaper therapeutics due to high cost and trepidation surrounding the adverse effects of modern/conventional drugs is inspiring awareness in the use of phytomedicine for treatment and management of diseases (Oladele et al., 2020e, f). Thus, this chapter focuses on the phytochemistry, health benefits, and pharmacological activities of some of the selected tropical food spices also used in Nigeria.

16.2 CINNAMON

16.2.1 SOURCES OF CINNAMON

Cinnamon is a tropical plant and one of the most prevalent and significant spices used globally not only for culinary but also for traditional and contemporary medicines. About 250 species have been documented among the *Cinnamomum* genus across the globe (Vangalapati et al., 2012). Taxonomically, cinnamon belongs to family Lauraceae, and four species are of economic significance and are traded globally. These are Vietnamese cinnamon (*Cinnamomum loureiroi*), Indonesian cinnamon (*Cinnamomum burmannii*), Chinese cinnamon (*Cinnamomum cassia* or *Cinnamomum aromaticum*), and Ceylon cinnamon (*Cinnamomum verum* or *Cinnamomum*

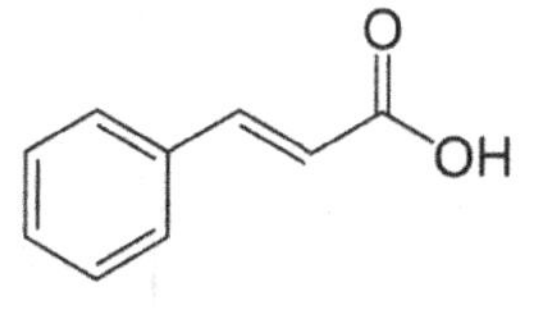

FIGURE 16.1 Resinous compounds isolated from cinnamon.

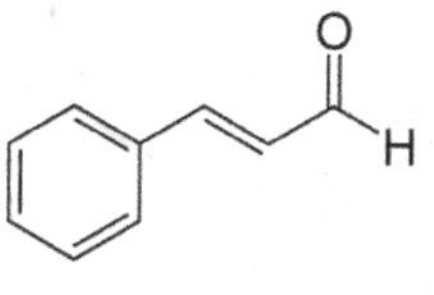

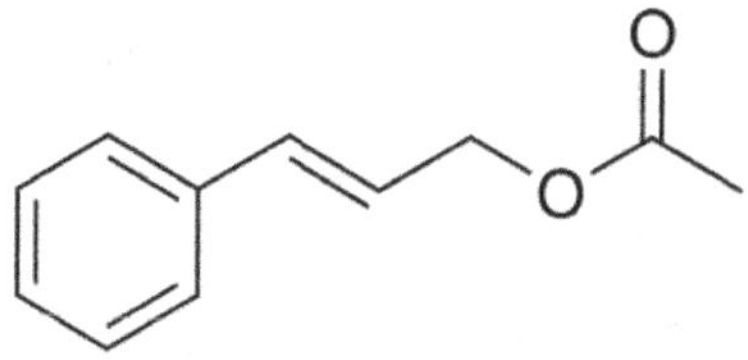

zeylanicum) from India and Sir Lanka (Ribeiro-Santos et al., 2017). The inner bark is mainly used as spice, and cinnamon is used mainly in the essence and aroma industries owing to its outstanding fragrance, which can be integrated into diverse kinds of perfumes, deodorants, food items, and medicinal products (Huang et al., 2007).

16.2.2 CHEMICAL CONSTITUTES OF CINNAMON

The most vital components of cinnamon are *trans*-cinnamaldehyde and cinnamaldehyde. They are constituents of the essential oil, which imparts the fragrance and contributes to the different biological activities performed by cinnamon (Yeh et al., 2013). Experimental studies have reported that *trans*-cinnamaldehyde is present in different species of *Cinnamomum* (Chang et al., 2008), and (E)-cinnamaldehyde is one of the main components of essential oil extracted from *C. Zeylanicum*. (E)-cinnamaldehyde has been reported to display anti-tyrosinase activity (Marongiu et al., 2007), having cinnamaldehyde as the major compound accountable for the activity (Chou et al., 2013).

Polyphenols and volatile phenols are the two major classes of chemical compounds isolated from cinnamon. Cinnamon comprises different kinds of resinous compounds, such as cinnamic acid, cinnamate, cinnamaldehyde, and some essential oils (Figure 16.1) (Muchuweti et al., 2007). Furthermore, isolated polyphenols in cinnamon includes ferulic, p-coumaric, protocatechuic, gallic, caffeic, and vanillic acids (Figure 16.2) (Chou et al., 2013). Its bark consists of catechins and procyanidins. The constituents of procyanidins consist of procyanidin A-type and B-type linkages (Figure 16.3) (Peng et al., 2008). Procyanidins isolated from cinnamon has been reported to have antioxidant activities (Tanaka et al., 2008). The presence of cinnamaldehyde is responsible for the spicy taste and fragrance of cinnamon (Maata-Riihinen et al., 2005). Cinnamon's colour darkens as it increases in age—a process that enhances the resinous compounds present in the spice. Some of the essential oils that have been isolated from cinnamon include α-thujene, terpinolene, α-terpineol, eugenol, cinnamyl acetate, *trans*-cinnamaldehyde, L-bornyl acetate, α-cubebene, caryophyllene oxide, e-nerolidol, b-caryophyllene, and L-borneol (Figure 16.4) (Tung et al., 2010).

16.2.3 BIOLOGICAL AND MEDICINAL SIGNIFICANCE OF CINNAMON

16.2.3.1 Anti-Inflammatory Activity of Cinnamon

Studies have established the anti-inflammatory activity of constituents of cinnamon and its essential oils (Tung et al., 2010). Phytocompounds with proven anti-inflammatory activities have been isolated from cinnamon; these include quercetin, oroxindin, hypolaetin, hibifolin, hesperidin, gnaphalin,

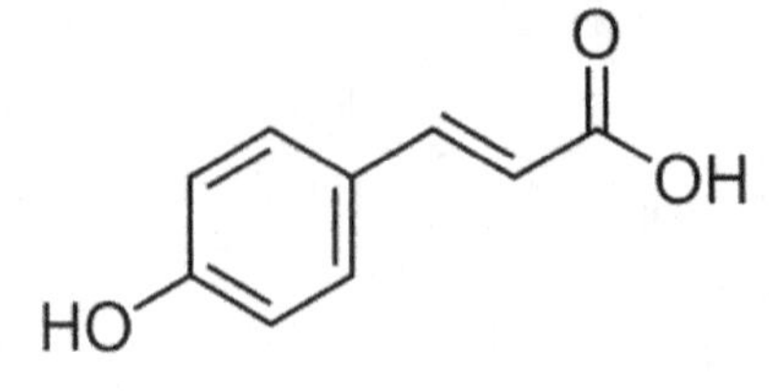

P-coumaric acid

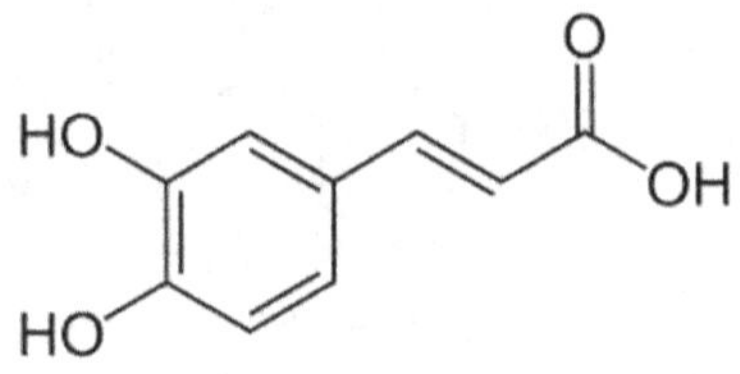

Caffeic acid

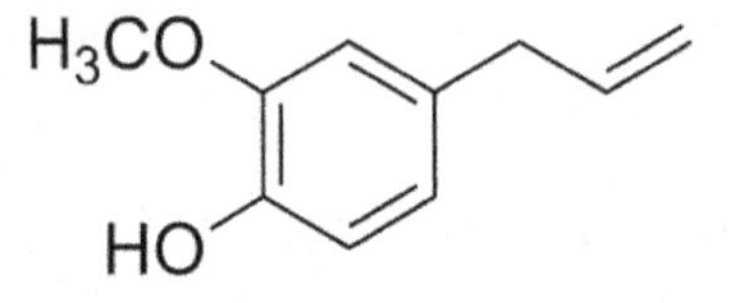

Eugenol

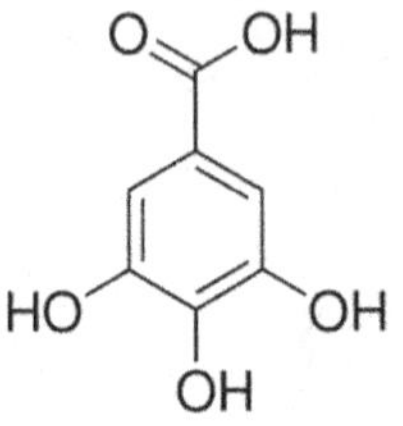

Gallic acid

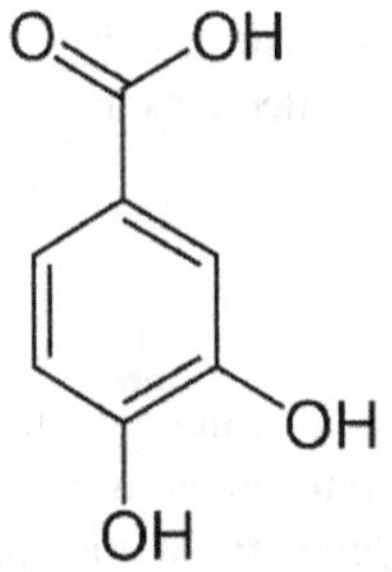

Protocatechuic acid

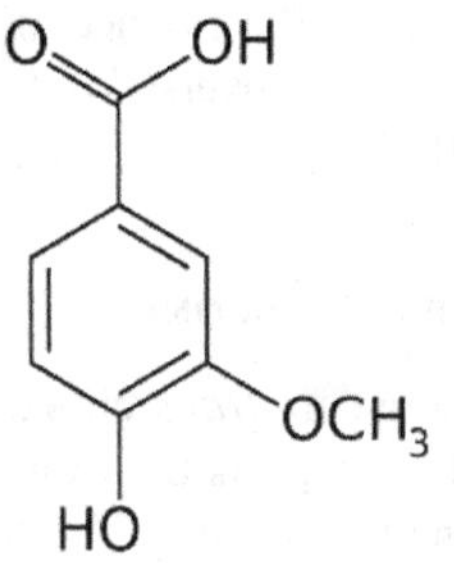

Vanillic acid

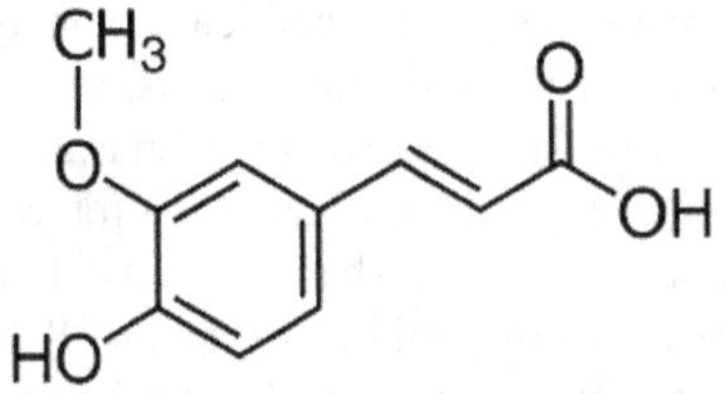

Ferulic acid

FIGURE 16.2 Phenolic compounds isolated from cinnamon.

and gossypin (Cho et al., 2013). Furthermore, 2′-hydroxycinnamaldehyde isolated from *C. cassia* bark was reported to suppress nitric oxide production through the inactivation of the nuclear factor kappa-light-chain-enhancer of activated B cells (NF-κB), thus suggesting 2′-hydroxycinnamaldehyde as a potential anti-inflammatory mediator (Lee et al., 2005). Similarly, phytochemicals contained in *C. ramulus* exhibited anti-inflammatory activities through inhibition of nitric oxide (NO) production, cyclooxygenase-2 (COX-2), and inducible nitric oxide synthesis (iNOS) expression in the brain, suggesting it has therapeutic potential for the prevention and/or treatment of inflammation-induced neurodegenerative diseases (Hwang et al., 2009). Furthermore, the aqueous extract of

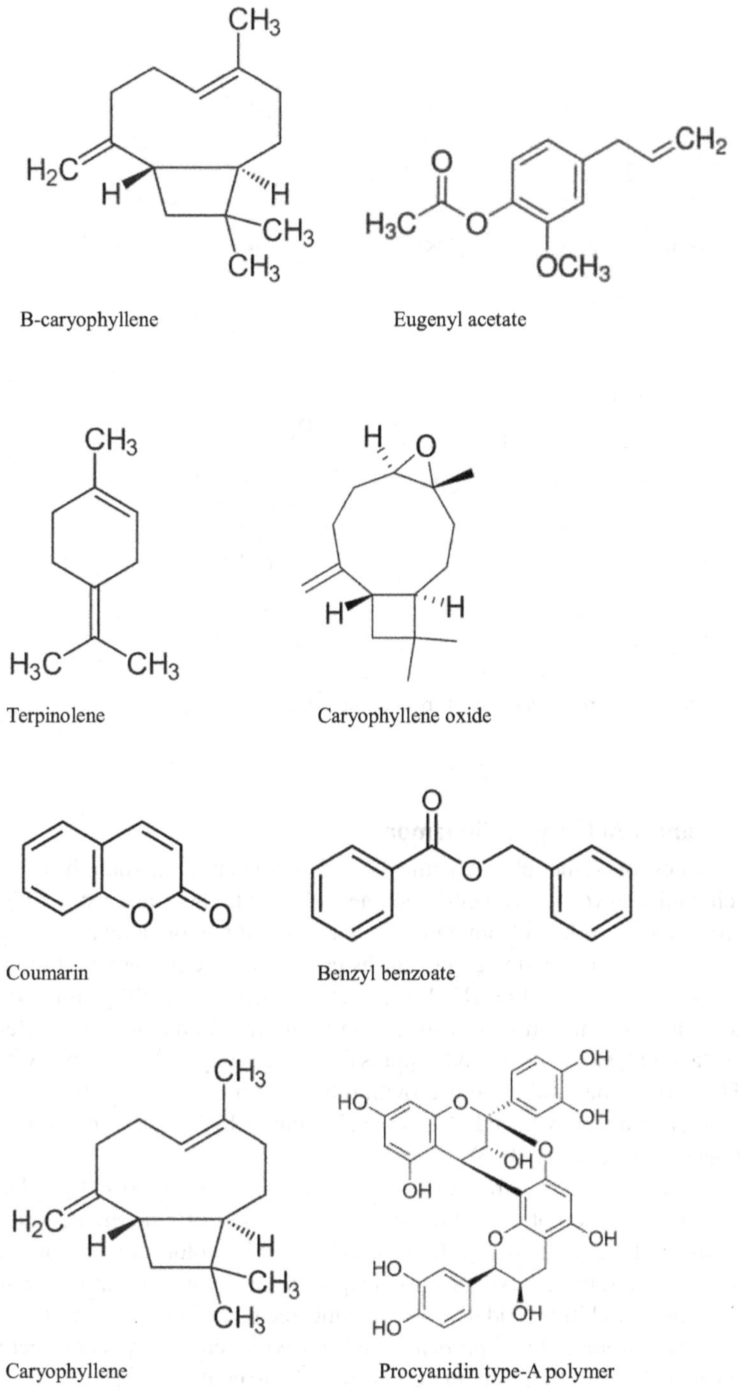

FIGURE 16.3 Other compounds isolated from cinnamon.

cinnamon reverses serum level of tumour necrosis factor-α mediated by lipopolysaccharide (Hong et al., 2012). The ethanol extract of *C. cassia* exhibited anti-inflammatory activities by suppressing nuclear factor kappa-light-chain-enhancer of activated B cells (NF-κB)-induced activation of Src/spleen-tyrosine kinase (Src/Syk) (Yu et al., 2012).

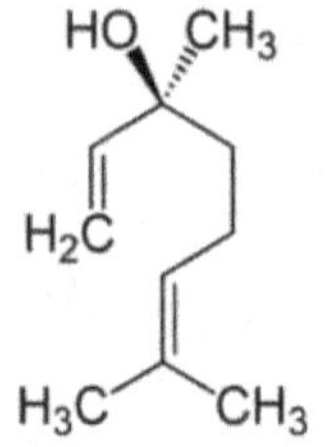

FIGURE 16.4   Hydroxyl compounds isolated from cinnamon.

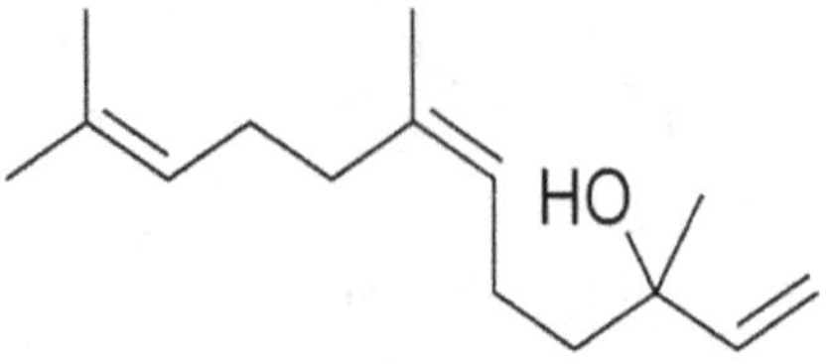

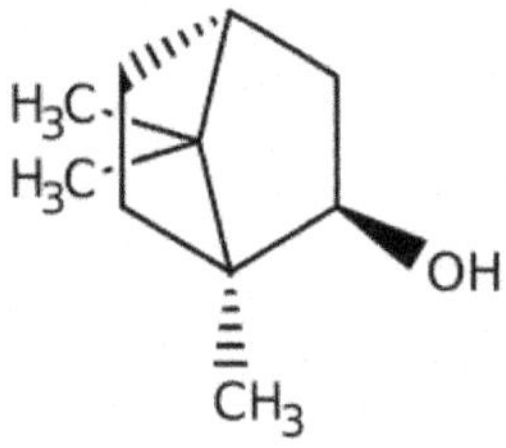

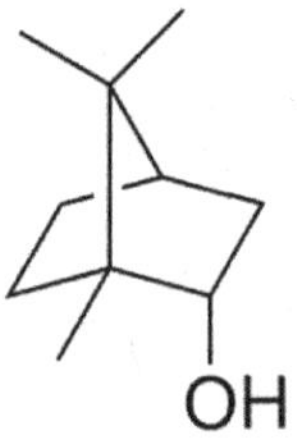

16.2.3.2   Anti-cancer Activity of Cinnamon

Anti-cancer activities of some phytochemicals isolated from cinnamon have been reported. For instance, cinnamaldehydes have been documented as inhibitors against angiogenesis (Kwon et al., 1997). Procyanidins isolated from aqueous extract of cinnamon demonstrated protective role against carcinogenesis by suppressing angiogenesis process via the inhibition of the vascular endothelial growth factor subtype 2 (VEGFR2) kinase activity (Lu et al., 2010). Cinnamic aldehyde has been suggested as a potential anti-cancer agent inhibiting the production of interleukin-8 (IL-8), tumour necrosis factor alpha (TNF-α), and suppressing the activity of NF-κB in A375 cells (Cabello et al., 2009). The anti-tumour and tumour growth inhibitory properties of CB403—a phytochemical obtained from cinnamaldehyde and 2'-hydroxycinnamaldehyde further potentiate cinnamon as an anti-cancer agent (Jeong et al., 2003).

Anti-cancer effect of aqueous extract of cinnamon has also been reported via the enhancement of the antioxidant enzyme activities and detoxification process of glutathione-s-transferase, with associated decrease in lipid peroxidation level in animals with colon cancer (Bhattacharjee et al., 2007). The essential oils from *C. Cassia* suppress alpha melanocyte-stimulating hormone-mediated melanin secretion, thus inhibiting oxidative stress in murine B16 melanoma cells (Chou et al., 2013). *Trans*-cinnamaldehyde from *C. Osmophloeum* has also exhibited anti-cancer effect via increasing tumour cell apoptosis and limiting tumour cell growth (Fang et al., 2004).

16.2.3.3   Antioxidant Activity of Cinnamon

Spices and medicinal plants have gained special interest as reservoirs of helpful antioxidant compounds that can be used to mitigate the pathogenesis of several diseases (Shaj, 2006). Foods containing antioxidant compounds perform crucial functions in human health, helping to prevent diseases and serving as health-enhancing agents. They help to prevent damage in metabolic diseases, attacks from free radicals, and age-related syndromes in animals and humans (Halliwell, 2011).

The essential oils, eugenol, (E)-cinnamaldehyde, and linalool have been reported to effectively inhibit peroxynitrite-induced nitration and lipid peroxidation (Chericoni et al., 2005). Another study has also documented the antioxidant activities of different extracts of cinnamon, which includes methanol, aqueous, and ether extracts (Mancini-Filho, 1998). Cinnamon displayed the highest antioxidant activity among 26 other spices used in a comparative study, suggesting cinnamon as a source of antioxidant (Shan et al., 2005).

Similarly, Lin et al. revealed the antioxidant activity of hot water and ethanol extracts of the dry bark of *C. cassia*. The ethanol extract of *C. cassia* exhibited substantial inhibition (96.3%) compared to α-tocopherol (93.74%) (Lin et al., 2003). Taken together, cinnamon showed higher antioxidant activities compared to other dessert spices (Murcia et al., 2004). Rats treated with bark powder of *C. verum* (10%) for 90 days exhibited antioxidant activities similar to that shown by glutathione (GSH), lipid conjugate dienes, and hepatic and cardiac antioxidant enzymes (Dhuley, 1999). A study examined the effect of different concoctions of spices on antioxidant and oxidative stress markers in high fructose-fed insulin-resistant rats. The concoction that contained 1 g/100 g cinnamon bark exhibited marked antioxidant activity (Suganthi et al., 2007). Flavonoids isolated from cinnamon have antioxidant and free-radical-scavenging activities (Okawa et al., 2001), while mixture of alcohol and aqueous extract (1:1) of cinnamon in an *in vitro* study inhibited lipid peroxidation and fatty acid oxidation (Shobana and Naidu, 2000).

Volatile oils from *C. zeylanicum* exhibited substantial antioxidant activities (Jayaprakasha and Rao, 2011). Lee et al. (2002) reported the inhibitory effects of cinnamaldehyde against nitric oxide production as well as the expression of inducible nitric oxide. The highest inhibitory effects were documented as 41.2%, 71.7%, and 81.5% at 0.1, 0.5, and 1.0 $\mu g/\mu L$, respectively (Lee et al., 2002). Cinnamon oil has been reported to display superoxide-dismutase- (SOD-) like activity, as showed by the inhibition of the inhibiting capacity of pyrogallol autoxidation (Kim et al., 1995). A study on extracts (aqueous, alcohol, and *n*-hexane extracts) of *C. malabatrum* leaves to evaluate the phytochemical contents and antioxidant activities showed that all of the extracts contained phenolic compounds and exhibited inhibition against free radicals such as nitric oxide, lipid peroxide, and hydrogen peroxide (Aravind et al., 2012). Studies on the antioxidant activities of various parts (buds, leaf, and barks) of *C. cassia* found that ethanol extract of all the plant parts exhibited significant antioxidant properties (Yang et al., 2012).

16.2.3.4 Anti-diabetic Activity of Cinnamon

A comparative study on the insulin-potentiating effects of many spices reported that the aqueous extract of cinnamon was 20-fold higher than some other spices (Broadhurst et al., 2000). A compound named 'insulin-potentiating factor' (IPF) has been isolated from cinnamon, and various studies have established the anti-diabetic effects of cinnamon extracts (Crawford, 2009). Naphthalene methyl ester—a new phytocompound characterised from hydroxycinnamic acid derivatives has been shown to have anti-diabetic properties via its ability to lower blood glucose level (Kim et al., 2006). Furthermore, beneficial effects of cinnamon bark against streptozotocin-induced diabetes in rats has been reported (Onderoglu et al., 1999).

Cinnamon has been reported to improve glycaemic control in diabetics due to its ability to boost insulin secretion (Lee et al., 2013). The antioxidant and anti-inflammatory activities of cinnamon in the pancreas may confer protection to pancreatic β cells (Lee et al., 2013). Isolation and characterisation of polyphenol type-A polymers from cinnamon with insulin-like activity has also been reported as possible anti-diabetic agents of cinnamon. Some of the polyphenols present are isorhamnetin, kaempferol, quercetin, catechin and rutin (Li et al., 2008; Yang et al., 2012), and aqueous extract of cinnamon (rich in these polyphenols) exhibited insulin-like activity (Cao et al., 2007). The aqueous extract also significantly reduced the absorption of alanine in intestine of rats. Alanine plays a critical function in gluconeogenesis, acting as a substrate for gluconeogenesis (Kreydiyyeh et al., 2000).

16.2.3.5   Neuroprotective Activity of Cinnamon

Parkinson disease (PD)—the second most common neurodegenerative disease after Alzheimer disease is becoming widespread among aged individuals of 65 years and above with prevalence rate of 2% (Oladele et al., 2020f). A study reported that cinnamon metabolite (sodium benzoate) upregulates *DJ-1* via modulation of mevalonate metabolites (Khasnavis et al., 2012). Alterations in the *DJ-1* gene induced PD protein 7 (PARK7), which is an autosomal recessive form of early-onset Parkinsonism (Bonifati et al., 2004). PARK7 is one of the major neuroprotective proteins in the brain that defend cells against free radical attack and harmful effect of oxidative stress (Khasnavis et al., 2012). In another study, the two major pathological features of Alzheimer disease—filament formation and tau aggregation were significantly reduced following administration of aqueous extract of *C. zeylanicum*. The extract also enhanced complete fragmentation of recombinant tau filaments and facilitated alteration on the morphology of paired helical filaments from Alzheimer disease brain (Peterson et al., 2009). Cinnamon and its sodium benzoate metabolite also enhance the activities of neurotrophin-3 (NT-3) and brain-derived neurotropic factors (BDNF) (Jana et al., 2013).

A phytochemical isolated from cinnamon has been reported to significantly decrease the development of β-amyloid polypeptide (Aβ) oligomers and ameliorates its toxic effect on neuronal pheochromocytoma (PC12) cells (Frydman-Marom et al., 2011). This study revealed that the phytochemical enhances movement and completely eliminated the tetrameric species of Aβ in the brain of the fly model of Alzheimer disease, decreases plaques, and enhances the cognitive functions of transgenic mice models (Frydman-Marom et al., 2011). Cinnamophilin—a novel thromboxane A2 receptor antagonist isolated from *C. philippinensis* (Yu et al., 1994) has been documented to protect against ischaemic damage in rat brain, thus enhanced neurobehavioral outcomes and has a positive effect on abridged brain infarction (Lee et al., 2009).

16.3   ONION

16.3.1   Sources of Onion

Onion (*Allium cepa* L.) has been treasured as a medicinal plant and food since time immemorial. Onion is documented to be one among the quondam vegetables and was mentioned in some ancient scriptures (Singh, 2008). It is a vegetable bulb plant common to most cultures and consumed globally (Figure 16.5). Onion is extensively cultivated (with global production of 74,250,809 tonnes from an area of 4,364,000 hectares), second only to tomato (FAO, 2012). In 2015, the global onion production is estimated to be 78.31 million tons, with the average productivity of 19.79 t/ha (FAO, 2015). The name is possibly derived from the Latin *unus* meaning 'one', and it is introduced to

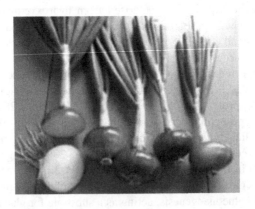

FIGURE 16.5   Onion bud.

Britain by the Romans, where it is then transported to the Americas (Burnie et al., 1999). Onion is a member of the Amaryllidaceae family; it smells when crushed and is either biannual or perennial plant (depending on the cultivar) (WHO, 1999). The plant has tubular leaves, bulb, and shallow adventitious fibrous roots (Ranjitkar, 2003). It is frequently referred to as 'Queen of the kitchen' due to its highly cherished aroma, unique taste, flavour, and the medicinal activities of its phytochemicals (Griffiths et al., 2002). Onion is a rich source of dietary flavonoids, fructans, and organosulphur compounds, and has much usefulness, for instance, it is used as condiment, as spice, in curries, in salads, or in food processing such as in flakes, paste, or pickles.

16.3.2 Chemical Constitutes of Onion

Phytochemical investigations on *A. cepa* showed that it contains many phytocompounds that confer its unique taste, flavour, and therapeutic properties. Sulphur-containing compounds have been isolated from onion such as iso-alliin, propiin, methiin, dipropyl disulfide, dipropyl trisulfide, allicin, S-alk(en)yl-L-cysteine sulfoxides (such as alliin and c-glutamylcysteine), and lipid-soluble sulphur compounds (e.g., diallyl sulfide, diallyl disulfide), all of which contribute to the unique taste and smell of fresh onion (Vazquez-Armenta et al., 2014). Thiopropal S-oxide—a compound from the sulphur volatiles is an irritating lachrymatory factor distinctively found in onions, which is converted by the action of the enzyme alliinase to methylpentanols (Thomas and Parkin, 1994).

Moreover, different phenolic compounds such as kaempferol, quercetin, protocatechuic acid, gallic acid, and ferulic acid have been isolated from four kinds (red, violet, white, and green) of *A. cepa* (Prakash et al., 2007). Similarly, different types of anthocyanins have been isolated from onion. These anthocyanins include: 5-carboxypyranocyanidin 3-O-β-glucopyranoside, 5-carboxypyranocyanidin 3-O-(6′-O-malonyl-β- glucopyranoside, malvidin 3′ -glucoside, peonidin 3′ -glucoside petunidin 3′ -glucoside acetate, peonidin 3-O-(6′-O-malonyl-β-glucopyranoside), peonidin 3-O-(6′-O-malonyl-β-glucopyranoside)-5-O-β-glucopyranoside, cyanidin 3,4′-di- O-β-glucopyranoside, cyanidin 7-O-(3′-O-β-glucopyranosyl-6′-O-malonyl-β-glucopyranoside)-4′-O-β-glucopyranoside, and cyanidin3-O-(3′-O-β-glucopyranosyl-6′-O-malonyl-β-glucopyranoside)-4′-O-β-glucopyranoside(Fredotovic et al., 2017; Perez-Gregorio et al., 2010).

Various kinds of flavonoids such as quercetin 3,4′-diglucoside, isorhamnetin 3,4′-diglucoside, quercetin 7,4′-diglucoside, quercetin 3,7,4′-triglucoside, querctin 3-glycosides, delphinidin 3,5-diglycosides, quercetin-3-monoglucosid, quercetin-4′-monoglucoside, quercetin-3,4′-diglucoside, and quercetin aglycon have been reported in different types of onions (Figure 16.6) (Zhang et al., 2016). Non-reducing sugars and pyruvic acid have been confirmed to be present in onions (Dhumal et al., 2007). The pyruvic acid content, which is a product of alkenyl-cysteine sulfoxide enzymatic degradation, is used as an indirect measure of the pungency of *A. Cepa* (Yoo et al., 2006). Oxalic, tartaric, succinic, malic, citric, and ascorbic acids are the organic acids identified in the bulb extracts. Aldehydes and ketones have also been detected in onion (Liguori et al., 2017).

16.3.3 Biological and Medicinal Significance of Onion

16.3.3.1 Antioxidant Activity of Onion

Oxidation of lipids, proteins, and DNA by the activities of free radicals and reactive oxygen/nitrogen species (ROS/RNS) is one of the processes that contribute significantly to the pathogenesis of some disease conditions, such as neurodegenerative diseases (e.g., Parkinson's disease, Alzheimer's disease), inflammatory diseases (e.g., Ulcerative colitis), cardiovascular disorders, cancer, and ageing (Oladele et al., 2020f, g, h). Experimental studies have demonstrated that phytochemical-based diets, especially those rich in fruits and vegetable, supply the body with enough phytochemicals and antioxidants (such as glutathione, vitamins C, and E, flavonoids, phenolic compounds) that can help the cells mitigate oxidative stress-induced cellular damage (Dimitrios, 2006).

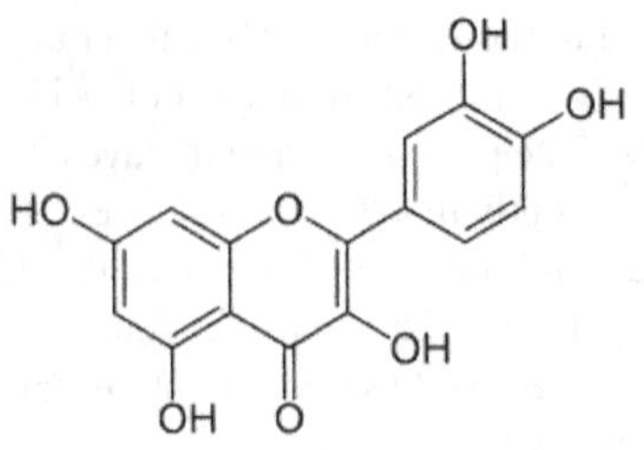

Quercetin

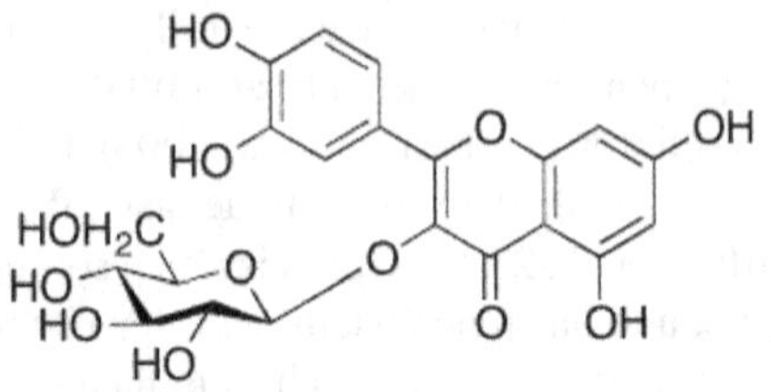

Quercetin-3-o-beta-glucoside

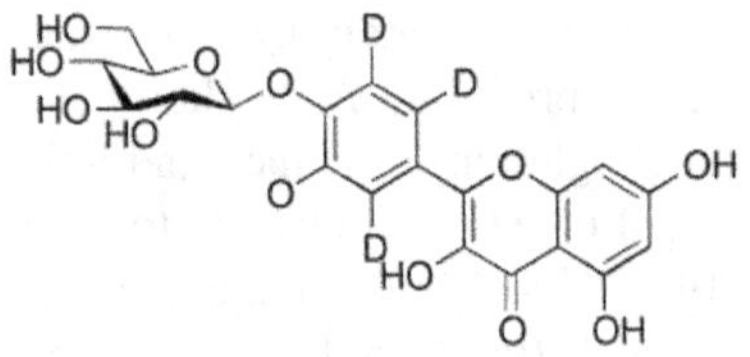

Quercetin-4'-o-beta-glucoside

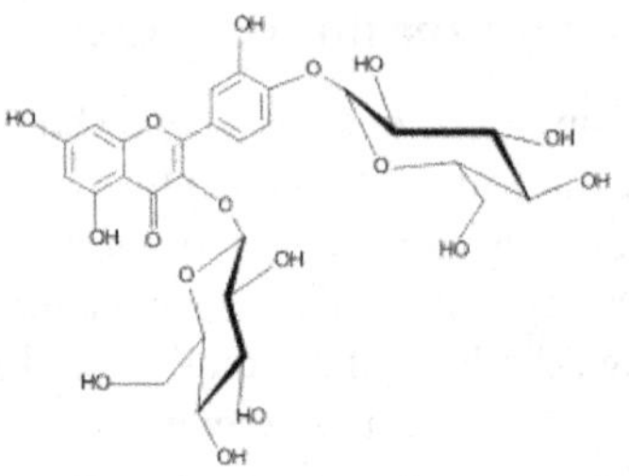

Quercetin-3,4'-o-beta-glucoside

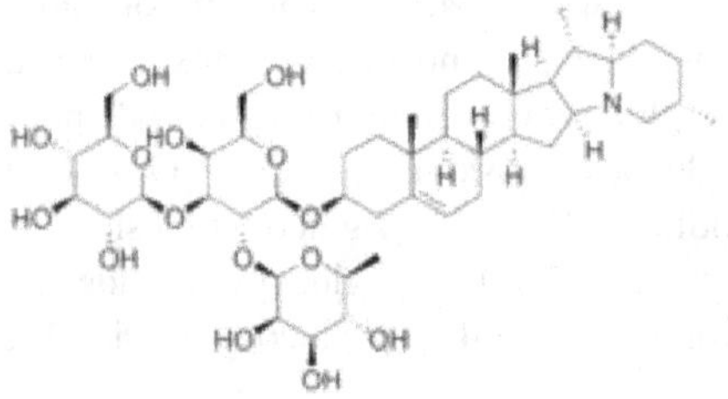

Saponin

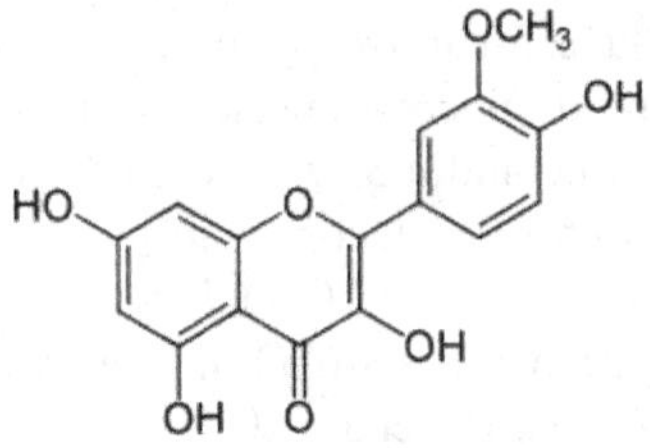

Isorhamnetin

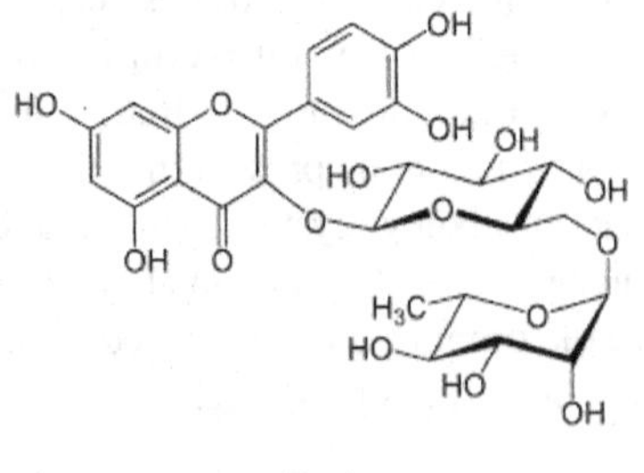

Rutin

FIGURE 16.6 Compounds isolated from onion.

Onion contains flavonoids and anthocyanins such as kaempferol and quercetin in the conjugate form (quercetin 3,7,4'-O-β-triglycopyranoside, quercetin 3,4'-O-β-diglycopyranoside, and quercetin 4'-O-β-glycopyranoside), which are effective antioxidants against oxidation of low-density lipoproteins (LDL) and non-enzymatic lipid peroxidation. Quercetin and its dimerized compounds have demonstrated great antioxidant activities compared to α-tocopherol (Park et al., 2007). The top–bottom and brown skin of industrial onion wastes have been suggested to be a source of functional ingredients, because it contained high levels of flavonoids and phenolic compounds with high antioxidant activity (Benitez et al., 2011). Similarly, the antioxidant activity of different extracts of red onion peel were examined using many methods. It was documented that the ethyl acetate extract has the highest amounts of polyphenols and could be a natural antioxidant in nutraceutical products (Singh et al., 2009).

16.3.3.2 Anti-carcinogenic and Anti-mutagenic Activities of Onion

Anti-carcinogen activities of onion and related *Allium* species have been documented in different experimental studies (Galeone et al., 2006). Chemoprevention of onion has been reported against skin cancer (Byun et al., 2010), endometrial cancer (Galeone et al., 2009a), ovarian cancer (Shen et al., 1999), breast cancer (Levi et al., 1993), liver cancer (Fukushima et al., 2001), bladder (Malaveille et al., 1996), prostate cancer (Hsing et al., 2002), brain cancer (Hu et al., 1999), lung cancer (Le Marchand et al., 2000), colorectal cancer (Tache et al., 2007), and gastric cancer (Gonzalez et al., 2006).

The anti-carcinogenic mechanisms of onion have been suggested to involve induction of apoptosis and inhibition of cell division to prevent cellular proliferation (Brisdelli et al., 2007); suppression of oxidative damage due to its antioxidant action (Raso et al., 2001); inhibition of gene transcription (Bora and Sharma, 2009); repression of bioactivating enzymes of procarcinogens (Platt et al., 2010); modification of carcinogen pathogenic pathways by inducing phase II enzymes such as UDP-glucuronosyl transferase, NAD(P)H-dependent quinine reductase, and glutathione S-transferase (GST) (Tsuda et al., 2004), which enhances the carcinogen hydrophilicity, thus easing its excretion from the body (Brisdelli et al., 2007), and suppression of cyclooxygenase and lipoxygenase activities (anti-inflammatory effect) (Rosa et al., 2001).

S-methylcysteine, S-allyl cysteine, *N*-acetylcysteine, dipropyl disulphide, dipropyl sulphide, and diallyl disulphide are some of the isolated phytocompounds in onion that have shown anti-cancer effects against kidney, liver, lung, mammary gland, oesophagus, forestomach, and colon cancers (Bora and Sharma, 2009). Similarly, luteolin (3,4,5,7-tetrahydroxyflavone) and quercetin are flavonoids isolated from onions with anti-inflammatory (Ueda et al., 2002; Zayachkivska et al., 2016), anti-angiogenic (Bagli et al., 2004), antioxidative (Manju et al., 2005), anti-metastatic, and anti-proliferative (Lee et al., 2006) effects in cancer cell assay.

16.4 RED CHILI PEPPER

16.4.1 Sources of Red Chili Pepper

A major commercial crop that is grown worldwide is chili (*Capsicum* spp.). It is a dicotyledonous flowering plant with various names that belongs to the family Solanaceae. It has a variety of nutritional and medicinal values, such as spicy pepper, chili pepper, and bell pepper etc., *Capsicum baccatum*, *Capsicum pubescens*, *Capsicum chinense*, *Capsicum annuum*, and *Capsicum frutescens* are the five domesticated types of chili pepper. One of the commercially significant species of the genus *Capsicum* is red chili pepper (C. *frutescens* L.), also identified as bird pepper (local name in Yoruba, Ata wewe) (Figure 16.7) (Forero et al., 2008). The fruit has a conical shape of about 2 cm in length and 0.5–1 cm in diameter. In the culinary world, chili peppers (*Capsicum* spp.) with their distinctive pungent taste are an essential ingredient. Chili pepper is endowed with vivid, appealing colour and attractive fragrance. Although these fruits are gastronomically demanding but are frequently used in cuisine by people from all over the world (Bosland et al., 2012; Russo, 2012). Scientific research has documented the flavour and nutritional components of chili (C. *frutescens*) peppers (Bosland et al., 2012; Russo, 2012). Like their purposeful benefits, chili peppers are entirely used due to their taste, which depends on the compounds associated with the aroma and flavour.

16.4.2 Chemical Constitutes of Red Chili Pepper

Red chilis are rich source of vitamin A and consist of significant quantity of antioxidants, including vitamin A, vitamin B-complex vitamin groups such as thiamin (vitamin B1), riboflavin, pyridoxine

FIGURE 16.7   Red chili pepper.

(vitamin B6), and niacin and flavonoids such as lutein, alpha-carotene, zea-xanthine, beta-carotene, and cryptoxanthine. It also comprises various minerals such as magnesium, iron, manganese, and potassium. Violaxanthin, ferulic acid, sinapic acid, and lutein are other bioactive compounds found in Chili (Chopan and Littenberg, 2017). The key bioactive compound in chilli is capsaicin, and is responsible for its incomparable taste and various fitness benefits. Capsaicin is traditionally used to treat and manage headaches, mastectomy pain, arthritis pain, psoriasis, osteoarthritis, post-herpetic neuralgia, and diabetic neuropathy; it is also used to protect against stroke, improves heart health, heal intestinal problems, and lowers blood sugar level (Jin et al., 2009). There are also several patents for herb or animal repellents, insecticides, and capsaicinoid-containing pesticides (Eich, 2008). Scientific research on humans has confirmed that capsaicin has the anti-cancer effect against prostate cancer cells (Diaz-Laviada, 2010). The pungency of *C. frutescens* pods are recognised for their dietetic, neurological, and pharmacological effectiveness. This pungency is due to the presence of Capsaicinoids—non-volatile alkaloids that are acid amides of C_9-C_{11} branched-chain fatty acids and vanillylamine (Figure 16.8). They have been reported to decrease the blood cholesterol level when used at low levels in the regular diet and displayed significant antibiotic activity (Govindarajan et al., 1991).

16.4.3 Biological and Medicinal Significance of Red Chili Pepper

16.4.3.1 Immunomodulatory Properties of Red Chili Pepper

Chilli pepper contains phytocompounds that can improve the body's immune system such as β-carotene or pro-vitamin A, and vitamin C is present abundantly in green chilli peppers. Vitamin A, also known as the anti-infection vitamin, has been reported to boost immunity in healthy mucous membranes, which serve as the first line of defence of the body against attack of pathogens and line

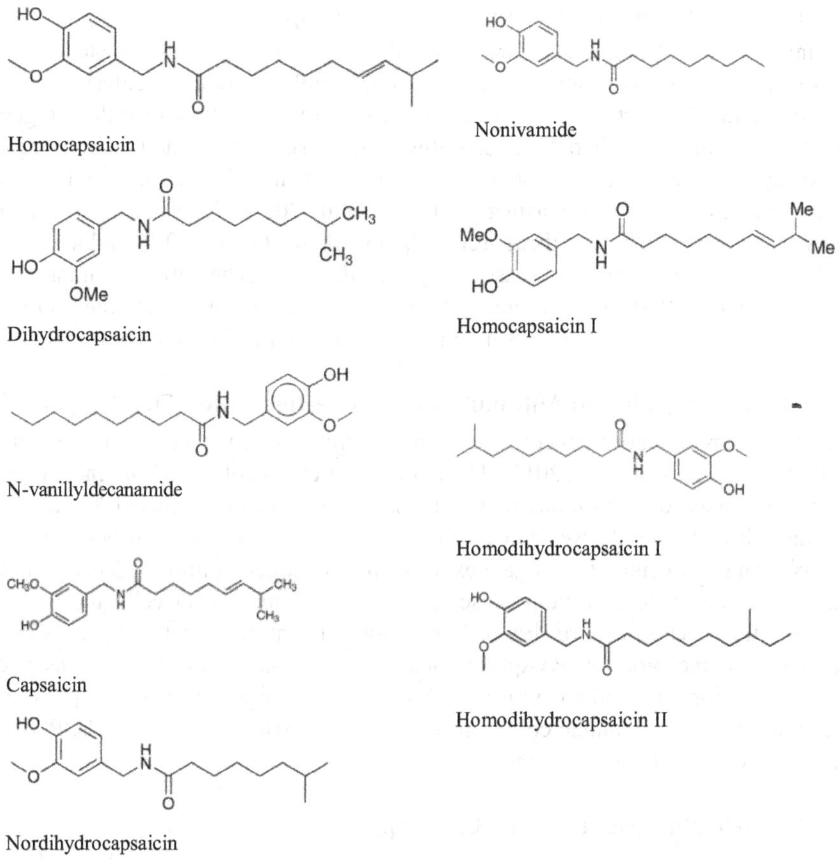

FIGURE 16.8 Chemical structures of phytochemicals in chili pepper.

the intestinal tract, lungs, nasal passages, and urinary tract. Capsaicin also improves the defence mechanism of the body (Payan et al., 1984). Vitamin C in *Capsicum* is a vital immune supportive bioactive compound that enhances the repair process of injured brain tissues, enhances bone health, and decreases the risk of cancer, paediatric asthma, and oxidative stress (Sancho et al., 2002; Ahuja et al., 2006). Capsaicinoids are also known for their immunosuppressive activity. Capsiates and their dihydro derivatives are the major capsaicinoids of sweet pepper that do not stimulate the vanilloid receptor type1 (VR1). It, however, shares some biological processes mediated in a VR1-independent manner with capsaicin, such as inhibition of early and late processes involved in T cell activation.

16.4.3.2 Effects of Red Chili Pepper on Cardiovascular Diseases

Red chili (cayenne) has been documented to decrease platelet aggregation, triglyceride levels, and blood cholesterol, as well as enhance the formation of blood clots via dissolution of fibrin in the body (Payan et al., 1984). Similarly, regular diet containing hot chili have been reported to prevent stroke and pulmonary embolism, improve rate of heart attack, and lower blood pressure (Wead et al., 1987). Studies of flavonoids in *Capsicums* have suggested that they are beneficial towards preventing coronary heart disease, as they are vasodilative and can improve hypotension and blood flow. This is most likely due to the tachykinins, which are known bioactive ingredients in *Capsicum* (Clement et al., 2012). Spicing meals with chili peppers have also been documented to prevent the deposition of fats in blood vessel walls caused by free radicals—a first step in the pathogenesis of atherosclerosis (Ahuja and Ball, 2006).

16.4.3.3 Anti-Inflammatory Properties of Red Chili Pepper

The potent inhibitor of substance P—a neuropeptide associated with inflammatory processes is capsaicin (8-methyl-*N*-vanillyl-6-nonenamide). It reduces inflammation by enhancing blood flow to the site. Capsaicin is important in the treatment of sensory nerve fibre disorders, together with pain associated with arthritis, diabetic neuropathy, and psoriasis. Capsaicin-containing diet was reported to delay the onset of arthritis and reduced paw inflammation in animals injected with a substance that induces inflammatory arthritis (Clement et al., 2012). *Capsicum's* anti-inflammatory activity has been shown to relieve arthritis-related pain (Clement et al., 2012) and neurogenic pain such as that associated with Crohn's disease (Sancho et al., 2002). The anti-inflammatory action of capsaicin can interfere with the oxygen radical transfer mechanism common to the cyclooxygenase and lipoxygenase pathways. As such, capsaicin is sold as a very potent ointment that relieves pain.

16.4.3.4 Anti-carcinogenic and Anti-mutagenic Properties of Red Chili Pepper

The prevention and treatment of cancer is another bioactive consequence of tannins found in the *Capsicum* molecule (Clement et al., 2012). The antioxidant effects of *Capsicum* indicate that other forms of cancer such as some hormone-related prostate cancers can be treated. It can also inhibit cancer cell growth and degrade weakened cellular structures in the human body (Ernst, 2011). Capsaicin has been documented to mitigate experimentally induced tumourigenesis and mutagenesis (Oyagbemi et al., 2010). Capsaicin causes apoptosis of some cancer cells and has a putative role in the prevention of cancer (Surh, 1999). A ligand-dependent transcription factor is Peroxisome proliferator-activated receptor γ (PPARγ)—a member of the super family of nuclear receptors. Peroxisome proliferator-activated receptor γ (PPARγ) pathway may be linked to capsaicin-induced apoptotic cell death in HT-29 human colon cancer cells. In a variety of cancer cells, PPARγ activation results in growth arrest and/or apoptosis.

16.4.3.5 Other Health Benefits of Red Chili Pepper

Capsicum comprises astringent tannins and is being used to treat gastrointestinal disorders (clement et al., 2012). When ingested, *Capsicum* operates as mucilage by enhancing the secretion of gastric mucus. Many studies have shown the use of *Capsicum* in the treatment of peptic ulcer disease with its anti-inflammatory properties (Ernest, 2011). Similarly, many clinical studies have demonstrated the efficacy of *Capsicum* when applied topically to treat fibromyalgia symptoms (Kim et al., 2008). The *Capsicum* chili pepper vitamins are known to have antioxidant effect on cell tissue that can suppress ageing (Clement et al., 2012). Cayenne's production of capsaicin has the known effect of dilating the arteries, thereby helping to reduce blood pressure. Furthermore, regular diet containing chili pepper has been found to lower insulin requirements and suppress the risk of hypernsulinaemia associated with type 2 diabetes. C-peptide/insulin is another vital clinical tool in diagnosis of diabetes, which reveals the rate of hepatic insulin clearance in the body. Diet containing chili pepper has been reported to increase the ratio of C-peptide/insulin (Oyagbemi et al., 2010). Phytochemicals in chili pepper such as carotenoids and vitamin C can also enhance insulin regulation (Kiran et al., 2006).

16.5 CONCLUSION AND PROSPECTS

This chapter showed that spices are natural plants, vegetable materials, or food products used for seasoning, flavouring, and improving/imparting aroma in foods, and possess great beneficial potentials not only as foods but also as traditional and modern medicine. These spices (cinnamon, onion, and red chili pepper) are regarded as essential spices serving as very rich sources of variety of phytochemicals, most of which are biologically active. The practice of application of these spices in diverse traditional systems of medicine expands their importance from culinary (in cooking) to therapeutic agents as well as use in cosmetics, and suggests their significance in economy and agriculture. Scientific studies have documented that several bioactive compounds in these spices: sesquiterpenes, aromatics, volatile constituents, essential oils, resinous compounds, sulphur-containing compounds, and phenolic compounds have both pharmacological and biological activities that enhance their use in traditional medicine. Cinnamon, onion, and red chili pepper have been used in the treatment and/or management of headaches, cancer, mastectomy pain, hypertension, diarrhoea, arthritis pain, psoriasis, obesity, osteoarthritis, post-herpetic neuralgia, diabetic neuropathy, and strokes. These three spices have also been reported to improve heart health, heal intestinal problems, as well as lower blood sugar levels.

Despite the massive migration of people internationally, local taste desires and culinary customs will persist. Therefore, there is a growing need to fashion a new cultural ethnographic market for these spices and their products. In addition to the demand due to taste and culinary practices, scientific studies have increased the awareness of the nutraceutical benefits of these spices. To meet this continuous demand and trade of these spices, biotechnology is one of the most promising approaches that can enhance their production and protection while also maintaining the quality.

REFERENCES

Ademiluyi, A.O., Oyeniran, O.H., and Oboh, G. (2020). Tropical food spices: a promising panacea for the novel coronavirus disease (COVID-19). *eFood*, *1*(5), 347–356. doi: 10.2991/efood.k.201022.001.

Ahuja, K.D. and Ball, M.J. (2006). Effects of daily ingestion of chili on serum lipoprotein oxidation in adult men and women. *British Journal Nutrition*, *96*(2), 239–242.

Ahuja, K.D., Robertson, I.K., Geraghty, D.P., and Ball, M.J. (2006). Effects of chili consumption on postprandial glucose, insulin, and energy metabolism. *The American Journal of Clinical Nutrition*, *84*(1), 63–69.

Aravind, R., Aneesh, T., Bindu, A., and Bindu, K. (2012) Estimation of phenolics and evaluation of antioxidant activity of *Cinnamomum malabatrum* (Burm. F). *Blume, Asian Journal of Research in Chemistry*, *5*(5), 628–632.

Bagli, E., Stefaniotou, M., Morbidelli, L., Ziche, M., Psillas, K., Murphy, C., and Fotsis, T. (2004). Luteolin inhibits vascular endothelial growth factor-induced angiogenesis; inhibition of endothelial cell survival and proliferation by targeting phosphatidylinositol 3′–kinase activity. *Cancer Research*, *64*, 7936–46.

Bang, M.A. and Kim, H.A. (2010). Dietary supplementation of onion inhibits diethylnitrosamine-induced rat hepatocellular carcinogenesis. *Food Science of Biotechnology*, *19*(1), 77–82.

Benítez, V., Mollá, E., Martín-Cabrejas, M.A., Aguilera, Y., López-Andréu, F.J., Cools, K., Terry, L.A., and Esteban, R.M. (2011). Characterization of industrial onion wastes (*Allium cepa* L.): dietary fibre and bioactive compounds. *Plant Foods Human Nutrition*, *66*, 48–57.

Bhattachar, S.J., Rana, T., and Sengupta, A. (2007) Inhibition of lipid peroxidation and enhancement of GST activity by cardamom and cinnamon during chemically induced colon carcinogenesis in Swiss albino mice. *Asian Pacific Journal of Cancer Prevention*, *8*(4) 578–582.

Bonifati, V., Oostra, B.A., and Heutink, P. (2004) Linking DJ-1 to neurodegeneration offers novel insights for understanding the pathogenesis of Parkinson's disease. *Journal of Molecular Medicine*, *82*(3), 163–174.

Bora, K.S. and Sharma, A. (2009). Phytoconstituents and therapeutic potential of *Allium cepa* Linn. – A Review. *Pharmacognosy Review*, *3*(5), 170–180.

Bosland, P.W. and Votava, E.J. (2012). *Peppers: Vegetable and Spice Capsicums*. 2nd edition, CABI, London, UK, pp. 1–12.

Brisdelli, F., Coccia, C., Cinque, B., Cifone, M.G., and Bozzi, A. (2007). Induction of apoptosis by quercetin: different response of human chronic myeloid (K562) and acute lymphoblastic (HSB-2) leukemia cells. *Molecular and Cellular Biochemistry*, *296*, 137–149.

Broadhurst, C.L., Polansky, M.M., and Anderson, R.A. (2000) Insulin like biological activity of culinary and medicinal plant aqueous extracts *in vitro*. *Journal of Agricultural and Food Chemistry*, *48*(3), 849–852.

Burnie, G., Forrester, S., and Greig, D., et al. (1999). *Botanica: The Illustrated A-Z of Over 10,000 Garden Plants*, 3rd edition, Random House, New South Wales, p. 74.

Byun, S., Lee, K.W., Jung, S.K., Lee, E.J., Hwang, M.K., and Lim, S.H., et al. (2010). Luteolin inhibits protein kinase Cε and c-Src activities and UVB-induced skin cancer. *Cancer Research*, *70*, 2415–2423.

Cabello, C.M., Bair III, W.B., and Lamore, S.D., et al. (2009). The *cinnamon*-derived Michael acceptor cinnamic aldehyde impairs melanoma cell proliferation, invasiveness, and tumor growth. *Free Radical Biology and Medicine*, *46*(2), 220–231.

Cao, H., Polansky, M.M., and Anderson, R.A. (2007). Cinnamon extract and polyphenols affect the expression of tristetraprolin, insulin receptor, and glucose transporter 4 in mouse 3T3-L1 adipocytes. *Archives of Biochemistry and Biophysics, 459*(2), 214–222.

Chang, C.W., Chang, W.L., Chang, S.T., and Cheng, S.S. (2008). Antibacterial activities of plant essential oils against *Legionella pneumophila. Water Research, 42*(1–2), 278–286.

Chericoni, S., Prieto, J.M., Iacopini, P., Cioni, P., and Morelli, I. (2005). *In vitro* activity of the essential oil of *Cinnamomum zeylanicum* and eugenol in peroxynitrite-induced oxidative processes. *Journal of Agricultural and Food Chemistry, 53*(12), 4762–4765.

Cho, N., Lee, K.Y., and Huh J., et al. (2013). Cognitive-enhancing effects of *Rhus verniciflua* bark extract and its active flavonoids with neuroprotective and anti-inflammatory activities. *Food and Chemical Toxicology, 58*, 355–361.

Chopan, M. and Littenberg, B. (2017). The association of hot red chili pepper consumption and mortality: a large population-based cohort study. *PLoS One, 12*(1), 0169876.

Chou, S.T., Chang, W.L., Chang, C.T., Hsu, S.L., Lin, Y.C., and Shih, Y. (2013). *Cinnamomum* cassia essential oil inhibits α-MSHinducedmelanin production and oxidative stress inmurine B16 melanoma cells. *International Journal of Molecular Sciences, 14*(9), 19186–19201.

Clément, K.K., Zinzendorf, Y.N., Serge, J.L., Solange, A., and Rose, K.N. (2012). Bioactive compounds and some vitamins from varieties of pepper (*Capsicum*) grown in Côted'Ivoire. *Pure and Applied Biology, 1*(2), 40–47.

Crawford, P. (2009). Effectiveness of cinnamon for lowering hemoglobin A1C in patients with type 2 diabetes: a randomized, controlled trial. *The Journal of the American Board of Family Medicine, 22*(5), 507–512.

Dhuley, J.N. (1999). Anti-oxidant effects of cinnamon (*Cinnamomum verum*) bark and greater cardamon (*Amomumsubulatum*) seeds in rats fed high fat diet. *Indian Journal of Experimental Biology, 37*(3), 238–242.

Dhumal, K., Datir, S., and Pandey, R. (2007). Assessment of bulb pungency level in different Indian cultivars of onion (*Allium cepa* L.). *Food Chemistry, 100*(4), 1328–30.

Díaz-Laviada, I. (2010). *Effect of capsaicin on prostate cancer cells, Future Oncol, 6*(10), 1545–1550.

Dimitrios, B. (2006). Sources of natural phenolic antioxidants. *Trends in Food Science and Technology, 17*(9), 505–512.

Eich, E., (2008). *Solanaceae and Convolvulaceae: Secondary Metabolites.* Springer, Germany, pp. 282–292.

Ernst, E. (2011). Herbal medicine in the treatment of rheumatic diseases. *Rheumatic Diseases Clinics of North America, 37*(1), 95–102.

Fang, S.H., Rao, Y.K., and Tzeng, Y.M. (2004). Cytotoxic effect of trans-cinnamaldehyde from cinnamomum osmophloeum leaves on Human cancer cell lines. *International Journal of Applied Science and Engineering, 2*(2) 136–147.

FAO (2012). World onion production. *Food and Agriculture Organization of the United Nations*, http://faostat.fao.org. Accessed 27 February 2017.

FAO (2015). *Food and Agriculture Organization Statistical Pocketbook on World Food and Agriculture*, ISBN 978-92-5-108802-9.

Forero, M.D., Quijano, C.E., and Pino, J.A. (2008). Volatile compounds of chile pepper (*Capsicum annuum* L.var. *gabriusculum*) at two ripening stages. *Flavour and Fragrance Journal, 24*(1), 25–30.

Fredotovic, Z., Sprung, M., Soldo, B., Ljubenkov, I., Budic-Leto, I., Bilusic, T., Cikes-Culic, V., and Puizina, J. (2017). Chemical composition and biological activity of *Allium cepa* L. and allium cornutum (clementi ex visiani 1842) Methanolic extracts. *Molecules, 22*(3), 448.

Frydman-Marom, A., Levin, A., and Farfara, D., et al. (2011). Orally administered cinnamon extract reduces β-amyloid oligomerization and corrects cognitive impairment in Alzheimer's disease animal models, *PLoS One, 6*(1), e16564.

Fukushima, S., Takada, N., Wanibuchi, H., Hori, T., Min, W., and Ogawa, M. (2001). Suppression of chemical carcinogenesis by water-soluble organo-sulphur compounds. *Journal of Nutrition, 131*(3), 1049s–1053s.

Galeone, C., Pelucchi, C., Levi, F., Negri, E., Franceschi, S., and Talamini, R., et al. (2006). Onion and garlic use and human cancer. *American Journal of Clinical Nutrition, 84*, 1027–1032.

Galeone, C., Pelucchia, P., Dal Masoa, L., Negria, E., Montellaa, M., and Zucchettoa, A., et al. (2009a). *Allium* vegetables intake and endometrial cancer risk. *Public Health Nutrition, 12*, 1576–1579.

Gonzalez, C.A., Pera, G., Agudo, A., Bueno-De-Mesquita, H.B., Ceroti, M., and Boeing, H., et al. (2006). Fruit and vegetable intake and the risk of stomach and oesophagus adenocarcinoma in the European prospective investigation into cancer and nutrition (EPIC-EURGAST). *International Journal of Cancer, 118*, 2559–2566.

Govindarajan, V.S. and Sathyanarayana, M.N. (1991). *Capsicum* production, technology, chemistry and quality. part v. impact on physiology, pharmacology, nutrition and metabolism: structure, pungency, pain and desensitization sequences, *Critical Reviews in Food Science and Nutrition, 29*, 435–474.

Griffiths', G, Trueman, L., Crowther, T., Thomas, B., and Smith, B. (2002). Onions: a global benefit to health. *Phytother Research, 16*(7), 603–615.

Gutteridge, J.M.C. (1993). Free radicals in disease processes: a complication of cause and consequence. *Free Radical Research Communications, 19*(3), 141–158.

Halliwell, B. (2011). Free radicals and antioxidants—*quo vadis? Trends in Pharmacological Sciences, 32*(3), 125–130.

Hong, J.W., Yang, G.E., Kim, Y.B., Eom, S.H., Lew, J.H., and Kang, H. (2012). Anti-inflammatory activity of *cinnamon* water extract *in vivo* and *in vitro* LPS-inducedmodels. *BMC Complementary and Alternative Medicine, 12*(1), 237.

Howard, L.R., Talcott, S.T., Brenes, C.H., and Villalon, B. (2000). Changes in phytochemical and antioxidant activity of selected pepper cultivars (*Capsicum* species) as influenced by maturity. *Journal of Agricultural and Food Chemistry, 48*(5), 1713–1720.

Hsing, A.W., Chokkalingam, A.P., Gao, Y.T., Madigan, M.P., Deng, J., and Gridley, G., et al. (2002). *Allium* vegetables and risk of prostate cancer: a population-based study. *Journal of the National Cancer Institute, 94*(21), 1648–1651.

Hu, J., La Vecchia, C., Negri, E., Chatenoud, L., Bosetti, C., and Jia, X., et al. (1999). Diet and brain cancer in adults. A case-control study in northeast China. *International Journal of Cancer, 81*(1), 20–23.

Huang, T.C., Fu, H.Y., Ho, C.T., Tan, D., Huang, Y.T., and Pan, M.H. (2007). Induction of apoptosis by cinnamaldehyde from indigenous cinnamon *Cinnamomum osmophloeum* Kaneh through reactive oxygen species production, glutathione depletion, and caspase activation in human leukemia K562 cells. *Food Chemistry, 103*(2), 434–443.

Hwang, S.H., Choi, Y.G., Jeong, M.Y., Hong, Y.M., Lee, J.H., and Lim, S. (2009). Microarray analysis of gene expression profile by treatment of *Cinnamomi ramulus* in lipopolysaccharide stimulated BV-2 cells. *Gene, 443*(1–2), 83–90.

Iyer, A., Panchal, S., Poudyal, H., and Brown, L. (2009). Potential health benefits of Indian spices in the symptoms of the metabolic syndrome: a review. *Indian Journal of Biochemistry and Biophysics, 46*(6), 467–481.

Jana, A., Modi, K.K., Roy, A., Anderson, J.A., Van Breemen, R.B., and Pahan, K. (2013). Up-regulation of neurotrophic factors by *cinnamon* and its metabolite sodium benzoate: therapeutic implications for neurodegenerative disorders. *Journal of Neuroimmune Pharmacology, 8*(3), 739–755.

Jaya, G., prakasha K., and Rao, L.J.M. (2011). Chemistry, biogenesis, and biological activities of *Cinnamomum zeylanicum. Critical Reviews in Food Science and Nutrition, 51*(6), 547–562.

Jeong, H.W., Han, D.C., and Son, K.H., et al. (2003). Antitumor effect of the cinnamaldehyde derivative CB403 through the arrest of cell cycle progression in the G2/M phase, *Biochemical Pharmacology, 65*(8), 1343–1350.

Jin, R., Pan, J., Xie, H., Zhou, B., and Xia, X., (2009). Separation and Quantitative analysis of capsaicinoids in chili peppers by reversed-phase argentation LC. *Chromatographia, 70* (5–6), 1011–1013.

Kaefer, C.M. and Milner, J.A. (2011). Herbs and spices in cancer prevention and treatment. Chapter 17. In: Benzie, I.F.F., Wachtel-Galor, S. (eds.) *Herbal Medicine: Biomolecular and Clinical Aspects.* CRS Press/Taylor and Francis, Boca Raton, FL.

Kennedy, D.O. and Wightman, E.L. (2011). Herbal extracts and phytochemicals: plant secondary metabolites and the enhancement of human brain function. *Advances in Nutrition, 2*(1), 32–50.

Khasnavis, S. and Pahan, K. (2012). Sodium benzoate, a metabolite of cinnamon and a food additive, upregulates neuroprotective parkinson disease protein DJ-1 in astrocytes and neurons, *Journal of Neuroimmune Pharmacology, 7*(2), 424–435.

Kim, S., Ha, T.Y., and Park, J. (2008). Characteristics of pigment composition and colour value by the difference of harvesting times in Korean red pepper varieties (*Capsicum annuum* L.). *International Journal of Food Science and Technology, 43*(5), 915–920.

Kim, S.H., Hyun, S.H., and Choung, S.Y. (2006). Anti-diabetic effect of cinnamon extract on blood glucose in db/db mice. *Journal of Ethnopharmacology, 104*(1–2), 119–123.

Kim, S.J., Han, D., Moon, K.D., and Rhee, J.S. (1995) Measurement of superoxide dismutase-like activity of natural antioxidants. *Bioscience, Biotechnology, and Biochemistry, 59*(5), 822–826.

Kiran, D.K.A, Robertson, I.K., Geraghty, D.P., and Ball, M.J. (2006). Effects of chili consumption on postprandial glucose, insulin, and energy metabolism, *The American Journal of Clinical Nutrition, 84*(1), 63–69.

Kreydiyyeh, S.I., Usta, J., and Copti, R. (2000). Effect of cinnamon, clove and some of their constituents on the Na+ -K+ -ATPase activity and alanine absorption in the rat jejunum. *Food and Chemical Toxicology, 38*(9), 755–762.

Kwon, B.M., Lee, S.H., and Cho, Y.K., et al. (1997). Synthesis and biological activity of cinnamaldehydes as angiogenesis inhibitors. *Bioorganic and Medicinal Chemistry Letters*, *7*(19), 2473–2476.

Le Marchand, L., Murphy, S.P., Hankin, J.H., Wilkens, L.R., and Kolonel, L.N. (2000). Intake of flavonoids and lung cancer. *Journal of the National Cancer Institute*, *92*(2), 154–160.

Lee, E.J., Chen, H.Y., and Hung Y.C., et al. (2009). Therapeutic window for cinnamophilin following oxygen-glucose deprivation and transient focal cerebral ischemia. *Experimental Neurology*, *217*(1), 74–83.

Lee, H.S., Kim, B.S., and Kim, M.K. (2002). Suppression effect of *Cinnamomum cassia* bark-derived component on nitric oxide synthase. *Journal of Agricultural and Food Chemistry*, *50*(26), 7700–7703.

Lee, S.C., Xu, W.X., Lin, L.Y., Yang, J.J., and Liu, C.T. (2013). Chemical composition and hypoglycemic and pancreas-protective effect of leaf essential oil fromindigenous *cinamon* (*Cinnamomumosmophloeum* Kanehira). *Journal of Agricultural and Food Chemistry*, *61*(20), 4905–4913.

Lee, S.H., Lee, S.Y. Son, D.J., et al. (2005). Inhibitory effect of 2'- hydroxycinnamaldehyde on nitric oxide production through inhibition of NF-κB activation inRAW264.7 cells. *Biochemical Pharmacology*, *69*(5), 791–799.

Lee, W.J., Wu, L.F., Chen, W.K., Wang, C.J., and Tseng, T.H. (2006). Inhibitory effect of luteolin on hepatocyte growth factor/scatter factor-induced HepG2 cell invasion involving both MAPK/ERKs and PI3K-Akt pathways. *Chemico-Biological Interactions*, *160*(2), 123–133.

Levi, F., La Vecchia, C., Gulie, C., and Negri, E. (1993). Dietary factors and breast cancer risk in Vaud, Switzerland. *Nutrition and Cancer*, *19*(3), 327–335.

Li, H.B., Wong, C.C., Cheng, K.W., and Chen, F. (2008). Antioxidant properties *in vitro* and total phenolic contents in methanol extracts from medicinal plants. *LWT-Food Science and Technology*, *41*(3), 385–390.

Liguori, L., Califano, R., Albanese, D., Raimo, F., Crescitelli, A., and Di Matteo, M. (2017). Chemical composition and antioxidant properties of five white onion (Allium cepa L.) landraces. *Journal of Food Quality*, *2017*, 1–9.

Lin, C.C., Wu, S.J., Chang, C.H., and Ng, L.T. (2003). Antioxidant activity of *Cinnamomum cassia*. *Phytotherapy Research*, *17*(7), 726–730.

Lu, J., Zhang, K., Nam, S., Anderson, R.A., Jove, R., and Wen, W. (2010). Novel angiogenesis inhibitory activity in cinnamon extract blocks VEGFR2 kinase and downstream signaling. *Carcinogenesis*, *31*(3), 481–488.

M¨a¨att¨a-Riihinen, K.R., K¨ahk¨onen, M.P., T¨orr¨onen, A.R., and Heinonen, I.M. (2005). Catechins and procyanidins in berries of vaccinium species and their antioxidant activity. *Journal of Agricultural and Food Chemistry*, *53*(22), 8485–8491.

Malaveille, C., Hautefeuille, A., Pignatelli, B., Talaska, G., Vineis, P., and Bartsch, H. (1996). Dietary phenolics as anti-mutagens and inhibitory of tobacco related DNA adduction in the urothelium of smokers. *Carcinogenesis*, *17*(10), 2193–2200.

Mancini-Filho, J., van-Koiij, A., Mancini, D.A.P., Cozzolino, F.F., and Torres, R.P. (1998) Antioxidant activity of *cinnamon* (*Cinnamomum zeylanicum*, breyne) extracts. *Bollettino Chimico Farmaceutico*, *137*(11), 443–447.

Manju, V., Balasubramaniyan, V., and Nalini, N. (2005). Rat colonic lipid peroxidation and antioxidant status: the effects of dietary luteolin on 1,2-dimethylhydrazine challenge. *Cellular and Molecular Biology Letters*, *10*(3), 535–51.

Marongiu, B., Piras, A., and Porcedda, S., et al. (2007). Supercritical CO_2 extract of *Cinnamomum zeylanicum*: chemical characterization and antityrosinase activity. *Journal of Agricultural and Food Chemistry*, *55*(24), 10022–10027.

Muchuweti, M., Kativu, E., Mupure, C.H., Chidewe, C., Ndhlala, A.R., and Benhura, M.A.N. (2007). Phenolic composition and antioxidant properties of some spices. *American Journal of Food Technology*, *2*(5), 414–420.

Murcia, M.A., Egea, I., Romojaro, F., Parras, P., Jimènez, A.M., and Martínez-Tomè, M. (2004). Antioxidant evaluation in dessert spices compared with common food additives. Influence of irradiation procedure. *Journal of Agricultural and Food Chemistry*, *52*(7), 1872–1881.

Oboh, G., and Rocha, J.B.T. (2007). *Antioxidant in Foods: A New Challenge for Food processors*. Leading Edge Antioxidants Research, Nova Science Publishers Inc. New York US, 35–64

Okawa, M., Kinjo, J., Nohara, T., and Ono, M. (2001). DPPH (1,1-diphenyl-2-Picrylhydrazyl) radical scavenging activity of flavonoids obtained from some medicinal plants. *Biological and Pharmaceutical Bulletin*, *24*(10), 1202–1205.

Oladele, J.O., Adewale, O.O., Oyeleke, O.M., Oyewole, I.O., Salami, M.O., and Owoade, G. (2020h). Annona muricata protects against cadmium mediated oxidative damage in brain and liver of rats. Acta *Facultatis Medicae Naissensis*, *37*(3), 252–260.

Oladele, J.O., Ajayi, E.I.O., Oyeleke, O.M., Oladele, O.T., Olowookere, B.D., Adeniyi, B.M., Oyewole, O.I., Oladiji, and A.T. (2020). A systematic review on COVID-19 pandemic with special emphasis on curative potentials of medicinal plants. *Heliyon, 6*(9), 1–17. doi: 10.1016/j.heliyon.2020.e04897.

Oladele, J.O., Ajayi, E.I.O., Oyeleke, O.M., Oladele, O.T., Olowookere, B.D., Adeniyi, B.M., and Oyewole, O.I., (2020d). Curative potentials of Nigerian medicinal plants in COVID-19 treatment: a mechanistic approach. *Jordan Journal of Biological Sciences, 13*, 681–700.

Oladele, J.O., Oyeleke, O.M., Awosanya, O.O., and Oladele, T.O. (2020a). Effect of *Curcuma longa* (Turmeric) against potassium bromate-induced cardiac oxidative damage, hematological and lipid profile altera-tions in rats. *Singapore Journal of Scientific Research, 10*(1), 8–15. doi: 10.3923/sjsres.2020.8.15

Oladele, J.O., Oyeleke, O.M., Awosanya, O.O., Olowookere, B.D., and Oladele, O.T. (2020e). Fluted pump-kin (*Telfaira occidentalis*) protects against phenyl hydrazine-induced anaemia in rats. *Advances in Traditional Medicine, 21*, 739–745. doi: 10.1007/s13596-020-00499-7.

Oladele, J.O., Oyeleke, O.M., Oladele, O.T., and Olaniyan, M.D. (2020f). Neuroprotective mechanism of *Vernonia amygdalina* in a rat model of neurodegenerative diseases. *Toxicology Report, 7*, 1223–1232.

Oladele, J.O., Oyeleke, O.M., Oladele, O.T., Babatope, O.D., and Awosanya, O.O. (2020g). Nitrobenzene-induced hormonal disruption, alteration of steroidogenic pathway, and oxidative damage in rat: protec-tive effects of *Vernonia amygdalina*. *Clinical Phytoscience, 6*, 1–13.

Oladele, J.O., Oyeleke, O.M., Oladele, O.T., Olowookere, B.D., Oso, B.J., and Oladiji, A.T. (2020b). Kolaviron (kolaflavanone), apigenin, fisetin as potential coronavirus inhibitors: in *silico* investigation. *Research Square Preprints*, doi: 10.21203/rs.3.rs–51350/v1.

Onderoglu, S., Sozer, S., Erbil, K.M., Ortac, R., and Lermioglu, F. (1999). The evaluation of long-termeffects of cinnamon bark and olive leaf on toxicity induced by streptozotocin administration to rats. *Journal of Pharmacy and Pharmacology, 51*(11), 1305–1312.

Opara, E.I. and Chohan, M. (2014). Culinary herbs and spices: their bioactive properties, the contribution of polyphenols and the challenges in deducing their true health benefits. *International Journal of Molecular Sciences, 15*(10), 19183–19202.

Oyagbemi, A.A., Saba, A.B., and Azeez, O.I. (2010). Capsaicin: a novel chemo preventive molecule and its underlying molecular mechanisms of action. *Indian Journal of Cancer, 47*(1), 53–58.

Park, J., Kim, L., and Kim, M.K., (2007). Onion flesh and onion peel enhance antioxidant ssstatus in aged rats. *Journal of Nutrition Science and Vitaminology, 53*(1), 21–29.

Payan, D.G. and Levine, J.D., and Goetzl, E.J. (1984). Modulation of immunity and hypersensitivity by sen-sory neuropeptide. *The Journal of Immunology, 132*(4), 1601–1604.

Peng, X., Cheng, K.W., and Ma J., et al. (2008). Cinnamon bark proanthocyanidins as reactive carbonyl scav-engers to prevent the formation of advanced glycation endproducts. *Journal of Agricultural and Food Chemistry, 56*(6), 1907–1911.

Perez-Gregorio, R.M., Garcia-Falcon, M.S., Simal-Gandara, J., Rodrigues, A.S., and Almeida, D.P. (2010). Identification and quantification of flavonoids in traditional cultivars of red and white onions at harvest. *Journal of Food Composition and Analysis, 23*(6), 592–598.

Peter, K.V. (2001). *Handbook of Herbs and Spices*. Woodhead Publishing Limited, Abington Hall, Abington, Cambridge CB1 6AH, England. www.woodhead-publishing.com.

Peterson D.W., George, R.C., and Scaramozzino, F., et al. (2009). Cinnamon extract inhibits tau aggregation associated with alzheimer's disease *in vitro*. *Journal of Alzheimer's Disease, 17*(3), 585–597.

Platt, K.L., Eden Harderb, R., Aderholda, S., Muckelc, E., and Glattc, H. (2010). Fruits and vegetables protect against the genotoxicity of heterocyclic aromatic amines activated by human xenobiotic-metabolizing enzymes expressed in immortal mammalian cells. *Mutation Research, 703*(2), 90–98.

Prakash, D., Singh, B.N., and Upadhyay, G. (2007). Antioxidant and free radical scavenging activities of phe-nols from onion (Allium cepa). *Food Chemistry, 102*(4), 1389–1393.

Ranjitkar, H.D. (2003). *A Hand book of Practical Botany*. ArunKumar Ranjitkar, Kathmandu.

Raso, G.M., Meli, R., Di Carlo, G., Pacilio, M., and Di Carlo, R. (2001). Inhibition of inducible nitric oxide synthase and cyclooxygenase-2 expression by flavonoids in macrophage J774A.1. *Life Sciences, 68*, 921–931.

Ribeiro-Santos, R., Andrade, M., Madella, D., Martinazzo, A.P., Moura, L.A.G., Melo, N.R., and Sanches-Silva, A. (2017). Revisiting an ancient spice with medicinal purposes: cinnamon. *Trends in Food Science and Technology, 62*, 154–169.

Russo, V.M. (2012). *Peppers: Botany, Production and Uses*. CABI, London, UK, pp xix–xx.

Sancho, R., Lucena, C., and Macho, A., et.al. (2002). Immuno suppressive activity of capsaicinoids: capsi-ate derived from sweet peppers inhibits NF-κB activation and is a potentanti-inflammatory compound invivo. *European Journal of Immunology, 32*(6), 1753–1763.

Scalbert, A. and Williamson, G. (2000). Dietary intake and bioavailability of polyphenols. *The Journal of Nutrition, 130*, 2073S–2085S.

Shan, B., Cai, Y.Z., Sun, M., and Corke, H. (2005). Antioxidant capacity of 26 spice extracts and characterization of their phenolic constituents. *Journal of Agricultural and Food Chemistry, 53*(20), 7749–7759.

Shen, F., Herenyiova, M., and Weber, G. (1999). Synergistic down-regulation of signal transduction and cytotoxicity by tiazofurin and quercetin in human ovarian carcinoma cells. *Life Sciences, 64*(21), 1869–1876.

Shobana, S. and Akhilender Naidu, K. (2000). Antioxidant activity of selected Indian spices. *Prostaglandins Leukotrienes and Essential Fatty Acids, 62*(2), 107–110.

Singh, B.N., Singh, B.R., Singh, R.L., Prakash, D., Singh, D.P., and Sarma, B.K., et al. (2009). Polyphenolics from various extracts/fractions of red onion (*Allium cepa*) peel with potent antioxidant and antimutagenic activities. *Food and Chemical Toxicology, 47*(6), 1161–1167.

Singh, U. (2008). *A History of Ancient and Early Medieval India: From the Stone Age to the 12th Century.* Pearson Education India, New Delhi, p. 677.

Srinivasan, K. (2005). Role of spices beyond food flavoring: nutraceuticals with multiple health effects. *Food Reviews International, 21*(2), 167–188.

Suganthi, R., Rajamani, S., Ravichandran, M.K., and Anuradha, C.V. (2007). Effect of food seasoning spices mixture on biomarkers of oxidative stress in tissues of fructose-fed insulin-resistant rats. *Journal of Medicinal Food, 10*(1), 149–153.

Suhaj, M. (2006). Spice antioxidants isolation and their antiradical activity: a review. *Journal of Food Composition and Analysis, 19*(6–7), 531–537.

Surh, Y.J. (1999). Molecular mechanisms of chemo preventive effects of selected dietary and medicinal phenolic substances. *Mutation Research/Fundamenta Land Molecular Mechanisms of Mutagenesis, 428*(1), 305–327.

Taché, S., Ladam, A., and Corpet, D.E. (2007). Chemoprevention of aberrant crypt foci in the colon of rats by dietary onion. *European Journal of Cancer, 43*(2), 454–458.

Tanaka, T., Matsuo, Y., Yamada, Y., and Kouno, I. (2008) Structure of polymeric polyphenols of cinnamon bark deduced from condensation products of cinnamaldehyde with catechin and procyanidins. *Journal of Agricultural and Food Chemistry, 56*(14), 5864–5870.

Thomas, D.J. and Parkin. K.L. (1994). Quantification of alk (en) yl-L-cysteine sulfoxides and related amino acids in alliums by high performance liquid chromatography. *Journal of Agricultural and Food Chemistry, 42*(8), 1632–1638.

Tsuda, H., Ohshima, Y., Nomoto, H., Fujita, K.I., Matsuda, E., and Iigo, M., et al. (2004). Cancer prevention by natural compounds. *Drug Metabolism and Pharmacokinetics, 19*(4), 245–263.

Tung, Y.T., Yen, P.L., Lin, C.Y., and Chang, S.T. (2010). Antiinflammatory activities of essential oils and their constituents from different provenances of indigenous *cinnamon* (*Cinnamomum osmophloeum*) leaves. *Pharmaceutical Biology, 48*(10), 1130–1136.

Ueda, H., Yamazaki, C., and Yamazaki, M. (2002). Luteolin as an anti-inflammatory and anti-allergic constituent of *Perilla frutescens*. *Biological and Pharmaceutical Bulletin, 25*(9), 1197–1202.

Vangalapati, M., Sree Satya, N., Surya Prakash, D., and Avanigadda, S. (2012). A review on pharmacological activities and clinical effects of cinnamon species. *Research Journal of Pharmaceutical, Biological and Chemical Sciences, 3*(1), 653–663.

Vazquez-Armenta, F., Ayala-Zavala, J., Olivas, G., Molina-Corral, F., and Silva-Espinoza, B. (2014). Anti-browning and antimicrobial effects of onion essential oil to preserve the quality of cut potatoes. *Acta Alimentaria, 43*(4), 640–649.

W.H.O (1999). *W.H.O Monographs on Selected Medicinal Plants*, vol. 1, pp. 5–12. World Health Organization, Geneva.

Wead, W.B., Cassidy, S.S., Coast, J.R., Hagler, H.K., and Reynold, R.C. (1987). Reflex cardiorespiratory responses to pulmonary vascular congestion. *Journal of Applied Physiology, 62*(3), 870–9.

Yang, C.H., Li, R.X., and Chuang, L.Y. (2012). Antioxidant activity of various parts of *Cinnamomum cassia* extracted with different extraction methods. *Molecules, 17*(6), 7294–7304.

Yeh, H.F., Luo, C.Y., Lin, C.Y., Cheng, S.S., Hsu, Y.R., and Chang, S.T. (2013). Methods for thermal stability enhancement of leaf essential oils and their main Constituents from Indigenous Cinnamon (*Cinnamomum osmophloeum*). *Journal of Agricultural and Food Chemistry, 61*(26), 6293–6298.

Yoo, K.S., Pike, L., Crosby, K., Jones, R., and Leskovar, D. (2006). Differences in onion pungency due to cultivars, growth environment, and bulb sizes. *Scientia Horticulturae, 110* (2), 144–149.

Yu, S.M., Ko, F.N., Wu, T.S., Lee, J.Y., and. Teng, C.M. (1994). Cinnamophilin, a novel thromboxane A2 receptor antagonist, isolated from *Cinnamomum philippinense*. *European Journal of Pharmacology, 256*(1), 85–91.

Yu, T., Lee, S., and Yang W.S., et al.(2012). The ability of an ethanol extract of *Cinnamomum cassia* to inhibit Src and spleen tyrosine kinase activity contributes to its anti-inflammatory action. *Journal of Ethnopharmacology, 139*(2), 566–573.

Zayachkivska, O.S., Konturek, S.J., Drozdowich, D., Konturek, P.C., Brzozowski, T., and Ghegotsky, M.R. (2005). Gastroprotective effects of flavonoids in plant extract. *Journal of Physiology and Pharmacology, 56* (suppl 1), 219–231.

Zhang, S.L., Peng, D., Xu, Y.C., Leu, S.W., and Wang, J.J. (2016). Quantification and analysis of anthocyanin and flavonoids compositions, and antioxidant activities in onions with three different colors. *Journal of Integrative Agriculture, 15*(9), 2175–2181.

17 Commercialization of Medicinal Plants

Opportunities for Trade and Concerns for Biodiversity Conservation

Mahboob Adekilekun Jimoh, Muhali Olaide Jimoh,
Sefiu Adekilekun Saheed, Samuel Oloruntoba Bamigboye,
Charles Petrus Laubscher and Learnmore Kambizi

CONTENTS

17.1 INTRODUCTION

Medicinal plants include all flora that have been a source of medicine to man and animals from the ancient times due to their proven therapeutic effects that could be found in orthodox drugs (Ahn, 2017). During that time, plants were only used empirically with no mechanistic knowledge of their pharmacological activities or bioactive chemicals. The link between man and his search for drugs in nature dates to ancient times, as evidenced by writings, historic monuments, and even plant-based

DOI: 10.1201/9781003206620-17

medication. In his search for natural drugs and struggle against illnesses, man learned to hunt for drugs in leaves, barks, roots, seeds, fruits, and other parts of plants, leading to the evolution of ideas about the use of medicinal plants (Petrovska, 2012). Only in the 18th century did Anton von Störck, who studied poisonous herbs like aconite and colchicum, and William Withering, who studied foxglove for oedema treatment, lay the groundwork for rational clinical investigation of medicinal herbs (Atanasov et al., 2015).

Through metabolic processes, plants synthesize various phytoconstituents that are implicated as carriers of biologically active compounds of high medicinal importance thought to be responsible for different pharmacological effects attributed to plants (Ahn, 2017; Patwardhan & Mashelkar, 2009). Medicinal plants may be administered to heal a particular illness, different kinds of diseases, or used to maintain normal health condition (Smith-Hall et al., 2012; Sneader, 2005). Modern science has recognized the bioactive potential of these phytoconstituents, and it has incorporated them in modern pharmacotherapy—a variety of plant-based drugs known to ancient civilizations and used for ages (Petrovska, 2012). The evolution of thoughts about the use of plants as medicine and food, as well as the propagation of the medicinal potential of plants, has created different layers of medicine–food continuum required to advance the biocultural adaptation of humans to the occurrence of diseases and prevention (Jennings et al., 2015; Júnior et al., 2015).

Information on the use of plants lacks the state-of-knowledge synopsis needed to harness various documents that are scattered across a wide range of disciplines, sectors, and countries around the world. This delays the systematic integration of herbal medicine into national healthcare systems in many nations, though there are some remarkable examples of herbal medicine integration into national health legislation (Smith-Hall et al., 2012). In a fact sheet entitled "The Millennium Development Goals and the United Nations Role", the United Nations' Department of Public Information emphasized the significance of medicinal plants in health care and as major contributors to the realization of the millennium development goals (Smith-Hall et al., 2012; United Nations, 2002). Nowadays, trade in botanical drugs and herbal products has elicited interest across the world having created a multi-billion-dollar market, which was projected to even exceed 7 trillion dollars by 2050 (Mosihuzzaman, 2012; Sen et al., 2011), although not without omissions in terms of financial records, global database of plant families or species involved, business routes, registered trade channels, and partners, retailers, and final consumers of herbal products. The potential market in herbal medicine has forced leading pharmaceutical industries to diversify their products by producing and upselling herbal products using their trademarks and retail channels to get involved fully in the business (Cunningham, 1989; Fokunang et al., 2011), thereby propagating the commercialization and consumption of plant-based drugs.

It is thus glaring that the commercialization of medicinal plants has reconstructed the market structure of drugs production and marketing globally with attendant wealth creation, and deliberate and unintended destruction of biodiversity. This chapter reviews the opportunities that are created by large-scale harvesting, growing, and trade in medicinal plants as well as the threats it portends to biodiversity to reassess conservation priorities and implement various recovery plans proposed by conservation botanists.

17.2 DIVERSITY OF MEDICINAL PLANTS

The distribution of life on earth is characterized by striking latitudinal variation in species richness based on palaeontological, biogeographic, ecological, and geospatial indices (Brown, 1990; Rocchini et al., 2015). Assessment of species richness is important to macroecology, conservation planning, and comparative community studies to estimate the number of endemic species in a geographic area detected from spatial and temporal changes in species distribution (Bye et al., 1995; Iknayan et al., 2014).

According to the Leipzig Catalogue of Vascular Plants (LCVP; version 1.0.2), there are 351,176 plants in the world with accepted names spread within 13,422 genera, 561 families, and 84 orders derived from relevant plant taxonomic databases and coherently versioned to accommodate future updates (Freiberg et al., 2020). The global distribution of medicinal plants is not uniform (Jamshidi-Kia et al., 2018) largely due to environmental inequalities; however, more than 72,000 plants are used worldwide as botanical drugs and in cosmeceutical applications (Chen et al., 2016b; Hassanpouraghdam et al., 2022; Jimoh & Kambizi, 2022; Schippmann et al., 2006).

The African continent is endowed with a huge diversity of plant species numbering over 50,000, a quarter of which are being used in the treatment of diseases since ancient times (Iwu, 2014). Of recent, African medicinal plants have gained some prominence as dietary supplement with the rising prevalence of polyherbacy, polypharmacy, and multi-drug therapy. The increased usage of herbal products as drugs and nutritional supplements in developed nations has created a new vista in the use of medicinal plants in global health care delivery (Benzie & Wachtel-Galor, 2011; Iwu, 2014; Ritchie, 2007)

17.2.1 Misidentification of Medicinal Plants

Inaccurate identification of medicinal plants has caused a major setback to the use, adoption, and promotion of herbal drugs. This has compromised further the quality of remedies received by consumers and predisposed them to higher levels of medical hazards in several ways (Street et al., 2008). In the developing countries and all over the world, most of the trade in medicinal plants takes place indoor or on the streets where wrong prescription of botanical drugs is done by poorly trained vendors who identify plants based on visual features without reference to the morphology, chemical diversity, and ethnopharmacology of the species (Snyman et al., 2005).

The consequences of misidentification of medicinal plants may be fatal, as it may lead to poisoning due to overconsumption, abuse, or misapplication (Abdul et al., 2018; Ali et al., 2021). In some cases, uses and abuses of drugs derived from misidentified plants may ultimately induce certain behavioural changes occasioned by variations in protein–protein interactions, synaptogenesis and neurogenesis, neuro-dendritic networks, gene expression, and protein synthesis depending on the molecular and cellular reaction to drug exposure (Kreek et al., 2005). In the long run, changes in the genetic factors due to misidentified and misapplied plant drugs may alter the pharmacodynamics and pharmacokinetics of other drugs administered for the treatment of certain chronic diseases (Kreek et al., 2005), although the relative health risk of botanical drugs may have been overemphasized (Kroll et al., 2006; Wu et al., 2020).

Nevertheless, the recent advances in deoxyribonucleic acid (DNA) biotechnology have offered a panacea to identifying plants accurately at intraspecific and interspecific levels via DNA barcoding techniques (Olatunji & Afolayan, 2019; Vigliante et al., 2019). This approach proves to be effective, as it provides a baseline for proper delimitation of plants using their molecular characteristics, enhances botanical authentication, and promotes effective management of adulteration in the production and administration of herbal drugs, which minimizes the concerns of quality assurance of plant-based drugs (Simmler et al., 2018; Snyman et al., 2005; Street et al., 2008).

17.2.2 Overharvesting of Medicinal Plants and Threats to Plant Diversity

The number of wild medicinal plants available globally continues to decline due to large-scale overharvesting and destructive methods of plant collection being used by profit-making plant hoarders (Groner et al., 2022; Williams et al., 2013). This is evident in the fact that these indiscriminate harvesters do not take into consideration conventional conservation practices and seasonal limitations that influence the availability of plant species (van Wyk & Prinsloo, 2018).

The dynamics of supply and demand for medicinal plants is a major factor that drives over-harvesting of wild stock, as most harvesters are lured into the business for economic gains (Liu et al., 2019). This is so because a premium price is placed on wild plants in the market, and they are most sought-after (Liu et al., 2014). Once a plant species or a group of plants is noted to possess certain pharmacological benefits, plant poachers go after such a plant and its close relatives wherever they exist to meet the market demand. This portends danger to the wild population of the plant species, as its natural population may be artificially driven into extinction due to excessive harvesting, thereby resulting in genetic erosion and biodiversity loss (Jimoh & Jimoh, 2021; Liu et al., 2014).

Other contributing factors to the excessive harvesting and destruction of wild population of medicinal plants include the competing demand for land use for industrialization, animal grazing, road construction, recreational centres, schools, hospitals, aesthetic, sport facilities, and a rising interest in herbal drug production and trade (Castle et al., 2014; Jimoh & Jimoh, 2021). Thus, increased harvesting pressure has caused severe consequences on the ecological integrity of biodiversity and may have accelerated the declining status of overexploited plants, species rarity, and, sometimes, their extinction (Chen et al., 2016a; van Andel & Havinga, 2008).

17.2.3 Opportunities for Local Trade and Foreign Direct Investment in Medicinal Plants

Exportation of the American medicinal plants was part of the international drug trade by the second-half of the 18th century across the Western and Northern Europe (Gänger, 2015). During this period, the Peruvian balsam, *sassafras, cinchona*, jalap root, *guaiacum, sarsaparilla, ipecacuanha,* and other plant products that formed part of the standard medical repertoire were supplied as standard medical remedies by pharmacies in Europe, Iberian Peninsula, Italian Peninsula, England, Russia, and major territories of the Holy Roman Empire (Curth, 2006; Gänger, 2015).

Nowadays, over 3.5 billion people depend on medicinal plants for healthy living, well-being, and economic survival worldwide (Cock et al., 2018; Khasim et al., 2020; Rajasekharan & Wani, 2020; Singh, 2015). This has translated to economic fortune for individual households and a source of foreign direct investment for developing economies. Although, records of market flows and governance of inter-border trade for medicinal plants are poorly documented (He et al., 2018), about 702,813 tonnes of medicinal plants worth USD 3.60 billion was traded globally in the year 2014 (Vasisht et al., 2016).

The United State of America, Europe, and Japan are the major consumers of medicinal plants products in the world. These include phytopharmaceuticals, tannins, spices, gums, essential oils, crude extracts, cosmetic ingredients, and all kinds of exudations from trees or products derived from plants (Vasisht et al., 2016). The annual demand for herbal products and plant-based medicine grows at the rate of about 25%, and, with this pace, the World Health Organization has estimated that the market value of this demand may worth more than 5 trillion USD by the year 2050 (Rajasekharan & Wani, 2020; Sen et al., 2011).

Thus, commercialization of medicinal plants is important to the global economy, as it significantly improves the local economies of indigenous people in different regions of Sub-Sahara Africa, Asia, Latin America, and many horticultural industries that develop and exchange cultivation protocols, grow, and trade in medicinal plants (He et al., 2018; Schippmann et al., 2006; van Andel & Havinga, 2008; van Wyk & Prinsloo, 2018; Vasisht et al., 2016). As a result, medicinal plants are crucial to community livelihoods, economic growth, and human capital development, as they serve as a vehicle for redistribution of wealth, knowledge, and emerging medical approaches for the treatment of chronic diseases. Some medicinal plants that are traded globally are listed in Table 17.1.

TABLE 17.1

Medicinal Plants Traded Globally

S/N	Botanical Names	Common Names	Family Names	Bioactive Compounds	Application	References
1	*Actaea racemosa*	Black cohosh	Ranunculaceae	Formononetin	Treatment of climacteric complaints, menopausal symptoms	Huntley (2004); Jiang et al. (2006); Rajasekharan & Wani (2020); Wuttke et al. (2014)
2	*Allium sativum* L.	Garlic	Alliaceae	Allicin	Arteriosclerosis, antimicrobial effect, cardiovascular disease, and antiplatelet activities	Agustina et al., (2022); Kendler (1987)
3	*Aloe ferox* Mill.	Cape aloe, bitter aloe	Asphodelaceae	Aloin	Anti-cancer agent, laxative agent, and treatment of loperamide-induced constipation	Chen et al. (2012); Dagne et al. (2005); Esmat et al. (2006); Wintola et al. (2010)
4	*Aloe barbadensis* Miller.	Aloe Vera	Asphodelaceae	C-glucosyl chromone, cinnamonic acid, phenols, lupeol, urea nitrogen, salicylic acid, and sulphur	Anti-inflammation, *anti-viral, and anti-tumour activity, generation of metallothionein, moisturizing, and anti-ageing effect, antiseptic effect, and potential inhibitor of COVID-19 virus main protease (M^{pro})*	Aslani et al. (2015; Klein & Penneys (1988); Mpiana et al. (2020); Surjushe et al. (2008)
5	*Aquilaria agallocha* Roxb.	Agarwood	Thymelaeaceae	Aquillochin, cucurbitacins	Hypersensitivity, acroparalysis, asthma, cough, sedative, anti-nociceptive, anti-cancer, and anti-diabetic	Alam et al. (2015); Bhandari et al. (1982); Chen et al. (2014); Kim et al. (1997)
6	*Asparagus racemosus* Willd.	Shatavari	Asparagaceae	Asparagamine, 8-methoxy-5,6,4'-trihydroxyisoflavone 7-o-β-D-glucopyranoside, saponin, diosgenin, Shatavarins I–IV, sarsasapogenin, quercetin-3 glucuronide, hyperoside	Diabetes, amenorrhoea, dysentery, and diarrhoea, diuretic, rheumatism, jaundice, hair protection, and phytoestrogenic properties	Asmari et al. (2004); Bala et al. (2010); Bopana & Saxena (2007); Sekine et al. (1994)

(Continued)

TABLE 17.1 (Continued)
Medicinal Plants Traded Globally

S/N	Botanical Names	Common Names	Family Names	Bioactive Compounds	Application	References
7	*Atropa belladonna*	Deadly nightshade	Solanaceae	Hyoscyamine 6β-hydroxylase, alkaloids	Asthmatic sedatives, anti-cholinergic, analgesic, Parkinson disease, rheumatism, psychiatric disorders, and neuralgia	Kamada et al., (1986); Kennedy (2014); Kwakye et al. (2018); Li et al. (2012)
8	*Carapichea ipecacuanha* (Brot) L. Andersson	poaia or ipecac	Rubiaceae	isoquinoline alkaloids (emetine and cephalin)	Asthma, anti-inflammation, and respiratory diseases	Ferreira Júnior et al. (2012); Machado Perucci Pereira dos Santos et al. (2022)
9	*Centella asiatica* (L) Urban	Gotu kola	Apiaceae	Asiatic acid, asiaticoside, madecasosside, centellasaponins, isothankunic acid, terminolic acid, β-caryphyllene, and madecassic acid	Skin diseases, neurodegenerative disorders, chronic venous insufficiency, and cancer	Bylka et al. (2014); Raimi et al. (2021)
10	*Commiphora wightii* (Arnott) Bhandari	Guggulu	Burseraceae	Guggultetrols, sesquiterpenoids, sterols, monoterpenoids, ferulic acid, lignans, and tetraols	Rheumatism, inflammation, obesity, gout, and hypolipidaemic activity	Sarup et al. (2015); Urizar & Moore (2003); Zhu et al. (2001)
11	*Echinacea purpurea* (L.) Moench	Purple coneflower	Asteraceae	Caffeic acid and alkamides	Chemotherapy, immunomodulation, inflammation, seizure, skin diseases, cancer, snake bite, toothache, bowel pain, and chronic arthritis	Manayi et al. (2015)
12	*Echinacea angustifolia* DC.	blacksamson echinacea	Asteraceae	Alkamides, cichoric acid, caftaric acid, cynarin, and echinacoside	Immuno-stimulation and common flu	Binns et al. (2002); Clarenc Aarland et al. (2016); Turner et al. (2005)
13	*Garcinia indica* Choisy	Kokum	Guttiferae	Isogarcinol, hydroxycitric acid, hydroxycitric acid, lactone, citric acid, cyanidin-3-sambubioside; and cyanidin-3-glucoside	Anti-cancer, anti-ulcerogenic, chemopreventive, anti-obesity, and cardioprotective	Baliga et al. (2011); Kadam et al. (2012); Nayak et al. (2010a, b); Prabhu et al. (2020)

(Continued)

TABLE 17.1 (Continued)
Medicinal Plants Traded Globally

S/N	Botanical Names	Common Names	Family Names	Bioactive Compounds	Application	References
14	Ginkgo biloba L.	Ginkgo, maidenhair-tree	Ginkgoaceae	Ginkgolides, ginkgetin, carotenoids, kaempferol, aglycones, and prodelphinidin	Nerve and neuro cells protection, blood circulation, strengthening capillary walls, treatment of dementia, asthma, and memory impairment in in Alzheimer patient	Diamond et al. (2000); Singh et al. (2008)
15	Glycyrrhiza glabra L.	Liquorice, black Sugar	Fabaceae	Glycyrrhizin, glycosyl-glycyrrhizin, stigmasterol, and β-sitosterol, sapogenins, licoricidin, rhamno-isoliquiritin, liqcoumarin, liquiritin, licoflavonol, licoricone, and formononetin	Oestrogenic activity and gastric ulcer	Dhingra et al. (2004); Fenwick et al. (1990); Gupta et al. (2008); Lim (2016); Pastorino et al. (2018)
16	Griffonia simplicifolia (Vahl ex DC.) Baill.	Griffonia	Fabaceae	GS Isolectin B4, 5-hydroxytryptamine, and 5-hydroxytryptophan	Management of overweight and obesity	Balogun et al. (2020); Benton et al. (2008); Fellows & Bell (1970); Rondanelli et al. (2013)
17	Hippophae rhamnoides L.	Sea buckthorn	Elaeagnaceae	Dulcioic acid, 19-alpha-hydroxyursolic acid 5-hydroxymethyl-2-furancarbox-aldehyde, cirsiumaldehyde hippophae cerebroside, oleanolic acid, ursolic acid, and isorhamnetin	Immunomodulatory, hepatoprotective, anti-atherogenic, radioprotective, anti-stress, tissue repair, skin diseases, asthma, gastric ulcers, and pulmonary disorders	Ciesarová et al. (2020); Sleyman et al. (2001); Suryakumar & Gupta (2011)
18	Hydrastis canadensis L.	Goldenseal	Ranunculaceae	Isoquinoline alkaloids, β-hydrastin, and berberine	Gastrointestinal ailments	Barceloux (2008); Hoshino et al. (2000); Mahady et al. (2003); Mahady & Chadwick (2001)

(Continued)

TABLE 17.1 (*Continued*)
Medicinal Plants Traded Globally

S/N	Botanical Names	Common Names	Family Names	Bioactive Compounds	Application	References
19	*Matricaria chamomilla* L.	Chamomile, scented mayweed	Asteraceae	Apigenin	Sedative and spasmolytic properties	Avallone et al. (2000); Singh et al. (2011)
	Mucuna pruriens L.	Velvet bean, cowhage, cowitch	Fabaceae	L-3,4-dihydroxyphenylalanine, isoflavanones, lectin, and 1,2,3,4-tetrahydroisoquinoline alkaloids	Parkinson disease, neurorestoration, epilepsy, scorpion stings, snakebite, malaria, cancer, elephantiasis, and infertility	Jimoh et al. (2020); Misra & Wagner (2004)
20	*Oenothera biennis* L.	Evening primrose	Onagraceae	γ-Linolenic acid and eicosanoids	Anti-tumour effect and cell cycle arrest	Dalla Pellegrina et al. (2005); Timoszuk et al. (2018)
21	*Papaver somniferum* L.	Opium poppy	Papaveraceae	Papaverine, Magnoflorine, 6-acetonyldihydrosanguinarine, norsanguinarine, codeine, morphine, oxysanguinarine, protopine sanguinarine, dihydrosanguinarine, cryptopine, sanguinarine, berberine, and tubocurarine	Anti-cancer, analgesics, vasodilator, cough suppressant, the muscle relaxant, antimicrobial agents, and cholesterol-lowering	Furuya et al. (1972); Labanca et al. (2018)
22	*Pelargonium sidoides* DC	South African geranium	Geraniaceae	Coumarin, 5,6-dimethoxy, prodelphinidins, 7-hydroxy-coumarin, and benzopyranones.	Gastrointestinal disorders, infections of respiratory tract, common cold, rhinopharyngitis, acute bronchitis, tuberculosis, tonsllitis, and sinusitis, rhinosinusitis	Germer et al. (2007); Kayser et al. (1995); Kolodziej et al. (2003); Krone et al. (2001); Moyo & van Staden (2014); Timmer et al. (2013)
23	*Rauvolfia serpentina* (L.) Benth. ex Kurz	Serpentine wood, devil peppier	Apocynaceae	β-Carbolinium anhydronium, 3,4,5,6-tetradehydrogeissoschizol, 3,4,5,6-tetradehydrogeissoschizine-17-O-β-d-glucopyranoside, 3,4,5,6-tetradehydroyohimbine, 3,4,5,6-tetradehydro-(Z)-geissoschizol	Treatment of cardiovascular diseases, anti-malarial, and anti-cancer and activity	Khan et al. (2018); Wachsmuth & Matusch (2002a, b)

(*Continued*)

TABLE 17.1 (Continued)

Medicinal Plants Traded Globally

S/N	Botanical Names	Common Names	Family Names	Bioactive Compounds	Application	References
24	*Sabal serrulate* (Michx.) Schult.f.))	Saw palmetto	Arecaceae	Polyoxethylene glycol and liposterolic extract	Antiphlogistic, anti-oedematous, spasmolytic effect, benign prostatic hyperplasia, and human prostate cancer	Bondarenko et al. (2003); Booker et al. (2014); MacLaughlin et al. (2006); Wagner et al. (1981)
27	*Silybum marianum* (L.) Gaertn.	Milk thistle	Asteraceae	Silymarin, dehydrosilybin, silidianin, silicristin, silybin, apigenin, flavonolignans, silibinin A and B, deoxysilyn dianin, and deoxysilyn cristin	Hepatoprotection, cardiovascular protection, intoxication, renal protection, Alzheimer prevention, and digestive disorders	Bahmani et al. (2015); Hikino et al. (1984); Shaker et al. (2010)
28	*Tanacetum parthenium* (L.) Schultz Bip.	Feverfew	Asteraceae	Apigenin, flavonols 6-hydroxykaempferol and quercetagetin, luteolin 7-glucuronides, and tanetin (6-hydroxykaempferol 3,6,4'-trimethyl ether)	Treatment of arthritis and migraine	Lopresti et al. (2020); Palevitch et al. (1997); Pfaffenrath et al. (2002); Williams et al. (1995, 1999)
29	*Taxus wallichiana* Zucc.	Himalayan yew	Taxaceae	**Taxines, 7-O-acetyltaxine A (2) and 2α-acetoxy-2β-deacetylaustrospicatine**	Antibiotics, anti-inflammatory, anti-leukaemic, and analgesic	Juyal et al. (2014); Khan et al. (2006); Kovács et al. (2007); Nisar et al. (2008); Prasain et al. (2001); Qayum et al. (2012)
30	*Ulmus rubra* Muhl.	**Slippery elm**	*Ulmaceae*	Sesquiterpenes, 7-hydroxycalamenenal, 7-hydroxycadalenal, 7-hydroxycadalene, 7-hydroxy-3methoxycadalenal, D-galacturonic acid, and L-rhamnose	Sore throat and pharyngitis, irritable bowel syndrome, and constipation	Banerjee et al. (2006); Brunet et al. (2016); Edwards et al. (2015); Qayum et al. (2012); Rowe et al. (1972)
31	*Vaccinium macrocarpon* Aiton	Cranberries	Ericaceae	Quercetin-3-O-(6'-benzoyl)-β-galactoside, quercetin-3-α-arabinopyranoside, 3'-methoxyquercetin-3-α-xylopyranoside, quercetin-3-β-glucoside, quercetin-3-O-(6'-p-coumaroyl)-β-galactoside, and myricetin-3-β-xylopyranoside,	Stomach ulcers, prevention of urinary tract infections, and gum disease	McKay & Blumberg (2007a, b); Rajasekharan & Wani (2020); Wu et al. (2008)

17.3   A SYSTEMATIC REVIEW OF RESEARCH TREND
ON TRADED MEDICINAL PLANTS

In this systematic review, data from scientific publications such as journals, conference proceedings, and books were extracted using the bibliometric techniques (Jimoh et al., 2022). The trend of research outputs on medicinal plants was monitored with a comprehensive technology to assess the scientific significance of authors' contribution to knowledge, the academic community, and the impact of research findings to the industrial sector via various publication outlets (Jimoh et al., 2022; Jimoh & Kambizi, 2022).

Since time immemorial, medicinal plants have been the fulcrum of human civilization having aided the survival of man in terms of health and wealth, and have attracted interest from industries, households and communities, although certain societal, psychological, and biochemical barriers limit the effective utilization of herbal medicine (Jimoh et al., 2020; Jimoh & Kambizi, 2022; Srivastava et al., 2019). The main focus of this bibliometric analysis was to assess and synchronize various contributions made by researchers on commercially important plants that are traded globally for their medicinal values with a view to identify notable researchers and the impact of their contributions as documented in the literature, given the rising demand for natural products from plants as drugs, food, and diet fortification in the last 10 years, and the need to conserve the wild relatives of the affected plants and preserve their natural reserve (Balogun et al., 2020; Jimoh et al., 2021; Khasim et al., 2020). This may prevent overexploitation of medicinal plants being harvested indiscriminately for commercial purposes and enhance biodiversity conservation, as attention may be shifted to other plants having similar or equivalent bioactive constituents.

17.3.1   MATERIAL AND METHODS

17.3.1.1   Data Sourcing

Data used for the analysis were extracted from the Web of Science platform, which is a leading multi-disciplinary academic databases globally. It provides access to review articles, research articles, abstracts, conference proceedings, books, short communications, and technical papers across 171 million platforms of which 79 million are core collection (Liu, 2019). Keywords such as *'plant-based natural products', 'medicinal plants', herbal products', 'medicinal plant trade', 'Sales of herbal products', 'Sales of medicinal plants', 'Sales of botanical drugs', 'medicinal plants trade', 'trade in plant-based natural products', 'trade in herbal products', 'commercial medicinal plants', 'natural products market', 'commercial herbal products', 'commercial botanical drugs', 'botanical drugs trade', 'medicinal plant market', 'herbal products market', 'botanical drugs market', 'natural products market', 'green plants market', 'flora market', 'flora species trade', 'herbal trade', 'herbal market', 'herbal commerce', 'flora commerce', 'trade in medicinal plants', 'plant product trade', and 'natural products commerce'* that are directly associated with commercialization of medicinal plants were used. The keywords were separated with 'OR' otherwise known as the Boolean operator. The title search was used to avoid accumulation of false results from unrelated data (Jimoh et al., 2022).

The analysis was based on 7,425 articles obtained in plain text file format from the Web of Science database (Tenen, 2017) since the beginning of this millennium (2000–2021) after which documents from early access and unrelated publications were selectively excluded (Jimoh et al., 2022; Smith et al., 2021). The extracted files in plain text file format were exported to R-studio (version 4.0.2) in which the Biblioshyny tool was used to map the retrieved datasets over the web interface (Aria & Cuccurullo, 2017) using bibliometric indices like annual scientific production, most productive journals/source, collaboration network, most productive countries, most productive authors, bibliographic coupling, and thematic map. The relative productivity of authors was evaluated based on the citation matrix of their publications (Adetunji et al., 2022; Smith et al., 2021).

17.3.2 RESULTS AND DISCUSSION

17.3.2.1 Main Data Summary from Web of Science

The main data summary of all documents retrieved from Web of Science is presented in Table 17.2. This includes main information about data, document types, document contents, authors, and authors' collaboration.

TABLE 17.2

Main Data Summary from Web of Science

Description	Results
Main information about data	
Timespan	2000:2021
Sources (Journals, Books, etc)	1,422
Documents	7,425
Average years from publication	8.99
Average citations per documents	22.51
Average citations per year per doc	2006
References	23,6292
Document types	
Article	6,132
Article; book chapter	2
Article; proceedings paper	102
Biographical item	1
Book review	13
Correction	58
Editorial material	130
Letter	66
News item	17
Review	896
Review; book chapter	8
Document contents	
Keywords Plus (ID)	11,556
Author's Keywords (DE)	16,400
Authors	
Authors	22,425
Author Appearances	31,956
Authors of single-authored documents	402
Authors of multi-authored documents	22,023
Authors collaboration	
Single-authored documents	485
Documents per Author	0.331
Authors per Document	3.02
Co-Authors per Documents	4.3
Collaboration Index	3.17

17.3.2.2   The Annual Scientific Production

The number of articles produced on medicinal plants and trade in herbal medicine grew steadily from year 2000 to 2021 from 130 to 573 articles, respectively. During this period, the lowest number of articles (116) was published in the year 2001, while the highest number of publications (573) was recorded in 2021, representing 7.72% of the total number of articles published within 22 years under review (Figure. 17.1). Articles published in the year 2002 had the highest mean total citation per article of 67.23 generated by 124 articles, indicating that these articles are of higher impact than those published in 2000 and 2001. However, article citation depends on publishing journals, cited articles, authors' collaborative network, authors' impact, and journal visibility (Aksnes, 2006; Onodera & Yoshikane, 2015).

17.3.2.3   Twenty Most Productive Authors

In the past 22 years, a total number of 22,425 authors have published articles on medicinal plant-related topics. This includes multi-authored papers where about 98.21% of authors participated, whereas 1.79% of the papers were single-authored (Table 17.1). The high volume of multi-authored papers may be attributed to successful collaboration among medicinal plant researchers. Likewise, it could be deduced from the bibliometric data that Van Staden J—a leading South African-based phytomedicine expert was the most active and topmost researcher having contributed the highest number of articles and citation matrix in medicinal plant-related research for the period under review (Figure. 17.2; Table 17.3). This justifies earlier claims that South Africa is a major hotspot of plant diversity, some of which are exploited for pharmacological and commercial purposes (Fennell et al., 2004; Street et al., 2008; van Wyk, 2008; Williams et al., 2013). Other leading authors are

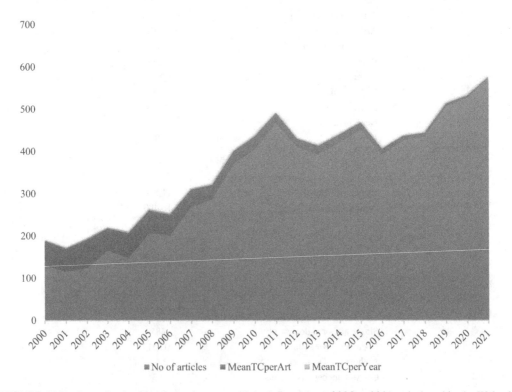

FIGURE 17.1   Annual scientific production on medicinal plants from 2000 to 2021 as indexed in the Web of Science database. Note: MeanTCperYear (Mean total citation per year); MeanTCperArt (Mean total citation per article).

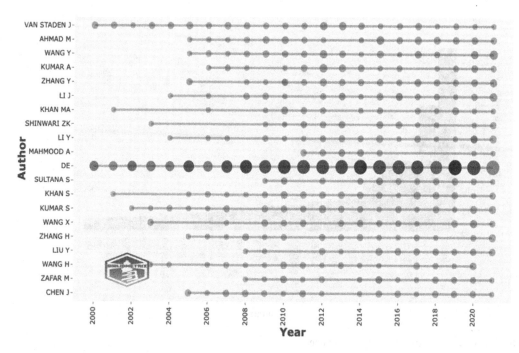

FIGURE 17.2   Twenty topmost authors' production over time.

TABLE 17.3

A Bibliometric Impact of Top 20 Authors on Herbal Trade and Medicinal Plant Related Research

Authors	Articles	Articles Fractionalized	h_Index	g_Index	m_Index	TC	PY_Start
Van Staden J	64	19.66	30	52	1.304	2,770	2000
Ahmad M	42	7.31	15	30	0.833	939	2005
Wang Y	42	8.46	16	25	0.941	645	2006
Kumar A	37	10.02	14	26	0.824	714	2006
Zhang Y	33	5.35	9	29	0.5	855	2005
Li J	31	5.40	11	21	0.579	481	2004
Khan MA	28	5.08	11	23	0.786	553	2009
Shinwari ZK	28	6.79	15	27	0.75	1,029	2003
Li Y	27	4.26	11	22	0.688	519	2007
Mahmood A	27	6.51	17	24	1.417	855	2011
DE	26	3.81	13	25	0.565	903	2000
Sultana S	26	4.38	15	26	1.071	808	2009
Khan S	25	5.04	10	18	0.588	349	2006
Kumar S	24	5.66	10	17	0.476	308	2002
Wang X	24	3.83	9	14	0.563	244	2007
Zhang H	24	4.97	12	18	0.667	362	2005
Liu Y	23	3.67	10	17	0.667	324	2008
Wang H	23	3.82	8	18	0.471	344	2006
Zafar M	22	3.13	13	22	0.929	687	2009
Chen J	21	4.42	9	15	0.5	241	2005

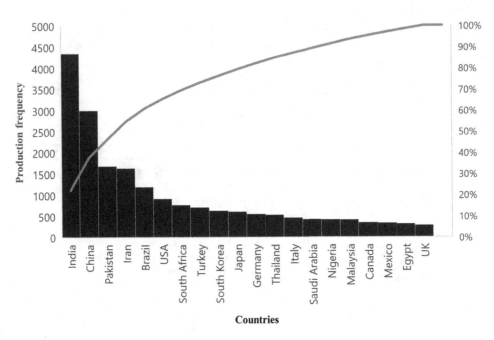

FIGURE 17.3 Countries production frequency.

Ahmad M and Wang Y whose publication years started in 2005 and 2006, respectively, and have contributed 42 papers each (Figure. 17.2; Table 17.3).

17.3.2.4   Twenty Most Productive Countries

Based on these bibliometric data, India leads other countries in herbal medicine- and medicinal plant-related research that was conducted from 2000 to 2021 (Figure. 17.3). However, South Africa, Nigeria, and Egypt were ranked 7th, 15th, and 19th positions, respectively, ahead of the United Kingdom (UK).

17.3.2.5   Twenty Topmost Affiliations

The Quaid-i-Azam University based in Islamabad, Pakistan was the most relevant affiliation on herbal medicine and medicinal plant research, as supported by the bibliometric data. Two universities from Africa namely—the University of Kwa-Zulu Natal and University of Witwatersrand made the 11th and 19th positions, respectively (Figure. 17.4).

17.3.2.6   Journal Dynamics

The Journal of Ethnopharmacology was the most sought-after source for publication of herbal medicine-related research (Figure. 17.5). This may not be unconnected with the journal's scope of publishing experimental observation of the biological activities of plants utilized in folk medicine and its openness to a well-informed ethnobotanical approach to the study of indigenous drugs. Other important journals are *Journal of Medicinal Plant Research, Phytotherapy Research Journal, South African Journal of Botany, African Journal of Biotechnology, Pharmaceutical Biology,* among others (Figure. 17.5).

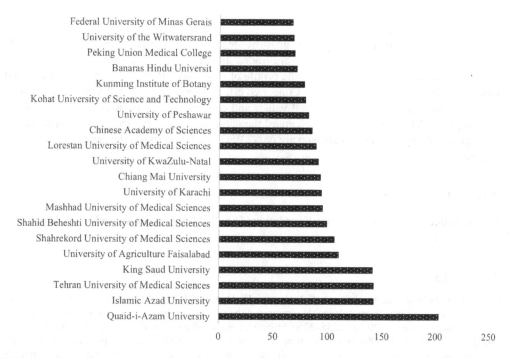

FIGURE 17.4 Twenty most relevant affiliations.

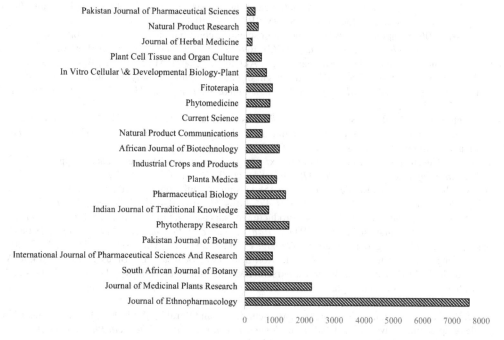

FIGURE 17.5 Journal dynamics.

17.4 CONCLUSION

The economic value of medicinal plants is growing steadily, as evidenced by the two-decade bibliometric analysis of medicinal plants that are traded globally. Increasing demands for medicinal plants has significantly improved the income profile of individuals that harvest plant for trade and economies of some developing countries. Concerted efforts should be made to retrieve the statistics of lost species due to their high medicinal values and capture the economic potential of trade in medicinal plants. This is important to safeguard the natural plant reserve from incessant depletion orchestrated by desperate individuals who are only after monetary gains at the expense of biodiversity conservation.

ACKNOWLEDGEMENTS

We acknowledge the National Research Foundation of South Africa: Grant No. SRUG210326591092, for supporting this work.

REFERENCES

Abdul, M.I.M., Siddique, S., Ur Rahman, S.A., Lateef, D., Dan, S., Mandal, P., and Bose, A. (2018). A critical insight of modern herbal drugs therapy under the purview of toxicity and authenticity. *Article in Biomedical Research*, *29*(16), 3255–3260. https://doi.org/10.4066/biomedicalresearch.29-18-968.

Adetunji, T.L., Olawale, F., Olisah, C., Adetunji, A.E., and Aremu, A.O. (2022). Capsaicin: a two-decade systematic review of global research output and recent advances against human Cancer. *Frontiers in Oncology*, *12*, 908487. https://doi.org/10.3389/FONC.2022.908487.

Agustina, L., Gan, E., Yuliatiq, N., and Sudjarwo, G.W. (2022). *In vitro* antiplatelet activities of aqueous extract of garlic (*Allium sativum*) and black garlic in human blood. *Research Journal of Pharmacy and Technology*, *15*(4), 1579–1582. https://doi.org/10.52711/0974-360X.2022.00263.

Ahn, K. (2017). The worldwide trend of using botanical drugs and strategies for developing global drugs. *BMB Reports*, *50*(3), 111. https://doi.org/10.5483/BMBREP.2017.50.3.221.

Aksnes, D.W. (2006). Citation rates and perceptions of scientific contribution. *Journal of the American Society for Information Science and Technology*, *57*(2), 169–185. https://doi.org/10.1002/ASI.20262.

Alam, J., Mujahid, M., Badruddeen, Rahman, M.A., Akhtar, J., Khalid, M., Jahan, Y., Basit, A., Khan, A., Shawwal, M., and Iqbal, S.S. (2015). An insight of pharmacognostic study and phytopharmacology of *Aquilaria agallocha*. *Journal of Applied Pharmaceutical Science*, *5*(8), 173–181. https://doi.org/10.7324/JAPS.2015.50827.

Ali, S., Rech, K.S., Badshah, G., Soares, F.L.F., and Barison, A. (2021). 1H HR-MAS NMR-Based metabolomic fingerprinting to distinguish morphological similarities and metabolic profiles of *Maytenus ilicifolia*, a *brazilian* medicinal plant. *Journal of Natural Products*, *84*(6), 1707–1714.

Aria, M., and Cuccurullo, C. (2017). Bibliometrix: an r-tool for comprehensive science mapping analysis. *Journal of Informetrics*, *11*(4), 959–975. https://doi.org/10.1016/j.joi.2017.08.007.

Aslani, A., Ghannadi, A., and Raddanipour, R. (2015). Design, formulation and evaluation of *Aloe vera* chewing gum. *Advanced Biomedical Research*, *4*, 175. https://doi.org/10.4103/2277-9175.163999.

Asmari, S., Zafar, R., and Ahmad, S. (2004). Production of sarsasapogenin from tissue culture of *Asparagus racemosus* and its quantification by HPTLC. *Iranian Journal of Pharmaceutical Research*, *3*(2), 66–67.

Atanasov, A.G., Waltenberger, B., Pferschy-Wenzig, E.-M., Linder, T., Wawrosch, C., Uhrin, P., Temml, V., Wang, L., Schwaiger, S., Heiss, E.H., Rollinger, J.M., Schuster, D., Breuss, J.M., Bochkov, V., Mihovilovic, M.D., Kopp, B., Bauer, R., Dirsch, V.M., and Stuppner, H. (2015). Discovery and resupply of pharmacologically active plant-derived natural products: a review. *Biotechnology Advances*, *33*(8), 1582–1614. https://doi.org/10.1016/j.biotechadv.2015.08.001.

Avallone, R., Zanoli, P., Puia, G., Kleinschnitz, M., Schreier, P., and Baraldi, M. (2000). Pharmacological profile of apigenin, a flavonoid isolated from *Matricaria chamomilla*. *Biochemical Pharmacology*, *59*(11), 1387–1394. https://doi.org/10.1016/S0006-2952(00)00264-1.

Bahmani, M., Shirzad, H., Rafieian, S., and Rafieian-Kopaei, M. (2015). *Silybum marianum*: beyond Hepatoprotection. *Journal of Evidence-Based Complementary and Alternative Medicine*, *20*(4), 292–301. https://doi.org/10.1177/2156587215571116.

Bala, B.K., Hoque, M.A., Hossain, M.A., and Uddin, M.B. (2010). Drying characteristics of *Asparagus* Roots (*Asparagus* racemosus Wild.). *Drying Technology*, *28*(4), 533–541. https://doi.org/10.1080/07373931003618899.

Baliga, M.S., Bhat, H.P., Pai, R.J., Boloor, R., and Palatty, P.L. (2011). The chemistry and medicinal uses of the underutilized Indian fruit tree *Garcinia indica Choisy* (*kokum*): a review. *Food Research International*, *44*(7), 1790–1799. https://doi.org/10.1016/J.FOODRES.2011.01.064.

Balogun, M.M., Jimoh, M.O., and Ogundipe, O.T. (2020). Conservation of a rare medicinal plant: a case study of *Griffonia simplicifolia* (Vahl ex DC) Baill. *European Journal of Medicinal Plants*, *31*(10), 152–160.

Banerjee, A.K., Laya, M.S., and Poon, P.S. (2006). Sesquiterpenes classified as phytoalexins. *Studies in Natural Products Chemistry*, *33*(PART M), 193–237. https://doi.org/10.1016/S1572-5995(06)80028I-1.

Barceloux, D.G. (2008). *Medical Toxicology of Natural Substances:* Foods, fungi, medicinal herbs, plants, and venomous animals. John Wiley & Sons, 1–1157. https://doi.org/10.1002/9780470330319.

Benton, R.L., Maddie, M.A., Minnillo, D.R., Hagg, T., and Whittemore, S.R. (2008). *Griffonia simplicifolia* isolectin B4 identifies a specific subpopulation of angiogenic blood vessels following contusive spinal cord injury in the adult mouse. *Journal of Comparative Neurology*, *507*(1), 1031–1052. https://doi.org/10.1002/CNE.21570.

Benzie, I.F.F., and Wachtel-Galor, S. (2011). *Herbal Medicine: Biomolecular and Clinical Aspects*, 2nd edition. CRC Press, Francis.

Bhandari, P., Pant, P., and Rastogi, R.P. (1982). Aquillochin, a coumarinolignan from *Aquilaria agallocha*. *Phytochemistry*, *21*(8), 2147–2149. https://doi.org/10.1016/0031-9422(82)83075-6.

Binns, S., Arnason, J., and Baum, B. (2002). Phytochemical variation within populations of *Echinacea angustifolia* (*Asteraceae*). *Biochemical Systematics and Ecology*, *30*(9), 837–854.

Bondarenko, B., Walther, C., Funk, P., Schlafke, S., and Engelmann, U. (2003). Long-term efficacy and safety of PRO 160/120 (a combination of Sabal and Urtica extract) in patients with lower urinary tract symptoms (LUTS). *Phytomedicine: International Journal of Phytotherapy & Phytopharmacology*, *10*(4), S53–S53.

Booker, A., Suter, A., Krnjic, A., Strassel, B., Zloh, M., Said, M., and Heinrich, M. (2014). A phytochemical comparison of saw palmetto products using gas chromatography and 1H nuclear magnetic resonance spectroscopy metabolomic profiling. *Journal of Pharmacy and Pharmacology*, *66*(6), 811–822. https://doi.org/10.1111/JPHP.12198.

Bopana, N., and Saxena, S. (2007). *Asparagus racemosus*—Ethnopharmacological evaluation and conservation needs. *Journal of Ethnopharmacology*, *110*(1), 1–15. https://doi.org/10.1016/J.JEP.2007.01.001.

Brown, J.H. (1990). Species diversity. *Analytical Biogeography*, Springer, Dordrecht. pp. 57–89. https://doi.org/10.1007/978-94-009-0435-4_3

Brunet, J., Zalapa, J., and Guries, R. (2016). Conservation of genetic diversity in slippery elm (*Ulmus rubra*) in Wisconsin despite the devastating impact of Dutch elm disease. *Conservation Genetics*, *17*(5), 1001–1010. https://doi.org/10.1007/S10592-016-0838-1/TABLES/6.

Bye, R., Linares, E., and Estrada, E. (1995). Biological diversity of medicinal plants in México. *Phytochemistry of Medicinal Plants,* Springer, Boston, MA, 65–82. https://doi.org/10.1007/978-1-4899-1778-2_4.

Bylka, W., Znajdek-Awizeń, P., Studzińska-Sroka, E., Dańczak-Pazdrowska, A., and Brzezińska, M. (2014). *Centella asiatica* in dermatology: an overview. *Phytotherapy Research*, *28*(8), 1117–1124. https://doi.org/10.1002/PTR.5110.

Castle, L., Leopold, S., Craft, R., and Kindscher, K. (2014). Ranking tool created for medicinal plants at risk of being overharvested in the wild. *Ethnobiology Letters*, *5*, 77–88. https://doi.org/10.14237/ebl.5.2014.169.

Chen, C.H., Kuo, T.C.Y., Yang, M.H., Chien, T.Y., Chu, M.J., Huang, L.C., Chen, C.Y., Lo, H.F., Jeng, S.T., and Chen, L.F.O. (2014). Identification of cucurbitacins and assembly of a draft genome for *Aquilaria agallocha*. *BMC Genomics*, *15*(1), 1–11. https://doi.org/10.1186/1471-2164-15-578/FIGURES/5.

Chen, S.L., Yu, H., Luo, H.M., Wu, Q., Li, C.F., and Steinmetz, A. (2016a). Conservation and sustainable use of medicinal plants: problems, progress, and prospects. *Chinese Medicine (United Kingdom)*, *11*(1), 1–10. https://doi.org/10.1186/s13020-016-0108-7.

Chen, S.L., Yu, H., Luo, H.M., Wu, Q., Li, C.F., and Steinmetz, A. (2016b). Conservation and sustainable use of medicinal plants: problems, progress, and prospects. *Chinese Medicine (United Kingdom)*, *11*(1). https://doi.org/10.1186/S13020-016-0108-7.

Chen, W., van Wyk, B.E., Vermaak, I., and Viljoen, A.M. (2012). Cape aloes- A review of the phytochemistry, pharmacology and commercialisation of *Aloe ferox*. *Phytochemistry Letters*, *5*(1), 1–12. https://doi.org/10.1016/J.PHYTOL.2011.09.001.

Ciesarová, Z., Murkovic, M., Cejpek, K., Kreps, F., Tobolková, B., Koplík, R., Belajová, E., Kukurová, K., Daško, Ľ., Panovská, Z., Revenco, D., and Burčová, Z. (2020). Why is sea buckthorn (*Hippophae rhamnoides* L.) so exceptional? a review. *Food Research International*, *133*, 109170. https://doi.org/10.1016/J.FOODRES.2020.109170.

Clarenc Aarland, R., Ernesto Bañuelos-Hernández, A., Fragoso-Serrano, M., del Carmen Sierra-Palacios, E., Díaz de León-Sánchez, F., Josefina Pérez-Flores, L., Rivera-Cabrera, F., Alberto Mendoza-Espinoza, J., Ernesto Ba, A., andez, nuelos-H., ıaz de Le on-S anchez, F.D., Josefina erez-Flores, L.P., and Alberto

Mendoza-Espinoza, J. (2016). Studies on phytochemical, antioxidant, anti-inflammatory, hypogly-caemic and antiproliferative activities of *Echinacea purpurea* and *Echinacea angustifolia* extracts. *Pharmaceutical Biology*, *55*(1), 649–656. https://doi.org/10.1080/13880209.2016.1265989.

Cock, I.E., Selesho, M.I., and van Vuuren, S.F. (2018). A review of the traditional use of southern African medicinal plants for the treatment of selected parasite infections affecting humans. *Journal of Ethnopharmacology*, *220*, 250–264. https://doi.org/10.1016/J.JEP.2018.04.001.

Cunningham, T. (1989). Herbal medicine trade: a hidden economy. *Indicator South Africa- Rural Trends*, *6*(3), 51–54.

Curth, L.H. (ed.). (2006). *From Physick to Pharmacology: five Hundred Years of British Drug Retailing*. Bath Spa University, Ashgate Publishing Ltd, UK.

Dagne, E., Bisrat, D., Viljoen, A., and van Wyk, B.E. (2005). Chemistry of *Aloe species*. *Current Organic Chemistry*, *4*(10), 1055–1078. https://doi.org/10.2174/1385272003375932.

Dalla Pellegrina, C., Padovani, G., Mainente, F., Zoccatelli, G., Bissoli, G., Mosconi, S., Veneri, G., Peruffo, A., Andrighetto, G., Rizzi, C., and Chignola, R. (2005). Anti-tumour potential of a gallic acid-containing phenolic fraction from Oenothera biennis. *Cancer Letters*, *226*(1), 17–25. https://doi.org/10.1016/J.CANLET.2004.11.033.

Dhingra, D., Parle, M., and Kulkarni, S.K. (2004). Memory enhancing activity of Glycyrrhiza glabra in mice. *Journal of Ethnopharmacology*, *91*(2–3), 361–365.

Diamond, B., Shiflett, S., Feiwel, N., Matheis, R., Noskin, O., Richards, J.A., and Schoenberger, N.E. (2000). Ginkgo biloba extract: mechanisms and clinical indications. *Archives of Physical Medicine and Rehabilitation*, *81*(5), 668–678.

Edwards, S.E., Da Costa Rocha, I., Williamson, E.M., and Heinrich, M. (2015). Slippery Elm. *Phytopharmacy: An Evidence-Based Guide to Herbal Medical Products,* John Wiley & Sons, Ltd. pp. 360–362. https://doi.org/10.1002/9781118543436.CH102.

Esmat, A.Y., Tomasetto, C., and Rio, M.C. (2006). Cytotoxicity of a natural anthraquinone (Aloin) against human breast cancer cell lines with and without ErbB-2-topoisomerase IIa coamplification. *Cancer Biology and Therapy*, *5*(1), 97–103. https://doi.org/10.4161/CBT.5.1.2347

Fellows, L.E., and Bell, E.A. (1970). 5-hydroxy-l-tryptophan, 5-hydroxytryptamine and l-tryptophan-5-hydroxylase in griffonia simplicifolia. *Phytochemistry*, *9*(11), 2389–2396. https://doi.org/10.1016/S0031-9422(00)85745-3.

Fennell, C.W., Light, M.E., Sparg, S.G., Stafford, G.I., and van Staden, J. (2004). Assessing African medicinal plants for efficacy and safety: agricultural and storage practices. *Journal of Ethnopharmacology*, *95*(2–3), 113–121. https://doi.org/10.1016/j.jep.2004.05.025.

Fenwick, G.R., Lutomski, J., and Nieman, C. (1990). Liquorice, *Glycyrrhiza glabra* L.– composition, uses and analysis. *Food Chemistry*, *38*(2), 119–143. https://doi.org/10.1016/0308–8146(90)90159-2.

Ferreira Júnior, W.S., Cruz, M.P., Dos Santos, L.L., and Medeiros, M.F.T. (2012). Use and importance of quina (*Cinchona spp.*) and ipeca (*Carapichea ipecacuanha* (Brot.) L. Andersson): plants for medicinal use from the 16th century to the present. *Journal of Herbal Medicine*, *2*(4), 103–112. https://doi.org/10.1016/J.HERMED.2012.07.003

Fokunang, C.N., Ndikum, V., Tabi, O.Y., Jiofack, R.B., Ngameni, B., Guedje, N.M., Tembe-Fokunang, E.A., Tomkins, P., Barkwan, S., Kechia, F., Asongalem, E., Ngoupayou, J., Torimiro, N.J., Gonsu, K.H., Sielinou, V., Ngadjui, B., Angwafor, I., Nkongmeneck, A., Abena, O.M., et.al. (2011). Traditional medicine: past, present, and future research and development prospects and integration in the national health system of cameroon. *African Journal of Traditional, Complementary, and Alternative Medicines*, *8*(3), 284. https://doi.org/10.4314/AJTCAM.V8I3.65276.

Freiberg, M., Winter, M., Gentile, A., Zizka, A., Muellner-Riehl, A.N., Weigelt, A., and Wirth, C. (2020). LCVP, The Leipzig catalogue of vascular plants, a new taxonomic reference list for all known vascular plants. *Scientific Data*, *7*(1), 1–7. https://doi.org/10.1038/s41597-020-00702-z.

Furuya, T., Ikuta, A., and Syono, K. (1972). Alkaloids from callus tissue of Papaver somniferum. *Phytochemistry*, *11*(10), 3041–3044. https://doi.org/10.1016/0031–9422(72)80101-8.

Gänger, S. (2015). World trade in medicinal plants from spanish America, 1717–1815. *Medical History*, *59*(1), 44–62. https://doi.org/10.1017/mdh.2014.70.

Germer, S., Hauer, H., Erdelmeier, C., and Schoetz, K. (2007). A detailed view on the constituents of EPs® 7630. *Planta Medica*, *73*(09). https://doi.org/10.1055/S-2007-986775.

Groner, V.P., Nicholas, O., Mabhaudhi, T., Slotow, R., Akçakaya, R.H., Mace, G.M., and Pearson, R.G. (2022). Climate change, land cover change, and overharvesting threaten a widely used medicinal plant in South Africa. *Ecological Applications*, *32*(4), e2545, Wiley Online Library. https://doi.org/10.1002/eap.2545.

Gupta, V.K., Fatima, A., Faridi, U., Negi, A.S., Shanker, K., Kumar, J.K., Rahuja, N., Luqman, S., Sisodia, B.S., Saikia, D., and Darokar, M.P. (2008). Antimicrobial potential of *Glycyrrhiza glabra* roots. *Journal of Ethnopharmacology*, *116*(2), 377–380.

Hassanpouraghdam, M.B., Ghorbani, H., Esmaeilpour, M., Alford, M.H., Strzemski, M., and Dresler, S. (2022). Diversity and distribution patterns of endemic medicinal and aromatic plants of iran: implications for conservation and habitat management. *International Journal of Environmental Research and Public Health*, *19*(3), 1552. https://doi.org/10.3390/ijerph19031552.

He, J., Yang, B., Dong, M., and Wang, Y. (2018). Crossing the roof of the world: trade in medicinal plants from Nepal to China. *Journal of Ethnopharmacology*, *224*, 100–110. https://doi.org/10.1016/J.JEP.2018.04.034.

Hikino, H., Kiso, Y., Wagner, H., and Fiebig, M. (1984). Antihepatotoxic actions of flavonolignans from *Silybum marianum* fruits. *Planta Medica*, *50*(3), 248–250. https://doi.org/10.1055/S-2007-969690/BIB.

Hoshino, O., Murakata, M., Ishizaki, M., Kametani, T., Honda, T., Fukumoto, K., and Ihara, M. (2000). Alkaloids, *Ullmann's Encyclopedia of Industrial Chemistry*. https://doi.org/10.1002/14356007.A01_353.

Huntley, A. (2004). The safety of black cohosh (*Actaea racemosa, Cimicifuga racemosa*). *Expert Opinion on Drug Safety*, *3*(6), 615–623. https://doi.org/10.1517/14740338.3.6.615.

Iknayan, K.J., Tingley, M.W., Furnas, B.J., and Beissinger, S.R. (2014). Detecting diversity: emerging methods to estimate species diversity. *Trends in Ecology & Evolution*, *29*(2), 97–106. https://doi.org/10.1016/J.TREE.2013.10.012.

Iwu, M.M. (2014). Handbook of African medicinal plants. *Handbook of African Medicinal Plants*, 2nd edition, CRC Press, Taylor & Francis Group. https://doi.org/10.1201/B16292.

Jamshidi-Kia, F., Lorigooini, Z., and Amini-Khoei, H. (2018). Medicinal plants: past history and future perspective. *Journal of Herbmed Pharmacology*, *7*(1), 1–7. https://doi.org/10.15171/jhp.2018.01.

Jennings, H.M., Merrell, J., Thompson, J.L., and Heinrich, M. (2015). Food or medicine? the food-medicine interface in households in Sylhet. *Journal of Ethnopharmacology*, *167*, 97–104. https://doi.org/10.1016/J.JEP.2014.09.011.

Jiang, B., Kronenberg, F., Balick, M.J., and Kennelly, E.J. (2006). Analysis of formononetin from black cohosh (*Actaea racemosa*). *Phytomedicine*, *13*(7), 477–486. https://doi.org/10.1016/J.PHYMED.2005.06.007.

Jimoh, M.A., Idris, O.A., and Jimoh, M.O. (2020). Cytotoxicity, phytochemical, antiparasitic screening, and antioxidant activities of *Mucuna pruriens* (Fabaceae). *Plants*, *9*(9), 1249. https://doi.org/10.3390/plants9091249.

Jimoh, M.A. and Jimoh, M.O. (2021). Economic consequences of plant biodiversity loss. In: Aliero, A.A., Agboola, D.A., and Vwioko, E.D. (eds.), *Plants and the Ecosystems*, FUK Press, Federal University of Kashere, Gombe State, Nigeria, pp. 397–411.

Jimoh, M.O., Afolayan, A.J., and Lewu, F.B. (2020). Toxicity and antimicrobial activities of *Amaranthus caudatus* L. (*Amaranthaceae*) harvested from formulated soils at different growth stages. *Journal of Evidence-Based Integrative Medicine*, *25*, 1–11. https://doi.org/10.1177/2515690X20971578.

Jimoh, M.O., and Kambizi, L. (2022). Aquatic phytotherapy: prospects, challenges and bibliometric analysis of global research output on medicinal aquatic plants from 2011 to 2020. In: Lall, N. (ed.), *Medicinal Plants for Cosmetics, Health and Diseases*, 1st edition, CRC Press, Taylor & Francis Group, pp. 507–522.

Jimoh, M.O., Okaiyeto, K., Oguntibeju, O.O., and Laubscher, C.P. (2022). A systematic review on *Amaranthus*-Related Research. *Horticulturae*, *8*(3), 239. https://doi.org/10.3390/HORTICULTURAE8030239.

Jimoh, M.O., Salami, S.O., and Laubscher, C.P. (2021). Essentials of plant physiology. In: Aliero, A.A., Agboola, D.A., and Vwioko, E.D. (eds.), *Plants and the Ecosystems*. FUK Press, Federal University of Kashere, Gombe State, Nigeria, pp. 20–31.

Júnior, W.S.F., de Oliveira Campos, L.Z., Pieroni, A., and Albuquerque, U.P. (2015). Biological and cultural bases of the use of medicinal and food plants. *Evolutionary Ethnobiology*, 175–184. https://doi.org/10.1007/978-3-319-19917-7_13.

Juyal, D., Thawani, V., Thaledi, S., and Joshi, M. (2014). Ethnomedical properties of taxus *Wallichiana Zucc.* (Himalayan Yew). *Journal of Traditional and Complementary Medicine*, *4*(3), 159–161. https://doi.org/10.4103/2225-4110.136544.

Kadam, M.P.V., Yadav, K.N., Patel, A.N., Navsare, V.S., Bhilwade, S.K., and Patil, M.J. (2012). Phytopharmacopoeial specifications of *Garcinia indica* fruit rinds. *Pharmacognosy Journal*, *4*(31), 23–28. https://doi.org/10.5530/PJ.2012.31.5.

Kamada, H., Okamura, N., Satake, M., Harada, H., and Shimomura, K. (1986). Alkaloid production by hairy root cultures in *Atropa belladonna*. *Plant Cell Reports*, *5*(4), 239–242. https://doi.org/10.1007/BF00269811.

Kayser, O., and Kolodziej, H. (1995). Highly oxygenated coumarins from Pelargonium sidoides. *Phytochemistry*, *39*(5), 1181–1185.

Kendler, B.S. (1987). Garlic (*Allium sativum*) and onion (*Allium cepa*): a review of their relationship to cardio-vascular disease. *Preventive Medicine*, *16*(5), 670–685. https://doi.org/10.1016/0091-7435(87)90050-8.

Kennedy, D. (2014). The deliriants- the nightshade (Solanaceae) family. *Plants and the Human Brain*, Oxford University Press, pp. 131–137.

Khan, M., Verma, S.C., Srivastava, S.K., Shawl, A.S., Syamsundar, K.V., Khanuja, S.P.S., and Kumar, T. (2006). Essential oil composition of *Taxus wallichiana* Zucc. from the Northern Himalayan region of India. *Flavour and Fragrance Journal*, *21*(5), 772–775. https://doi.org/10.1002/FFJ.1682.

Khan, S., Banu, T., Akter, S., Goswami, B., Islam, M., Hani, U., and Habib, A. (2018). *In vitro* regeneration protocol of *Rauvolfia serpentina* L. *Bangladesh Journal of Scientific and Industrial Research*, *53*(2), 133–138. https://doi.org/10.3329/BJSIR.V53I2.36674.

Khasim, S.M., Chunlin Long, K., and Thammasiri, H.L. (2020). Medicinal Plants: Biodiversity, Sustainable Utilization ,and Conservation. In: Khasim, S.M., Chunlin Long, K. and Thammasiri, H.L. (eds.), *Medicinal Plants: Biodiversity, Sustainable Utilization and Conservation*, Springer, Singapore. https://doi.org/10.1007/978-981-15-1636-8.

Kim, Y.C., Lee, E.H., Lee, Y.M., Kim, H.K., Song, B.K., Lee, E.J., and Kim, H.M. (1997). Effect of the aqueous extract of *Aquilaria agallocha* stems on the immediate hypersensitivity reactions. *Journal of Ethnopharmacology*, *58*(1), 31–38. https://doi.org/10.1016/S0378-8741(97)00075-5.

Klein, A.D., and Penneys, N.S. (1988). *Aloe vera. Journal of the American Academy of Dermatology*, *18*(4), 714–720. https://doi.org/10.1016/S0190-9622(88)70095-X.

Kolodziej, H., Kayser, O., Radtke, O.A., Kiderlen, A.F., and Koch, E. (2003). Pharmacological profile of extracts of Pelargonium sidoides and their constituents. *Phytomedicine*, *10*(4), 18–24. https://doi.org/10.1078/1433-187X-00307.

Kovács, P., Csaba, G., Pállinger, É., and Czaker, R. (2007). Effects of taxol treatment on the microtubular system and mitochondria of *Tetrahymena. Cell Biology International*, *31*(7), 724–732. https://doi.org/10.1016/j.cellbi.2007.01.004.

Kreek, M., Nielsen, D., Butelman, E., and LaForge, K. (2005). Genetic influences on impulsivity, risk taking, stress responsivity and vulnerability to drug abuse and addiction. *Nature Neuroscience*, *8*(11), 1450–1457. https://doi.org/10.1038/nn1583.

Kroll, D.J., Shaw, H.S., Mathews, J.M., and Oberlies, N.H. (2006). Simple and inexpensive methods to prevent botanical misfortune. *Acta Horticulturae*, *720*, 137–147. https://doi.org/10.17660/ACTAHORTIC.2006.720.13.

Krone, D., Mannel, M., Pauli, E., and Hummel, T. (2001). Immunomodulatory principles of *Pelargonium sidoides. Phytotherapy Research*, *15*(2), 122–126. https://doi.org/10.1002/PTR.785.

Kwakye, G.F., Jiménez, J., Jiménez, J.A., and Aschner, M. (2018). *Atropa belladonna* neurotoxicity: Implications to neurological disorders. *Food and Chemical Toxicology*, *116*(Part B), 346–353. https://doi.org/10.1016/J.FCT.2018.04.022.

Labanca, F., Ovesnà, J., and Milella, L. (2018). *Papaver somniferum* L. taxonomy, uses and new insight in poppy alkaloid pathways. *Phytochemistry Reviews*, *17*(4), 853–871. https://doi.org/10.1007/S11101-018-9563-3/FIGURES/6.

Li, J., van Belkum, M.J., and Vederas, J.C. (2012). Functional characterization of recombinant hyoscyamine 6β-hydroxylase from *Atropa belladonna. Bioorganic & Medicinal Chemistry*, *20*(14), 4356–4363. https://doi.org/10.1016/J.BMC.2012.05.042.

Lim, T.K. (2016). Glycyrrhiza glabra. *Edible Medicinal and Non-Medicinal Plants*, Springer, vol. 10, pp. 1–659. https://doi.org/10.1007/978-94-017-7276-1/COVER.

Liu, H., Gale, S.W., Cheuk, M.L., and Fischer, G.A. (2019). Conservation impacts of commercial cultivation of endangered and overharvested plants. *Conservation Biology*, *33*(2), 288–299. https://doi.org/10.1111/COBI.13216.

Liu, H., Luo, Y.B., Heinen, J., Bhat, M., and Liu, Z.J. (2014). Eat your orchid and have it too: a potentially new conservation formula for Chinese epiphytic medicinal orchids. *Biodiversity and Conservation*, *23*(5), 1215–1228. https://doi.org/10.1007/S10531-014-0661-2.

Liu, W. (2019). The data source of this study is Web of Science Core Collection? Not enough. *Scientometrics*, *121*(3) 1815-1824. https://doi.org/10.1007/s11192-019-03238-1

Lopresti, A.L., Smith, S.J., and Drummond, P.D. (2020). Herbal treatments for migraine: a systematic review of randomised-controlled studies. *Phytotherapy Research*, *34*(10), 2493–2517. https://doi.org/10.1002/PTR.6701.

Machado Perucci Pereira dos Santos, C., Moll Hüther, C., Borella, J., Santos Ribeiro, F.N., Alves Duarte, G.C., Ferreira de Carvalho, L., de Oliveira, E., Alves Lameira, O., Ferreira de Pinho, C., de Barros Machado, T., and Rodrigues Pereira, C. (2022). Season and shading affect emetine and cephalin production in *Carapichea ipecacuanha* plants. *Official Journal of the Societa Botanica Italiana, 156*(1), 51–60. https://doi.org/10.1080/11263504.2020.1832602.

MacLaughlin, B.W., Gutsmuths, B., Pretner, E., Jonas, W.B., Ives, J., Kulawardane, D.V., and Amri, H. (2006). Effects of homeopathic preparations on human prostate cancer growth in cellular and animal models. *Integrative Cancer Therapies, 5*(4), 362–372. https://doi.org/10.1177/1534735406295350.

Mahady, G.B. and Chadwick, L.R. (2001). Goldenseal (*Hydrastis canadensis*): is There enough scientific evidence to support safety and efficacy? *Nutrition in Clinical Care, 4*(5), 243–249. https://doi.org/10.1046/J.1523–5408.2001.00004.X.

Mahady, G.B., Pendland, S.L., Stoia, A., and Chadwick, L.R. (2003). *In vitro* susceptibility of *Helicobacter pylori* to isoquinoline alkaloids from *Sanguinaria canadensis* and Hydrastis canadensis. *Phytotherapy Research, 17*(3), 217–221. https://doi.org/10.1002/PTR.1108.

Manayi, A., Vazirian, M., and Saeidnia, S. (2015). *Echinacea purpurea*: pharmacology, phytochemistry and analysis methods. *Pharmacology, Phytochemistry and Analysis Methods, 9*(17), 63–72. https://www.ncbi.nlm.nih.gov/pmc/articles/PMC4441164/.

McKay, D.L. and Blumberg, J.B. (2007a). Cranberries (*Vaccinium macrocarpon*) and cardiovascular disease risk factors. *Nutrition Reviews, 65*(11), 490–502. https://doi.org/10.1111/J.1753-4887.2007.TB00273.X.

McKay, D.L. and Blumberg, J.B. (2007b). Cranberries (*Vaccinium macrocarpon*) and cardiovascular disease risk factors. *Nutrition Reviews, 65*(11), 490–502. https://doi.org/10.1111/J.1753-4887.2007.TB00273.X.

Misra, L., and Wagner, H. (2004). Alkaloidal constituents of *Mucuna pruriens* seeds. *Phytochemistry, 65*(18), 2565–2567.

Mosihuzzaman, M. (2012). Herbal medicine in healthcare-an overview. *Natural Product Communications, 7*(6), 807–812.

Moyo, M. and van Staden, J. (2014). Medicinal properties and conservation of *Pelargonium sidoides* DC. *Journal of Ethnopharmacology, 152*(2), 243–255.

Mpiana, P.T., Ngbolua, K.T.N., Tshibangu, D.S.T., Kilembe, J.T., Gbolo, B.Z., Mwanangombo, D.T., Inkoto, C.L., Lengbiye, E.M., Mbadiko, C.M., Matondo, A., Bongo, G.N., and Tshilanda, D.D. (2020). Identification of potential inhibitors of SARS-CoV-2 main protease from Aloe vera compounds: a molecular docking study. *Chemical Physics Letters, 754*, 137751. https://doi.org/10.1016/J.CPLETT.2020.137751.

Nayak, C.A., Rastogi, N.K., and Raghavarao, K.S.M.S. (2010a). Bioactive constituents present in *Garcinia Indica* Choisy and its potential food applications: a review. *International Journal of Food Products, 13*(3), 441–453. https://doi.org/10.1080/10942910802626754.

Nayak, C.A., Srinivas, P., and Rastogi, N.K. (2010b). Characterisation of anthocyanins from *Garcinia Indica* Choisy. *Food Chemistry, 118*(3), 719–724. https://doi.org/10.1016/J.FOODCHEM.2009.05.052.

Nisar, M., Khan, I., Ahmad, B., Ali, I., Ahmad, W., and Choudhary, M.I. (2008). Antifungal and antibacterial activities of *Taxus wallichiana* Zucc. *Journal of Enzyme Inhibition and Medicinal Chemistry, 23*(2), 256–260. https://doi.org/10.1080/14756360701505336.

Olatunji, T.L., and Afolayan, A.J. (2019). Evaluation of genetic relationship among varieties of *Capsicum annuum* L. and *Capsicum frutescens* L. in West Africa using ISSR markers. *Heliyon, 5*(5), e01700. https://doi.org/10.1016/j.heliyon.2019.e01700.

Onodera, N., and Yoshikane, F. (2015). Factors affecting citation rates of research articles. *Journal of the Association for Information Science and Technology, 66*(4), 739–764. https://doi.org/10.1002/ASI.23209.

Palevitch, D., Earon, G., and Carasso, R. (1997). Feverfew (*Tanacetum parthenium*) as a prophylactic treatment for migraine: a double-blind placebo-controlled study. *Phytotherapy Research, 11*(7), 508–511. https://doi.org/10.1002/(SICI)1099-1573(199711)11.

Pastorino, G., Cornara, L., Soares, S., Rodrigues, F., and Oliveira, M.B.P.P. (2018). Liquorice (*Glycyrrhiza glabra*): a phytochemical and pharmacological review. *Phytotherapy Research, 32*(12), 2323–2339. https://doi.org/10.1002/PTR.6178.

Patwardhan, B., and Mashelkar, R.A. (2009). Traditional medicine-inspired approaches to drug discovery: can Ayurveda show the way forward? *Drug Discovery Today, 14*(15–16), 804–811. https://doi.org/10.1016/J.DRUDIS.2009.05.009.

Petrovska, B.B. (2012). Historical review of medicinal plants' usage. *Pharmacognosy Reviews, 6*(11), 1. https://doi.org/10.4103/0973–7847.95849.

Pfaffenrath, V., Diener, H.C., Fischer, M., Friede, M., and Henneicke-Von Zepelin, H.H. (2002). The efficacy and safety of *Tanacetum parthenium* (feverfew) in migraine prophylaxis - a double-blind, multicentre, randomized placebo-controlled dose-response study. *Cephalalgia, 22*(7), 523–532. https://doi.org/10.1046/J.1468–2982.2002.00396.X.

Prabhu, P.R., Prabhu, D., and Rao, P. (2020). Analysis of *Garcinia indica* Choisy extract as eco-friendly corrosion inhibitor for aluminum in phosphoric acid using the design of experiment. *Journal of Materials Research and Technology, 9*(3), 3622–3631. https://doi.org/10.1016/J.JMRT.2020.01.100.

Prasain, J.K., Stefanowicz, P., Kiyota, T., Habeichi, F., and Konishi, Y. (2001). Taxines from the needles of *Taxus wallichiana*. *Phytochemistry, 58*(8), 1167–1170. https://doi.org/10.1016/S0031–9422(01)00305–3.

Qayum, M., Nisar, M., Shah, M.R., Adhikari, A., Kaleem, W.A., Khan, I., Khan, N., Gul, F., Khan, I.A., Zia-Ul-Haq, M., and Khan, A. (2012). Analgesic and antiinflammatory activities of taxoids from *Taxus wallichiana* Zucc. *Phytotherapy Research, 26*(4), 552–556. https://doi.org/10.1002/PTR.3574.

Raimi, I., Mugivhisa, L., Lewu, F., Amoo, S., and Olowoyo, J. (2021). Levels of trace metals in five anticancer medicinal plants harvested from soil treated with organic and inorganic fertilizers. *Current Topics in Toxicology, 17*, 41–58.

Rajasekharan, P.E., and Wani, S.H. (2020). Distribution, diversity, conservation and utilization of threatened medicinal plants. In: Rajasekharan, P.E. and Wani, S.H. (eds.), *Conservation and Utilization of Threatened Medicinal Plants*. 1st edition, Springer, Cham. pp. 3–30. https://doi.org/10.1007/978-3-030-39793-7_1.

Ritchie, M.R. (2007). Use of herbal supplements and nutritional supplements in the UK: what do we know about their pattern of usage? *Proceedings of the Nutrition Society, 66*(4), 479–482. https://doi.org/10.1017/S0029665107005794.

Rocchini, D., Hernández-Stefanoni, J.L., and He, K.S. (2015). Advancing species diversity estimate by remotely sensed proxies: a conceptual review. *Ecological Informatics, 25*, 22–28. https://doi.org/10.1016/J.ECOINF.2014.10.006.

Rondanelli, M., Opizzi, A., Faliva, M., Bucci, M., and Perna, S. (2013). Relationship between the absorption of 5-hydroxytryptophan from an integrated diet, by means of Griffonia simplicifolia extract, and the effect on satiety in overweight females after oral spray administration. *Eating and Weight Disorders – Studies on Anorexia, Bulimia and Obesity, 17*(1), 22–28. https://doi.org/10.3275/8165.

Rowe, J.W., Seikel, M.K., Roy, D.N., and Jorgensen, E. (1972). Chemotaxonomy of ulmus. *Phytochemistry, 11*(8), 2513–2517. https://doi.org/10.1016/S0031–9422(00)88527–1.

Sarup, P., Bala, S., and Kamboj, S. (2015). Pharmacology and phytochemistry of oleo-gum resin of *Commiphora wightii* (Guggulu). *Scientifica, 2015*, 1–14. https://doi.org/10.1155/2015/138039.

Schippmann, U., Leaman, D., and Cunningham, A. (2006). A comparison of cultivation and wild collection of medicinal and aromatic plants under sustainability aspects. In: Bogers, R.J., Craker, L.E., and Lange, D. (eds.), *Medicinal and Aromatic Plants,* Springer, Netherlands, pp. 75–95. https://library.wur.nl/ojs/index.php/frontis/article/view/1225.

Sekine, T., Fukasawa, N., and Kashiwagi, Y. (1994). Structure of asparagamine, a novel polycyclic alkaloid from *Asparagus racemosus*. *Chemical and Pharmaceutical Bulletin, 42*(6), 1360–1362.

Sen, S., Chakraborty, R., and De, B. (2011). Challenges and opportunities in the advancement of herbal medicine: India's position and role in a global context. *Journal of Herbal Medicine, 1*(3–4), 67–75. https://doi.org/10.1016/J.HERMED.2011.11.001.

Shaker, E., Mahmoud, H., and Mnaa, S. (2010). Silymarin, the antioxidant component and *Silybum marianum* extracts prevent liver damage. *Food and Chemical Toxicology, 48*(3), 803–806. https://doi.org/10.1016/J.FCT.2009.12.011.

Simmler, C., Graham, J.G., Chen, S.N., and Pauli, G.F. (2018). Integrated analytical assets aid botanical authenticity and adulteration management. *Fitoterapia, 129*, 401–414. https://doi.org/10.1016/J.FITOTE.2017.11.017.

Singh, B., Kaur, P., Singh, R., and Ahuja, P.S. (2008). Biology and chemistry of *Ginkgo biloba*. *Fitoterapia, 79*(6), 401–418.

Singh, O., Khanam, Z., Misra, N., and Srivastava, M.K. (2011). Chamomile (*Matricaria chamomilla* L.): an overview. *Pharmacognosy Reviews, 5*(9), 82. https://doi.org/10.4103/0973–7847.79103.

Singh, R. (2015). Medicinal plants: a review. *Journal of Plant Sciences, 3*(1), 50–55. https://doi.org/10.11648/j.jps.s.2015030101.18.

Sleyman, H., Demirezer, L., Bykokuroglu, M.E., Akcay, M.F., Gepdiremen, A., Banoglu, Z.N., and Ger, F. (2001). Antiulcerogenic effect of *Hippophae rhamnoides* L. *Phytotherapy Research, 15*(7), 625–627. https://doi.org/10.1002/PTR.831

Smith, H.H., Idris, O.A., and Maboeta, M.S. (2021). Global trends of green pesticide research from 1994 to 2019: a bibliometric analysis. *Journal of Toxicology, 2021*, Article ID: 6637516. Hindawi Limited. https://doi.org/10.1155/2021/6637516.

Smith-Hall, C., Larsen, H.O., and Pouliot, M. (2012). People, plants and health: a conceptual framework for assessing changes in medicinal plant consumption. *Journal of Ethnobiology and Ethnomedicine, 8*(1), 1–11. https://doi.org/10.1186/1746-4269-8-43/FIGURES/2.

Sneader, W. (2005). *Drug Discovery: A History,* 1st edition. In: Schedler, D.J.A. (ed.), John Wiley and Sons Ltd, University of Strathcyde, Glasgow, UK.

Snyman, T., Stewart, M.J., Grove, A., and Steenkamp, V. (2005). Adulteration of South African traditional herbal remedies. *Therapeutic Drug Monitoring, 27*(1), 86–89. https://doi.org/10.1097/00007691-200502000-00015.

Srivastava, A., Srivastava, P., Pandey, A., Khanna, V.K., and Pant, A.B. (2019). Phytomedicine: a potential alternative medicine in controlling neurological disorders. In: *New Look to Phytomedicine: Advancements in Herbal Products as Novel Drug Leads*. Elsevier, Imam Abdulrahuman Bin Faisal University, Saudi Arabia, pp. 625–655. https://doi.org/10.1016/B978-0-12-814619-4.00025–2.

Street, R.A., Stirk, W.A., and van Staden, J. (2008). South African traditional medicinal plant trade—challenges in regulating quality, safety and efficacy. *Journal of Ethnopharmacology, 119*(3), 705–710. https://doi.org/10.1016/J.JEP.2008.06.019.

Surjushe, A., Vasani, R., and Saple, D. (2008). *Aloe vera*: a short review. *Indian Journal of Dermatology, 53*(4), 163. https://doi.org/10.4103/0019–5154.44785.

Suryakumar, G., and Gupta, A. (2011). Medicinal and therapeutic potential of sea buckthorn (*Hippophae rhamnoides* L.). *Journal of Ethnopharmacology, 138*(2), 268–278. https://doi.org/10.1016/J.JEP.2011.09.024.

Tenen, D. (2017). *Plain Text: The Poetics of Computation,* Stanford University Press, Stanford, CA. Hardcover ISBN: 9781503601802.

Timmer, A., Günther, J., Motschall, E., Rücker, G., Antes, G., and Kern, W.V. (2013). Pelargonium sidoides extract for treating acute respiratory tract infections. *Cochrane Database of Systematic Reviews, 10*, CD006323. https://doi.org/10.1002/14651858.CD006323.PUB3.

Timoszuk, M., Bielawska, K., and Skrzydlewska, E. (2018). Evening primrose (*Oenothera biennis*) biological activity dependent on chemical composition. *Antioxidants, 7*(8), 108. https://doi.org/10.3390/ANTIOX7080108.

Turner, R.B., Bauer, R., Woelkart, K., Hulsey, T.C., and Gangemi, J.D. (2005). An evaluation of *Echinacea angustifolia* in experimental rhinovirus infections. *New England Journal of Medicine, 353*(4), 341–348.

United Nations. (2002). *United Nations: The Millennium Development Goals and the United Nations role-Fact Sheet*. United Nations Department of Public Information. https://www.un.org/millenniumgoals/MDGs-FACTSHEET1.pdf.

Urizar, N.L. and Moore, D.D. (2003). Gugulipid: a natural cholesterol-lowering agent. *Annual Review of Nutrition, 23*, 303–313.

van Andel, T. and Havinga, R. (2008). Sustainability aspects of commercial medicinal plant harvesting in Suriname. Forest Ecology and Management, *256*(8), 1540–1545. https://doi.org/10.1016/J.FORECO.2008.06.031

van Wyk, A. and Prinsloo, G. (2018). Medicinal plant harvesting, sustainability and cultivation in South Africa. *Biological Conservation, 227*, 335–342.

van Wyk, B.E. (2008). A broad review of commercially important southern African medicinal plants. *Journal of Ethnopharmacology, 119*(3), 342–355. https://doi.org/10.1016/J.JEP.2008.05.029

Vasisht, K., Sharma, N., and Karan, M. (2016). Current perspective in the International trade of medicinal plants material: an update. *Current Pharmaceutical Design, 22*(27), 4288–4336.

Vigliante, I., Mannino, G., and Maffei, M.E. (2019). Chemical characterization and DNA fingerprinting of griffonia simplicifolia baill. *Molecules, 24*(6), 1032 https://doi.org/10.3390/molecules24061032.

Wachsmuth, O., and Matusch, R. (2002a). Anhydronium bases from *Rauvolfia serpentina*. *Phytochemistry, 61*(6), 705–709. https://doi.org/10.1016/S0031–9422(02)00372–2.

Wachsmuth, O., and Matusch, R. (2002b). Anhydronium bases from *Rauvolfia serpentina*. *Phytochemistry, 61*(6), 705–709. https://doi.org/10.1016/S0031–9422(02)00372–2.

Wagner, H., Flachsbarth, H., Medica, G.V.P. (1981). A new antiphlogistic principle from *Sabal serrulata*. *Planta Medica, 41*(3), 244–251.

Williams, C.A., Harborne, J.B., Geiger, H., and Hoult, J.R.S. (1999). The flavonoids of *Tanacetum parthenium* and T. vulgare and their anti-inflammatory properties. *Phytochemistry, 51*(3), 417–423. https://doi.org/10.1016/S0031-9422(99)00021-7.

Williams, C.A., Hoult, J.R.S., Harborne, J.B., Greenham, J., and Eagles, J. (1995). A biologically active lipophilic flavonol from *Tanacetum parthenium. Phytochemistry, 38*(1), 267–270. https://doi.org/ 10.1016/0031-9422(94)00609-W.

Williams, V.L., Victor, J.E., and Crouch, N.R. (2013). Red Listed medicinal plants of South Africa: status, trends, and assessment challenges. *South African Journal of Botany, 86*, 23–35. https://doi.org/10.1016/J. SAJB.2013.01.006.

Wintola, O.A., Sunmonu, T.O., and Afolayan, A.J. (2010). The effect of Aloe ferox Mill. In the treatment of loperamide-induced constipation in Wistar rats. *BMC Gastroenterology, 10*(1), 1–5. https://doi. org/10.1186/1471-230X–10–95.

Wu, C., Lee, S.L., Taylor, C., Li, J., Chan, Y.M., Agarwal, R., Temple, R., Throckmorton, D., and Tyner, K. (2020). Scientific and regulatory approach to botanical drug development: a U.S. FDA perspective. *Journal of Natural Products, 83*(2), 552–562.

Wu, V.C.H., Qiu, X., Bushway, A., and Harper, L. (2008). Antibacterial effects of American cranberry (*Vaccinium macrocarpon*) concentrate on foodborne pathogens. *LWT - Food Science and Technology, 41*(10), 1834–1841. https://doi.org/10.1016/J.LWT.2008.01.001.

Wuttke, W., Jarry, H., Haunschild, J., Stecher, G., Schuh, M., and Seidlova-Wuttke, D. (2014). The non-estrogenic alternative for the treatment of climacteric complaints: black cohosh (*Cimicifuga or Actaea racemosa*). The Journal of Steroid Biochemistry and Molecular Biology, *139*, 302–310. https://doi. org/10.1016/J.JSBMB.2013.02.007.

Zhu, N., Rafi, M.M., DiPaola, R.S., Xin, J., Chin, C.K., Badmaev, V., Ghai, G., Rosen, R.T., and Ho, C.T. (2001). Bioactive constituents from gum guggul (*Commiphora wightii*). *Phytochemistry, 56*(7), 723–727. https:// doi.org/10.1016/S0031-9422(00)00485-4

18 Ethnomedicinal and Other Ecosystem Services Provided by Threatened Plants
A Case Study of Cycads

Terence Nkwanwir Suinyuy, Peter Tshepiso Ndhlovu,
John Awungnjia Asong, Bongani Petros Kubheka,
Babajide Charles Falemara and Wilfred Otang Mbeng

CONTENTS

18.1 INTRODUCTION

Ecosystem services (ESs) are the benefits that humans derived from ecosystems (Nicholson et al., 2009). The Millennium Ecosystem Assessment (MEA) highlights the impact of natural ecosystems to human well-being and determines the trend and status of ESs globally (Millennium Ecosystem Assessment, 2005). Four categories of ESs have been identified, namely, provisioning services (food, medicines, fuel wood), cultural services (aesthetic value, recreation and tourism, spiritual, and religious use), supporting services (ecosystem processes like nitrogen fixation), and regulating services (regulation of climate, diseases, and water) (Millennium Ecosystem Assessment, 2005).

Natural ecosystems such as forests, grasslands, and other natural areas provide a variety of ESs to society. Such important ecosystems are to be carefully managed to secure the provision of ESs in the short and long term. However, these ecosystems have witnessed changes that have been accompanied by loss of biodiversity such as plants that are vital to the provision of ESs. The loss of plants will eventually lead to them being threatened with extinction.

The tenth meeting of the Conference of Parties (COP 10) to the CBD in 2010 adopted a global strategic plan for biodiversity from 2011 to 2020. This strategic plan includes, besides strategic

DOI: 10.1201/9781003206620-18

goals, also 20 targets, known as the Aichi targets. The Aichi targets of the CBD include the prevention of extinction of threatened species and enhancement of the benefit of all from ESs (Marques et al., 2014). This is because threatened plants may provide important ES, and a decline in their abundance and probably extinction will eventually lead to a loss in provision of ESs. Eventually, such a loss will negatively affect the people whose livelihoods are strongly tied to local ES. The conservation of threatened plants is, therefore, imperative for ensuring sustainable development and provision of essential ESs. Many plant species are subject to extinction due to anthropogenic and natural causes, and there is a concern that this may lead to a collapse in their natural ecosystem and ESs. One such plant group that is threatened with extinction and may provide important ESs is the cycads. Currently, the provisioning of ESs is not included and rarely discussed in cycad conservation planning, as it is considered a threat to the existence of the plants. Nonetheless, these extraordinary plants have provided avenues for some of the most important studies in ecological and evolutionary biology (Suinyuy and Johnson, 2021), as conservation efforts will not only conserve the cycads but both the ecological and evolutionary processes that are important for the persistence of the plants (Cowling et al., 1999) for continuous provision of ESs. There is, therefore, a need for the inclusion of ESs in the formulation of conservation strategies of such plants, especially, as these systems that provide ES have to be carefully managed for the provision of the present and the future ES.

Cycads are gymnosperms that originated about 300 million years ago and are considered the oldest extant seed plants (Norstog and Nicholls, 1997). They are regarded as flagship species of conservation importance, as they have a higher proportion of threatened species with more than 50% being threatened globally and are listed as Rare or Endangered in the International Union for Conservation of Nature (IUCN) Red List (Millennium Ecosystem Assessment, 2005). Once widely distributed when the world's climate was mild and permitted larger ranges, today their ranges have reduced to tropical and subtropical habitats. They have attracted much interest from diverse groups and are collected for their educational, economic, medicinal, and aesthetic value. Cycads are of particular biological interest, providing important information for understanding the evolution of angiosperms and gymnosperms because their morphological characteristics are intermediate between ancient plants such as ferns and more derived plants including angiosperms (Brenner et al., 2003). They are slow-growing, long-lived, perennial, and dioecious plants that have developed specific interactions with insect visitors having highly specialised mutualisms with the plants (Suinyuy and Johnson, 2021). Their irregular coning nature may disrupt these mutualisms and affect reproduction. Most species are range-restricted and occur in fragmented populations, experiencing much localised seed dispersal, low recruitment and seedling survival (Raimondo and Donaldson, 2003), and high seed death rate due to herbivory, fire, and other natural and anthropogenic disturbances. Consequently, the impact of climate change (extreme weather conditions) together with these factors could promote the decline of cycads, and disrupt the ecosystem function and the provision of ESs.

In addition to their biological, conservation, and evolutionary significance, there have been several studies on the importance of cycads to humans throughout history. This has led to the development of several conservation actions and monitoring programmes to keep the status of this group of plants in check. Several studies on the uses of cycads have focused on their significance as an important food source in time of famine (Thieret, 1958; Whitelock, 2002), ethnobotany, ethnological and economic importance (Thieret, 1958), and its neurotoxicity and toxicity. Furthermore, other studies have documented key information on benefits, importance, and uses of cycads in North and Central America (Blake, 2012; Bonta, 2010), Australia, South Africa (Bamigboye and Tshisikhawe, 2019; Cousins et al., 2011, 2012) India (Krishnamurthy et al., 2013; Varghese and Ticktin, 2006), and Japan (Hayward and Kuwahara, 2013). Currently, there is no published synthesis on the importance of cycads to human well-being, making it difficult for researchers and conservation practitioners working on cycads to access such information, which may be critical in conservation planning. The purpose of this review is, therefore, to bring together the different studies on the benefits, importance, and use of cycads to humans and to highlight the importance of cycads in providing ESs with the hope that this aspect will be incorporated into cycad conservation programmes. Furthermore, the review seeks to highlight opportunities for further study and to provide an additional aspect to be incorporated in the future research and cycad conservation-related work.

18.2 MATERIAL AND METHODS

Following procedure from the previous studies (Wang et al., 2016), different search engines such as Web of Science, Science Direct, and Research Gate, as well as Google Scholar, PubMed, Science Direct, Scopus, and JSTOR, were used to look for relevant studies. In addition, thesis, dissertations, books, and conference proceedings were retrieved from the University of Mpumalanga (UMP) library. Since ES is not yet discussed in cycads, searches on related papers and research works on ES used keywords and terminologies like benefits, importance, uses, and ecosystem functions, ecological functions and services, environmental functions and services, and ethnobiology and ethnobotany of cycads. These terms are sometimes used interchangeably with ES, which is the benefit that humans derived from either natural (Daily, 1997) or human-modified (Costanza et al., 1997; Millennium Ecosystem Assessment, 2005) ecosystems. The bibliographies from the retrieved papers were also checked and saved using Endnote index manager. All the scientific plant names were validated in references to The Plant List (www.theplantlist.org) and International Plant Names Index (IPNI) (https://www.ipni.org/), South African National Biodiversity Institute (SANBI) Red List of South African Plants (redlist.sanbi.org/species).

18.3 RESULTS AND DISCUSSION

18.3.1 PROVISIONING SERVICES

18.3.1.1 Ethnomedicinal Services of Cycad

Cycas species are one of the largest groups of gymnosperms (Wang et al., 2016). They provide different services to both humans and the environment, including educational and ethnomedicinal services. They have enormous ethnomedicinal uses recorded among indigenous communities globally due to the fact that gymnosperms have evolved for many years as a result. Humans have had a long history of interaction with the *Cycas* species. As a result of their evolutionary history of more than 280 million years ago (Haynes, 2011) and human interaction, cycas provide a variety of ESs, including education and ethnomedicine. Among these different ecosystem species, ethnomedicine services are the highest percentage of services provided, accounting for 34.92% (Table 18.1 and Figure 18.1) of the total ESs rendered by *Cycas* species. Currently, 25 species grouped in six genera of *Cycas* are being used to treat over 32 different health conditions in seven countries around the world (Table 18.2).

18.3.1.1.1 Ailments Cycas is Used on

Cycas is reported to be used in many ailments. Some are used to cure boils and skin diseases using young leaves. Diabetic people use it as a food supplement (Radha and Singh, 2008). Cycads are used for sexual potency, as they are aphrodisiac. They are also known to improve sperm count. For this purpose, it is mainly used as an invigorating and nutritive tonic made from the pith of the stem, dry seeds, and flour from the seeds (Varghese et al., 2012). The male cones are widely used to cure rheumatoid, arthritis, and muscle pains. Powdered endosperm is used to relieve burning sensations and general debility, while the tender leaves relieve flatulence and vomiting. A poultice made from bark and seeds is used to treat swellings and sores. A piece of young female cone is eaten daily to treat painful urination, paralysis, piles, ulcers, other stomach ailments, indigestion, and snake bite (Hayward and Kuwahara, 2013). Some people use *cycas* stem for disinfection of the umbilical cord during childbirth (Bonta et al., 2019 and Vovides et al., 2010).

TABLE 18.1
Classification of Ecosystem Services by Cycad spp

Ecosystem Services	Species	Citations (N)	Citations (%)
Aesthetic	All cycads	1	
	Ceratozamia fuscoviridis	1	
	Cycas spp	1	
	Dioon spp	1	
	Dioon spp, *Zamia* spp	1	
	Encephalartos transvenosus	1	
	Stangeria, Zamia, some Encephalartos	1	
Aesthetic total		7	5.56%
Agricultural	*Ceratozamia latifolia*	1	
	Cycas micronesica	1	
	Cycas revolute	1	
	Cycas thouarsi	1	
	Dioon edule	1	
	Dioon sonorense	3	
	Dioon spp	1	
	Encephalartos spp	2	
	Macrozamia riedlei	2	
	Stangeria eriopus	2	
	Zamia loddigesi	1	
	Zzmia integrifolia	1	
Agricultural total		17	13.49%
Carbon sequestration	*Cycas panzhihuaensis*	4	
	Cycas revolute	1	
	D. sonorense	1	
Carbon sequestration total		6	4.76%
Educational	*All cycads*	1	
	D. mejiae	1	
Educational total		2	1.59%
Ethnomedicinal	*Zamia* sp.	1	
	Ceratomzamia mexicana	1	
	Cycas beddomei	6	
	Cycas circinalis	8	
	Cycas media	1	
	Cycas pectinata	5	
	Cycas revolute	2	
	Cycas rumphii	1	
	D. sonorense	1	
	Encephalarots villosus	1	
	Encephalartos ferox	1	
	Encephalartos longifolius	1	
	Encephalartos natalensis	1	
	E. transvenosus	2	

(Continued)

TABLE 18.1 (*Continued*)
Classification of Ecosystem Services by Cycad spp

Ecosystem Services	Species	Citations (N)	Citations (%)
	S. eriopus	1	
	Zamia aff. Nesophila	1	
	Zamia cunaria	1	
	Zamia dressleri	1	
	Zamia ipetiensis	1	
	Zamia loddigesii	1	
	Zamia manicata	1	
	Zamia neurophyllida	2	
	Zamia oblique	1	
	Zamia pseudoparasitica	1	
	Zamia sp.	1	
Ethnomedicinal total		44	34.92%
Food	*Ceratozamia latifolia*	1	
	Ceratozamia tenuis	1	
	C. beddomei	2	
	C. circinalis	1	
	Cycas media	1	
	Cycas pectinata	1	
	Cycas revolute	1	
	Cycas sphaerica	1	
	Cycas spp	1	
	Dioon edule	1	
	D. mejiae	1	
	D. sonorense	1	
	Dioon spinulosum	1	
	Encephalartos hildebrandtii	1	
	Encephalartos manikensis	1	
	E. transvenosus	1	
	Macrozamia communis	1	
	Zamia oblique	1	
Food total		19	15.08%
Goods	*C. circinalis*	1	
	Cycas seemannii	1	
	D. mejiae	2	
	Dioon spinolosum	2	
	E. transvenosus	1	
	Zamia floridana	1	
Goods total		8	6.35%
Insect repellent	*Ceratozamia tenuis*	1	
	Cycas bedommei	1	
	C. circinalis	1	
Insect repellent total		3	2.38%

(*Continued*)

TABLE 18.1 (*Continued*)
Classification of Ecosystem Services by Cycad spp

Ecosystem Services	Species	Citations (N)	Citations (%)
Spiritual and religious use	*Ceratozamia fuscoviridis*	1	
	Cycas angulate	1	
	C. circinalis	1	
	Cycas hainanensis	1	
	Cycas pectinata	1	
	Cycas revolute	3	
	C. seemannii	3	
	Cycas szechuansis	1	
	Dioon holmgrenii	1	
	D. mejiae	1	
	Dioon merolae	1	
	Dioon spp	1	
	E. natalensis	1	
	E. transvenosus	1	
	Encephalartos villosus	1	
	S. eriopus	1	
Spiritual and religious use total		20	15.87%
Grand total		126	100.00%

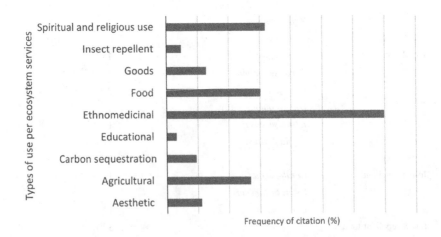

FIGURE 18.1   Distribution of types of uses per ecosystems services (*n* = 126).

TABLE 18.2

Ecosystem Services Provided by Cycads. The Botanical Names Were Verified Using the Plant List

Species	Uses	Ecosystem Services	Plant Parts Used	Country	Reference
All cycads	Cycads are grown in botanical gardens for educational purposes	Educational	Grown in botanical gardens	Global	Vovides et al. (2010); Keppel (2009); Singh and Singh (2011)
All cycads	Grown in home and public gardens for landscaping	Aesthetic	Cycads are used for landscaping	Global	Marler and Moore (2010)
Cycas angulata R.Br. Cycadaceae	Community is named after this cycad	Spiritual and religious use	Whole plant	Australia	Kearney and Bradley (2009)
C. beddomei Dyer Cycadaceae	Make a paste from it to cure boils and skin diseases, cure arthritis, use it to increase sexual potency (aphodisiac), used as a food supplement and rejuvenator, and for diabetes. Pith is used in processing flour for cooking, and young leaves are cooked and eaten as vegetables.	Ethnomedicinal, food, and insect repellent	Young leaves, pith, Bark, male and female cones	India	Radha and Singh (2008)
Ceratozamia becerrae Pérez-Farr., Vovides & Schutzman Zamiaceae	Seeds are ground and used as insecticide against flies. Seeds are ground and mixed with jam and marmalade.	Insect repellent and Food	Seeds	Mexico	Vovides et al. (2010)
Cycas thouarsi Syn: Cycas aculeata K.D.Hill & H.T.Nguyen Cycadaceae	Plant roots fix nitrogen	Agricultural	*Encephalartos* spp, *Cycas thouarsi* and *S. eriopus* fix significant amount of nitrogen	South Africa	Grobbelaar et al. (1989)

(Continued)

TABLE 18.2 (Continued)
Ecosystem Services Provided by Cycads. The Botanical Names Were Verified Using the Plant List

Species	Uses	Ecosystem Services	Plant Parts Used	Country	Reference
Ceratozamia latifolia Miq. Zamiaceae	Planted on the edges of maize field to protect the growing maize	Agricultural	Whole plant	Honduras	Bonta et al. (2019)
C. circinalis L. Cycadaceae	Dry seeds, flour from the seeds, and the pith of the stem are for general medicinal use, such as an invigorating and nutritive tonic for men and children, which improves sperm count, narcotics, and aphrodisiac. The powdered endosperm is used to relieve burning sensations and general debility. While the tender leaves relieve flatulence and vomiting. A poultice made from bark and seeds is used to treat swellings and sores. Leaves are cooked as vegetables, seeds are used to produce flour for cooking, and mature leaves are used for roof thatching, constructing 'pandhals' structures erected for special rituals.	Ethnomedicinal, food, goods, spiritual, and religious use	The pith of the stem, leaves, seeds, roots, male-female cones, and male bracts	India, Guam	Cheruku et al. (2012); Varghese et al. (2012)
Cycas hainanensis subsp. changjiangensis (N.Liu) N.Liu Cycadaceae	It is called 1,000-year-old phoenix grass and symbolises the Mythical Fenghuang bird——the 'Chinese phoenix'	Spiritual and religious use	Whole plant	China	Bonta (2012)
Cycas media R.Br. Cycadaceae	Nuts are dried, soaked, fermented, and then roasted Before eating	Food	Seeds	Australia	Kaur (2006)
Cycas micronesica K.D.Hill Cycadaceae	Plant roots fix nitrogen	Agricultural	Adult Cycas micronesica fix soil nitrogen, enhance soil carbon and phosphorus build-up	USA	Marler and Calonje (2020)
C. circinalis L Cycadaceae	Male cones are placed in rice paddy fields to repel insects that attack the young plants	Insect repellent	Male cones	India	Varghese and Ticktin (2006); Varghese et al. (2012)

(Continued)

TABLE 18.2 (Continued)
Ecosystem Services Provided by Cycads. The Botanical Names Were Verified Using the Plant List

Species	Uses	Ecosystem Services	Plant Parts Used	Country	Reference
Cycas revoluta Thunb. Cycadaceae	Leaves are used on Palm Sunday, which are kept for a year, burn, and use the ash-on-Ash Wednesday in the Catholic church and by other religious groupings. Leaves are taken to church during Holy Week for blessing, which are then displayed in the house to protect it from natural disasters. A piece of young female cone is eaten daily to treat painful urination, paralysis, indigestion, and snake bite. Seeds and stems are used in making flour for bread and porridge It is called 1,000-year-old phoenix grass and symbolises the Mythical Fenghuang bird—'the Chinese phoenix'.	Spiritual, religious use, ethnomedicine, food, and carbon sequestration	Female cone and whole plant, Seed, stem, and the leaves	The Americas and Philippines, India, Bangladesh Japan and China Spain	Hayward and Kuwahara (2013); Rout et al. (2012); Mollik et al. (2010)
Cycas pectinata Buch.-Ham. Cycadaceae	Enhance sexual potency, treat arthritis or joint pains, piles, ulcers, and other stomach ailments such as stomach aches and ulcers, and repel snakes from entering the house. Leaves are used in Hindu shrines for deities.	Ethnomedicine, food, spiritual, and religious use	Leaves, seeds and male cones	India	Singh and Singh (2011); Bonta (2012)
Cycas panzhihuaensis L.Zhou & S.Y.Yang Cycadaceae	Carbon sequestration: Total carbon storage of *Cycas panzhihuaensis* communities is 246.507 ± 3.495 t C/ha. Soil carbon storage is 233.822 t C/ha Vegetation carbon storage (11.092 ± 4.834 t C/ha). Litter carbon storage (1.593 ± 1.339 t C/ha).	Carbon sequestration	Whole plant	China	Yuandan et al. (2009)
C. seemannii A.Br. Cycadaceae	Emblem of the national flag. Dry sclerotesta of the seeds use for making cloth pins and toys.	Goods, spiritual, and religious use	Whole plant, leaves, and seeds	Vanuatu	Keppel (2009); Ala, and Osborne (2009); Bonta (2012)

(Continued)

TABLE 18.2 (Continued)
Ecosystem Services Provided by Cycads. The Botanical Names Were Verified Using the Plant List

Species	Uses	Ecosystem Services	Plant Parts Used	Country	Reference
Cycas spp Cycadaceae	Seeds are processed into flour for making different types of bread, and leaves are harvested for decoration	Food and aesthetic	Seeds and leaves	Australia, Fiji, Tonga, Vanuatu and India	Hayward and Kuwahara (2013); Radji and Kokou (2015)
Cycas sphaerica Roxb. Cycadaceae	Leaves are sliced, fried, and eaten	Food	Fresh and young leaves	India	Radji and Kokou (2015)
Cycas szechuansis Syn: C. aculeata K.D.Hill & H.T.Nguyen Cycadaceae	It is called 1,000-year-old phoenix grass and symbolises the Mythical Fenghuang bird—'the Chinese phoenix'	Spiritual and religious use	Whole plant	China	Bonta (2012)
Dioon holmgrenii De Luca, Sabato & Vázq.Torres Zamiaceae	Leaves are used to make crowns for Good Friday celebrations, and decorate churches during weddings and other feast days.	Spiritual and religious use	Leaves	Mexico	Chemnick (2013); Bonta (2012)
Dioon spp Zamiaceae	Leaves of Dioon species are used in churches for decoration and on important feast days. They are also planted in graveyards.	Spiritual and religious use	Leaves	Mexico and Northen Central America	Bonta et al. (2019); Bonta (2012)
Dioon edule Lindl Zamiaceae	Seeds are edible. Prepared into flour/paste for making bread. Dioon edule are planted as fallows in maize fields.	Food and agricultural	Seeds and Whole plant	Mexico	Bonta (2012)
D. mejiae Standl. & L.O.Williams Zamiaceae	Dry sclerotesta of the seeds are used for making toys. The seeds are chopped up and cooked, ground into flour to make bread and cookies, and used in making drinks.	Goods, food educational, spiritual, and religious use	Seeds, whole plants and leaves	Mexico and Honduras	Bonta (2012);Bonta et al. (2006); Vovides et al. (2010)

(Continued)

TABLE 18.2 (Continued)
Ecosystem Services Provided by Cycads. The Botanical Names Were Verified Using the Plant List

Species	Uses	Ecosystem Services	Plant Parts Used	Country	Reference
	Dry sclertotesta is used to make pin-and-target toys and whistles.				
	Researchers used these plants to elucidate indigenous technical knowledge and customs surrounding local uses for and conservation of the cycad *D. mejiae*.				
	Wreaths are made from leaves and placed on graves on the Feast of All souls, leaves are used on alters for communal Catholic prayers, commemorating deaths of relatives. Leaves are used during Holy Week on Good Friday.				
	On the Day of the Cross, 3rd May, the Chiapas do a pilgrimage on foot to collect *D. merolae* leaves from sacred sites and distribute to people. The pilgrimage ends in an all-night vigil, which involves singing of a hymn to Christ.				
D. spinolosum Dyer ex Eichl Zamiaceae	Seeds are prepared and processed into flour for making bread/tortillas. Dry sclerotesta is used in making toys such as bullroarers.	Food and goods	Seeds	Mexico	Chamberlain (1909); Vovides et al. (2010)
D. sonorense (De Luca, Sabato & Vázq.Torres) Chemnick, T.Greg. & Salas-Morales Zamiaceae	Together with *Brahea aculeata* palm, total carbon storage of the rare and endangered cycad *D. sonorense* between 70 and 130 Mg C/ha. Plant roots fix nitrogen. Mucilage of the cycad used for treating pains.	Carbon sequestration, agricultural, and ethnomedicinal.	Whole plants Seedlings of *D. sonorense* fix 1.68±0.09 % N. Juveniles of *D. sonorense* fix 1.41±0.08%N. Adults of *D. sonorense* fix 1.47±0.08% of N.	Mexico	Álvarez-Yépiz and Dovčiak (2015); Álvarez-Yépiz et al. (2014); Vovides (2020)

(Continued)

TABLE 18.2 (*Continued*)

Ecosystem Services Provided by Cycads. The Botanical Names Were Verified Using the Plant List

Species	Uses	Ecosystem Services	Plant Parts Used	Country	Reference
Dioon spp *Zamiaceae*	Grown in home and public gardens for landscaping and intercropping with maize	Aesthetic and agricultural	Whole plant	Mexico and Northern Central America	Bonta et al. (2019); Vovides et al. (2010)
Encephalartos spp *Zamiaceae*	Plant roots fix nitrogen.	Agricultural	*Encephalartos* spp and *S. eriopus* fix nitrogen. Nitrogen from 12 *Encephalartos* spp cointained 0.189–5,490 atom % of 15N. *Encephalartos* spp, *Cycas thouarsi*, and *S. eriopus* fix significant amount of Nitrogen of up to 0.015% atom % of 15N.	South Africa	Grobbelaar et al. (1989); Grobbelaar et al. (1987)
Macrozamia riedlei (Gaudich.) C. A. Gardner *Zamiaceae*	Plant roots fix nitrogen.	Agricultural	*Macrozamia riedlei* fix 19 kg N/ha yr. *M. riedlei* fix 8 Kg N/ha yr.		Grove et al. (1980)

(Continued)

TABLE 18.2 (Continued)

Ecosystem Services Provided by Cycads. The Botanical Names Were Verified Using the Plant List

Species	Uses	Ecosystem Services	Plant Parts Used	Country	Reference
S. eriopus (Kunze) Baill. Zamiaceae	Keeps away bad spirit; and it is believed to make the human body invisible to malignant spirit when ingested. The roots serve plants by fixing nitrogen, and the pots and flowers are grown in private and public places for aesthetic	Spiritual, religious use, agriculture and aesthetic.	Roots, *Encephalartos* spp and *S. eriopus* fix nitrogen. Nitrogen from 12 *Encephalartos* spp contained 0.189 to 5,490 atom % of 15N. *Encephalartos* spp, *Cycas thouarsi* and *S. eriopus* fix significant amount of nitrogen of up to 0.015% atom % of 15N.	Global	Ndawonde et al. (2007); Marler and Moore (2010); Calonje et al. (2011); Grobbelaar et al. (1989); (Grobbelaar et al., (1987); Chen and Henny (2003); Heywood (2001); Bonta et al. (2019)
Zamia obliqua A.Braun Zamiaceae	They are used in processing flour for cooking	Food	Seed/stem	Panama	Blake (2012)
Zamia cunaria Dressler & D.W.Stev. Zamiaceae	Symbolises the authority of the traditional chiefs, planted along graves, and in dance grounds, pigs are tight to them before ceremonial feasts.	Ethnomedicinal, Spiritual, and religious use.	Stem mixed with sundry herbs and leaves.	Panama and Vanuatu	Blake (2012); Keppel (2009)

(Continued)

TABLE 18.2 (Continued)

Ecosystem Services Provided by Cycads. The Botanical Names Were Verified Using the Plant List

Species	Uses	Ecosystem Services	Plant Parts Used	Country	Reference
	Leaves have high cultural and spiritual value: they indicate taboos or restrictions, are a symbol of peace, are involved in conferring status to a recipient in an act of initiation, are held up during guard-of-honour in funeral processions for high-ranking chiefs, people guilty of severe crimes are punished by burning them with hot leaves, and pinnae on leaves are used for keeping records.				
Z. floridana A.DC *Zamiaceae*	Starch is produced from the stem	Goods	Stem		Seth (2003)
Zamia integrifolia L.f. ex Aiton *Zamiaceae*	Plant roots fix nitrogen.	Agricultural	Adult *Zamia integrifolia* fix soil nitrogen, enhance soil carbon, and phosphorus build-up.	USA	Marler and Calonje (2020)
Zamia loddigesi Syn: *Zamia acuminata Oerst. ex Dyer Zamiaceae*	Planted as fallows in maize fields, and stem is used for disinfection of the umbilical cord during childbirth	Agricultural and ethnomedicinal	Whole plant and stem	Mexico and Northern Central America	Bonta et al. (2019); Vovides et al. (2010)

Source: http://www.theplantlist.org/.

Different parts of *Cycas* species have been employed for the treatment of different health conditions. Among the parts used, the stems ($n=12$), male cones ($n=11$), and roots ($n=8$) are the most utilised parts. The application of the various *Cycas* species for medicinal purposes is an indication of huge deposits of phytochemicals present in the cells of this unique group of plants. It will not be an overstatement to recommend a multidisciplinary approach to further explore this plant species to produce affordable pharmaceuticals for all mankind.

18.3.1.1.2 Species that are Mainly Used

Among the six genera, the most represented genus is *Zamia* ($n=11$), followed by *Cycas* ($n=6$), *Encephalartos* ($n=5$), while the other genera *Stangeria, Ceratomzamia, Dioon,* are represented by a single species each. Despite *Zamia* being the most represented genus, *Cycas circinalis* ($n=8$) is the most used species used for treating different ailments in children and adults, ranging from general well-being, reproductive, and excretory problems (Tables 18.2 and Figure 18.1). Among the different health problems treated with *Cycas* species, aphrodisiac ($n=9$), stomach ailments, muscle and joint pain ($n=8$), and wounds ($n=5$) are the most popular health problems treated using the *Cycas* species (Table 18.1). Among the seven countries, India followed by Panama and South Africa are the leading countries in terms of the number of disease-treated *Cycas* species and the number of species used (Table 18.3).

18.3.1.1.3 Countries that Developed Ethnomedicines from Cycas

According to available data, *Cycas* species are used in seven countries for medicinal purposes with India ($n=19$), Panama ($n=11$), and South Africa ($n=7$) being the top three countries where *Cycas* species have been used for various health conditions representing 36.07% of the total ESs, as compared to the rest. It is not a coincidence that the three countries top the list for the exploitation of *Cycas* species for medicinal purposes. For instance, India is reported to have one of the oldest indigenous medicinal systems of the world, while South Africa has approximately 371 years of recorded use of plants for medicine (Gurib-Fakim, 2006; Haynes, 2011). The Indian traditional medicine system, '*Ayurveda*' is one of the oldest indigenous medical systems in the world that recorded the use of medicinal plants as far back as 500 BC (Gurib-Fakim, 2006; Haynes, 2011).

Although, *Cycas* spp are used for health purposes in many countries, very little has been documented for educational purposes. *Cycas* have been planted in most botanic gardens around the globe, where they serve as useful educational tools, especially in conservation studies. Globally, all cycads species and *Dioon sonorense* have been spotted in botanic gardens where they are used for educational purposes. Their usage accounts for approximately 1.59% of the services provided by *Cycas* in the ecosystem and approximately 1.64% when evaluated among countries (Table 18.3). Most of the cycads have been reintroduced in some parks to shield them from extinction and at the same time serving as study materials for naturalist, boosting ecotourism, and thereby generating natural capital (Maunder, 1992).

18.3.1.2 Food Ecosystem Services

According to the results obtained, the food ESs provided by cycad constitutes 15.08% of the total ESs (Table 18.2). Many communities in countries rely on it as a staple food providing starch or carbohydrates (Martínez et al., 2020). This includes countries like Japan, Panama, Zanzibar, Zimbabwe, Mexico, India, and Mozambique (Calonje et al., 2011; Varghese and Ticktin, 2006; Vovides et al. (2010). The only country that uses the whole plant for food is Mexico. Probably the reason Mexico is using the whole plant is the fact that it is one of the countries that started consuming cycads as far back as the 16th century; so they learnt different ways of using the whole plant. However, the only species that is used as the whole plant is *D. sonorense*. It is used in brewing alcohol as one of the ingredients (Vovides et al., 2010). Among the countries that use food ESs of cycad, Mexico and India are leading (Table 18.3).

TABLE 18.3

Classification of Cycads Ecosystem Services by Countries

Ecosystem services	Country	Citations (N)	Citations (%)
Aesthetic	Fiji, Tonga, Vanuatu, India	1	
	Global	2	
	Mexico	2	
	Mexico and Northern Central America	1	
	South Africa	1	
Aesthetic total		7	5.74
Agricultural	Honduras	1	
	Japan	1	
	Mexico	3	
	Mexico	1	
	Mexico and Northern Central America	2	
	South Africa	5	
	USA	2	
Agricultural total		15	12.30
Carbon sequestration	China	4	
	Mexico	1	
	Spain	1	
Carbon sequestration total		6	4.92
Educational	Global	1	
	Honduras	1	
Educational total		2	1.64
Ethnomedicinal	Australia	1	
	Bangladesh	2	
	Guam	1	
	India	19	
	Mexico	3	
	Panama	11	
	South Africa	7	
Ethnomedicine total		44	36.07
Food	Australia	3	
	Honduras	1	
	India	5	
	Japan	1	
	Mexico	5	
	Mozambique/Zimbabwe	1	
	Panama	1	
	South Africa	1	
	Zanzibar	1	
Food total		19	15.57
Goods	Honduras	1	
	India	1	
	Mexico	3	
	South Africa	1	

(Continued)

TABLE 18.3 (*Continued*)
Classification of Cycads Ecosystem Services by Countries

Ecosystem services	Country	Citations (N)	Citations (%)
Goods total		6	4.92
Insect repellent	India	2	
	Mexico	1	
Insect repellent total		3	2.46
Spiritual and religious use	Australia	1	
	China	3	
	Honduras	1	
	India	2	
	Mexico	2	
	Mexico and North and Central America	2	
	Philippines	1	
	South Africa	4	
	The Americas, SE Asia,	1	
	Vanuatu	3	
Spiritual and religious use total		20	16.39
Grand total		122	100.00

Other people use just plant parts of cycads, not the whole plant. It is reported that some communities use leaves as vegetables and grind seeds to make flour; others are using stems to make flour (Table 18.2). Most people who use leaves as cooked vegetables are in India (Krishnamurthy et al., 2013; Radha and Singh, 2008; Varghese and Ticktin, 2006). The other countries that use seeds use them to make flour for porridge and bread. Some communities detoxify seeds before making the flour by drying or smoking them; others repeatedly rinse them before grinding (Bonta et al., 2019; Martínez et al., 2020). The main reason for detoxification is that consumption of cycads as food is associated with neurological diseases in humans (Giménez-Roldán et al., 2021; Spencer, 2020). Other studies reported that consumption of flour from cycads seed increases liver toxicity due to toxins such as cycasins (Giménez-Roldán et al., 2021). The association of cyanobacteria and cycad also increases brain diseases due to neurotoxic effect caused by the combination of the cycad toxins and compounds produced by cyanobacteria (Giménez-Roldán et al., 2021)

Among the cycads genera that provide food ESs, *Cycas* (42%) dominates (Table 18.3). It dominates because all its plant parts (seeds, stem, and leaves) can be used in various foods. Stem and seeds are used to make flour for different kinds of bread and porridge (Hayward and Kuwahara, 2013). Moreover, seeds may be consumed raw or roasted as a source of starch (Kaur, 2006). Some people slice leaves dry and fry them before eating, while others just cook them and eat as vegetables (Bonta et al., 2019). *Zamia* and *Macrozamia* genera are the least consumed (Table 18.2). Zamia is only used in Panama to make flour out of the seeds and stem, while *Macrozamia* is used only in Australia to make flour out of the seeds. The dominating species of *Cycas* in providing food services is *Cycas beddomei*. It is reported more than others in India.

18.3.1.3 Ecosystem Services Provided by Cycads: Goods

Cycads provide many ecosystem goods; surface fibres, obtainable from the evergreen, strong, and leathery leaves of different species of cycads (*Cycas revoluta, C. circinalis, Encephalartos trans-venosus, Macrozamia spiralis*), are reportedly used for roof thatching (Lal, 2003; Varghese and

Ticktin, 2006), preparation of twines, ropes, cloths, mats, baskets, paper, brooms, yarns, stuffing pillows, producing mattresses, and many other purposes (Lal, 2003; Zarchini et al., 2011). Seeds of some species like *Cycas seemannii, Dioon spinolosum,* and *Dioon mejiae* are used in making cloths, pins, toys, bracelets, ornaments, and whistles (Bonta et al., 2019; Keppel, 2009; Vovides et al., 2010). Oil can also be extracted from the seeds of cycads, especially *C. revoluta*, and processed into cooking oils as well as biodiesel (Pasaribu et al., 2020). Starch is similarly obtainable from the processed stem of *C. revolute*, which is utilised as food during the starvation period (Lal, 2003; Zarchini et al., 2011). The stem is also used in making clothing and ropes (Lal, 2003). A total of 6.35% of studies were conducted on goods ESs, as provided by cycad plant species. These were focused majorly on species, including *D. mejiae, D. spinolosum, C. circinalis, C. seemannii, E. transvenosus, and Zamia floridana* (Table 18.2). Most of the studies investigated on ESs provided by cycads in form of goods services (4.92%) were conducted in Mexico, Honduras, India, and South Africa (Tables 18.2 and 18.3).

18.3.2 Cultural Services

18.3.2.1 Aesthetic Services of Cycad

In the current study, all cycad species have aesthetic values (Table 18.2). Furthermore, *E. transvenosus, Ceratozamia fuscoviridis, Dioon* spp, *and Zamia* were used for multiple aesthetic services, such as growing in homes and public gardens for landscaping. In Fiji, Tonga, Vanuatu, and India, the leaves of *C. circinalis, Cycas pectinata, C. revoluta,* and *C. seemannii* are harvested for floriculture. The whole plant of *Dioon* spp is being used in Mexico and Northern Central America for landscaping (Bonta et al., 2019), and *Dioon* spp *and Zamia* spp are being used in home and public gardens for landscaping in Mexico (Vovides et al., 2010). In South Africa, Ravele and Makhado (2010) reported that the leaves of *E. transvenosus* are harvested for decoration.

18.3.2.2 Ecosystem Services Provided by Cycads: Spiritual and Religious Uses

The spiritual and religious ESs provided by cycads transcends time and historical diversity with age-long spiritual and traditional norms (Bonta et al., 2019). This is one of the reasons why the plant is well known and appreciated with significant conservation value all over the world. Bonta et al. (2019) recorded 16 cycad species dedicated for religious uses, including ceremonies such as death, birth, and concept of afterlife and transcendence during death. The spines of the leaves are worn as crowns during Holy Week processions, Good Friday, or Palm Sundays to connote painful crowns of thorns. The leaves used during these religious periods are kept in the church and burnt after a year for ash-on-ash Wednesdays in the Catholic church (Bonta et al., 2019). In addition, the leaves of this multipurpose plant species are also made into wreaths and placed on graves in commemoration of the dead. Cycad leaves are also consumed as sacraments in churches (Bonta et al., 2019) due to the durable and malleable nature of the leaves. Catholics use the leaves for decoration during wedding and significant feasting days (Bonta et al., 2019). According to Ala and Osborne (2009), this unique plant species is usually planted around graves and dance grounds, as well as it symbolises authority of traditional chiefs, while the Hindus used the leaves in shrines for deities (Bonta, 2010). Particularly, *Encephalartos natalensis* species is believed to drive away evil spirits when planted in homesteads (Zukulu et al., 2012), while *Stangeria eriopus* species keep bad spirits away, and it is believed to make the human body invisible when ingested orally (Ndawonde et al., 2007). A total of 15.87% of research papers reported spiritual and religious ESs provided by cycads, while more studies were focused on *C. revoluta* and *C. seemannii* (Table 18.2).

18.4 CONCLUSIONS AND PROSPECTS

The number of cycads and their benefits to humans reported in this review provides evidence that cycads harbour a high diversity of plant species that will continue to play an important role in ESs. Medicinal services are the most cited, followed by food and goods. Cultural services are the second most cited ESs provided by cycads. The high citation of medicinal uses is not surprising, because cycads occur mostly in least developed countries, where local indigenous people rely on plants to treat their illnesses. It is, therefore, imperative that local people should be included and involved in the development of cycad management strategies.

FUNDING

S.T.N received funding from the National Research Foundation in Pretoria, South Africa (Grant UID: 129403).

CONFLICTS OF INTEREST

The authors declare that there is no conflict of interest. The opinions, conclusions/recommendations herein this study is based on the findings of the authors; therefore, the funders (NRF) accept no liability whatsoever in this regard.

REFERENCES

Ala, P. and Osborne, R. (2009). Some observations on *Cycas seemannii* in Vanuatu. The Cycad Newsletter.

Álvarez-Yépiz, J.C. and Dovčiak, M. (2015). Enhancing ecosystem function through conservation: threatened plants increase local carbon storage in tropical dry forests. *Tropical Conservation Science, 8*(4), 999–1008.

Álvarez-Yépiz, J.C., Búrquez, A., and Dovčiak, M. (2014). Ontogenetic shifts in plant–plant interactions in a rare cycad within angiosperm communities. *Oecologia, 175*(2), 725–735.

Bamigboye, S. and Tshisikhawe, M.P. (2019). The impacts of bark harvesting on a population of *Encephalartos transvenosus* (Limpopo cycad), in Limpopo Province, South Africa. *Biodiversitas Journal of Biological Diversity, 21*(1), 117–121.

Blake, A. (2012). Notes on the ethnobotany of panamanian cycads. In: *Proceedings of Cycad 2008. The 8th International Conference on Cycad Biology, Panama City, Panama, 13–15 January 2008.* New York Botanical Garden Press, Panama City, Panama, 165–177.

Bonta, M. (2010). Human Geography and Ethnobotany of Cycads in Xi'ui, Teenek, and Nahuatl Communities of Northeastern Mexico, Final Report. Cleveland.

Bonta, M. (2012). Cycads and human life cycles: outline of a Symbology In: *Proceedings of Cycad 2008. The 8th International Conference on Cycad Biology, Panama City, Panama, 13–15 January 2008.* New York Botanical Garden Press, 133–150.

Bonta, M., Pulido-Silva, M.T., Diego-Vargas, T., Vite-Reyes, A., Vovides, A.P., and Cibrián-Jaramillo, A. (2019). Ethnobotany of Mexican and northern central American cycads (Zamiaceae). *Journal of Ethnobiology and Ethnomedicine, 15*(4), 1–34.

Brenner, E.D., Stevenson, D.W., McCombie, R.W., Katari, M.S., Rudd, S.A., Mayer, K.F.X., Palenchar, P.M., Runko, S.J., Twigg, R.W., and Dai, G. (2003). Expressed sequence tag analysis in *Cycas*, the most primitive living seed plant. *Genome Biology, 4,* 1–11.

Calonje, M., Kay, J., and Griffith, M.P. (2011). Propagation of cycad collections from seed: applied reproductive biology for conservation. *The Journal of Botanic Garden Horticulture, 9,* 77–96.

Chamberlain, C.J. (1909). Spermatogenesis in *Dioon edule. Botanical Gazette, 47*(3), 215–236.

Chemnick, J. (2013). The Dioons of Oaxaca. *Cactus and Succulent Journal, 85*(1), 19–27.

Chen, J. and Henny, R.J. (2003). ZZ: a unique tropical ornamental foliage plant. *HortTechnology, 13*(3), 458–462.

Cheruku, A., Nimmanapalli, Y., and Dandu, C. (2012). Physicochemical analysis and antimicrobial activity of *Cycas beddomei* Dyer. *Vegetative parts. Journal of Pharmacy Research, 5,* 2553–2558.

Cowling, R.M., Pressey, R.L., Lombard, A.T., Desmet, P.G., and Ellis, A.G. (1999). From representation to persistence: requirements for a sustainable system of conservation areas in the species-rich mediterranean-climate desert of Southern Africa. *Diversity and Distributions, 5,* 51–71.

Daily, G.C. (1997). *Nature's Services: Societal Dependence on Natural Ecosystems*. In: Gretche C. Daily (ed.), Island Press, Washington, DC.

Donaldson, J.S. (2003). Cycads: status survey and conservation action plan. IUCN--the World Conservation Union.

Giménez-Roldán, S., Steele, J.C., Palmer, V.S., and Spencer, P.S. (2021). Lytico-bodig in Guam: historical links between diet and illness during and after Spanish colonization. *Journal of the History of the Neurosciences, 30*(4), 335–374.

Grobbelaar, M.C., Scott, W.E., Hatting, W., and Marshall, J. (1987). The indentification of the coraloid root endophytes of the Southern Africa cycads and the ability of the endophytes to fix dinitrogen. *South Africa Journal of Botany, 53*(2), 111–118.

Grobbelaar, N., Meyer, J.J.M., and Burchmore, J. (1989). Coning and sex ratio of *Encephalantos transvenosus* at the Modjadji Nature Reserve. *South African Journal of Botany, 55*(1), 79–82.

Grove, T., O'connell, A., and Malajczuk, N. (1980). Effects of fire on the growth, nutrient content and rate of nitrogen fixation of the *cycad Macrozamia riedlei*. *Australian Journal of Botany, 28*(3), 271–281.

Gurib-Fakim, A. (2006). Medicinal plants: traditions of yesterday and drugs of tomorrow. *Molecular Aspects of Medicine, 27*(1), 1–93.

Haynes, J.L. (2011). World list of cycads: a historical review. IUCN/SSC Cycad Specialist Group, pp. 8–35.

Hayward, P. and Kuwahara, S. (2013). *Cycads*, sustenance and cultural landscapes in the Amami islands. *The Islands of Kagoshima*, Chapter 5. Kohima University Research Center for the Pacific Islands, pp. 29–37.

Heywood, V. (2001). Conservation and sustainable use of wild species as sources of new ornamentals. *International Symposium on Sustainable Use of Plant Biodiversity to Promote New Opportunities for Horticultural Production, 598,* 43–53.

Kaur, K. (2006). The role of ecosystem services from tropical savannas in the well-being of aboriginal people: a scoping study. *A Report for the Tropical Savannas Cooperative Research Centre*, Darwin, NT. pp. 1–78.

Kearney, A. and Bradley, J.J. (2009). Too strong to ever not be there: place names and emotional geographies. *Social & Cultural Geography, 10*(1), 77–94.

Keppel, G. (2009). Morphological variation, an expanded description and ethnobotanical evaluation of *Cycas seemannii* A. Braun (Cycadaceae). *The South Pacific Journal of Natural and Applied Sciences, 27*(1), 20–27.

Krishnamurthy, V., Mandle, L., Ticktin, T., Ganesan, R., Saneesh, C.S., and Varghese, A. (2013). Conservation status and effects of harvest on an endemic multi-purpose cycad, *Cycas circinalis* L, western Ghats, India. *Tropical Ecology, 54*(3), 309–320.

Lal, J.J. (2003). Sago palm, In: Caballero, B. (ed.,) *Encyclopedia of Food Sciences and Nutrition*. Elsevier, The Netherlands.

Marler, T.E. and Calonje, M. (2020). Stem branching of cycad plants informs horticulture and conservation decisions. *Horticulturae, 6*(4), 65.

Marler, T.E. and Moore, A. (2010). Cryptic scale infestations on *Cycas revoluta* facilitate scale invasions. *HortScience, 45*(5), 837–839.

Marques, A., Pereira, H.M., Krug, C., Leadley, P.W., Visconti, P., Januchowski-Hartley, S.R., Krug, R.M., Alkemade, R., Bellard, C., and Cheung, W.W.L. (2014). A framework to identify enabling and urgent actions for the 2020 Aichi Targets. *Basic and Applied Ecology, 15*(8), 633–638.

Martínez, E.T., Martínez, J.F., and Bonta, M. (2020). Toxic harvest: *Chamal* Cycad (*Dioon edule*) Food culture in Xi'Iuy indigenous communities of San Luis Potosi, Mexico. *Journal of Ethnobiology, 40*(4), 519–534.

Maunder, M. (1992). Plant reintroduction: an overview. *Biodiversity & Conservation, 1*, 51–61.

Millennium Ecosystem Assessment (2005). *Ecosystems and Human Well-Being: Biodiversity Synthesis*. World Resources Institute, Washington, DC.

Mollik, A.H., Hassan, A.I., Paul, T.K., Sintaha, M., Khaleque, H.N., Noor, F.A., Nahar, A., Seraj, S., Jahan, R., and Chowdhury, M.H. (2010). A survey of medicinal plant usage by folk medicinal practitioners in two villages by the rupsha river in bagerhat district, Bangladesh. *American-Eurasian Journal of Sustainable Agriculture, 4*(3), 349–357.

Ndawonde, B.G., Zobolo, A.M., Dlamini, E.T., and Siebert, S.J. (2007). A survey of plants sold by traders at Zululand muthi markets, with a view to selecting popular plant species for propagation in communal gardens. *African Journal of Range and Forage Science, 24*(2), 103–107.

Nicholson, E., Mace, G.M., Armsworth, P.R., Atkinson, G., Buckle, S., Clements, T., Ewers, R.M., Fa, J.E., Gardner, and T.A., Gibbons, J. (2009). Priority research areas for ecosystem services in a changing world. *Journal of Applied Ecology, 46*(6), 1139–1144.

Norstog, K.J. and Nicholls, T.J. (1997). *The Biology of the Cycads.* Cornell University Press, Ithaca, New York.

Pasaribu, B., Fu, J.H., and Jiang, P.L. (2020). Identification and characterization of caleosin in *Cycas* revoluta pollen. *Plant Signaling & Behavior, 15*(8), 1779486.

Radha, P. and Singh, R. (2008). Ethnobotany and conservation status of Indian *Cycas* species. *Encephalartos, 93,* 15–21.

Radji, R. and Kokou, K. (2015). Distribution of the horticultural plants in Togo according to decorative parts and medicinal value. *International Journal of Current Research and Academic Review, 3*(4), 249–265.

Raimondo, D.C. and Donaldson, J.S. (2003). Responses of cycads with different life histories to the impact of plant collecting: simulation models to determine important life history stages and population recovery times. *Biological Conservation, 111*(3), 345–358.

Ravele, A.M. and Makhado, R.A. (2010). Exploitation of *Encephalartos transvenosus* outside and inside *Mphaphuli cycads* nature reserve, limpopo province, South Africa. *African Journal of Ecology, 48*(1), 105–110.

Rout, J., Sajem, A.L., and Nath, M. (2012). Medicinal plants of north cachar hills district of Assam used by the dimasa tribe. *Indian Journal of Traditional Knowledge, 11*(3), 520–527.

Seth, M., (2003). Trees and their economic importance. *The Botanical Review, 69*(4), 321–376.

Singh, R. and Singh, K.J. (2011). The importance of Odisha Cycas in India. *Biodiversity, 12*(1), 21–27.

Spencer, P.S. (2020). Consumption of unregulated food items (false morels) and risk for neurodegenerative disease (amyotrophic lateral sclerosis). *Health Risk Analysis, 2020*(3), 94–99.

Suinyuy, T.N. and Johnson, S.D. (2021). Evidence for pollination ecotypes in the African cycad *Encephalartos ghellinckii* (Zamiaceae). *Botanical Journal of the Linnean Society, 195*(2), 233–248.

Thieret, J.W. (1958). Economic botany of the cycads. *Economic Botany*, Springer, *12,* 3–41.

Varghese, A. and Ticktin, T. (2006). *Harvest, Trade, and Conservation of the Endemic Multiuse Cycad, Cycas circinalis L., in the Nilgiri Biosphere Reserve, South India People and Plants*, International and Keystone Foundation.

Varghese, A., Krishnamurthy, V., and Ticktin, T. (2012). Harvest, use, and ecology of *Cycas circinalis* L.–a case study in the Nilgiri biosphere reserve area, Western Ghats, India, *Proceedings of Cycad 2008. The 8th International Conference on Cycad Biology, Panama City, Panama, 13–15 January 2008.* New York Botanical Garden Press, pp. 178–191.

Vovides, A.P. (2020). The mesoamerican use of Cycads as food, medicine and ritual. *EC Clinical and Experimental Anatomy, 3*(1), 01–02.

Vovides, A.P., Pérez-Farrera, M.A., and Iglesias, C. (2010). Cycad propagation by rural nurseries in Mexico as an alternative conservation strategy: 20 years on. *Kew Bulletin, 65,* 603–611.

Wang, W., Ma, L., Becher, H., Garcia, S., Kovarikova, A., Leitch, I.J., Leitch, A.R., and Kovarik, A. (2016). Astonishing 35S rDNA diversity in the gymnosperm species *Cycas revoluta* Thunb. *Chromosoma, 125*(4), 683–699.

Whitelock, L.M. (2002). *The Cycads.* Timber Press Inc, Portland, Oregon, p. 374.

Yuandan, M., Jiang, H., Wang, B., Zhou, G., Yu, S., Peng, S., Hao, Y., Wei, X., Liu, J., and Yu, Z. (2009). Carbon storage of cycad and other gymnosperm ecosystems in China: implications to evolutionary trends. *Polish Journal of Ecology, 57*(4), 635–646.

Zarchini, M., Hashemabadi, D., Kaviani, B., Fallahabadi, P.R., and Negahdar, N. (2011). Improved germination conditions in *Cycas revoluta* L *by using sulfuric acid and hot water, Plant Omics Journal, 4*(7), 350–353.

Zukulu, S., Dold, T., Abbott, T., and Raimondo, D. (2012). *Medicinal and Charm Plants of Pondoland.* South African National Biodiversity Institute (SANBI), South Africa.

19 Discovery of Antibacterial Lead Compounds from Three South African Marine Algae

Omolola Afolayan

CONTENTS

19.1 INTRODUCTION

Marine organisms such as algae have been documented in the literature as producers of secondary metabolites with a broad spectrum of therapeutic activities (Carroll et al., 2020). Species of the genus of brown algae, *Dictyota*, are known to produce bioactive natural products including terpenes. At least 233 diterpenes have been reported (Chen et al., 2018). The biological activity of some of the metabolites produced exhibit biological activities that include antimicrobial, anti-inflammatory, anti-plasmodial, anti-viral, and cytotoxicity (Ioannou et al., 2013). Likewise, the

DOI: 10.1201/9781003206620-19

members of red algae of the *Laurencia* genus have been widely studied with over 700 isolated natural products (Wang et al., 2013). The major metabolites produced comprise halogens, typically bromine and some chlorine atoms chamigrane-type sesquiterpenes (Suzuki et al., 2009). However, there are few reports of diterpenes, triterpenes, and C-15 acetogenins from this genus. This class of natural products serves as promising drug candidate or as lead in discovery, because they are structurally unique and diverse, and exhibit significant pharmacological activities. Considering that the species of these genera are known to produce bioactive compounds, especially antibacterial agents, the present study aimed to explore the chemical constituents of three marine algae namely, *Dictyota naevosa*, *Laurencia pumila*, and *Laurancia sodwaniensis* for antibacterial lead compounds.

19.2 MATERIALS AND METHODS

19.2.1 GENERAL EXPERIMENTAL PROCEDURES

Column chromatography was carried out using silica gel 60 (0.040–0.063 mm) purchased from Merck (Darmstadt, Germany). High-performance liquid chromatography (HPLC) separations were performed using Agilent technologies equipped with a UV 100 detector at 250 nm and a Whatman 10 μm semi-preparative column (50 cm). All NMR spectra were measured with an AvanceCore Bruker spectrometer operating at 400 MHz for1H-NMR at 100 MHz for 13C NMR in CDCl$_3$. All the solvents that were used for extraction were redistilled before use. Deuterated chloroform, dichloromethane, ethyl acetate, hexane, methanol, and silica gel 60 were purchased from Merck (Darmstadt, Germany). Silica gel 60 (0.040–0.063 mm) was obtained from Merck KGaA (Germany). Diaion® HP20SS was purchased from Supelco (USA). Nuclear magnetic resonance spectroscopy (NMR) data were recorded on a AvanceCore Bruker spectrometer at 400 MHz in deuterated chloroform. Chemical shifts were measured in parts per million (ppm) and referenced to undeuterated solvent signals at δ_H 7.26.

19.2.2 BIOLOGICAL MATERIAL

A sample of *D. naevosa* (collection code: SB20140302-3), *L. pumila* (collection code: KZN13107-1), and *Laurencia sodwanienisis* (collection code: KZN1311-1) were collected by hand at the coast of KwaZulu-Natal in March 2014, October 2013, and November 2013, respectively. Botanical identification was done by Professor John J. Bolton at the University of Cape Town, South Africa. Voucher specimens were deposited at the Marine Biodiscovery collection, University of the Western Cape. Samples were stored at –20°C until further use. *D. naevosa* is a brown alga with flat, broad lamina, and dichotomy axes (Anderson, et al., 2016). *L. pumila* is green with several branches emerging from a rhizoidal holdfast, sparsely branched thallus, and club-shaped apices (Francis, 2014). *L. sodwaniensis* is a yellowish-brown alga connected by discoid holdfast with sparsely branched fronds (Francis, 2014; De Clerck and Tronchin, 2005).

19.2.3 EXTRACTION AND ISOLATION

Frozen specimens of *D. naevosa* (3.3 g), *L. pumila* (2.3 g), and *L. sodwaniensis* (23.2 g) were submerged in CH$_3$OH (75 mL) at 4°C overnight. The solvents were decanted, and the algae was re-extracted three times with CH$_3$OH-CH$_2$Cl$_2$ (1:2, 75°mL) at 36°C. The two organic fractions were combined, and the solvents were dried under reduced pressure to yield 529.6, 105.7, and 180.7 mg of organic extracts, respectively.

   D. naevosa crude extract (388.7 mg) was separated by silica gel column chromatography and eluted in *n*-hexane (100%), *n*-hexane–EtOAc (90:10; 80:20; 60:40; 40:60; 20:80) EtOAc (100%), EtOAc–MeOH (50:50), and MeOH (100%) to obtain nine fractions (Fr1–Fr9). Fraction 2 (40.5 mg) eluted with *n*-hexane–EtOAc (90:10) was further purified with a semi-preparative normal phase

HPLC (Agilent Technologies phenomenox silica gel 10 μm column 50 mm) and eluted with n-hexane–EtOAc (90:10) at a rate of 3 mL/min resulting in the isolation of **1** (4.90 mg).

L. pumila crude extract (105.7 mg) was chromatographed by silica gel column and eluted in n-hexane (100%), n-hexane–EtOAc (90:10; 80:20; 60:40; 40:60; 20:80), EtOAc (100%), EtOAc–MeOH (50:50), and MeOH (100%) to obtain nine fractions (Fr1–Fr9). Fraction 3 (7.0 mg) eluted with n-hexane–EtOAc (80:20) was further purified with a semi-preparative normal phase HPLC (Agilent technologies phenomenox silica gel 10 μm column 50 mm) and eluted with n-hexane–EtOAc (80:20) at a rate of 3 mL/min resulting in the isolation of **2** (1.10 mg).

L. sodwaniensis crude extract (180.7 mg) was purified by silica gel column chromatography with n-hexane (100%), a mixture of n-hexane–EtOAc (90:10; 80:20; 60:40; 40:60; 20:80), EtOAc (100%), EtOAc–MeOH (50:50), and MeOH (100%). As a result, nine fractions were obtained. Compound **3** (86.9 mg) was identified in fraction 3 [(n-hexane–EtOAc (80:20)].

19.2.3.1   Pachydictyol A

Pachydictyol A (**1**): colourless oil, 1H-NMR (CDCl$_3$, 400 MHz), and 13C NMR (CDCl$_3$ 100 MHz) (See Tables 19.1 and 19.2.).

19.2.3.2   C-15 Acetogenin

C-15 acetogenin (**2**): colourless oil, 1H-NMR (CDCl$_3$, 400 MHz), and 13C NMR (CDCl$_3$ 100 MHz) (See Tables 19.1 and 19.2.).

19.2.3.3   Cartilagineol

Cartilagineol (**3**): white solid,1H-NMR (CDCl$_3$, 400 MHz), and 13C NMR (CDCl$_3$ 100 MHz) (See Tables 19.1 and 19.2.).

19.2.4   Culturing of MRSA 33591 Test Strain

Culturing of MRSA 33591 test strain Luria-Bertani (LB) broth consisting of 0.5% tryptone, 0.5% sodium chloride, and 0.25% yeast extract in 500 mL distilled water at pH 7.0 was prepared and auto-claved at 121°C for 30 min. MRSA ATCC 33591 was inoculated into 5 mL LB broth and incubated at 37°C for 24 hours, shaking at 160 rpm. The 5 mL cell culture was diluted into 250 mL LB broth and incubated as per the aforementioned growth parameters. The optical density (OD) of the culture for use in TLC-bioautography was measured at 600 nm, and the culture OD was adjusted to 0.5 for TLC-bioautography.

19.2.5   Thin-Layer Bioautography Assay

A 20 μg of compounds (**1–3**) was spotted on a 1×1 cm^2 partitioned block on a 20×10 cm TLC silica gel plate in duplicate. An antibiotic standard (10 μL) vancomycin with a concentration of 10 mg/mL was spotted on each silica TLC gel plate as a positive control, and 20 μg of dichloromethane (DCM) was used as a negative control. A sterile cotton wool swab was used to apply MRSA 33591 onto the silica gel plates. The plates were incubated at 37°C in an enclosed plastic container containing a moist paper towel for 24 hours. Thereafter, the plates were treated with 20 μL of 0.25% (w/v) MTT 3-(4,5-dimethylthiazol-2-yl)-2,5-diphenyltetrazolium bromide (tetrazolium dye) and re-incubated for 3 hours, then evaluated for the appearance of purple spots (indicative of organism growth) or white spots (indicative of organism growth inhibition) (Suleimana et al., 2009).

19.2.6   Microdilution Assay

A 500 μg/mL stock solution of each algae extract and compound was prepared by dissolving in dimethyl sulfoxide (DMSO). Serial dilution to concentrations of 0.39, 0.78, 1.56, 3.12, 6.25, 12.5, 25, and 50 μg/mL was established for each sample. A 20 μL of the serial dilution was added to 96-well

TABLE 19.1
1H NMR Data in CDCl$_3$ (400 and100 MHz) of Compounds 1–3

Position	δH (Multiplicity, J, Hz) Literature[a]	1	δH (Multiplicity, J, Hz) Literature[b]	2	δH (Multiplicity, J, Hz) Literature[c]	3
1	2.6 (1H, m)	2.62 (m)	3.16 (brd, 1.5)	3.16 (m)	2.05 (dt, 13.6, 4.2) 1.70 (dd, 6.0, 3.2)	2.06 (td, 13.6, 3.8) 1.74 (m)
2a	2.2 (2H, m)	2.20 (2H, m)	—	—	2.38 (dd, 13.6, 2.8) 1.82 (dd, 6.0, 3.2)	2.39 (m) 1.82 (m)
B	2.4 (2H, m)	2.48 2H, m)				
3	5.2 (1H, brs)	5.33 (brs)	5.55 (brd, 9.5)	5.56 (m)	—	—
4	—	—	6.03 (dtd, 10.8, 7.6, 0.9)	6.33 (dt, J = 6.8, 15.6)	4.45 (m)	4.42 (ddd, 11.7, 4.8, 2.5)
5	2.2 (1H, m)	2.20 (m)	2.77 (m) 2.95 (m)	2.83 (m) 2.87 (m)	3.08 (d, 16.0) 2.78 (dd, 16.0, 4.0)	3.08 (d, 16.4) 2.78 (m)
6	3.8 (1H, d, J=7)	3.91 (d, J=6.9)	3.59 (brdddd, 8.8, 5.6, 3.3)	3.59 (m)	—	—
7	1.5 (1H, m)	1.55 (m)	4.07 (ddd, 10.6, 5.4, 3.3)	4.06 (m)	—	—
8	1.8 (2H, m)	1.80 (m)	2.46 (brddd, 13.0, 5.4, 5.4) 3.18–3.49 (m)	2.48 (m) 3.22 (m)	2.70 (t, 11.9) 2.45 (m)	2.70 (t) 2.45 (m)
9	2.25 (2H, m)	2.23 (m)	5.52–5.69 (m)	5.62 (m)	3.68 (dt, 11.9, 4)	3.67 (dt, 11.9, 4.1)
10	—	—	5.52–5.69 (m)	5.61 (m)	4.41 (q, 3.7, 1.6)	4.42 (m)
11	1.4 (1H, m)	1.52 (m)	2.58 (brddd, 13.0, 5.4, 5.4) 3.18–3.49 (m)	1.75 (m) 1.53 (m)	—	—
12a	1.1 (2H, m)	1.19 (m)	4.27 (ddd, 10.5, 5.4, 3.1)	4.25 (m)	1.31 (s)	1.32 (s)
b	1.4 (2H, m)	1.52 (m)				
13a	1.6 (2H, m)	1.60 (m)	3.23 (ddd, 8.4, 5.9, 3.1)	2.68 (m)	1.04 (s)	1.06 (s)
b	1.8 (2H, m)	1.80 (m)				
14	5.02 (1H, br. t, J=7.5)	5.10 (brt, J=7.1)	1.82–1.98 (m)	1.88 (m)	5.20 (sl) 4.97 (sl)	5.19 (sl) 4.95 (sl)
15	—	—	0.86 (t, 7.5)	0.86 (m)	1.75 (sl)	1.74 (sl)
OH	—	—	—	—	2.27 (d, 11.3)	2.25 (m)
16	1.61 (3H, s)	1.60 (s)	—	—	—	—
17	1.75 (3H, brs)	1.76 (brs)	—	—	—	—
18	4.6 (2H, br)	4.74 (br)	—	—	—	—
19	0.94 (3H, d, J=6.0)	0.99 (d, J=5.8)	—	—	—	—
20	1.54 (3H, s)	1.60 (s)	—	—	—	—

[a] Gedara et al. (2003) (1H NMR measured in CDCl3 at 400 or 500 MHz).
[b] Awakura et al. (1999).
[c] Da Silva Machado et al. (2011).

TABLE 19.2

13C NMR Data in CDCl$_3$ (400 and 100 MHz) of Compounds 1–3

Position	δ_C (Literature)[d]	1	δ_C (Literature)[b]	2	δ_C (Literature)[c]	3
1	46.9	46.8	82.5	82.3	24.3	24.3
2a	34.7	34.7	—	81.6	32.5	32.4
2b						
3	124.0	124.0	111.3	111.9	73.3	73.3
4	141.4	141.4	140.3	141.6	57.2	57.1
5	60.4	60.4	34.6	34.3	33.8	33.8
6	75.0	75.0	81.4	81.6	44.0	43.9
7	47.8	47.7	62.2	63.2	147.2	147.1
8	23.5	23.5	33.5	33.3	39.4	39.3
9	40.7	40.6	128.5	129.3	69.6	69.6
10	152.6	152.5	130.0	130.0	76.4	76.3
11	35.0	35.0	34.3	29.8	43.7	43.7
12a	35.4	35.4	54.7	63.5	24.8	24.8
12b						
13a	25.8	25.6	84.0	63.2	25.3	25.3
13b						
14	124.6	124.7	27.9	29.1	114.7	114.7
15	131.6	131.5	9.7	10.4	33.1	33.1
OH	25.6	25.6	—	—	—	—
16	15.9	15.9	—	—	—	—
17	107.1	107.1	—	—	—	—
18	17.5	17.5	—	—	—	—
19	17.7	17.7	—	—	—	—
20	46.9	46.8	—	—	—	—

[b] Awakura et al. (1999).
[c] Da Silva Machado et al. (2011).
[d] Ayyad et al. (2011) (13C measured in CDCl3 at 150 MHz).

plate containing 180 µL of bacterial inoculum per well for a final volume of 200 µL per well, with each one tested in triplicates. Positive control included 180 µL of culture plus 20 µL of vancomycin (10 mg/mL). Negative control was 180 µL of culture and 20 µL of DMSO, and the sterile control was made up of 180 µL of broth and 20 µL of DMSO. Following dispensing of samples and controls into the plate as per setup, the plate was covered with a Breathe-Easy® sealing membrane and was incubated for 24 hours at 37°C. Following incubation, 20 µL of 0.25% (w/v) MTT was added to each well and further incubated at 37°C for 3 hours. The plate was placed in a microplate reader, and the optical density was read at OD$_{570}$. The readings were used to calculate the percentage growth of each sample using the following formulae outlined:

$$\% \text{ growth} = \frac{\text{Final value (mean of triplicate dilution concentration)} \times 100}{\text{MRSA (mean of triplicate negative control)}}$$

$$\% \text{ growth positive control} = \frac{(\text{mean of triplicate vancomycin dilution concentration}) \times 100}{\text{MRSA (mean of triplicate negative control)}}$$

$$\% \text{ growth negative control} = \frac{(\text{mean of triplicate DMSO dilution concentration}) \times 100}{\text{MRSA (mean of triplicate negative control)}}$$

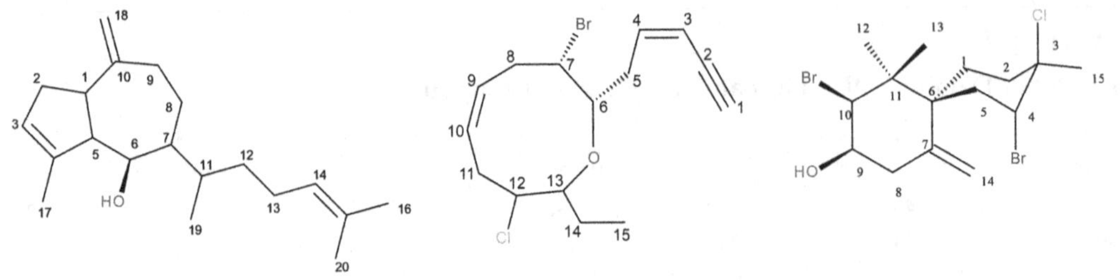

(1)   (2)   (3)

FIGURE 19.1   Chemical structures of compounds (**1–3**) isolated from *Dictyota naevosa, Laurencia pumila,* and *Laurencia sodwaniensis*, respectively.

19.3   RESULTS AND DISCUSSION

19.3.1   ISOLATION OF **1**

The brown alga *D. naevosa* was collected by hand at the coast of KwaZulu-Natal, South Africa. A series of column chromatographic separations of the organic crude extract and further purification through semi-preparative HPLC afforded compound **1** (Figure 19.1) were performed. Following detailed analysis of 1D and 2D NMR spectroscopy and comparison with literature data, the structure of **1** was elucidated.

19.3.2   STRUCTURE ELUCIDATION OF **1**

Compound **1** showed a protonated molecular ion peak at m/z 289.2535 corresponding to a molecular formula of $C_{20}H_{32}O$ (calcd for 289.2533). Following the evaluation of the 1D and 2D NMR spectral data for **1** in $CDCl_3$ and comparison with literature data, the spectroscopic values for all the signals were assigned. The 1H NMR spectrum of **1** showed the presence of two methine signals at δ 5.33 (H-3) and 5.01 (H-14), four methyl signals at δ 1.76 (H-17), 1.60 (H-16), 1.55 (H-20), 0.94 (H-19), and six methylene signals at δ 4.70 (H-18), 2.23 (H-9), 2.20/2.48 (H-2), 1.60/1.80 (H-13), 1.80 (H-8), and 1.19/1.52 (H-12) (Table 19.1). The 13C NMR spectrum revealed 20 carbon signals consistent with a diterpene skeleton (Table 19.2). The HSQC spectrum showed that olefinic protons at δ 5.3 (H-3) and 5.01 (H-14) were attached to carbons at δ 124.0 (C-3) and 124.7 (C-14), respectively. The COSY spectrum showed correlations of 2.62 (H-1) –2.20/2.48 (H-2), 2.20 (H-5), 2.20/2.48 (H-2)– 5.33 (H-3), 1.55 (H-7)–3.91 (H-6), and 1.52 (H-11). The pentene ring is established based on a COSY correlation between 2.20 (H-5) and 3.91 (H-6). The isoprenyl is then linked to the ring through a COSY bond between 1.19/1.52 (H-12) and 1.60/1/80 (H-13). The methylene proton at 4.70 (H-18) showed HMBC correlation of 40.6 (C-9), and 2.23 (H-9) showed a correlation of 47.7 (C-7). The HMBC correlations of the methyl proton at 1.76 (H-17)–124.0 (C-3) and 60.4 (C-5) completed the methylated pentene ring of the diterpene. All spectroscopic data for compound **1** are in agreement with literature data reported for pachydictyol A (Gedara et al., 2003; Ayyad et al., 2011). Compound **1** was first isolated from brown alga *Pachydictyon coriaceum* by Hirschfeld et al. (1973). Its occurrence in other algae species has been reported from *Dictyota flabellate* (De Andrade Moura et al., 2014), *Cystoseira myrica* (Ayyad et al., 2003), and *P. coriaceum* (Robertson and Fenical, 1977).

19.3.3   ISOLATION OF **2**

A specimen of *L. pumila* was collected manually from the coast of KwaZulu-Natal, South Africa. The compound was extracted by column chromatography and subsequently subjected to a semi-preparative normal phase HPLC for further separation and purification to yield **2**. The 2D NMR spectroscopic results for structural elucidation of **2** (Figure 19.1) are discussed and compared with those reported in the literature.

19.3.4 STRUCTURE ELUCIDATION OF 2

The 1H NMR spectrum of **2** indicated the presence of a terminal methyl signal at δ 0.89 (H-15) and four methylene signals at δ 2.83 (H-5), 2.48 (H-8), 2.58 (H-11), and 1.88 (H-14). An acetylenic signal at δ 3.16 (H-1), and further upfield, four olefinic proton signals were exhibited at δ 5.61 (H-10), 5.62 (H-9), 5.56 (H-3), and 6.33 (H-4) (Table 19.1). The 13C NMR spectrum displayed fifteen carbon signals (Table 19.2). This included four methylene signals at δ 34.3 (C-5), 33.3 (C-8), 29.8 (C-11), and 29.1 (C-14), an alkyne carbon signal δ 81.6 (C-2), and nine methine carbon signals at δ 82.3 (C-1), 111.9 (C-3), 141.6 (C-4), 81.6 (C-6), 63.2 (C-7), 129.3 (C-9), 130.0 (C-10), 63.5 (C-12), and 63.2 (C-13). The presence of four methine carbons signals δ 111.9 (C-3), 141.6 (C-4), 129.3 (C-9), and 130.0 (C-10) is indicative of two double bonds for which 111.9 (C-3) and 141.6 (C-4) forms part of a linear chain, while 129.3 (C-9) and 130.0 (C-10) is indicative of a double bond in a ring. The methine carbons 81.6 and 63.5 confirm an ether linkage. A nine-membered ring was established based on the HMBC correlation of δ 2.68 (H-13) to the carbon δ 81.6 (C-6). The enyne side chain of the ring was established by long-range correlation of δ 6.33 (H-4)– δ 81.6 (C-2) and a COSY correlation of δ 6.02 (H-4) and δ 5.56 (H-3). The skeleton is indicative of a C-15 acetogenin, which is not uncommon in the *Laurencia* species. A literature search of the spectral data revealed a close relationship to 3Z, 12R, 13R-obtusenyne, as reported by Awakura et al. (1999) and Noite et al. (1991). The major difference observed between **2** and 3Z, 12R, 13R-obtusenyne was the13C chemical shift at δ C-11, C-12, and C-13 (Table 19.2). This suggests the presence of a different electronegative compound or halogen at position 12 other than bromine.13C NMR spectrum prediction of compound **2** on ChemDraw 18.2, that is, substituting bromine at position 12 for chlorine gave a similar spectrum to the obtained compound **2**. A literature search for compound **2**—a dichloronated version of 3Z, 12R, 13R-obtusenyne did not yield any result. Further experiments to confirm the structure could not be carried out due to the small quantity of material available and the fact it requires further purification. This could, in fact, generate a novel halogenated C-15 acetogenin compound from *L. pumila*.

19.3.5 ISOLATION OF 3

The alga *L. sodwaniensis* was collected by hand from the shoreline in KwaZulu-Natal, South Africa. The compound was purified through the application of the organic extracts to a silica gel column for chromatographic separation to yield **3** (Figure 19.1). Compound **3** was analysed by 1D and 2D NMR spectroscopy.

19.3.6 STRUCTURE ELUCIDATION OF 3

In compound **3**, the 1H NMR showed three distinct methyl signals at δ 1.03 (H-13), 1.32 (H-12), and 1.74 (H-15). Five overlapping methylene proton signals were observed at δ 1.74/2.06 (H-1), 1.82/2.39 (H-2), 2.78/3.08 (H-5), 2.45/2.70 (H-8), 4.95/5.19 (H-14) as well as three methine signals at δ 4.42 (H-10), 3.67 (H-9), and 4.42 (H-4) (Table 19.1). The 13C NMR spectrum showed 15 carbon signals consistent with sesquiterpenes (Table 19.2). Five of these carbon signals were confirmed as methylene carbons with a13C DEPT-135 NMR spectrum showing signals at δ 24.3 (C-1), 32.4 (C-2), 33.8 (C-5), 39.3 (C-8), and 114.7 (C-14). Three methyl signals were observed at δ 24.8 (C-12), 25.3 (C-13), and 33.1 (C-15). The presence of halogenated carbons was established by the three-carbon signals observed at 57.1 (C-4), 73.3 (C-3), and 76.3 (C-10). The deshielding of the methyl group at δ 33.1 (C-15) is indicative of the close proximity of a halogen to the methyl group. The HMBC spectrum showed a correlation of 4.42 (H-4) –33.1 (C-15). Since a brominated methine appears generally

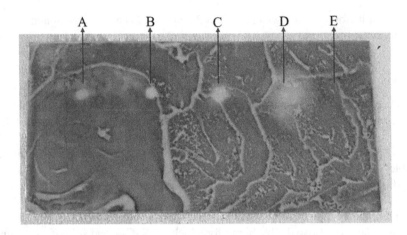

FIGURE 19.2 Thin-layer bioautographic assay of compounds **1, 2,** and **3** against MRSA. (a) 1, (b) 2, (c) 3, (d) Vancomycin, and (e) Dichloromethane. All the assays were in duplicates; clear zone-inhibition of MRSA by compounds.

upfield from its chlorinated analog (Crews et al., 1984), the 13C signal at 57.1 (C-4) is indicative of a bromomethine at 57.1 (C-4) and chlorination at 73.3 (C-3). All spectroscopic data for compound **3** were in agreement with literature data reported for cartilagineol by Da Silva Machado et al. (2011) which was isolated from the Brazilian red alga *Laurencia dendroidea*. The HR-LCMS of fraction 3 from which **3** was derived showed a base peak at m/z 305.1527, which is not consistent with any of the fragments expected from the ionisation of **3**. The molecular ion of **3** was also expected at 414.6 g/mol; however, it is not present in the HR-LCMS.

19.3.7 Biological Activity Observed for Compounds 1–3

The compounds (**1–3**) displayed antibacterial activities against MRSA at a concentration of 20 μg/mL from the TLC bioautographic assay (Figure 19.2). The active extracts and compounds (**1–3**) were represented by the presence of clear spots on the TLC plates that effectively inhibited the growth of MRSA. Similar observations were made by Suleimana et al. (2009) who used the TLC bioautographic procedure to screen hexane, acetone, dichloromethane, and methanol extracts of seven South African plants.

19.3.8 The Susceptibility Assessment of Extracts and Compounds 1–3 against MRSA by Broth Microdilution

The susceptibility of MRSA to *D. naevosa*, *L. pumila,* and *L. sodwaniensis* extracts and isolated compounds is shown in Figure 19.3. Neither the extracts nor the compounds showed striking activity against MRSA when tested by the microdilution method. This is in direct contrast to the results obtained from the TLC-bioautography assay. The result shows that the dilution concentration at which *D. naevosa* extract appears to show some activity against MRSA growth was at 1.56 μg/mL and with the least percentage growth of 93.74%. MRSA growth was only slightly suppressed at 3.12 μg/mL and 0.78 μg/mL. Ideally, it is expected that the higher concentration would show increased MRSA inhibition; however, these concentrations appear to enhance the growth of the organism. Similar observation was reported by Luziatelli et al. (2019) who observed that vegetal-derived bioactive compounds and extracts from a tropical lettuce stimulated the growth of epiphytic bacteria with biological control activity against pathogens. In addition, the extract may not be fully

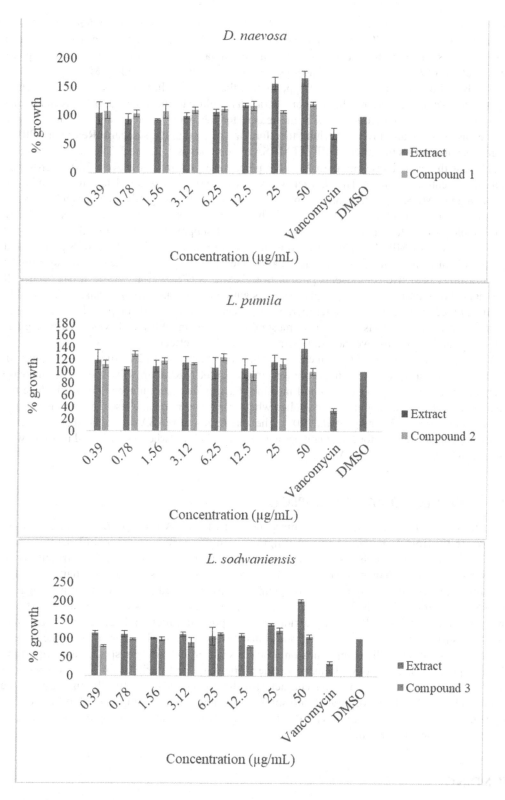

FIGURE 19.3 Evaluation of susceptibility of MRSA to extracts and compounds of *Dictyota naevosa*, *Laurencia pumila*, and *Laurencia sodwaniensis* by broth microdilution method.

solubilised in liquid medium (Delgado et al. 2005). The result also showed that MRSA growth was not susceptible to *L. pumila* and *L. sodwaniensis* extracts, as the percentage growth was above 100% when tested within the set concentration range evaluated. Similar to *D. naevosa*, the highest concentrations of *L. pumila* and *L. sodwaniensis* extracts resulted in the highest MRSA percentage growth. Based on these observations, it appears that the susceptibility of MRSA to algal extracts is not dose-responsive. Perhaps, higher concentrations could have been tested, but it is likely that the poor solubility of the compounds is the reason for the poor activity. Vancomycin has been reported as effective anti-mycobacterium; therefore, it was used as the standard (Rens et al., 2014). Figure 19.3 also shows the susceptibility of MRSA to compounds **1**, **2**, and **3**. The result shows that MRSA growth was not susceptible to **1**, as the percentage growth was above a 100% across the dilution concentration ranges. The *D. naevosa* extract from which **1** was isolated shows inhibitory activity against MRSA growth at 3.12 µg/mL, 1.56 µg/mL, and 0.78 µg/mL, whereas an increase in growth was observed at these concentrations with the purified compound. This suggests that **1** may be insoluble, or that the dilution concentration at which **1** suppresses MRSA growth is at 12.5 µg/ mL. The growth of MRSA is also susceptible to **3** at 3.12 µg/mL, 1.56 µg/mL, 0.78 µg/mL, and 0.39 µg/mL. It is also quite possible that the antibacterial activity is dependent on the synergistic action of other bioactive components present in the extracts (Pérez et al., 2016). It has been well documented that taking the whole plant or extracts with no isolation of components, as practiced in traditional medicine, produces a better therapeutic effect than individual compounds. According to Thomford et al. (2018), this is important as most of the plant metabolites likely work in a synergistic fashion or concurrently to give the plant extract its therapeutic effect.

A contrast in the activity of these metabolites in nutrient broth and TLC silica plate is observed. On the TLC silica plates, the extracts showed inhibition of MRSA growth, whereas an increase in growth rate was detected in the liquid medium. Factors that can influence growth include the medium in which the metabolites were applied whether solid or liquid and also the nutrients present in the liquid medium (Wiegand et al., 2008). The application of MRSA on a solid surface such as the silica gel plate permits for direct contact with the metabolites, whereas the liquid medium with nutrients may have caused MRSA to thrive.

19.4   CONCLUSION AND PROSPECTS

This study describes the isolation of a diterpene pachydictyol A (**1**) from *D. naevosa*—a C-15 acetogenin (**2**) from *L. pumila,* and a chamigrane sesquiterpene cartilagineol (**3**) from *L. sodwaniensis*. These compounds (**1–3**) were isolated for the first time in these species. Structural elucidation of the compounds was based on a detailed analysis of the spectroscopic data generated with reference to literature data. The compounds (**1–3**) displayed antibacterial activity against MRSA, following a TLC-bioautograpy assay. Broth microdilution assay of the algae extracts and compounds against MRSA was carried out, and poor antimicrobial activity was observed. Due to insufficient material availblity, additional tests could not be carried out. The extraction and re-isolation of the compounds on a larger scale from the algae species are important. Greater quantity of extracts will enable further separation and purification of the compounds by chromatographic means to better characterise the compounds structurally. In addition, the compounds can then be incorporated into solid lipid nanoparticles to improve solubility, following which further biological and cytotoxicity assays can take place to gain clarity on the mode of action of isolated compounds.

FUNDING

This study was supported by the National Research Foundation of South Africa (NRF, Grant No: 94922).

ACKNOWLEDGEMENTS

I am grateful to E.M. Antunes (Department of Chemistry, University of the Western Cape, South Africa) for the nuclear magnetic resonance spectroscopy data. I further extend my gratitude to D.R. Beukes (School of Pharmacy, University of the Western Cape, South Africa) for supervising this study. In addition, I am grateful to J.J. Bolton (Department of Biological Sciences, University of Cape Town, South Africa) for the identification of the algae specimen used in the study. I acknowledge M. Le Roes-Hill (Institute of Biomedical and Microbial Biotechnology, Cape Peninsula University of Technology, South Africa) for providing the facilities required to carry out bioactivity assays. Finally, I am grateful to M. Trindade (Institute for Microbial Biotechnology and Metagenomics, University of the Western Cape, South Africa) for co-supervising this study.

REFERENCES

Anderson, R., Stegenga, H., and Bolton, J. (2016). *Seaweeds of the South African south coast.* World Wide Web electronic publication, University of Cape Town. http://southafrseaweeds. uct.ac.za.

Awakura, D., Fujiwara, K., and Murai, A. (1999). Determination of the absolute configurations of norte's obtusenynes by total syntheses of (12 R, 13 R)-(−)-and (12 S, 13 R)-(+)- obtusenynes. *Chemistry Letters*, 28(6), 461–462.

Ayyad, S.E., Makki, M.S., Al-Kayal, N.S., Basaif, S.A., El-Foty, K.O., Asiri, A.M., and Badria, F.A. (2011). Cytotoxic and protective DNA damage of three new diterpenoids from the brown alga Dictyota dichotoma. *European Journal of Medicinal Chemistry*, 46(1), 175-182.

Ayyad, S.E.N., Abdel-Halim, O.B., Shier, W.T., and Hoye, T.R. (2003). Cytotoxic hydroazulene diterpenes from the brown alga Cystoseira myrica. *Zeitschrift für Naturforschung* C, 58(1–2), 33–38.

Carroll, A.R., Copp, B.R., Davis, R.A., Keyzers, R.A., and Prinsep, M.R. (2020). Marine natural products. *Natural Product Reports, 37*(2), 175–223.

Chen, J., Li, H., Zhao, Z., Xia, X., Li, B., Zhang, J., and Yan, X. (2018). Diterpenes from the marine algae of the genus Dictyota. *Marine Drugs*, 16(5), 159.

Crews, P., Naylor, S., Hanke, F.J., Hogue, E.R., Kho, E., and Braslau, R. (1984). Halogen regiochemistry and substituent stereochemistry determination in marine monoterpenes by carbon-13 NMR. *The Journal of Organic Chemistry*, 49(8), 1371–1377.

da Silva Machado, F.L., Pacienza-Lima, W., Rossi-Bergmann, B., de Souza Gestinari, L.M., Fujii, M.T., de Paula, J.C., and Soares, A.R. (2011). Antileishmanial sesquiterpenes from the Brazilian red alga Laurencia dendroidea. *Planta Medica*, 77(07), 733–735.

de Andrade Moura, L., Marqui de Almeida, A.C., Domingos, T.F.S., Ortiz-Ramirez, F., Cavalcanti, D.N., Teixeira, V.L., and Fuly, A.L. (2014). Antiplatelet and anticoagulant effects of diterpenes isolated from the marine alga, Dictyota menstrualis. *Marine Drugs*, 12(5), 2471–2484.

De Clerck, O., Tronchin, E., and Schils, T. (2005). Red Algae *Guide to the Seaweeds of Kwazulu-Natal, National Botanic Garden of Belgium*, 131–267.

Delgado, S., Flórez, A.B., and Mayo, B. (2005). Antibiotic susceptibility of Lactobacillus and Bifidobacterium species from the human gastrointestinal tract. *Current Microbiology, 50*(4), 202–207.

Francis, C.M. (2014). Systematics of the Laurencia complex (Rhodomelaceae, Rhodophyta). University of Cape Town, Southern Africa.

Gedara, S.R., Abdel-Halim, O.B., El-Sharkawy, S.H., Salama, O.M., Shier, T.W., and Halim, A.F. (2003). Cytotoxic hydroazulene diterpenes from the brown alga Dictyota dichotoma. *Zeitschrift für Naturforschung, C, 58*(1–2), 17–22.

Hirschfeld, D.R., Fenical, W., Lin, G.H.Y., Wing, R.M., Radlick, P., and Sims, J.J. (1973). Marine natural products. VIII. Pachydictyol A, an exceptional diterpene alcohol from the brown alga, Pachydictyon coriaceum. *Journal of the* American Chemical Society, 95(12), 4049–4050.

Ioannou, E., Vagias, C., and Roussis, V. (2013). Isolation and structure elucidation of three new dolastanes from the brown alga Dilophus spiralis. *Marine Drugs*, 11(4), 1104–1112.

Luziatelli, F., Ficca, A.G., and Colla, G., Švecová, E. B., and Ruzzi, M. (2019). Foliar application of vegetal-derived bioactive compounds stimulates the growth of beneficial bacteria and enhances microbiome biodiversity in lettuce. *Frontiers in Plant Science, 10*, 60.

Noite, M., Gonzalez, A.G., Cataldo, F., Rodríguez, M.L., and Brito, I. (1991). New examples of acyclic and cyclic C-15 acetogenins from laurencia pinnatifida. Reassignment of the absolute configuration for E and Z pinnatifidienyne. *Tetrahedron*, 47(45), 9411–9418.

Pérez, M.J., Falqué, E., and Domínguez, H. (2016). Antimicrobial action of compounds from marine seaweed. *Marine Drugs, 14*(3), 52.

Rens, C., Laval, F., Daffé, M., Denis, O., Frita, R., Baulard, A., and Fontaine, V. (2016). Effects of lipid-lowering drugs on vancomycin susceptibility of mycobacteria. *Antimicrobial Agents and Chemotherapy, 60*(10), 6193–6199.

Robertson, K.J., and Fenical, W. (1977). Pachydictyol-A epoxide, a new diterpene from the brown seaweed Dictyota flabellata. *Phytochemistry, 16*(7), 1071–1073.

Suleimana, M.M., McGaw, L.J., Naidoo, V., and Eloff, J.N. (2009). Detection of antimicrobial compounds by bioautography of different extracts of leaves of selected South African tree species. *African Journal of Traditional, Complementary and Alternative Medicines, 7*(1), 64–78.

Suzuki, M., Takahashi, Y., Nakano, S., Abe, T., Masuda, M., Ohnishi, T., and Seki, K.I. (2009). An experimental approach to study the biosynthesis of brominated metabolites by the red algal genus Laurencia. *Phytochemistry, 70*(11), 1410–1415.

Thomford, N.E., Senthebane, D.A., Rowe, A., Munro, D., Seele, P., Maroyi, A., and Dzobo, K. (2018). Natural products for drug discovery in the 21st century: innovations for novel drug discovery. *International Journal of Molecular Sciences, 19*(6), 1578.

Wang, B.-G., Gloer, J.B., Ji, N.-Y., and Zhao, J.-C. (2013). Halogenated organic molecules of Rhodomelaceae origin: chemistry and biology. *Chemical Reviews, 113*(5), 3632–3685.

Wiegand, I., Hilpert, K., and Hancock, R.E. (2008). Agar and broth dilution methods to determine the minimal inhibitory concentration (MIC) of antimicrobial substances. *Nature Protocols, 3*(2), 163.

20 An Endophytic *Beauveria bassiana* (Hypocreales) Strain Enhances the Flavonol Contents of *Helichrysum petiolare*

Ninon Geornest Eudes Ronauld Etsassala, Neo Macuphe, Ilyaas Rhoda, Fanie Rautenbach and Felix Nchu

CONTENTS

DOI: 10.1201/9781003206620-20

20.1   INTRODUCTION

Beauveria bassiana is an endophytic entomopathogenic fungus that occurs naturally and ubiquitously in the soil (Espinoza et al., 2019). *B. bassiana* can form a mutually beneficial symbiotic relationship with a plant. It can live in a plant's tissues without causing infection while strengthening plant defence, and the plant serves as the host in exchange (Behie et al., 2014). While some fungal species live as latent or inactive pathogens that become active and multiply when their host plants are under stressful environmental conditions, others are neutral and offer no benefit (Sikora et al., 2007). *B. bassiana* has been reported by several authors to enhance the production of bioactive compounds that improve plant resistance to pests and diseases, the rate of nutrient uptake tolerance to abiotic and biotic stresses, and growth (Gana et al., 2022). Furthermore, endophytic fungi could induce plant production of antioxidants, and protect host plants from pathogens and environmental stresses (White and Mónica, 2010). Hence, the endophytic fungi–plant relationship could be explored for the cultivation of targeted high-valued medicinal plant species to optimise their medicinal properties by increasing the quantity and quality of secondary metabolites, which are responsible for the bioactivities demonstrated by plants (Dutta et al., 2014; Caruso et al., 2020). Several studies are ongoing to investigate alternative ways to improve the phytochemical constituents in plants by manipulating the biotic and abiotic environmental factors (Li et al., 2020). Hence, inoculating plants with an endophytic fungus such as *B. bassiana*— a biotic approach can be explored to enhance the production of secondary metabolites and induce the production of antioxidants in host plant species (Macuphe et al., 2021).

Soil cultivation of medicinal plants, another approach for cultivating medicinal plants, has many challenges, including exposure to pests, high pesticides, and fertilisers (Chen et al., 2016). Consequently, researchers are looking for other, more environment-friendly, cost-effective alternative methods. Hydroponics is a feasible alternative approach for cultivating medicinal plants (Giurgiu et al., 2014); the biotic and abiotic factors can be manipulated to optimise the secondary metabolite production and improve the bioactivities of plants (Vu et al., 2006). Plants' secondary metabolites play a vital ecological role in plants' defence, protection, and signalling mechanisms (Griesser et al., 2015).

Helichrysum petiolare Hilliard belongs to the plant family Asteraceae or Compositae, and it's commonly known as silver bush everlasting (English), akruie (Afrikaans), and imphepho (Xhosa). The species is endemic to South Africa and predominantly found in the Free State, Western Cape, Eastern Cape, and Northern Cape Provinces (Maroyi, 2019). The height of *H. petiolare* ranges from 30 to 120 cm. It grows better in drier inland and sheltered slopes in the fynbos biome, transition zones, and forest margins at an altitude that ranges from 120 to 1,420 m above the sea level (Maroyi, 2019). Van Wyk and Gericke (2013) recorded *H. petiolare* as one of the most popular *Helichrysum* species; it is utilised as herbal medicine against colds, coughs, headache, fever, menstrual pain, and infections in South Africa. Phytochemical studies showed that *H. petiolare* extracts and their isolated compounds have anti-fungal, anti-bacterial, anti-inflammatory, antioxidant, anti-tyrosinase, anti-genotoxicity, and cytotoxicity activities (Akinyede et al., 2021; Maroyi, 2019). Furthermore, the aerial parts (flowers, leaves) of *H. petiolare* are a rich source of terpenoid and non-terpenoid compounds (Giovanelli et al., 2018; Lourens et al., 2004; Ras, 2013). It has been reported that the ethanol and methanol extracts of *H. petiolare* exhibited good antioxidant (DPPH) activities with IC_{50} values of 44.3 μg/mL and 28.7 μg/mL, respectively (Akinyede et al. 2021, Maroyi, 2019). Hence, this research study aimed to primarily assess the effects of *B. bassiana* on secondary metabolite contents and antioxidant activity of *H. petiolare*.

20.2 MATERIALS AND METHODS

20.2.1 RESEARCH DESIGN

A greenhouse experiment of this research study was carried out at the Cape Peninsula University of Technology (CPUT), Bellville campus, Western Cape, South Africa. Potted *H. petiolare* plants were exposed to a fungus (*B. bassiana* [SM 3 strain]) or control treatment in a completely random design. The effect of the fungus inoculation on secondary metabolite contents and antioxidant activity was evaluated.

20.2.2 PLANT MATERIALS

H. petiolare (Silver Bush Everlasting) seedlings were purchased from Stodels Nurseries (Pty) Ltd. in Bellville, Western Cape Province, South Africa. *H. petiolare* seedlings were maintained under the following conditions: $28 \pm 2°C$, 60–80% RH, and 14/10 natural light/dark regime in a glasshouse at the Cape Peninsula University of Technology, Bellville campus, Western Cape, South Africa. Each plant was gently removed from the seedlings trays and transplanted into a substrate mix of silica sand, peat moss, vermiculite, and perlite at a 1:1:1:1 ratio. Prior to transplanting, the substrate was sterilised with 1% sodium hypochlorite for 15 minutes and rinsed with sterile distilled water three times.

20.2.3 FUNGUS PREPARATION

Cultures of a *B. bassiana* strain (SM 3) previously isolated from a vineyard and identified molecularly by Moloinyane and Nchu (2019) were used in this study. The method described by Moloinyane and Nchu (2019) was used to culture the fungus. Briefly, the *B. bassiana* strain was cultured on selective medium for fungus as follows: half strength (19.5 g/1,000 mL) of Potato Dextrose Agar (PDA) (Sigma-Aldrich Pvt. Ltd., Johannesburg, South Africa), 0.04 g streptomycin, and 0.02 g ampicillin sodium salt. The PDA was prepared on 9 cm diameter Petri dishes, and fungal cultures were incubated as described in Macuphe et al. (2021). The matured conidia were harvested using a sterile spatula and transferred into a 50 mL centrifuge tube containing 25 mL sterile water. The centrifuge tube was capped and shaken for 3 minutes and mixed vigorously for 2 minutes using a vortex mixer (MI0101002D Vortex Mixer, Silverson Machines, Inc., East Longmeadow, MA, USA) at 3,000 rpm to homogenise the conidial suspension. Furthermore, the homogenous conidial suspension was transferred into 1,000 mL bottles comprising 500 mL sterile distilled water and 0.05% Tween 80 (Polysorbate, Sigma-Aldrich, Johannesburg, South Africa). The desired conidial concentration, 1×10^8 conidia m/L, was determined using a haemocytometer (Bright-Line™, Sigma-Aldrich, Johannesburg, South Africa) and observed with a light microscope at 400× magnification. The method described by Latifian and Rad (2012) was employed to assess fungal viability. Germination percentage was evaluated on a 100-spores count at 40× magnification. Each plate was replicated four times, and over 90% conidial germination was observed.

20.2.4 GREENHOUSE EXPERIMENT

This experiment was carried out in a CPUT greenhouse, Bellville campus in the Department of Horticultural Sciences, Western Cape, South Africa. The following conditions were maintained in the greenhouse: $25 \pm 5°C$, $65\% \pm 5\%$ relative humidity, and the average light intensity was 31.77 kilo lux. Two weeks old *H. petiolare* seedlings were transferred into 15 cm pots containing a substrate mix of the substrates, silica sand, peat moss, vermiculite, and perlite, mixed at a ratio of 1:1:1:1. Twelve plants were placed into 15 cm pots individually. This experiment consisted of a control treatment

FIGURE 20.1   *Helycrysum petiolare* plants inoculated with *Beauveria bassiana conidia.*

and a *B. bassiana* treatment. Each plant was drenched with 100 mL of conidial suspension (1×10^8 conidia m/L) of *B. bassiana*, and the control plants were drenched with 100 mL of sterile distilled water with 0.05% Tween 20. Each treatment had six replicates ($n=6$) (Figure 20.1). The plants were fed using recommended commercial hydroponics fertiliser NUTRIFEED® hydroponic fertiliser (Starke Ayres Pvt. Ltd., Cape Town, South Africa) that had the following ingredients: N (65 mg/kg), P (27 mg/kg), K (130 mg/kg), Ca (70 mg/kg), Cu (20 mg/kg), Mo (10 mg/kg), Fe (1,500 mg/kg), Mg (22 mg/kg), S (75 mg/kg), B (240 mg/kg), Mn (240 mg/kg), and Zn (240 mg/kg). The fertiliser was diluted with sterile distilled water to a concentration of 10 g/ 5,000 mL, and 200 mL was added to each plant weekly. Moreover, each *H. petiolare* plant was watered twice a week with distilled water. Furthermore, each plant was watered with distilled water once a week for 6 weeks. This experiment was replicated three times.

20.2.5 Fungal Colonisation of Plant Tissue

The successful fungal colonisation of the tissues was determined using the method described in Moloinayne and Nchu (2019). Briefly, After 21 days post-treatment, fresh leaves were picked off plants and taken to the laboratory to assess fungal colonisation. Leaf sections were surfaced and sterilised in the following sequence: 0.5% sodium hypochlorite for 2 minutes, 70% ethanol for 2 minutes, and then rinsed with sterile distilled water for 1 minute. The sterilised leaf sections were placed on selective solid agar plates made up of half-strength PDA (19.5 g/1,000 mL of sterile water containing 0.04 g streptomycin and 0.02 g ampicillin sodium salt). They were incubated at 25 ± 2°C. Based on the fungal outgrowth on leaf sections, 83% of plants had successful fungal colonisation.

20.2.6 Sample Preparation for Chemical Analyses and Antioxidant Activities

Only plants that showed successful fungal colonisation (83%) were randomly selected to analyse secondary metabolite contents. Plants were oven dried at 35°C for 168 hours and were ground and transferred into plastic bags.

20.2.7 Analysis of Secondary Metabolites on Leaves of Inoculated Plants

20.2.7.1 Total Polyphenol

The Folin–Ciocalteu method was used to determine the total polyphenol content of the crude extracts (Swain and Hillis 1959; Singleton et al., 1999). A volume of 25 µL of the crude extract was mixed with 125 µL Folin–Ciocalteu reagent (diluted 1:10 with distilled water) (Merck, South Africa). A volume of 100 µL (7.5%) aqueous sodium carbonate (Na_2CO_3) (Sigma-Aldrich, South Africa) was added to each well after 5 minutes, followed by the absorbance reading of the solution in the microplates. The results were expressed as mg gallic acid equivalents per gram dry weight (mg GAE/g DW).

20.2.7.2 Total Flavonol

The flavonol content was determined using the protocol described by Daniels et al. (2015). Quercetin standard concentrations of 0, 5, 10, 20, 40, and 80 mg/L in 95% ethanol (Sigma-Aldrich, South Africa) were used. 12.5 µL of the crude extracts were mixed with 12.5 µL 0.1% hydrochloric acid (HCl) (Merck, South Africa) in 95% ethanol in the sample wells and then incubated for 30 minutes at room temperature. The results were expressed as mg quercetin equivalent per g dry weight (mg QE/g DW).

20.2.8 GC–MS Analysis

20.2.8.1 Sample Preparation

Twelve potted plants, six from each treatment, were used for this analysis. Only plants that showed fungal colonisation among the fungus-treated plants were used for GC–MS analysis.

20.2.8.2 GC-MS Analysis (Headspace)

The GC-MS method described by Moloinyane and Nchu (2019) was adopted for this research study. The whole leaves from the fresh lettuce plants were removed and freeze-dried at −80°C (overnight), and then crushed in liquid nitrogen. A mass of 1 g of the crushed leaves was transferred into a solid-phase microextraction (SPME) vial, and then 2 mL of 12% ethanol solution (v/v) at 3.5 and 3 mL of 20% NaCl were added to the vial. The samples were mixed vigorously using a vortex mixer. Finally, the headspace of the samples was analysed using a Divinylbenzene/Carboxen/Polydimethylsiloxane (DVB/CAR/PDMS) SPME fibre (grey).

20.2.8.3 Chromatographic Separation

The volatile compounds were identified and separated in the lettuce plants using a gas chromatograph (6890N, Agilent Technologies Network) coupled to an inert Xl EI/CI Mass selective detector (model 5975B, Agilent Technologies Inc., Palo Alto, CA). The protocol described in Moloinyane and Nchu (2019) was used. The GC–MS system used was combined with a CTC Analytics PAL autosampler. The volatiles were separated on a polar ZB-WAX (30 m, 0.25 mm ID, 0.25 μm film thickness) Zebron 7HG-G007-11 capillary column. Helium was used as the carrier gas at the flow rate maintained at 1 mL/min. The injector temperature was 250°C, with a split ratio of 5:1, and oven temperature was timed at 35°C for 6 minutes, at a rate of 3°C/min to 70°C for 5 minutes, then at 4°C/min to 120°C for 1 minute, and finally increased to 240°C at a rate of 20°C/min and maintained for 2.89 minutes. The mass selective detector was operated in full scan mode while maintaining the source, quad, and transfer temperatures at 230°C, 150°C, and 250°C, respectively. The electron impact mode at ionisation energy of the mass spectrometer was run below 70 eV, scanning from 35 to 500 m/z. Relative ratios were used to estimate quantities of volatiles, and they were determined using the expression (peak area/IS peak area)×IS concentration (IS = Internal standard). A cut-off match quality of at least 90% was used for organic volatile compound identification.

20.2.9 ANTIOXIDANTS

20.2.9.1 Sample Materials

The plant materials (*H. petiolare*) were oven dried at 35°C for 168 hours. The dried plant materials were ground, and the powered material was transferred into plastic bags. Twelve samples were weighed, and 0.1 g of powdered plant material was transferred into centrifuge tubes. The samples were extracted with 25 mL of 60% ethanol and placed inside the incubator for 24 hours.

20.2.9.2 Sample Preparation

The dried samples were ground, and the powdered material was transferred into plastic bags. Twelve samples per treatment were weighed, and 0.1 g of powdered *H. petiolare* samples were transferred into 50 mL centrifuge tubes. Tomato samples were extracted with 25 mL of 60% ethanol and placed inside the incubator for 24 hours (Macuphe et al., 2021).

20.2.9.3 Ferric-Reducing Antioxidant Power (FRAP)

The ferric-reducing antioxidant power assay method was adopted by Benzie and Strain (1996). The assay is based on the reduction of the ferric–tripyridyltriazine compound to its ferrous in the presence of antioxidants. The reagent used for this research study were as follow: 2.5 mL of a 10 mmol/L TPTZ (2,4,6- tripyridyl-s-triazine, Sigma-Aldrich, Johannesburg, South Africa) solution in 40 mmol/L HCl plus 2.5 mL of 20 mmol/L $FeCl_3$ and 25 mL of 0.3 mol/L acetate buffer and kept at pH 3.6 was freshly prepared and warmed at 37°C (Macuphe et al., 2021). The aliquots of 40 μL of the sample supernatant were mixed with 0.2 mL sterile distilled water and 1.8 mL FRAP reagent. Following the incubation at 37°C for 10 minutes, the spectrophotometric method was used to determine the absorbance of the reaction mixture at 593 nm. The standard solution was 1 mmol/L $FeSO_4$, and the result was represented as the concentration of antioxidants with a ferric-reducing capacity of 1 mmol/L $FeSO_4$.

20.2.9.4 Trolox Equivalent Antioxidant Capacity (TEAC)

The TEAC method was used to assess the antioxidants' potential to scavenge free radicals in tomatoes in a method described by Miller et al. (1993). The TEAC value is calculated by comparing the antioxidant's ability to scavenge the blue–green coloured 2,2′-azino-bis-(3-ethylbenzthiazoline-6-sulphonic acid) radical (ABTS+) radical cation to the water-soluble vitamin E analogue's ABTS•+ radical cation scavenging ability.

TABLE 20.1

Effect of *Beauveria bassiana* on Secondary Metabolites of *Helichrysum petiolare* Extracts ($n=24$)

Treatments	Polyphenols (mg GAE/L)	Flavonols (mg QE/L)	Alkaloids (mg AE/L)
Control	76.96±3.35a	31.94±1.69b	N.D
Fungal control	80.29±3.43a	42.73±2.56a	N.D

20.2.10 STATISTICAL ANALYSIS

The total phenolic contents are reported as Mean±SE, and the one-way analysis of variance (ANOVA) was used to compare the means among the treatments. All computations were completed with the software programmes STATISTICA® (13.5.0.17) and the Paleontological Statistics package for education and data analysis (PAST 3.14). Post-hoc analysis based on the Tukey test was used to separate the means. P-values of <0.05 were considered significant.

20.3 RESULTS

20.3.1 TOTAL FLAVONOL (MG QE/L)

Generally, the results showed that there was a significant difference between fungus-treated plants and control ($DF=1.22$; $F=12.35$; $P=0.002$), as shown in Table 20.1.

20.3.2 TOTAL POLYPHENOL (MG GAE/L)

Generally, the results showed that there was no significant difference between fungus-treated plants and the control ($DF=1.22$; $F=0.48$; $P=0.49$), as shown in Table 20.1.

20.3.3 GC–MS

The GC–MS results showed that there was no significant difference in the overall number of volatile compounds between *B. bassiana*-treated plants and control plants (Table 20.2), except for the significant variations in the quantities (area ratios) of a detected individual compound—p-anisic acid methyl ester that was observed ($DF=1.5$; $P<0.05$).

20.3.4 ANTIOXIDANTS

20.3.4.1 FRAP (μmol AAE/L)

The results showed that there was no significant difference ($DF=1.22$; $F=0.06$; $P=0.81$) between *B. basianna*-treated plants and control, as shown in Table 20.3.

20.3.4.2 Trolox Equivalent Antioxidant Capacity (μmol TE/L)

The results showed that there was no significant difference between the fungus-treated plants and the control plants ($DF=1.22$; $F=1.06$; $P=0.32$), as shown in Table 20.3.

20.4 DISCUSSION

The fungal strain of *B. bassiana* used in this study was successfully re-isolated from 83% of the leaf tissue of *H. petiolare*, suggesting that the fungus colonised the entire plant. The previous studies reported successful colonisation of many plant species following inoculation with *B. bassiana*

TABLE 20.2

The Mean Area Ratio of Volatile Compounds (Mean±SE) of Control and Fungal Treated of *Helichrysum petiolare*

Compounds Volatiles	Control	Fungal Treated	DF	F Value	P. Value
Alpha-pinene	6.51±0.98A	7.99±1.38A	1.10	0.77	0.40
Beta-pinene	14.88±3.32A	15.00±3.25A	1.10	0.01	0.98
Alpha-terpinene	1.21±0.17A	1.04±0.12A	1.10	0.66	0.43
Limonene	1.88±0.90A	1.98±0.99A	1.10	0.04	0.84
1,8 Cineole_(eucalyptol)	16.85±2.90A	23.26±4.09A	1.10	1.63	0.23
Dodecane	0.39±0.10A	0.77±0.19A	1.10	3.24	0.10
Trans-2-hexenal	0.78±0.08A	0.66±0.08A	1.10	1.05	0.33
Trans-beta-ocimene	84.88±23.74A	78.33±18.17A	1.10	0.05	0.83
Para-cymene	0.79±0.09A	0.83±0.10A	1.10	0.12	0.74
Terpinolene	0.78±0.12A	0.67±0.10A	1.10	0.48	0.50
Neo-allo-ocimene	1.53±0.30A	1.96±0.27A	110	1.11	0.31
Allo-ocimene	1.54±0.28A	1.94±0.19A	1.10	1.41	0.26
Tetradecane	1.56±0.22A	1.44±0.12A	1.10	0.22	0.65
m-di-tert-butyl_benzene	0.50±0.17A	0.30±0.04A	1.10	1.27	0.29
Mentha-1,4,8-triene	1.74±0.30A	1.40±0.23A	1.10	0.79	0.40
1-Octen-3-ol	10.17±0.68A	9.81±0.89A	1.10	0.10	0.76
Alpha-copaene	8.04±1.74A	7.32±1.20A	1.10	0.12	0.74
Alpha-gurjunene	1.90±0.59A	1.24±0.31A	1.10	0.98	0.35
6,7-Dimethyl-1,2,3,5,8,8a-hexahydronaphthaln	1.90±0.64A	2.43±1.30A	1.10	0.82	0.39
Beta-caryophyllene	4.36±1.34A	3.08±0.56A	1.10	0.78	0.40
Terpinen-4-ol	3.84±0.58A	3.55±0.38A	1.10	0.17	0.69
Lavandulyl_acetate	13.99±2.55A	14.27±1.94A	1.10	0.01	0.93
Ylangene	5.71±1.31A	4.68±0.76A	1.10	0.46	0.52
Longifolene	1.84±0.77A	0.73±0.19A	1.10	1.94	0.19
Aromadendrene	6.14±1.53A	4.70±0.92A	1.10	0.66	0.44
Beta-selinene	74.53±20.14A	51.97±9.66A	1.10	1.02	0.34
Alpha-chamigrene	20.22±2.89A	18.41±1.88A	1.10	0.27	0.61
Meryntyl_acetate	5.77±1.80A	3.64±0.79A	1.10	1.17	0.31
(-)-Aristolene	1.23±0.60A	0.36±0.20A	1.10	1.91	0.20
Alpha-Guaiene	2.48±0.55A	2.06±0.31A	1.10	0.45	0.52
Alpha-himachalene	5.13±2.14A	1.97±0.58A	1.10	2.04	0.18
Alpha-bulnesene	2.37±0.58A	1.33±0.18A	1.10	2.87	0.12
Valencene	7.05±1.76A	5.43±0.93A	1.10	0.69	0.43
Guaia-1(10),11-diene	13.77±4.23A	13.07±2.47A	1.10	0.57	0.47
Delta-cadinene	99.77±22.52A	76.72±12.07A	1.10	0.81	0.39
Dihydro-alpha-agarofuran	152.46±31.06A	107.75±13.30A	1.10	1.75	0.22
Myrentol	1.46±0.49A	0.86±0.15A	1.10	1.40	0.27
Trans-calamenene	1.32±0.30A	1.27±0.26A	1.10	0.01	0.92
1s,*cis*-calamenene	1.54±0.29A	1.32±0.13A	1.10	0.43	0.53
Patchouli alcohol	620.19±123.88A	433.31±45.47A	1.10	2.01	0.18
Phenylethyl valerate	0.70±0.11A	1.03±0.40A	1.10	0.84	0.38
Viridiflorol	0.65±0.14A	0.52±0.07A	1.10	0.76	0.41
Epiglobulol	0.91±0.23A	0.61±0.10A	1.10	1.37	0.17

(Continued)

TABLE 20.2 (*Continued*)

The Mean Area Ratio of Volatile Compounds (Mean±SE) of Control and Fungal Treated of *Helichrysum petiolare*

Compounds Volatiles	Control	Fungal Treated	DF	F Value	P. Value
Alpha-cadinol	2.75±1.38A	2.06±0.79A	1.10	1.13	0.31
Gleenol	0.88±0.27A	0.42±0.12A	1.10	2.46	0.15
p-Anisic acid, Methyl ester	0.75±0.07B	1.09±0.08A	1.10	9.20	0.01
Valeranon	13.99±2.37A	10.64±1.37A	1.10	1.49	0.25
Eudesm-4(14)-en-11-ol	0.40±0.12A	0.22±0.03A	1.10	1.20	0.19
Tau-Cadinol	1.49±0.27A	1.32±0.20A	1.10	0.26	0.62
T-Muurolol	1.45±0.34A	1.21±0.16A	1.10	0.78	0.40
(+)-epi-Bicyclosesquiphellandrene	0.82±0.22A	0.55±0.08A	1.10	1.32	0.28
Globulol	7.85±3.65A	2.37±0.30A	1.10	2.24	0.17
Trans-sesquisabinene hydrate	6.78±2.65A	2.75±0.63A	1.10	2.18	0.17
4-Isopropyl-1,6-dimethyl naphthalene	2.56±0.39A	2.09±0.11A	1.10	1.37	0.27
Geranyl-p-cymene	1.63±0.45A	1.01±0.16A	1.10	1.69	0.22
2,4-di-tert-butyl phenol	0.61±0.16A	0.38±0.04A	1.10	2.00	0.19

TABLE 20.3

Effect of *Beauveria bassiana* on Antioxidant Activities of *Helichrysum petiolare* Extracts (*n*=24)

Treatments	FRAP (µmol AAE/L)	TEAC (µmol TE/L)
Control	350.60±20.06a	478.49±19.02a
Fungal control	344.59±15.31a	506.49±19.51a

(Macuphe et al., 2021; Moloinyane and Nchu, 2019). The inoculation of *H. petiolare* with the conidial suspension of *B. bassiana* had various effects on secondary metabolites. The total flavonol contents were significantly influenced by the fungus-treated plants, while the total polyphenol contents were not affected. The phytochemical analysis also revealed that the fungus had no significant effects on the overall amount of volatile chemical constituents; however, significant variations in the quantity (area ratio) of a detected individual compound—p-anisic acid methyl ester was observed. Previously, Ludwig-Müller (2015) reported that some endophytic fungal strains could increase the synthesis of secondary metabolites such as flavonoids in host plants.

The antioxidant outcomes showed that there was no significant difference between fungus-treated plants and the control. However, acetone and methanol extracts of *H. petiolare* have demonstrated prominent antioxidant activity, which corroborated with our findings (Aladejana et al., 2021). Flavonoids are known to stabilise reactive oxygen species (ROS) via their scavenging actions through the oxidation of free radicals into more stable but less active or reactive radicals. Moreover, flavonoids with potent antioxidant activities have been shown to effectively modulate oxidative stress-related diseases through clearly defined mechanisms of action (Akinyede et al., 2022). Flavonoids and phenolics with potent antioxidant activities have been shown to effectively modulate oxidative stress-related diseases through clearly defined mechanisms of action (Akinyede et al., 2022). Numerous plants and their phytochemical constituents have been reported to be beneficial in preventing, alleviating, or treating oxidative stress-induced diseases (Etsassala et al., 2020).

While many studies have studied the effects of endophytic fungus on the protection of plants against abiotic and biotic stresses (Dutta et al. 2014; Agbessenou et al., 2020), few studies have been focused on the effects of endophytic fungi on medicinal plants (Espinoza et al., 2018), especially *H. petiolare*. Hence, the positive influence of *B. bassiana* inoculum on total flavonol contents in this study suggests that endophytic *B. bassiana* could be used to manipulate secondary metabolite productions during cultivation. Beyond their role in agricultural plant protection against pests and pathogens, endophytic entomopathogens, such as *B. bassiana*, could be used to optimise commercial cultivation of high-valued medicinal plants.

20.5 CONCLUSION AND PROSPECTS

B. bassiana inoculum significantly enhanced the flavonol contents in the fungus-treated plants. However, the *B. bassiana* inoculation had little effect on the volatile compounds and antioxidant activities. Endophytic *B. bassiana* could be used to optimise the cultivation of medicinal plants. Future studies on the effects of *B. bassiana* inoculation on other *Helichrysum* species are recommended.

ACKNOWLEDGEMENTS

We are grateful to the CPUT for awarding a Postdoctoral Fellowship to NGER.

REFERENCES

Agbessenou, A., Akutse, K.S., Yusuf, A.A. et al. (2020). Endophytic fungi protect tomato and nightshade plants against *Tuta absoluta* (Lepidoptera: Gelechiidae) through a hidden friendship and cryptic battle. *Scientific Report, 10*(1), 22195.

Akinyede, K.A. et al. (2021). Medicinal properties and *in vitro* biological activities of selected *Helichrysum* species from South Africa: a review. *Plants, 10*(8), 1566.

Akinyede, K.A. et al. (2022). Comparative study of the antioxidant constituents, activities and the GC-MS quantification and identification of fatty acids of four selected *Helichrysum* species. *Plants, 11*(8), 998.

Aladejana A.E., Bradley, G., and Afolayan, A.J. (2021). *In vitro* evaluation of the anti-diabetic potential of *Helichrysum petiolare* Hilliard & B.L. Burtt using HepG2 (C3A) and L6 cell lines. (version 2; peer review: 2 approved) *F1000Research, 9,* 1240.

Behie, S. W. and Bidochka, M. J. (2014). Nutrient transfer in plant–fungal symbioses. *Trends in Plant Science, 19*(11), 734–740.

Benzie, I. and Strain, J. (1996). The ferric reducing ability of plasma (FRAP) as a measure of antioxidant power: the FRAP assay. *Analytical Biochemistry, 239,* 70–76.

Caruso, G. et al. (2020). Biodiversity, ecology, and secondary metabolites production of endophytic fungi associated with Amaryllidaceae crops. *Agriculture, 10*(11), 533.

Dutta, D., Puzari, K. C., Gogoi, R., and Dutta, P. (2014). Endophytes: exploitation as a tool in plant protection. *Brazilian Archives of Biology and Technology, 57,* 621–629.

Espinoza, F., Vidal, S., Rautenbach, F., Lewu, F., and Nchu, F. (2019). Effects of *Beauveria bassiana* (Hypocreales) on plant growth and secondary metabolites of extracts of hydroponically cultivated chive (*Allium schoenoprasum* L. [Amaryllidaceae]). *Heliyon, 5*(12), e03038.

Etsassala, N.G., Badmus, J.A., Marnewick, J.L., Iwuoha, E.I., Nchu, F., and Hussein, A.A. (2020). Alpha-glucosidase and alpha-amylase inhibitory activities, molecular docking, and antioxidant capacities of *Salvia aurita* constituents. *Antioxidants, 9*(11), 1149.

Gana, L. P., Etsassala, N. G., and Nchu, F. (2022). Interactive effects of water deficiency and endophytic beauveria bassiana on plant growth, nutrient uptake, secondary metabolite contents, and antioxidant activity of *Allium cepa* L. *Journal of Fungi, 8*(8), 874.

Giovanelli, S. et al. (2018). Essential oil composition and volatile profile of seven *Helichrysum* species grown in Italy. *Chemistry & Biodiversity, 15*(5), e1700545.

Giurgiu, R.M., Morar, G.A., Dumitraş, A., Boancă, P., Duda, B.M., and Moldovan, C. (2014). Study regarding the suitability of cultivating medicinal plants in hydroponic systems in controlled environment. *Research Journal of Agricultural Science, 46*(2), 84–92.

Griesser, M., Weingart, G., Schoedl-Hummel, K., Neumann, N., Becker, M., Varmuza, K., and Forneck, A. (2015). Severe drought stress is affecting selected primary metabolites, polyphenols, and volatile metabolites in grapevine leaves (*Vitis vinifera* cv. Pinot noir). *Plant Physiology and Biochemistry*, *88*, 17–26.

Hillis, W. E. and Swain, T. (1959). The phenolic constituents of *Prunus domestica*. II.—The analysis of tissues of the Victoria plum tree. *Journal of the Science of Food and Agriculture*, *10*(2), 135–144.

Latifian, M., Rad, B., and Amani, M. (2014). Mass production of entomopathogenic fungi *Metarhizium anisopliae* by using agricultural products based on liquid-solid diphasic method for date palm pest control. *International Journal of Food and Agriculture Sciences*, *3*(4), 368–372.

Li, Y. et al. (2020). The effect of developmental and environmental factors on secondary metabolites in medicinal plants. *Plant Physiology and Biochemistry*, *148*, 80–89.

Lourens, A.C., Reddy, D., Başer, K.H., Viljoen, A.M, and Van Vuuren, S.F. (2004). *In vitro* biological activity and essential oil composition of four indigenous South African *Helichrysum* species. *Journal of Ethnopharmacology*, *95*, 253–258.

Ludwig-Müller, J. (2015). Plants and endophytes: equal partners in secondary metabolite production? *Biotechnology Letters*, *37*(7), 1325–1334.

Macuphe, N., Oguntibeju, O.O. and Nchu, F. (2021). Evaluating the endophytic activities of *Beauveria bassiana* on the physiology, growth, and antioxidant activities of extracts of lettuce (*Lactuca sativa* L.). *Plants*, *10*(6), 1178.

Maroyi, A. (2019). *Helichrysum petiolare* Hilliard and BL Burtt: a review of its medicinal uses, phytochemistry, and biological activities. *Asian Journal of Pharmaceutical Clinical Research*, *12*, 32–37.

Miller, N.J., Rice-Evans, C., Davies, M.J., Gopinathan, V., and Milner, A. (1993). A novel method for measuring antioxidant capacity and its application to monitoring the antioxidant status in premature neonates. *Clinical Science*, *84*(4), 407–412.

Moloinyane, S. and Nchu, F. (2019). The effects of endophytic *Beauveria bassiana* inoculation on infestation level of Planococcus ficus, growth and volatile constituents of potted greenhouse grapevine (*Vitis vinifera* L.). *Toxins*, *11*(2), 72.

Ras, A.M. (2013). Essential oil yield and composition of three *Helichrysum* species occurring in the eastern cape province of South Africa. *South African Journal of Botany*, *86*, 181.

Sikora, R.A., Schäfer, K., and Dababat, A.A. (2007). Modes of action associated with microbially induced in planta suppression of plant-parasitic nematodes. *Australasian Plant Pathology*, *36*(2), 124–134.

Singleton, G. R., Leirs, H., Hinds, L. A., and Zhang, Z. (1999). Ecologically-based management of rodent pests–re-evaluating our approach to an old problem. *Ecologically-Based Management of Rodent Pests*. Australian Centre for International Agricultural Research (ACIAR), Canberra, 17–29.

Van Wyk, B.E., Oudtshoorn, B.V., and Gericke, N. (2013). *Medicinal Plants of South Africa*. Briza Publications, Pretoria.

White Jr, J.F. and Torres, M.S. (2010). Is plant endophyte-mediated defensive mutualism the result of oxidative stress protection? *Physiologia Plantarum*, *138*(4), 440–446.

21 Ethnobotanical Study of Plants Used for the Management of Diabetes Mellitus in South Africa

Learnmore Kambizi and Callistus Bvenura

CONTENTS

21.1 INTRODUCTION

Diabetes mellitus is a multifaceted metabolic disorder characterised by hypoinsulinaemia or insulin resistance, and consequent hyperglycaemia and other cardiovascular complications (Jung et al., 2006). It manifests in two forms, namely insulin-dependent diabetes mellitus (IDDM), also known as type 1 diabetes, and non-insulin-dependent diabetes mellitus (NIDDM), also known as type 2 diabetes. The diabetic condition affects about 10% of the global population, and most cases are in developing countries (Wild et al., 2004). In South Africa, an estimated 1.8% of the population is living with diabetes (Azevedo and Alla, 2008). The South African National Burden of Disease estimates the age standardised death rate of diabetes to be between 40 and 80 deaths per 100,000 people (Rheeder, 2006). These figures warrant rigorous research into low-cost and readily available antidiabetic agents, such as medicinal plants.

An estimated 60% of the South African population uses traditional medicine to cure various ailments (Mander, 1998). Medicinal plants comprise a greater portion of traditional medicines. It is thus important to record the knowledge on these plants, as they are possible future sources of novel alternative treatments. The use of plant derivatives containing high concentrations of essential nutrients and other phytochemicals has been proposed for the treatment of diabetes (Oyedemi et al., 2009). Natural products, especially of plant origin are potential sources of novel antidiabetic principles (Marles and Farnsworth, 1995), hence the shift in global interest towards the use of plant-based medication (Erasto et al., 2005). Plant remedies are preferred, as they are readily available, inexpensive, and are perceived to be effective with fewer or no side effects compared to conventional medication (Farnsworth, 1994). Moreover, the use of plant remedies is part and parcel of culture among rural communities.

Ethnobotanical surveys are the method often used to search for important plant species (Farnsworth, 1994). Thus, an ethnobotanical survey is the first step towards prospecting for antidiabetic agents of plant origin, and these surveys have been conducted in the Eastern Cape Province (Erasto et al., 2005; Oyedemi et al., 2009). However, these surveys are inexhaustive due to the vastness of the region and varied vegetation, hence the survey among communities in the study area.

DOI: 10.1201/9781003206620-21

21.2 METHODOLOGY

The ethnobotanical survey was conducted to collect data from key informants who comprised traditional medical practitioners and knowledgeable community members. A total of 30 informants participated in the survey. All informants were requested to give verbal consent prior to participating in the survey. Only willing informants were interviewed.

Non-probability sampling techniques were employed, whereby the key informants were identified opportunistically by signposts, enquiries in the neighbourhood, and from other informants. At least two personal visits were arranged with each key informant. A rapid appraisal tool in the form of a structured questionnaire was used to obtain ethnomedical information. The questionnaire was administered through guided interviews and general conversations. To effectively communicate, conversations were conducted in the local language (Xhosa) and later translated into English. Subtle probing techniques were used to get the maximal information. Financial rewards and gifts were given to encourage the respondents to part with their information, not as a way of buying them but establishing a good rapport. Personal visits proved to be the most appropriate method due to the remoteness of the locations visited, the relatively low literacy levels of key informants, and their willingness to participate in the survey. The survey questions elicited for information that includes: the diagnosis of diabetes mellitus, local names of plants used for the treatment of diabetes mellitus, plant part(s) used, methods of preparation of remedies, dosage, and the most common age group treated.

Where possible, plant and picture specimens were collected either from informants or during walks in the forest accompanied by rural dwellers or traditional medical practitioners. Specimens were initially identified by their local names. The specimens were shown to at least two other informants to confirm identification and/or identify the plant(s) using alternative local names. Identification of specimens using binomial names was done at the Kei Herbarium with the aid of herbarium curators. Voucher specimens were prepared and deposited in the Kei Herbarium. An inventory of plants used for the management of diabetes mellitus was compiled. Only plants supported by at least two references were included in the inventory (Jayakumar et al., 2010).

21.3 RESULTS AND DISCUSSION

The study revealed 30 plant species used for the management of diabetes in the study area (Table 21.1). The diagnosis of diabetes among traditional medical practitioners is largely based on observation of clinical symptoms.

Analysis of the diagnostic methods used (Figure 21.1) revealed that 97% of the respondents depended wholly or in part on the presentation of clinical symptoms to diagnose diabetes. Only 3% of the respondents use divination alone to diagnose diabetes mellitus. The respondents claimed to diagnose diabetes in their patients by symptoms such as rapid weight loss, weakness, and blurred vision, ants on urine signifying the presence of sugar, slow healing wounds, and polyuria. Being a complex disorder, diabetes mellitus can only be treated by a few individuals among the traditional medical practitioners, and, even then, they treat the symptom and not the underlying cause of the symptom. According to Bryant (1966), indigenous knowledge systems work on the premise that the symptom is the disease, hence the symptoms are treated, which are not necessarily the root cause of the ailment. Traditional medical practitioners have limited knowledge of human anatomy and physiology, as their practice is based mainly on the experience acquired from trial-and-error methods.

Based on the number of plants used per family, Asphodelaceae tops the list with a 10% (three plants) representation, followed by Asteraceae, Apiaceae, and Hyacinthaceae with 6% (two plants) representation each, respectively. The rest of the families are represented by a single plant species. Compared to the previous studies conducted by Erasto et al. (2005) and Oyedemi et al. (2009) in the region, our findings reveal that there are more plants used traditionally for the management of diabetes mellitus. A total of 17 plant species (56.7%) are reported for the first time in the Eastern

TABLE 21.1
Plants Traditionally Used for the Management of Diabetes Mellitus

Family and Species Name	Local Name(s)	Plant Part Used	Preparation and Dosage
Hyacinthaceae			
Albuca sertosa	Inqwebeba	Corm	Half a cup of decoction made from bulbs of *Dioscorea* spp corm of *Albuca* spp and corm of
Bowiea volubilis Harv. ex Hook.F.	Umagaqana	Corm	*Bulbine* spp taken orally once a day. As discussed under *A. africanus*
Asphodelaceae			
Aloe ferrox Mill.	Salenxiweni	Root	One cup of root and leaf decoction or infusion taken three times a day.
		Leaves	Decoction of leaves of *A. arborescens* and roots of *Rubus* spp. Leftover concoction is thrown away as it becomes poisonous with time.
	Ikhala	Leaves	A leaf decoction taken three times a day before meals.
Bulbine spp	Ibhucu	Corm	As discussed under *Albuca* spp.
Apiaceae			
Alypedia amatymbica Eckl. & Zeyh	Iqwili	Roots	Teaspoon of root powder eaten or fresh root chewed, as discussed under *Curcubita* spp.
Daucus carota	Young carrot	Whole plant	Young fresh plants chewed after meals.
Polygonaceae			
Rumex spp	Idololenkonyane	Root, leaf	As discussed under *S. Africana*. Eaten as a vegetable. Quarter cup of root decoction taken orally before meals.
Asteraceae			
Artemisia afra Jacq. ex Wild	umhlonyane	Twigs	Decoction or infusion of twigs taken before meals can be mixed with *H. dorratissimum*.
Helichrysum odorratissimum (L.) Sweet	Impepho	Whole plant	Whole plant decoction taken orally three times a day before meals.
Mesmbyranthemaceae			
Carpobrotus edulis (L.) L.Bolus	Unomatyumtyum	Bulb	Bulb crushed and mixed with equal amount of crushed tuber of *R. digitata*. A decoction is made, and four tablespoons are taken orally in the evening.
Asparagaceae			
Asparagus africanus (Lam) Oberm.	Umayime	Rhizome	Decoction made from *A. africanus rhizome*, and bulbs of *H. albifos* and *B. volubilis* taken orally twice a day.
Cannabaceae			
Cannabis sativa L.	Umya	Leaves	One cup of leaf decoction taken before meals.
Curcubitaceae			
Curcubita spp	Uselentaka	Leaves	Decoction prepared from leaf and roots of *Curcubita* spp, root of *A. amatymbica*, and root of *R. multifidus*. Three tablespoons of the remedy are taken three times a day. Leaves are eaten as a vegetable.
Ebenaceae			
Euclea spp	Idungamuzi	Stem bark	Half cup stem bark tincture taken orally twice a day.
Gunneraceae			
Gunnera perpensa L	Uphuzi	Rhizome	Rhizome crushed and mixed with crushed root of *P. prunelloides*. A decoction is made from the mixture. Three tablespoons are taken orally two times a day.

(Continued)

TABLE 21.1 (*Continued*)
Plants Traditionally Used for the Management of Diabetes Mellitus

Family and Species Name	Local Name(s)	Plant Part Used	Preparation and Dosage
Amarallydaceae			
Haemanthus albifos	Umathunga	bulb	As discussed under *A. africanus*.
Anacardiaceae			
Harpephyllum caffrum Bernh	Umgwenya	Stem bark	Stem bark decoction or infusion taken before meals.
Dioscoreaceae			
Dioscorea spp	ingcolo	Tuber	Half a cup of decoction of *Dioscorea* spp, roots of *Albuca* spp, and *Bulbine* spp roots taken orally once a day.
Euphorbiaceae			
Spirostachys africana Sond.	Umthombothi	Stem bark	Infusion of inner part of bark taken before meals concurrently with root decoction of *Rumex* spp.
Loganiaceae			
Strychnos henningsii Gilg.	Umnonono	Roots	Decoction prepared from roots of *S. henningsii, R. multifidus*, and *G. perpensa* taken orally once a day.
Canellaceae			
Waburgia salutaris (Bertol.f.) Chiov	Isibharha	Stem bark	Half a teaspoon of ground bark is infused in a cup of boiling water and taken orally. Can be taken with *Hypoxis* spp decoction.
Hypoxidaeceae			
Hypoxis spp	Ugobeleweni,	Corms	Three tablespoons of decoction of crushed corms are taken orally.
Lamiaceae			
Leonotis leonurus (L.) R.Br.	Isikrakrisa	Stem and leaves	One cup stem and leaf decoction taken twice a day.
Rubiaceae			
Pentanisia prunelloides (Klotzsch ex Eckl. & Zeyh)	Icimamlilo	Rhizome	As discussed under *G. perpensa*.
Myrtaiceae			
Psidium guajava	Umgwava	Leaves	Crushed leaf infusion taken three times a day.
Ranunculaceae			
Ranunculus multifidus Forssk.	Umvuthuza	Root	As discussed under *S. henningsii*.
Myrsinaceae			
Rapanea melanophloes (L.) Mez	Umaphipha	Stem bark	A decoction prepared from stem bark mixed with *P. prunelloides* root and *Aloe* spp leaves is taken daily.
Vitaceae			
Rhoicissus digitata (*L.f.*) Gilg & Brandt	Uchithibhunga	Tuber	As discussed under *C. edulis*.
Rosaceae			
Rubus spp	Umbhimbi	Roots	As discussed under *A. arborescens*.

Cape as used for the treatment of diabetes mellitus. These findings suggest that narrowing the study area in these kinds of surveys may help researchers to document more indigenous knowledge as far as medicinal plants are concerned. The families Asteraceae, Hypoxidaceae, and Asphodelaceae appear to be the common contributors of plants used for the treatment of diabetes mellitus in the region. Similar to our findings, both Erasto et al. (2005) and Oyedemi et al. (2009) reported the use of *Aloe, Bulbine,* and *Helichrysum* species. The species used, however, seem to depend on its availability in a given area, as evidenced by the use of different species in the three different study areas. Erasto et al. (2005) also reported the use of *Artemisia afra* in the treatment of diabetes, as we

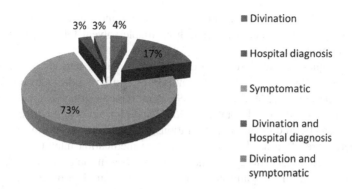

3% 3% 4%

17%

73%

- Divination
- Hospital diagnosis
- Symptomatic
- Divination and Hospital diagnosis
- Divination and symptomatic

FIGURE 21.1 Methods used by respondents to diagnose diabetes mellitus.

did. Similarly, Oyedemi et al. (2009) reported the uses of *Albuca sertosa, Strychnos heningsii,* and *Leonotis leonorus* in the management of diabetes. The repeated citation of similar plant genera and species may imply the effectiveness of the plants in the treatment of the diabetic condition. Thus, such plants must be considered for pharmacological evaluation.

Antidiabetic properties are distributed among various plants from different families. However, no general conclusions can be drawn from families that exhibit antidiabetic activity. The presence of similar phytochemicals, however, may give direction in which to search for antidiabetic principles (Jayakumar, 2010). The general consensus by respondents was that bitter plant remedies are effective in the treatment of diabetes mellitus. This concurs with the findings by Thring and Weitz (2006) on their study of medicinal plant use in the Bredasdorp/Elim region of Southern Overberg in the Western Cape Province of South Africa, in which they reported that bitter plants were implicated by most of their respondents in the treatment of diabetes mellitus. In their study entitled: Evaluation of Virgin Olive Oil Bitterness by Total Phenol Content Analysis, Beltran et al. (2007) assert that bitterness in plants is related to phenolic content. Phenolic and flavonoid compounds have been implicated as having various antidiabetic properties (Atangwho et al., 2009).

Underground plant parts were the most frequently cited, followed by the leaves, stem bark, and whole plants. Uprety et al. (2010) also made similar observations in their study on the indigenous use and bio-efficacy of medicinal plants in the Rasuwa District, Central Nepal. Underground plant parts include roots, tubers, bulbs, and corms. The preference for underground plant parts to prepare remedies concurs with scientific assertions that roots generally contain high concentrations of bioactive compounds (Baral and Kurmi, 2006; Uprety et al., 2010). It is noteworthy that these findings contradict other literature findings, where the leaves are usually cited as the most frequently used (Erasto et al., 2005; Thring and Weitz, 2006) and potent plant part (Muanda et al., 2011). Based on phytochemical distribution and variety, leaves have been demonstrated to be the richest compared to other parts of plants (Muanda et al., 2011), thus it makes sense that they are the most preferred plant part.

Remedies are made from a single plant or from a mixture of plants; however, polyherbal remedies are preferred. The preference for polyherbal remedies could be explained by the fact that polyherbal remedies contain a greater number of phytochemicals, minerals, vitamins, and increased dietary fibre amounts required to alleviate the diabetic condition (Atangwho et al., 2009). These components are believed to work singly or synergistically in alleviating diabetes symptoms. In some cases, animal products such as snake powder and common ingredients such as sugar and salt are added probably to enhance potency and/or palatability.

The choice of method of remedy preparation is influenced by the nature of the plant material and probably the ease with which the medicinal principles can be extracted. The fact that underground plant parts, which are the commonly used plant parts, are usually tough justifies decocting as the method of choice, which aggressively extracts the medicinal components from tough plant parts.

Moreover, rural dwellers have a limited choice with regard to more effective solvents for the extraction of medicinal principles. Water is the readily available solvent followed by alcohol, which is beyond the reach of many rural folks.

Remedies are frequently administered orally as teas for long periods of time, ranging from 2 weeks to a lifetime. This is consistent with findings of similar studies carried out in the region by Erasto et al. (2005) and Oyedemi et al. (2009). Plant remedies may also be taken as emetic or enematic preparations. Of the cited plants, *C. edulis* and *Harpephyllum caffrum* are also used in the preparation of emetic and enematic preparations, respectively. According to the respondents, enemas and emetics are often used as cleansers before medication intended to treat the diabetic condition is administered. The survey revealed inconsistency in the dosage of the remedies. With traditional medical practitioners, plant material is often measured in handfuls, and remedies are taken in cupfuls. There is no standardised dosage for taking remedies. Literature revealed that the cited plants are used for the treatment of various ailments, and some edible ones such as *Rumex* spp are eaten as vegetables (De Beer and Van Wyk, 2011).

Based on the frequency of citation by key informants, *H. caffrum* was observed as the most frequently used plant for the treatment of diabetes mellitus. Literature revealed that these plants are used for the treatment of other ailments besides the management of diabetes mellitus. In South Africa, *Helichrysum odorratissimum* smoke is used during traditional magical-medicinal healing ceremonies to invoke ancestral spirits.

The decoction from the stem bark of *H. caffrum* is implicated in the treatment of bacterial and fungal infections (Chinyama, 2009; Buwa and Van Staden, 2007;Van Wyk, 2010). An infusion of the stem bark is used as a wash for the treatment of acne and eczema, whereas a paste made from burnt bark is used in the treatment of fractures. The survey revealed that the plant has magical uses; it is often used as a love potion. *In vivo* screening of *H. odorratissimum* and *H. caffrum* crude extracts for hypoglycaemic properties is underway.

All respondents agreed that plant material was easier to find during the spring or summer season, when the plants had leaves and or flowers. At least two plants, for example *Aloe* species and *H. caffrum*, were available and could be easily identified all year round. Incidents of wrong identification and consequent herbal intoxication are common among rural communities. These are attributed to sending inexperienced persons to collect plant material and the absence of plant parts used for identification, which are usually fruits, flowers, and leaves.

All respondents strongly believe that their herbal remedies are equally effective or even better than the conventional medicine prescribed by modern medical practitioners. They, however, acknowledge that medicinal plant remedies may be harmful to patients. According to Street et al. (2008), it would be naive to assume that plant extracts are inevitably safe. Secondary plant metabolites are active chemicals intended for defence against competitors and predators. These are the chemicals that confer medicinal properties to plants. It would be dangerous to assume that plants are harmless because they are natural. Herbal intoxication is a common phenomenon among rural communities, and it may lead to death (De Villiers and Ledwaba, 2003). The westernised and traditional healing paradigms must recognise, accept each other, and work in unison to reduce herbal intoxication cases and to improve health delivery to many communities.

The survey revealed that diabetes mellitus is more prevalent in people aged 35 and above. The respondents, however, agreed that nowadays the prevalence of diabetes mellitus in the younger generation is increasing, as observed from the age groups of their patients and community members.

21.4   CONCLUSION AND PROSPECTS

The study has revealed 30 plant species used for the management of diabetes mellitus. This information may offer alternative and affordable ways to manage the condition. However, remedies must be taken with caution, since information on the safety of most plant species is scanty. Documentation of important plant species such as medicinal plants is imperative, as these may be a source of novel drug principles.

ACKNOWLEDGEMENTS

Research reported in this book chapter was supported by the South African Medical Research Council (SA MRC) under a Self-Initiated Research Grant. The views and opinions expressed are those of the authors and do not necessarily represent the official views of the SA MRC.

REFERENCES

Atangwho, I.J., Ebong, P.E., Eyong, E.U., Williams, I.O., Eteng, M.U., and Ebung, G.E. (2009). Comparative chemical composition of leaves of some antidiabetic medicinal plants: *Azadirachta indica*, *Vernonia amygdalina*, and *Gongronema latifolium*. *African Journal of Biotechnology*, *8*(18), 4685–4689.

Azevedo, M. and Alla, S. (2008). Diabetes in Sub-Saharan Africa: Kenya, Mali, Mozambique, Nigeria, South Africa, and Zambia. *International Journal of Diabetes in Developing Countries*, *28*(4), 101–108.

Baral, S.R. and Kurmi, P.P. (2006). A compendium of medicinal plants in Nepal. Rachana Sharma, Kathmandu. *American Journal of Plant Sciences*, *8*(6), 534.

Beltran, G., Ruano, M.T., Jimenez, A., Uceda, M., and Aguilera, M.P. (2007). Evaluation of virgin olive oil bitterness by total phenol content analysis. *European Journal of Lipid Science and Technology*, *109*(3), 193–197.

Bryant, A.T. (1966). *Zulu Medicine and Medicine Men*, Struik, Cape Town.

Buwa, L.V. and van Staden, J. (2007). Effects of collection time on the antimicrobial activities of *Harpephyllum caffrum* bark. *South African Journal of Botany*, *73*(2), 242–247.

Chinyama, R.F. (2009). Biological activities of medicinal plants traditionally used to treat septicaemia in the Eastern Cape, South Africa. MSc thesis. Nelson Mandela Metropolitan University.

De Beer, J.J.J. and Van Wyk, B.E. (2011). An ethnobotanical survey of the Agter-Hantam, Northern Cape Province, South Africa. *South African Journal of Botany*, *77*, 741–754.

De Villiers, F.P. and Ledwaba, M.J.P. (2003). Traditional healers and paedeatric care. *South African Medical Journal*, *93*(9), 664–665.

Erasto, P., Adebola, P.O., Grierson, D.S., and Afolayan, A.J. (2005). An ethnobotanical study of plants used for the treatment of diabetes in the Eastern Cape Province, South Africa, *African Journal of Biotechnology*, *4*(12) 1458–1460.

Farnsworth, N.R. (1994). Ethnopharmacology and drug development, *Ciba Found Symposium*, *185*, 42–59.

Jayakumar, G., Ajithabai, D.M., Sreedevi, S., Viswanathan, P.K., and Remeshkumar, B. (2010). Ethnobotanical survey of plants used in the treatment of diabetes. *Indian Journal of Traditional Knowledge*, *9*(1), 100–104.

Jung, M., Park, M., Lee, H.C., Kang, Y.H., Kang, E.S., and Kim, S.K. (2006). Antidiabetic agents from medicinal plants. *Current medicinal chemistry*, *13*(10), 1203–1218.

Mander, M. (1998). Marketing of indigenous medicinal plants in South Africa. In: *A Case Study in KwaZulu Natal*. Food and Agricultural Organisation of the United Nations, Rome.

Marles, R.J. and Farnsworth, N.R. (1995). Antidiabetic plants and their active constituents, *Phytomedicine*, *2*, 137–189.

Muanda, F., Koné, D., Dicko, A., Soulimani, R., and Younos, C. (2011). Phytochemical composition and antioxidant capacity of three Malian medicinal plant parts. *Evidence-Based Complementary and Alternative Medicine: eCAM*, *2011*, 674320.

Oyedemi, S.O., Bradley, G., and Afolayan, A.J. (2009). Ethnobotanical survey of medicinal plants used for the management of diabetes mellitus in the Nkonkobe municipality of South Africa. *Journal of Medicinal Plants Research*, *3*(12), 1040–1044.

Rheeder, P. (2006). Type 2 diabetes: the emerging epidemic. *South African Family Practice*, *48*(10), 20.

Street, R.A., Stirk, W.A., and Van Staden, J. (2008). South African traditional medicinal plant trade- Challenges in regulating quality, safety and efficacy. *Journal of Ethnopharmacology*, *119*, 705–710.

Thring, T.S.A. and Weitz, F.M. (2006). Medicinal plant use in the Bredasdorp/Elim region of Southern overberg in the Western Cape Province of South Africa. *Journal of Ethnopharmacology*, *103*(2), 261–275.

Uprety, Y., Aselin, H., Boon, E.K., Yadav, S., and Shrestha, K.K. (2010). Indigenous use and bio-efficacy of medicinal plants in the Rasuwa District, Central Nepal. *Journal of Ethnobiology and Ethnomedicine*, *6*, 3.

Van Wyk, C. (2010). Antifungal activity of medicinal plants against oral *Candida albicans* Isolates. *The Preliminary Program for IADR South African Division Meeting & Skills Transfer Workshop*, 13–14 September.

Wild, S., Roglic, G., Green, A., Sicree, R., and King, H. (2004). Gobal prevalence of diabetes: estimates for the year 2000 and projections for 2030. *Diabetes Care*, 27(5), 1047–1053.

22 Traditional Healthcare Practices of Herbal Drugs in Uttarakhand Himalayas, India

Sarla Saklani, Dinesh Prasad Saklani, Manisha Nigam, Chiara Sinisgalli and Abhay Prakash Mishra

CONTENTS

22.1 INTRODUCTION

The healthcare system in India is based on traditional medicine systems like Ayurveda, Unani, Siddha, Yoga, Naturopathy, and Homeopathy. As Ayurveda derivates from *Ayu*, which is life, and *Veda*, which is science, it means knowledge. Hence, Ayurveda refers to the science or knowledge of life. Ayurveda is the earliest Indian system of healthcare focused on views of man and his illnesses, where positive health means metabolically well-balanced human beings.

It is both a curative and a preventive form of therapy. Ayurvedic medicine is thousands of years old, but it is still as relevant as it was in ancient times. It is generally well known and agreed that it was utilized in India more than 5,000 years ago (Ros, 2011). Ayurvedic philosophy emerged between 10,000 and 2,500 B.C. in India. Rigveda and Atharvaveda (10,000 years–5,000 B.C.)—an ancient Indian knowledge—contains references to health and disease. Previously, they were mainly taught verbally from generation to generation. Ayurveda texts like Charak Samhita, which describes 341 plants and plant products for use in medicine, was documented in 1,900 B.C., whereas Sushruta Samhita (based on surgery) was documented in 600 B.C., which describes 395 medicinal plants, 57 drugs of animal origin, and 64 minerals and metals. Charak Samhita classified plant drugs into 50 groups based on their Sanskrit name (Mukherjee, 2001).

Twenty-one thousand plants used as medicinal remedies around the world have been listed by the World Health Organisation (WHO). Among these, there are about 150 species used commercially on a large scale (Zohary and Hopf, 2000). Since India produces the most medicinal herbs in the world, it is known as the botanical garden of the world. Approximately, 2,000 natural drugs are included in the Indian Materia Medica, almost all of which come from traditional systems and folklore practices in India (Narayana, Katayar, and Brindavanam, 1998). About 400 of these drugs come from mineral and animal sources, while the rest come from plants.

DOI: 10.1201/9781003206620-22

The Himalayas form a chain of mountains extending across 3,500 km from Afghanistan to China in a global context. Nearly 7,000 remedial plants are found there (Sheng-Ji, 2001). Uttarakhand—a state of India located in the Northwest Himalayas—harbors several wild medicinal, aromatic, edible fruits, vegetables, and spice plant species.

The extreme climate and different altitudinal zones of Uttarakhand create a remarkable range of habitats and micro-habitats, thus offering a great variety of useful plants to flourish and possibilities of exploring novel, active phytochemicals in the plants. Extreme cold, arid climate, scanty rainfall, high winds velocity, snowstorms, and high ultraviolet radiation characterize high-altitude regions. The vegetation adapts to the extreme conditions of these places.

Different environments and different climate conditions influence the biodiversity of the place. Several plant species are distributed in all Uttarakhand regions, and inhabitants use them for healthcare since ancient times. Indeed, inhabitants of Uttarakhand have developed a great traditional healthcare system by using various plant species.

Kirata and Tangan in ancient and Bhotiyas in medieval modern times had been dealing with the lucrative trade of herbal medicine (Sheng-Ji, 2001; Purohit, 1997; Semwal et al., 2010).

Most people use medicinal plants as medicines, and only the traditional healthcare system is used by the majority of the population, especially in high-altitude regions (Semwal et al., 2010; Dhar et al., 2000; Dhawan, 1997; Gaur et al., 1983). Nowadays, the application of plants in healthcare practices is decreasing due to a plethora of reasons such as:

- The younger generation does not have any interest in learning indigenous knowledge; instead, they prefer modern medicine.
- The majority of people know the economic value of medicinal plants, but they ignore their health-promoting properties and their use in the treatment of different ailments.
- The change in lifestyle by moving to a more Western one (Hazlett and Sawyer, 1998; Kalakoti and Pangety, 1988; Samal et al., 2004).
- The drug composition is considered a "family secret," and it is orally transmitted from generation to generation only in the Vaidya's families. According to ethical and popular belief, Vaidya does not reveal drug composition to common people to avoid misuse and earnings that could make it unsuccessful (Sharma et al., 1997; Ved et al., 1998).

People in some remote areas in the hills still live according to a traditional lifestyle, so they use the local plants to cure diseases and other purposes. As wild edibles are used by local inhabitants and play a significant role in their diet, they are especially beneficial in areas where the varieties and availability of marketed foods are limited. Wild edibles contain secondary metabolites in appreciable amounts and have the potential to be developed as functional food ingredients (Saklani et al., 2011).

Following important steps must be taken for the revival of the traditional healthcare system of Uttarakhand:

- To identify the medicinal plant resource base of Uttarakhand.
- To underline the activity and role of medicinal plants in the state healthcare system.
- Assess the value of documenting the conservation and sustainable utilization of medicinal plants by traditional communities.
- To evaluate the change of values and customs of the Uttarakhand inhabitants and their effects on the traditional healthcare system.
- Assess the present state of knowledge from every part of the state.
- Survey methods of study area for some unique cultural studies.
- The study of income-generation activities for some villagers.

22.2 STUDY AREA

The study area covers the present state of Uttarakhand (North-West Himalayas), located 28,043′ N to and 31° 27′ N latitude and 77° 34′E–81°02′ E longitude, covering an area of 53,483 km out of which 34,651 km^2 is a forest area. It borders China (Tibet) in the north, Nepal in the east, Himachal Pradesh in the northwest, and Uttar Pradesh in the south, which is enriched with an immense diversity of flora due to the region's eco-geographical and eco-climatic conditions. Plants in this region have high medicinal potential, including biological properties. It is rich in natural resources, especially, because of the several glaciers, rivers, dense forests, and snow-clad mountain peaks it nurses. Eco-geographical and eco-climatic conditions influence biodiversity and variety of plant species used in drug formulations. Several ethnic communities live in harmony with biodiversity. Among the tribes, Bhotias, Rajees, Tharus, and Boxas live in Uttarakhand (Figures 22.1 and 22.2), and they still practice their healthcare system (Saklani, 1998). A field survey was carried out in order to collect information from traditional healers and older men and women about the use of herbal species in the treatment of different diseases, especially methods and techniques adopted.

For this reason, the nearby forest and agricultural lands were explored with the help of local inhabitants and practitioners of herbal medicine.

22.3 RITUALS FOR COLLECTION OF MEDICINAL PLANTS

Over the years, local inhabitants of Uttarakhand harvest medicinal plants only during the "Nanda Ashtami" festival, during which selected villagers, the "Jagaryas," start to harvest flowers to offer to the goddess, Nanda. In this way, the collection begins. Harvesting medicinal plants before this ritual involves divine punishment, which results in natural disasters like flash flood, earthquake, landslip,

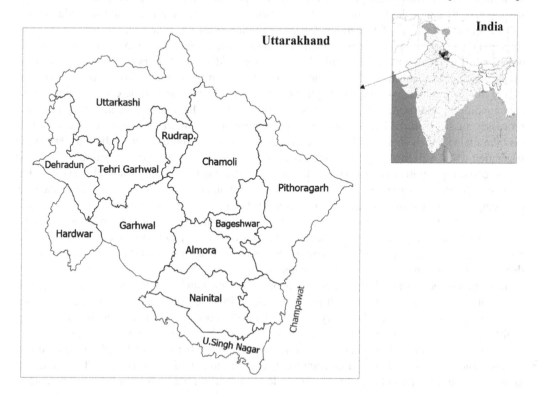

FIGURE 22.1 Map of Uttarakhand, India.

FIGURE 22.2   The landscape of Uttarakhand state of India (pictures taken by Dr. Sarla Saklani).

or drought. The festival takes place at the end of September and the beginning of October—the sowing period of most high-altitude medicinal plants. This practice avoids plant depredation and ensures the maintenance of variations. Other traditional rules are followed and respected in the Bughiyals for harvesting practices of medicinal plants. Men selected by Goddess herself are the only persons who can access the Bughiyals. They are considered mediators between the supernatural powers and the local inhabitants. Plant collection starts in the morning. To avoid over-exploitation of the natural habitat of medicinal plants, the selected men wear a white cotton dhoti, despite cold conditions, and they are fasting. Thus, the time spent on the land is less, and the pollination processes are not disturbed, as the activities are minimal in the morning. Moreover, local citizens visit the site with their shoes off during the growing season to be more cautious about the sprouting vegetation while walking. These precautions guarantee the maintenance of the natural habitat and the potential regeneration of the plant. A traditional mechanism for long-term wild harvesting has also been developed. Upon returning from the forest, inhabitants must leave a sample of the crop in a temple (located at the village entrance) as a gift to the goddess; violation of the rule results in punishment. In this way, wild resources are harvested sustainably, and seeds from harvesting were spread to establish re-growth. The inhabitants explained that the plant roots such as *Arnebia benthamii*, *Angelica glauca*, and *Dactylorhiza hatagirea* must be harvested after the seed is set to avoid decomposition and thus preserve the raw material.

In many villages besides *Vaidyas*, elderly people, both men and women, know different medicinal plants, especially species often used for common illnesses such as stomachache, headache, fever, cold, cough, viral fever, dysentery, diarrhoea, minor sores, and scratches, but they still believe in the efficacy of medicines given by *Vaidyas* (Table 22.1). However, individuals have preference to visit *Vaidyas* to identify their dilemma, and then select the medicinal herb based on the nature of diseases and the person because dose response is not the same in different persons. The method of administration depends on the plants, for example, some herbs are administered on an empty stomach because they are more effective, and the consumption of certain foods is restricted for those undergoing medication. It is not uncommon for some plant species to combine with many other herbs in precise quantities. Generally, the medicinal plants are powdered before the administration;

TABLE 22.1
Medicinal Plants Traditionally Used by *Vaidyas* in Uttarakhand

S. NO	Plants	Traditional Name	Family	Useful Part	Traditional Uses
1.	*Achyranthes aspera* L.	Latjiri, Apamarg	Amaranthaceae	Climber whole	Ringworm, asthma
2.	*Aconitum balfourii* Stapf	Mithabish	Ranunculaceae	Herb root	Leucoderma
3.	*Aconitum heterophyllum* Wall. ex Royle	Atis	Ranunculaceae	Herb root	Fever, stomachache
4.	*Acorus calamus* L.	Buch	Araceae	Herb whole	Dyspepsia, cough
5.	*Adhatoda vasica* Nees	Vashika	Acanthaceae	Shrub leaves	Bronchitis, fever
6.	*Allium cepa* L.	Pyaz	Liliaceae	Herb bulb	Stomach disease
7.	*Allium humile* Kunth	Jambu faran	Liliaceae	Herb leaves	Gastric, digestion
8.	*Allium sativum* L.	Lashun	Liliaceae	Herb bulb	Stomach disease
9.	*Aloe vera* (L.) Burm.f.	Ghirt kumara	Liliaceae	Herb leaves	Diabetes, skin disease
10.	*Angelica glauca* Edgew	Choru	Apiaceae	Herb root	Fever, gastric
11.	*Rhododendron arboreum* Sm.	Buransh	Ericaceae	Flowers, leaves, bark, root	Health tonic, rheumatism, gout, coughs, piles and liver disorders, early stage of cancer
12.	*Arisaema tortuosum* (Wall.) Schott	Bag-Mungri	Araceae	Herb fruit	Piles
13.	*Arnebia benthami* (Kitam.) Rech.f.& Riedl	Balchari	Boraginaceae	Herb root	Hair disease
14.	*Aesculus indica* (Wall. ex Cambess.) Hook.	Pangar, Indian Horse Chestnut	Hippocastanaceae	Tree root bark	Rheumatic pain
15.	*Asparagus racemosus* Willd.	Sataver, Jhirna	Liliaceae	Shrub root	Epilepsy
16.	*Berberis aristate* DC.	Chatru	Berberidaceae	Shrub root	Eye disease
17.	*Berberis lycium* Royle	Kinmor	Berberidaceae	Shrub root	Piles
18.	*Bergenia ciliate* (Haw.) Sternb.	Silphori	Saxifragaceae	Herb root	Kidney stone, piles, paralysis
19.	*Betula utilis* D. Don	Bhojpatra	Betulaceae	Tree bark, leaves	Gout, rheumatism, blood purify
20.	*Brassica campestris* L.	Sarsoo	Brassicaceae	Herb leaves, seed	Anaemia, skin disease
21.	*Cannabis sativa* L.	Bhang	Cannabaceae	Herb leaves	Piles
22.	*Capsicum annuum* L.	Mirch	Solanaceae	Herb fruit	Rabies, snake bite
23.	*Carica papya* L.	Papeeta	Cucurbitaceae	Tree fruit	Skin disease, anti-worm
24.	*Carum carvi* L.	Kala jeera	Apiaceae	Herb seed	Stomachache
25.	*Cedrus deodara* (Roxb. ex.D.Don) G.Don	Devdar	Pinaceae	Tree bark	Rheumatism, back pain
26.	*Centella asiatica* (L.) Urb.	Brahmi	Apiaceae	Herb leaves	Leucorrhoea, epilepsy, mental tonic

(Continued)

TABLE 22.1 (*Continued*)
Medicinal Plants Traditionally Used by *Vaidyas* in Uttarakhand

S. NO	Plants	Traditional Name	Family	Useful Part	Traditional Uses
27.	*Chlorophytum tuberosum* (Roxb.) Baker	Safed musli	Anthericaceae	Herb leaves	Cutaneous
28.	*Citrus aurantifolia* (Christm.) Swingle	Kaghzi Nimbu	Rutaceae	Tree fruit	Common cold
29.	*Coriandrum sativum* L.	Dhaniya	Apiaceae	Herb leaves	Stomach disorder
30.	*Cucumis hardwickii* Royel	Elaroo	Cucurbitaceae	Herb root	Urination
31.	*Cucumis sativus* L.	Kakree	Cucurbitaceae	Climber seed	Urinary disorder
32.	*Curcuma domestica* Valeton	Haldi	Zingiberaceae	Herb rhizome	Blood purifier, eye disease
33.	*Cymbopogon martini* (Roxb.) W. Watson	Mirchya ghass	Poaceae	Herb leaves	Itching
34.	*Cynodon dectylon* (L.) Pers.	Dub ghass	Poaceae	Creeper whole	Vomiting, dysentery,
35.	*Dactylorhiza hatagirea* (D. Don) Soo	Hatajari	Orchidaceae	Herb root	Cuts, wounds
36.	*Daucus carota* L.	Gajar	Apiaceae	Herb root, seed	Anaemia, abortifacient
37.	*Elettaria cardamomum* (L.) Maton	Badi elachi	Zingiberaceae	Herb seed	Heart disease, asthma
38.	*Eupatorium perfoliatum* L.	Bashya, Gandhel	Asteraceae	Shrub leaves	Cuts
39.	*Evolvulus alsinoides* (9L.) L.	Sankhpusphi	Convolvulaceae	Herb leaves	Bronchitis, cough, brain tonic
40.	*Ficus semicardata* Buch. - Hem.ex Sm.	Khina	Moraceae	Tree fruit	Milk, provide strength, baldness
41.	*Glycine max* (L.) Merr.	Kala bhatt	Fabaceae	Climber seed	Kidney stone
42.	*Hedychium spicatum* Sm.	Van-Haldi	Zingiberaceae	Herb rhizome	Asthma, energetic
43.	*Hippophae salicifolia* D. Don	Amesh	Elaeagnaceae	Tree fruit	Cold, cough, cancer
44.	*Juglans regia* L.	Akhrot	Juglandaceae	Tree embryo	Pregnancy
45.	*Lilium polyphyllum* D. Don	Cheerkaguli	Liliaceae	Herb flower	Fever
46.	*Lyonia ovalifolia* (Wall.) Drude	Anyar	Ericaceae	Tree buds	Itching
47.	*Macrotyloma uniflorum* (Lam.) Verdc.	Aheth	Fabaceae	Herb seed	Kidney stone
48.	*Megacarpaea polyandra* Benth. ex Madden	Bermula	Cruciferae	Herb root	Boils, wounds
49.	*Mentha arvensis* L.	Podina	Lamiaceae	Herb leaves	Stomach disorder
50.	*Microstylis muscifera* (Lindl.) Ridl.	Reebjak	Orchidaceae	Herb tuber	Tonic
51.	*Microstylis wallichii* Lindl.	Jeevak	Orchidaceae	Herb bulb	Bronchitis, tonic
52.	*Momordica charantia* L.	Karela	Cucurbitaceae	Climber fruit, seed	Rheumatic, stomachache

(Continued)

TABLE 22.1 (*Continued*)
Medicinal Plants Traditionally Used by *Vaidyas* in Uttarakhand

S. NO	Plants	Traditional Name	Family	Useful Part	Traditional Uses
53.	*Murraya koenigii* (L.) Spreng.	Curry Leaf, Kadhipatta	Rutaceae	Leaf, root, bark	Appetite, digestion, tonic, piles, leucoderma, blood disorder
54.	*Musa paradisca* L.	Kela	Musaceae	Tree spadix	Cough, cold
55.	*Myrica esculenta* Buch.- Ham. Ex. D.Don	Kafal	Myricaceae	Tree fruit	Cardiac disorder
56.	*Nardostachys grandiflora* DC.	Jatamansi	Valerianaceae	Herb rhizome	BP, jaundice, leprosy, heart
57.	*Ocimum sanactum* L.	Tulsi	Liliaceae	Herb leaves	Bronchitis, constipation
58.	*Paeonia emodi Royle*	Chandra	Paeoniaceae	Herb whole	Vomiting, epilepsy, dysentery
59.	*Panicum milaceum* L.	Cheena	Poaceae	Herb seed	Measles
60.	*Phyllanthus emblica* L.	Anowla	Euphorbiaceae	Tree fruit, bark	Blood purifier, sore throat
61.	*Picrorhiza kurrooa Royle*	Kutaki	Scrophulaceae	Herb root	Typhoid fever, jaundice
62.	*Pinus wallichiana* A. B. Jacks.	Kail	Pinaceae	Tree resin	Arthritis
63.	*Piper nigrum* L.	Kali mirch	Piperaceae	Herb fruit	Common cold
64.	*Pistacia integerrima* J.L. Steward ex Brandis	Kakersinghee	Anacardiaceae	Tree leaves	Jaundice, chronic, wounds
65.	*Plantago ovate* Forssk.	Isabgol	Plantaginaceae	Herb seed	Stomachache
66.	*Podophyllum hexandrum* Royle	Bankakri	Podophyllaceae	Herb root	Cancer
67.	*Polygonatum cerrhifolium* (Wall.) Royle	Bakrolu, Mahameda	Liliaceae	Herb root	Boils, wounds
68.	*Polygonatum verticillatum* (l.) All.	Salam misri, Meda	Liliaceae	Herb root	Anaemia, leucorrhoea
69.	*Potentilla fulgens* Wall. Ex Hook	Bajaradanti	Rosaceae	Herb leaves	Toothache
70.	*Prinsepia utilis Royle*	Bhekal	Rosaceae	Shrub fruit	Rheumatism
71.	*Prunus persica* (L.) Batsch	Aaru	Rosaceae	Tree leaves	Anti-worm
72.	*Pyracantha crenulata* (Roxb. ex D.Don) M.Roem.	Ghingharu	Rosaceae	Leaves	Diarrhoea, immunomodulator
73.	*Quercus leucotrichophora* A. Camus	Banj	Fagaceae	Tree seed	Snake bite
74.	*Raphanus sativus* L.	Muli	Brassicaceae	Herb root	Jaundice
75.	*Rauvolfia serpentina* (L.) Benth. ex Kurz	Sharpgandha	Apocynaceae	Herb root	Cough, heart disease

(*Continued*)

TABLE 22.1 (*Continued*)
Medicinal Plants Traditionally Used by *Vaidyas* in Uttarakhand

S. NO	Plants	Traditional Name	Family	Useful Part	Traditional Uses
76.	*Reinwardtia indica* Dumort.	Phiunli	Linaceae	Herbpetal	Tongue wash
77.	*Rhamnus virgatus* Roxb.	Chadolu	Rhamnaceae	Tree bark	Eczema, ringworm
78.	*Rheum australe* D. Don	Dolu	Polygonaceae	Herb root	Internal wounds
79.	*Rhododendron anthopogon* D. Don	Awon	Ericaceae	Tree leaves	Ringworm
80.	*Rumex hastatus* D. Don	Almoru	Polygonaceae	Herb leaves	Wounds, bleeding
81.	*Sapindus mukorossi* Gaertn.	Reetha	Sapindaceae	Tree fruit	Snakebite
82.	*Saussurea costus* (Falc.) Lipsch.	Kuth	Asteraceae	Herb root	Toothache, jaundice, snakebite
83.	*Saussurea obvallata* (DC.) Edgew.	Brahm kamal	Asteraceae	Herb flower	Leucorrhoea, mental Disorder
84.	*Selinum tenuifolium* Wall. ex C.B. Clarke	Bhutkeshi	Apiaceae	Herb root	Cough, asthma
85.	*Sesamum orientale* L.	Til	Pedaliaceae	Herb seed, leaves	Aphrodisiac, body Pain
86.	*Smilax aspera* L.	Kukurdara	Smilaceae	Climber root	Rheumatic, arthritis
87.	*Solanum nigrum* L.	Makoi	Solanaceae	Herb fruit	Liver, piles, diarrhoea
88.	*Spondias pinnata* (L.f.) Kurz	Amra	Anacardiaceae	Tree leaves	Ear disease
89.	*Swertia chirata* Buch.-Ham. ex Wall.	Cherayita	Gentianaceae	Herb whole	Fever, diabetes
90.	*Taxus baccata* L.	Thuner	Taxaceae	Tree bark	Anti-cancer, bone fracture
91.	*Terminalia arjuna* (Roxb. ex DC.) Wight & Arn.	Arjuna	Combretaceae	Tree bark	Bone fracture
92.	*Terminalia bellirica* (Gaertn.) Roxb.	Bahera	Combretaceae	Tree fruit	Provide strength
93.	*Terminalia chebula* Retz.	Haira	Combretaceae	Tree	Strengthened body, stomachic
94.	*Thalictrum javanicum* Blume	Peeli jari, Mameri	Ranunculaceae	Herb root	Diabetes, jaundice
95.	*Tinospora sinensis* (Lour.) Merr.	Gilai	Menispermaceae	Climber whole	Fever, leprosy, urinary
96.	*Trigonella foenum-graecum* L.	Methi	Fabaceae	Herb leaves	Pneumonia
97.	*Valeriana hardwickii* Wall.	Tagar	Valerianaceae	Herb root	Urinary disorder, joint pain
98.	*Verbascum Thapsus* L.	Akulbeer	Scrophulaceae	Herb whole	Bronchitis, asthma
99.	*Vigna mungo* (L.) Hepper	Kali dal	Fabaceae	Climber seed	Bone fracture
100.	*Withamia somnifera* (L.) Dunal	Ashwagandha	Solanaceae	Herb whole	Rheumatism, ulcer, carbuncle
101.	*Zanthxylum armatum* DC.	Timru	Rutaceae	Shrub bark	Toothache
	Zingiber officinale Roscoe	Adrak	Zingiberaceae	Herb rhizome	Paralysis, epilepsy

this form is considered more effective than others like pills or tablets by the local people. Many times, leaves, stems, fruits, and root/tuber were crushed by using mortar and pestle, and then they were boiled in water. Generally, the external application of medicine was employed for the treatment of burns, wounds, headaches, cuts, boils, and skin diseases. For example, concoctions were applied unswervingly on the bruise or the disease-ridden body part, or plant paste was useful to set bone dislocation or fracture and for muscular pain. In addition, to cure ailments like fever, garlands made from the stem or the roots were worn.

22.4 METHODS USED FOR THE FORMULATION

Ancient people together with *Vaidyas* know the healthy properties of plants and their use for common diseases such as headache, viral infection, cough, fever, stomachache, diarrhoea, and dysentery. Generally, people prefer to interface with *Vaidyas* who identifies the problem and the right treatment indicating the exact doses and times. The dose response of each drug depends on the person and on time.

The different formulations are guarded by *Vaidyas* in their diary in coded language.

The methods of formulation and uses of herbal drugs vary depending on the disease. In some cases, different plant species can be mixed in specific amounts. Medicinal plant associations are widely used in modern phytomedicine to achieve a synergistic effect between them; thus, it is considered a very important and effective method to use herbal drugs.

Some popular formulations (Tables 22.2 and 22.3) are:

- Powders
- Decoctions
- Juices
- Infusions
- Bhasma
- Paste

The powder form is the most used being the most effective, according to a popular thought. Regarding the decoctions, parts of the plants like leaves, fruits, seeds, and roots are crushed using mortar and pestle, and boiled with water. Then, the liquid is drunk cold or hot, or, in some cases, it is rubbed on the body in the case of wounds or infections. Bathing the patient directly in the decoction is recommended in case of skin diseases. Paste formulations are useful for bone dislocation or fracture and muscular pain. Roots and stems of the plant are woven into a garland for treating some ailments like fever.

22.5 EROSION OF TRADITIONAL KNOWLEDGE BASE

The influence of modernization and the limitations on medicinal plant extraction have reduced the use of traditional medicine in favour of allopathic drugs. In remote, high-altitude areas, the folk healthcare system is on the edge of vanishing. Earlier, there were many birth attendants, bonesetters, herbalists, and wandering monks; but, today, the number of them is gradually decreasing. As tribal folk maintains their oral tradition of passing on knowledge from generation to generation, they are slowly losing the wealth of information conserved as an unwritten materia medica. Many aspects of tribal knowledge remain secret. The information could relate to the occurrence, characteristics, therapeutic effects, processing methods, and the use of the plants themselves for treatment.

The poorest people still use the traditional healthcare system and contact the traditional *Vaidyas*; but they also prefer allopathic drugs for some ailments instead of the Ayurvedic system, as it has a slow effect and takes a longer time to cure the ailment.

TABLE 22.2

List of Uttarakhand Indigenous Medicinal Plants, Their Formulation, and Uses

S. NO	Scientific Name	Folk Name	Parts Used	Uses	Mode of Treatment
1.	*Achyranthes bidentata* Blume.	Dansh	Root	As laxative	Decoction in 1.0 L water is to be given two times.
2.	*Artemisia nilagirica* (C. B. Clarke) Pamp.	Patti, kunj	Whole plant	For urinary tract infection	Brew up in 1.0 L water; 1.0 cup is to be given with gur.
3.	*Artemisia sacrorum* Ledeb.	Kapar Patti, jholpatti	Leaf/Bud	For hair fall	Brew in 2.0 L of water; one cups is to be given twice a day.
4.	*Abies webbiana* Lindl.	Raisal barmi radha	Bud	In cough	Brew up in 3.0 L water is to be given thrice a day.
5.	*Abina cordifolia* Hook. F	Haldu	Bud & leaf	For wound & fever	Apply a paste of fresh buds to the sore. 0.5 L of leaves decoction is to be given three times a day during fever.
6.	*Acacia catechu* Wild.	Khair	Stem	In the urine problem, dysentery	Decoction in 0.5 L water; one cup is to be given four times a day.
7.	*Achyranthes aspera* L.	Chirchira	Whole plant	For teeth problem	Whole plant decoction in 0.5 L water.
8.	*Aconitum balfouria* Stapf.	Meethabish, bishjahar	Root	In wound	One matured root burn in 1.0 L of oil is to be given as ointment.
9.	*Acorus calamus* L.	Banj	Root	Fever, pain	Two matured roots with fibrous food are to be given daily.
10.	*Adiantum venusthum* G. Don.	Hanshraj	Seed	For chest problems and hair fall	One palmful seed to be given with fibrous food.
11.	*Aesculus indica* (Wall. ex Cambess.) Hook.	Pangar	Fruit	In stomach problem	The decoction of one palmful of fruit in 0.5 L water should be administered with gur.
12.	*Agrimonia pilosa* Ledeb.	Kafliya	Whole plant	For purification of blood	Half palmful whole plant decoction in 3.0–4.0 L water; 0.75 part is to be given with gur in the morning.
13.	*Ajuga parviflora* Benth.	Ratpatia	Whole plant	In arthritis	Brew up one palmsful whole plant part in 0.75 L water; one cup is to be given every day.
14.	*Allium stracheyi* Baker.	Jambu	Whole Plant	For stomach problem	Two palmful of whole plant part is to be given thrice a day.

(Continued)

TABLE 22.2 (*Continued*)
List of Uttarakhand Indigenous Medicinal Plants, Their Formulation, and Uses

S. NO	Scientific Name	Folk Name	Parts Used	Uses	Mode of Treatment
15.	*Allium wallichii* Kunth.	Jangali lasun	Root	In infection	Two nodes are given daily.
16.	*Aloe vera* L.	Patquar	Leaf	Stomach problem	Juice of leaves is to be given 0.5 cups a day.
17.	*Althaea officinalis* L.	Jangalihauli	Root	For termination of pregnancy	Three/four matured root decoction in 1.0 L water.
18.	*Anagallis arvensis* L.	Vish khaparia	Fruit/ Leaf	As pain killer	Two palmful fruit/ leaves are to be given daily.
19.	*Anemona obtusiloba* Don.	Kakaria	Leaf	In sinus	A cotton bud is dipped in a leaf paste using ghee to clean the sinuses.
20.	*Artemisia maritime* L.	-	Bud/ Leaf	For Indigestion	One palmful bud/leaves decoction in 1.0 L water; 1.0 cup is to be given daily.
21.	*Artemisia parviflora* Roxb.	Patti, dhopani	Leaf/ Bud	For roundworm	One palmful leaves/ bud brew up in 1.0 L water is to be given 0.125 L in 1h intervals.
22.	*Asparagus racemosus* Willd.	Kairuwa	Bud	In liver problem & to enhance lactation	One palmful bud is to be given twice a day.
23.	*Atropa belladonna* L.	Dhatur jahar	Leaf	In injury as a pain killer	Apply an ointment made from a paste of one handfuls of leaves burnt in oil.
24.	*Berberis aristata* DC	Kilmori	Root & stem	In fever, weakness	One palmful root/ stem brew up in 0.5 L water; one cup is to be given daily.
25.	*Bergenia ciliata* Moench.	Silfhora	Root	For hydrophobia	Two handful root decoction in 0.5 L water; 1.0 cup is to be given thrice a day.
26.	*Betula utilis* Don.	Bhuj, bhojpatra	Seed	To protect from worm	Two small pinches.
27.	*Boerhaavia diffusa* L.	Parnata	Leaf	In blood dysentery, in dropsy	Juice of leaves given thrice a day.
28.	*Brassica napus* L.	Kali sarso	Seed	In poor appetite	Twice a day, take two handfuls of seed with fibrous meals and gur.
29.	*Butea frondosa* Koen.	Dhank	Flower, seed	As painkiller	Paste

(*Continued*)

TABLE 22.2 (*Continued*)
List of Uttarakhand Indigenous Medicinal Plants, Their Formulation, and Uses

S. NO	Scientific Name	Folk Name	Parts Used	Uses	Mode of Treatment
30.	*Calendula officinalis* L.	Ganda (Tokar)	Leaf	In bleeding	Juice
31.	*Calotropis procera* R. Br.	Ank	Root	In indigestion	One palmful powder of root brew up in 1.0 L water; one cup is to be given twice a day.
32.	*Canna indica* L.	Kewara	Root	In disinterest, in afra	One handful of roots powder is to be administered with gur.
33.	*Capsella bursa-pastoris,* Moench.	Torighash	Whole plant	For sikka rog	Two handfuls of whole plant part brew up in the water are to be taken twice.
34.	*Capsicum annum* L.	Khusane, marac	Fruit	As oil massage.	One palmful fruit brew up in 3.0 L water; one cup is to be given twice a day.
35.	*Cardamine impatiens* L.	-	Whole plant	For Tantrika in calf	One handful whole plant concoction in 1.0 L water is to be given two times for vigor.
36.	*Cassoa absus* L.	Banar, chakwar	Seed	In urine problem	One palmful seeds concoction in 0.5 L water; one cup is to be given thrice a day.
37.	*Centella asiatica* (L.) Urban	Brahmi	Leaf	For brain fever	A paste made from green leaves is applied to the forehead.
38.	*Chenopodium album* L.	Bethuwa	Leaf/seed	For worm	Two palmful seed is to be given before breakfast.
39.	*Cinnamomum tamala*Ness.	Kiriya, karkiriya, dalchini	Leaf	In stomach problem, in gastric problem	Leaves and bark powder with half handful of fiber food.
40.	*Clerodendrum infortunatum* Gaertn.	Aranyo	Bark	In afra	One cup of powdered bark decoction in 2.0 L water is to be taken three times per day.
41.	*Cuminum cyminum* L.	Jeera	Seed	For indigestion	One palmful seed in 0.25 L water is to be given daily.
42.	*Cureuma angustifolia* Roxb.	Banhaldi	Root	In gastric problems, anti-worm	Paste of root.
43.	*Datura metel* L.	Dhatura	Seed	As a pain killer (for external use only)	25g roasted seed in 1.0 L oil used for massage.

(*Continued*)

TABLE 22.2 (*Continued*)
List of Uttarakhand Indigenous Medicinal Plants, Their Formulation, and Uses

S. NO	Scientific Name	Folk Name	Parts Used	Uses	Mode of Treatment
44.	*Datura stramonium* L.	Dhatura	Leaf	In injury as a pain killer	Paste of one palmful leaves is to be applied as an ointment.
45.	*Delphinium denudatum* Wall	Nirwishi, munel	Seed	In epilepsy	One palmful root decoction in 0.5 L water; two spoonsful is to be given thrice a day.
46.	*Digitalis purpurea* L.	Prawasit, degitelis, tilpushpi	Leaf	In burning	One palmful leaves roasted with oil are to be used as an ointment.
47.	*Emblica officinalis* Gaertn.	Aula, awla	Fruit	In eye disease/ good health	Two palmful fruits powdered with fibrous food.
48.	*Ephedra gerardiana* Wall.	Gidjing	Stem	In pain	One cup of decoction made from one bunch of stem pieces in 2.0 L of water is given in the morning.
49.	*Equisetum arvense* L.	Horsetel	Whole plant	For urinary problem	1/2 palmful whole plant decoction in 1.0 L water.
50.	*Euphorbia prolifera* Buch. Ham. ex. Don.	Duwila	Fruit	Used in dog bite	Powder
51.	*Foeniculum vulgare* Mill.	Saup	Seed	For hookworm	Before breakfast, a handful of seed in 0.125 L of water is administered.
52.	*Fragaria vesca* L.	Pudalia Kafal	Leaf	To protect abortion	Two palmful leaves are to be given daily.
53.	*Fumaria parviflora* Lamk.	Pitpapara	Whole plant	In skin etching(disease)	One palmful whole plant decoction in 1.0 L.
54.	*Gentiana tenella* (Roltb) H. Smith.	Kutuki, katuwi	Fruit	In hysteria, In weakness	25g of the bark of fruits decoction in 1.0 L water; 1.0 cup is to be given with honey daily.
55.	*Geranium ocellatum* Camb.	Bhiljari	Whole plant	As insecticide	Powder of the whole plant with fibrous food is to be given twice a day.
56.	*Hedychium spicatum* Hamex. Smith	Kapur Kachari	Root	For fever & cough	The root is to be given with gur.
57.	*Holarrhena antidysenterica*, Wall.	Quiar, indraw	Seed & bark	In fever, gastric & dysentery	One handful powder of bark/ seed brew up in 1.0 L water; 1.0 cup is to be given with gur.
58.	*Hyoscyamus niger* L.	Bran juwan	Leaf & Seed	As pain killer	As an ointment, a paste of leaves and seeds is used.

(*Continued*)

TABLE 22.2 (*Continued*)
List of Uttarakhand Indigenous Medicinal Plants, Their Formulation, and Uses

S. NO	Scientific Name	Folk Name	Parts Used	Uses	Mode of Treatment
59.	*Hypericum cernuum* Roxb.	Vaya, culi	Whole plant	For Hodgkin's disease, For wound	Two palmful whole plant decoctions in 1.0 L water are to be given two times.
60.	*Juglans regia* L.	Akhore	Leaf/ fruit	In stomach problems, as anti-worm	Two palmful leaves or two green fruits decoction in 1.0 L water; one cup with two spoons of honey is to be given thrice a day.
61.	*Juniperus communis* L.	Jhora, khichiya	Fruit	In liver disease	Twelve fruits daily.
62.	*Litsea umbrosa* Ness.	Circira	Leaf	In bone injury	Leaves paste in water as an ointment.
63.	*Linum usitatissimum* L.	Alsi	Whole plant	For strength	Two palmful whole plant decoction in 1.25 L water are to be given two times.
64.	*Litsaea polyantha* Juss.	Cirira	Leaf	In injury	Powder of bark & leaves in cold water as an ointment.
65.	*Lobelia pyramidalis* Wall.	Bran tambacoo	Whole Plant	For liver disease	Two palmful whole plant decoction in 0.75 L water; one spoon is to be given with honey thrice a daily.
66.	*Mallotus philippinensis* Muell. & Arg.	Roli, kasela	Fruit	To protect from worm	Fruit extract with one palmful fibrous food is to be given once a day.
67.	*Melilotus alba* Ledeb.	Banmethi	Whole plant	For stomach problem and indigestion	One palmful whole plant is to be given three times in a day for vigor.
68.	*Mentha arvensis* L.	Pudina, eliachi	Whole plant	In post-pregnancy problems	Two palmful whole plant decoction in 1.0 L water; 0.25 part is to be given thrice a day.
69.	*Ocimum sanctum* L.	Tulsi	Whole plant	In fever	Two palmful whole plant is to be given twice a day.
70.	*Origanum vulgare* L.	Jangali tulsi	Whole plant	Indigestion	04 palmful whole plant with fibrous food is to be given twice a day.
71.	*Plantago ovata* Forssk.	Isabgol	Seed	In dysentery	Semisolid paste of one palmful seed in 0.5 L water is to be given thrice a day.

(Continued)

TABLE 22.2 (Continued)
List of Uttarakhand Indigenous Medicinal Plants, Their Formulation, and Uses

S. NO	Scientific Name	Folk Name	Parts Used	Uses	Mode of Treatment
72.	*Paeonia emodi* Wall.	Bhoi Pawin	Root	In stomach problem	One matured root decoction in 0.75 L water; one cup is to be given with 100 g gur thrice a day.
73.	*Pimpinella diversifolia* DC	Dhanjari	Seed	For lactation	One palmful seed is to be given daily.
74.	*Piper longum* L.	Pipal	fruit	In low appetite, As oil massage	Powder of fruit or oil of fruit powder is used for massage.
75.	*Plantago major* L.	Vrantank	Leaf	In injury, teeth problem, fever	Paste of leaves in water (for teeth pain). Two bunch of leaves decoction in 1.0 L of water; 1/6 part is to be given thrice a day.
76.	*Potentilla argyrophylla* Wall. ex Lehm.	Danti, brajdanti	Leaf/ Root	For stomach problem	One palmful leaves/Two matured root decoction in 0.17 L water is to be given thrice in a day.
77.	*Primula denticulata* Smith.	Vish Khaparia	Fruit	In cough, useful for mammary glands	Two palmful flower are to be given with gur.
78.	*Punica granatum* L.	Darim	Skull of fruit	As antimicrobials	One palmful skull of fruit decoction in 0.5 L water; one cup is to be given three times a day with gur.
79.	*Quercus semecarpifolia* Sm.	-	Bark	In dysentery	Two palmful bark powdered decoction in one cup of water are to be given twice a day.
80.	*Quercus dilatata* Lindl.	Banj	Bark	In dysentery	Two palmful of bark powdered decoction in 1.0 L water; one cup is to be given twice a day.
81.	*Reinwardtia trigyna* Planch.	Pyuli	Root	In wound	One bunch of root decoction in 0.5 L water; one cup is to be given in a gap of two days.
82.	*Rhamnus virgata* Roxb.	Chaitula	Fruit	In leg swelling	05 matured fruits decoction in 0.25 L water is to be given daily.
83.	*Rheum emodi* Wall.	Dolu, archa	Root	For blood purification, for energy	One matured root decoction in 1.0 L water is to be given three times.

(Continued)

TABLE 22.2 (*Continued*)
List of Uttarakhand Indigenous Medicinal Plants, Their Formulation, and Uses

S. NO	Scientific Name	Folk Name	Parts Used	Uses	Mode of Treatment
84.	*Ribes grossularia* L.	Caktu	Whole plant	For preventing abortion	One palmful whole plant is to be given daily.
85.	*Ricinus communis* L.	Erind	Leaf	For internal injury	Oil
86.	*Rosa moschata* Herrm.	Kunj pani	Fruit	For leucorrhoea, bleeding, pregnancy termination.	Two palmful fruit with one spoon honey are to be given daily.
87.	*Rubus lasiocarpus* Sm.	Kala Hisalu	Leaf	In pregnancy	The leaf is useful for cows, especially in pregnancy pain.
88.	*Rubus paniculatus* Sm.	Kala Hisalu (Kadula)	Leaf	In pregnancy	Two palmful leaves decoctions in 0.5 L water; one cup is to be given twice a day.
89.	*Rumex hastatus* D. Don	Bhilmora	Whole plant	For skin disease, in fever	One palmful whole plant decoction in 0.75 L water; one cup is to be given thrice a day,
90.	*Senecio rufinervis* DC.	-	Seed	For wound	Three palmful seed is to be given twice a day.
91.	*Salix elegans* Wall.	Garbainsh	Fruit	In rickets	Three palmful fruits decoctions in 1.0 L water one cup is to be given thrice a day.
92.	*Salvia lanata* Roxb.	Sania, sunip	Whole plant	For vomiting, painkiller	Two palmful whole plant with gur and fibrous food thrice a day.
93.	*Satyrium nepalense* D. Don.	Salangmishri	Root	As tonic	Two palmful roots decoctions in 0.75 L liter water are to be given ½ parts twice a day.
94.	*Scutellaria angulosa* Benth.	Karuijhar	Whole plant	In acidity	One palmful whole plant decoction in 0.5 L water; one spoon is to be given with honey thrice a day.
95.	*Senecio chrysanthemoides* DC.	Ratpatia	Whole plant	For skin disease	Two palmful whole plant decoctions in 0.75 L water; one cup is to be given daily.
96.	*Swertia purpurascens* Wall.	Ciraita	Whole Plant	In fever, in weak appetite.	Two palmful whole plant decoctions in 1.0 L water; one cup is to be given thrice a day.
97.	*Tagetes erecta* L.	Hazari	Fruit	In vomiting, in healing wound	One palmful fruit is given with fibrous food at the time of vomiting. External use for filling wounds.

(Continued)

TABLE 22.2 (*Continued*)
List of Uttarakhand Indigenous Medicinal Plants, Their Formulation, and Uses

S. NO	Scientific Name	Folk Name	Parts Used	Uses	Mode of Treatment
98.	*Tanacetum nubigenum* Wall.	-	Leaf/ Fruit	As energy syrup, anti-microbial.	One palmful leaves/ fruit decoction in 1.0 L water; one spoon is to be given with honey.
99.	*Thymus serpyllum* L.	Van ajmain	Whole plant	In chest pain	One palmful whole plant decoction in 0.5 L water; one cup is to be given twice a day.
100.	*Trifolium repens* L.	Garila	Whole plant	For satrika (back pain/sciatica)	04 palmful whole plant is to be given Two times a day.
101.	*Urtica dioica* L.	Kandali	Leaf	Skin disease, for lactation	One palmful leaves is to be given with fibrous food in 1hour intervals.
102.	*Valeriana hardwichii* wall.	Samyo, Dhup	Root	For clostridium titaini (tetanus)	04 matured root decoctions in 2.0 L liter water; 0.25 L is to be given twice a day.
103.	*Verbascum thapsus* L.	Akalvir	Leaf	In bronchitis	One palmful leaves decoction in 0.75 L water; one cup is to be given thrice a day.
104.	*Viola serpens* Wall.	-	Root	For liver	Two palmful root decoctions in 1.0 L water is to be given three times with honey.
105.	*Viola biflora* L.	-	Whole plant	In calf for heart & faint problem	Two palmful whole plant Two times a day for attack. Three or four parts of two palmsful whole plant with a spoon of honey are to be given two times for heart and skin problems.
106.	*Viola patrinii* DC	-	Root	For liver	Two palmful root decoctions in 1.0 L water are to be given two times for vigor.
107.	*Viscum album* L.	Bana	Fruit	In pregnancy problem	Six fruits with milk twice a day.
108.	*Woodfordia floribunda* Salisb.	Dhow	Flower	As energy syrup	One palmful dry flower decoction in water.
109.	*Zingiber officinale* Roscoe	Banhaldi	Root	Internal injury, as anti-worm	Paste of root.

TABLE 22.3

Medicinal Plants of Uttarakhand Used in Various Ayurvedic Drugs

S.NO	Ayurvedic Formulation	Botanical Name	Local Name	Parts Used
1.	Abana	*Nepeta hindostana* (B. Heyne ex Roth) Haines	Billilotin	Whole plant
2.	Abana, Bonnisan, Geriforte, Koflet lozenge, Luol, Mentat, Renalka, Anxocare, Digyton, Bresol, Chyavanaprasha	*Elettaria cardamomum* (L.) Maton	Elaechi	Seed
3.	Abana, Diabecon, Diakof, Geriforte, Herbolax, Himcocid, Koflet, Menosan, Septillin, Rumalaya forte, Himpyrin, Regurin, Yastimadhu, Bleminor	*Glycyrrhiza glabra* L.	Muleti	Stem and root
4.	Abana, Diabecon, Diakof, Koflet, Ophthacare, Rumalaya, Canisep, Scavonvet, Hiora-K-mouthwash, Bresol, Dabecon	*Ocimum sanctum* L.	Tulsi	Whole plant
5.	Abana, Diakof, Mentat, Purim, Talekt, Dental cream (Himalaya), Anxocare, Appetonic forte,	*Embelia ribes* Burm.f.	Vidanga	Seed, fruit and bark
6.	Abana, Mentat, Anxocare, Galactin, Hiora-mouthwash	*Foeniculum vulgare* Mill	Sanuf	Whole plant
7.	Abana, Mentat, Rumalaya, Anxocare	*Nardostachys jatamansi* (D. Don) DC.	Jatamansi	Herb
8.	Agnikumar ras, Anand bhairavras, Pushyanug churn, Ativisha churn, Chandraprabha vati.	*Aconitum heterophyllum* Wall. ex Royle	Atis	Roots
9.	Amritpras ghrit, Dantasodhak, Yavih	*Juglans regia* L.	Akhrot	Leaves, stem bark, and fruits
10.	Appentonic fote, Lesuna	*Allium sativum* L.	Lasuna	Bulb
11.	Aragvadhadirsta, Aragvadhadighrita, Aragvadhadi tel, Agni- kumar churn, Kankayan gutika	*Cassia fistula* L.	Amaltas	Fruits and roots
12.	Arogyavardhi vati, Laxmi narayan ras, Mahayogray guggulu, Amritarista, Cruminillsyrup, Livertone, Carminex, Pigmeto, Livomyn, Hepatogard, Calcury Apimore (Teb& tonic)	*Picrorhiza kurrooa* Royle	Kutaki	Rhizomes and roots
13.	Arshonyte forte	*Shorea robusta* Gaertn.	Sal	Leaves and bark
14.	Arvindasava, Pushyanug churn, Sarivadi vati, drakshavleha.	*Nelumbo nucifera* Gaertn.	Kamal	Rhizome, flowers
15.	Arvindasava,Chandrakala ras, Chandanasava, Mahasudorshan kwath Livomyn drops	*Fumaria indica* (Hausskn.) Pugsley	Pit papra	Whole plant
16.	Asbwagandha churn, Ashwa gnadha rasayan, Ashwagan dharista, Kamdev ghrit, Chyaranpras, Phalaghrit Mahamash tel, Manoll, M_2 tone	*Withania somnifera* (L.) Dunal	Aswagandha	Roots
17.	Ashmeriher kwath, Pashanbhedadi kwath, Pushyanug churn, Vatvidhv ansak ras. Cystone tablets.	*Bergenia ligulata* Engl.	Silpara	Rhizomes

(Continued)

TABLE 22.3 (*Continued*)
Medicinal Plants of Uttarakhand Used in Various Ayurvedic Drugs

S.NO	Ayurvedic Formulation	*Botanical Name*	Local Name	Parts Used
18.	Bharmi churn, Bharmi ghrit, Brahmi rasayan	*Bacopa monnieri* (L.) Wettst.	Jalbirahmi	Whole plant
19.	Bharmi Vati, Panchtikta guggula, Madusnhi rasayan, Dantmanjan.	*Zanthoxylum armatum* DC	Timroo	Whole plants
20.	Bhringrajasava, Sutsekherras Anand bhairas ras, Acidocid syrup, Geriforte Mahabhringraj tel, Gandhak rasayana, Livomyn, Neo, Ojus, M$_2$- tone, Hepatogard.	*Eclipta prostrata* (L.) L.	Bhringraja (Asteraceae)	Whole plants
21.	Bilvadi kwath, Bilava churn, Bilva-majjadi yog, Dasamoo -larist, Tentex forte, Lukol, Diarex, Sumento (Tablets).	*Aegle marmelos* (L.) Corrêa	Bael	Fruit and roots bark
22.	Bonnisan, Geriforte, Liv 52, Herbolax	*Cassia occidentalis* L.	Kasondi	Herb
23.	Brahmi vati, Brahmi ghrit, Brahmi tel, Saraswatarista, Geriforte, Mental (Teb & syp)	*Centella asiatica* (L.) Urb.	Brahmi	Whole plant
24.	Brahtyadi kwath, Dasamoolarista, Dasamool kwath, Chyavanpras, Mahasudarshan churn, Mahanarayan tel	*Solanum anguivi* Lam.	Kandyari	Whole plant
25.	Canisep, Scavon	*Linum usitatissimum* L.	Atasi	Seed
26.	Chayavanpras, Amlaki churn, Triphala chrn, Triphala ghrit, Brahmi amla tel, Fresh fruit pulp	*Emblica officinalis* Gaertn.	Amla	Fruit
27.	Chayavanpras, Mahamash tel, Narsingh tel, Grabbdharin vati,	*Dioscorea bulbifera* L.	Genthi	Roots
28.	Chitrakadi vati, Chitrak haritake, Chitrak ghrit Mahasankh vati, Mahamritaunjay ras, Pigmento, Liv-52, Alarsin, Vigiroll	*Plumbago zeylanica* L.	Chitrak	Roots
29.	Chyavanpras, Triphala ghrit, Bramh rasayan, Mahanarayan tel	*Polygonatum verticillatum* (L.) All.	Kantula	Rhizomes
30.	Chyavranpras, Swanshar kwath, Kashhar kwath, Satpatradi churn, Tonali Livomyn, Livomap (tab & syrup)	*Phyllanthus fraternus* G. L. Webster	Jarmala	Whole plant
31.	Cold balm (Himalaya), Erina, Savon	*Eucalyptus globulus* Labill.	Lyptus	Leaf, bark, and oil
32.	Confido, Gerifort, Steman, Tentx forte	*Argyrcia nervosa*	Vriddadaru	Whole plant
33.	Confido, Lukol, Serpira	*Rauvolfia serpentina* (L.) Benth. ex Kurz	Sarpgandha	Root
34.	Confido, Lukol, Speman, Tentex royal, Spemanvet	*Hygrophila auriculata* (Schumach.) Heine	Talimkhana	Root, leaf, and seed
35.	Confido, Speman, Galactin, Chyvanprasaha	*Leptadenia reticulata* (Retz.) Wight & Arn.	Dori	Leaf and root
36.	Cyston, Diakof, Koflet, Nefrotec	*Onosma bracteatum* Wall.	Bal-jari	Root
37.	Cystone, Nefrotec	*Achyranthes aspera* L	Apamarga	Herb
38.	Cystone, Nefrotec	*Didymocarpus pedicellata* R.Br.	Shilapuspha	Leaf

(*Continued*)

TABLE 22.3 (*Continued*)
Medicinal Plants of Uttarakhand Used in Various Ayurvedic Drugs

S.NO	Ayurvedic Formulation	Botanical Name	Local Name	Parts Used
39.	Cystone, Nefrotec	*Vernonia cinerea* (L.) Less.	Kalijiri	Leaf, flower, and seed
40.	Dadimastak churn, Dadimadi ghrit, Lavanbhaskar churn Eladi churn.	*Punica granatum* L.	Dadim	Fruits
41.	Dasomoolarista, Dasamooll kwath, Pushyanug churn, Amritarista, Mahanarayan tel Chayavanpras etc.	*Oroxylum indicum* (L.) Kurz	Tantia	Root bark
42.	Devdaryadi kwath, Mahamash tel, Khadirarista Dasant lep, M₂ Tone	*Cedrus deodara* (Roxb. ex D. Don) G. Don	Devdara (Pinaceae)	Wood oil and bark
43.	Diabecon	*Abutilon indicum* (L.) Sweet	Atibalaa	Fruit, leaves, bark, seed, and roots
44.	Diabecon, Chyavanparash	*Gmelina arborea* Roxb.	Gambhari	Bark, root
45.	Diabecon, Diabetes, Nourishing cream (Himalaya)	*Pterocarpus marsupium* Roxb.	Bijesar	Gum leaf and flower
46.	Diakof, Koflet	*Hyssopus officinalis* L.	Jufa	Whole plant
47.	Diarex, Evecare, Dental cream (Himalaya)	*Acacia nilotica* (L.) Delile	Babool	Tree bark
48.	Erand pak, Erand tel Balarista, Pradrantak lauh, Bishgorbh tel, Cibolic capsules	*Ricinus communis* L.	Arandi	Roots, root bark, leaves, and seeds
49.	Erina plus	*Hibiscus rosa-sinensis* L.	Gudehal	Fruit
50.	Erina plus	*Pongamia pinnata* L.	Karanjua	Flower and seed
51.	Essential oil	*Tagetes minuta* L.	Jangali genda	Flowering tops
52.	Evecare, Renalka	*Hemidesmum indicus* (L.) R.Br.	Anantamul	Shrub
53.	Evercare, Menosan	*Sarca asoca* (Roxb.) Willd.	Ashoka	Bark and leaf
54.	Evercare, Styplon, Himfertin, M₂- Tone	*Symplocos racemosa* Roxb.	Lodhra	Bark
55.	Femiplex	*Amaranthus polygamus* L.	Chuali	Leaves
56.	Femiplex	*Lawsonia alba* Lam.	Mehandi	Leaves
57.	Femiplex, Crush	*Hygrophila spinosa* T. Anderson	Talimkhana	Leaves, seeds, and roots
58.	Femiplex, Livomyn	*Berberis aristata* DC.	Kingore	Roots, root bark, stem
59.	Femiplex, M₂-tone (Tablets & tonic)	*Bombax malabaricum* DC.	Semal	Bark, fruits, and seeds
60.	Femiplex, Neo, M2 Tone, Ojus, Himaplasia, Tentex Geriforte, Lukol, Renalka.	*Asparagus racemosus* Willd.	Satawari	Roots
61.	Femiplex.	*Ficus glomerata* Roxb.	Gular	Fruit, bark, and roots
62.	Foot cream (Himalaya), Revitalizing hair oil, Geriforte aqua, Immunol	*Trigonella foenum-graecum* L.	Methi	Seed and fruit
63.	Gangadher churn, Pushyanug churn, Abhyarista, Ashokarista, Arvidiasava, Femiplex	*Woodfordia fruiticosa* (L.) Kurz	Kurz	Flowers

(*Continued*)

TABLE 22.3 (*Continued*)
Medicinal Plants of Uttarakhand Used in Various Ayurvedic Drugs

S.NO	Ayurvedic Formulation	Botanical Name	Local Name	Parts Used
64.	Geriforte, Bonnisan, Liv-52	*Achillea millefolium* L.	Momoduru	Herb
65.	Geriforte, Diabecon	*Sphaeranthus indicus* L.	Mundi	Whole plant
66.	Giloy satva, Amritasava, Amritarista, Ashwaganda churn, Chandra, prabha Vati, Guduchi tel, Livonyn, Calcury, Pigmento Livosyp Dibecon, Mentat, Rumalaga Gasex, Geriforte, Liv- 52. Diakof, Manoll, Apimare	*Tinspora cardifolia* (Thunb.) Miers	Giloy	Stem
67.	Hair loss cream (Himalaya), Hairzone	*Butea parviflora* Roxb. Ex. DC.	Palashbheda	Bark, root, and stem
68.	Haritaki churn, Triphala churn, Agastgy haritaki aveleha, Abhyarista, Triphala ghrti, Chitrak, Haritaki, M2 tone, Hepatogard, Ojus, Mentat, Liv-52, Herbolax Pilex, Menosan, Gerforte, Menosan Gerforte, Abana	*Terminalia chebula* Retz.	Harrd	Frits
69.	Herbalax	*Operculina turpethum* (L.) Silva Manso	Triputa	Root
70.	Inflamin vet	*Nerium odorum* Aiton	Kaner	Leaf and root
71.	Jyotismasti tel, Himcotin cream, Sexotex cream Gerifort, Remem, Sumenta (Tab & syrup)	*Celastrus paniculatus* Willd.	Kaunya	Seeds and seeds oil
72.	Kafsina, Kofteb, Bans cough syrap, Cheston, Mehasudarsan ark.	*Viola pilosa* Blume	Vanfsa	Whole plants
73.	Kanchanar gugulu, Kanchnar kwath, Ushirasava, Chandanasava	*Bauhinia variegata* L.	Kanchnar	Stem bark
74.	Kankasava, Maha laxmi vilas ras, Asthma relief Alarex, Spasmolin	*Datura stramonium* L.	Dhatura	Leaves, flowering tops, and seeds
75.	Kantakariavleha, Kantakari ghrit, Vyaghiharitaki, Chyavanpras, Dasamoolarist, Dasamool kwath	*Solanum surattense* Burm. f.	Kantakari	Whole plants
76.	Katphaladi churn, Purhyanug churn, Khadiradi gutika, Irimedadi tel, Brihatphal ghrit.	*Myrica esculenta* Buch. -Ham. ex D. Don	Kaphal	Barks and fruits
77.	Kawanch pak, Kawanch churn, Kamdev ghrit, Ashwagandha churn, Musli pak, Dhatupaustic churn, Agstya haritaka, Badam pak, Strenex, Tentax forte, Speman tab.	*Mucuna pruriens* (L.) DC.	Kauch	Roots and hairs on pods
78.	Krimikuthar ras, Krimighatini vati, Kshisadi ghrit, Herbinol cream	*Mallotus philippensis* (Lam.) Müll.Arg.	Kmbhal	Ripe fruits
79.	Krmighna churn, Krimimudgar ras, Krmikutharras, Plashbeejj churn. Plash ghrit, Crush Lukol	*Butea monosperma* (Lam.) Taub.	Dhak	Tree Gum, seeds, and flowers
80.	Kutajarista, Kutajahan vati, Kutajawteha ras, Karpur ras	*Holarrhena antidysenterica* (Roth) Wall. ex A.DC.	Karu	Barks and seeds
81.	Lavanbhaskar churn,	*Rheum austral* D. Don.	Chukri	Rhizomes and roots

(Continued)

TABLE 22.3 (*Continued*)
Medicinal Plants of Uttarakhand Used in Various Ayurvedic Drugs

S.NO	Ayurvedic Formulation	Botanical Name	Local Name	Parts Used
82.	Lip balm, Oxitard	*Daucus carota* L.	Gajar	Seeds
83.	Liv-52	*Tecomella undulata* (Sm.) Seem.	Rohitaka	Bark
84.	Livomyn	*Coriandrum sativum* L.	Dhaniya	Fruit
85.	Livomyn, Pigmento	*Tephrosia purpurea* (L.) Pers.	Pal	Whole plant
86.	Livomyn, Pigmento, Hepatogard, Purim, Talekt	*Andrographis paniculata* (Burm. f.) Wall.	Kalmegh	Whole plant
87.	Livomyn, Pigmento, Livosyp, Diabecon, Herbolax, Reostra	*Aloe barbadensis* Mill.	Kumari	Leaves
88.	Livomyn, Vomiteb	*Zingiber officinale* Roscoe	Aadarak	Rhizome
89.	Livosyp, Hepatogard, Pilex	*Azadirachta indica* A. Juss.	Neem	Leaves and wood
90.	Lodhrasava, Vasasava, Sudarshan churh, Chandraprabha vati, Yograj guggulu, Dasamoolarista, Aswagandharisata	*Cinnamomum tamala* (Buch. -Ham.) T. Nees & Eberm.	Tejpatta	Leaves and bark
91.	M₂-Tone	*Sida cordifolia* L.	Balu	Leaves and roots
92.	M₂-Tone	*Eugenia jambolana* Lam.	Jamun	Bark and seed
93.	M₂-Tone	*Ficus benghalensis* L.	Bargad	Fruits, barks, and leaves
94.	M₂-tone	*Mangifera indica* L.	Aam	Fruits, seeds, and bark
95.	Maha laxadi tel	*Sapindus mukorossi* Gaertn.	Ritha	Fruits
96.	Mahamanjisthadyarista, Mahamanjisthadi ark, Manjisthadi tel, Mahanarayan tel, Ashwagandharista.	*Rubia cordifolia* L.	Majethi	Roots
97.	Manoll	*Spinacia oleracea* L.	Palak	Leaves and fruits
98.	Manoll, Calcury	*Tribulus terrestris* L.	Gukhru	Seeds
99.	Mohavishgarbh tel, Nirundi tel, Rumalaya cream, Langali rasayan.	*Gloriosa superba* L.	Kalihari	Rhizomes and seeds
100.	Muscle & Joint rub (Himalaya)	*Hyoscyamus niger* L.	Parasika yavani	Seed
101.	Neo, Gerifort	*Alternanthera sessilis* L.	Matsyakshi	Herb leaves
102.	Neo, Gerifort	*Allium cepa* L.	Pyaj	Whole plant
103.	Nirgundi ghrit, Nirgundi kwath, Nirgundi tel, Bishagarbh tel, Rumalagy cream, Himcolin cream.	*Vitex negundo* L.	Siwain	Leaves roots and fruits
104.	Ojus	*Vitis vinifera* L.	Angoor	Whole plant
105.	Ojus, Femiplex, M₂-Tone, Apimore	*Cuminum cyminum* L.	Zeera	Frit
106.	Parpaundrik kwath, Khadiradi vati, and Surma.	*Cassia absus* L.	Cheaksu	Seeds
107.	Pigmento	*Cyperus rotundus* L.	Motha	Tuberous roots

(Continued)

TABLE 22.3 (*Continued*)

Medicinal Plants of Uttarakhand Used in Various Ayurvedic Drugs

S.NO	Ayurvedic Formulation	*Botanical Name*	Local Name	Parts Used
108.	Pigmento	*Melia azedarach* L.	Daikan	Leaves, barks, and fruits
109.	Pigmento	*Solanum nigrum* L.	Makoi	Whole plant
110.	Pigmento and Diabacon	*Swertia chirayita* Roxb	Chiraita	Whole plant
111.	Pigmnto	*Acacia catechu* (L. F.) Willd.	Kher	Wood bark
112.	Pilex, Styplon	*Mimosa pudica* L.	Lajjalu	Root
113.	Priyangwadi tel, Ashwagandharista, Dasamoolarist, Chandana sava, Draksharista, Eladi churn.	*Callicarpa macrophylla* Vahl.	Daiya	Flowers buds
114.	Punarnava kwath, Punarnava -sova, Punarnavadi mandoor, Punarnavadi guggulu, Abana, Crush, Lukol, Geriforte, Livomyne,	*Boerhavia diffusa* L	Punarnava	Roots
115.	Pushyanug churn, Aswagantharista, Khadirarista, Patrangasava,	*Berberis lycium* Royle	Daruharidra	Roots, root bark, stem
116.	Pusyanug churn, Pathadi kwath, Mahayograj guggulu Agnimukh churn	*Cissampelos pareira* L.	Pahre	Roots
117.	Refreshing fruit pack (Himalaya)	*Ficus carica* L.	Fig	Fruit and latex
118.	Rumalaya forte, Rumalaya gel, Shallaki	*Boswellia serrata* Roxb. ex Colebr.	Shallaki	Bark and gum
119.	Rumalaya gel, Muscle & joint rub (Himalaya)	*Mentha arvensis* L.	Pudina	Leaves
120.	Rumalaya gel, Rumalaya liniment, Pain balm (Himalaya)	*Gaultheria fragrantissima* Wall.	Jalan-thrait	Leaf and fruit
121.	Rumalaya gel/ Rumalaya vet, Cold balm, Pain balm (Himalaya)	*Pinus roxburghii* Sarg.	Chir	Bark and resin
122.	Sanjivani vati, Saraswat Ashwagandharista, Laghu vishgarbha tel, Abana tebles, Sumento, Pigmento, Mentat, Alarsin, M_2-Tone. churn	*Acorus calamus* L.	Vacha	Rhizomes
123.	Septilin, Rumalaya, Pain massage oil (Himalaya), HimRop vet, Himfertin	*Moringa pterygosperma* Gaertn.	Shigru	Root
124.	Shatyadi chrun, Dasamoolarista, Kankayan gutika, Vomiteb, Chikara, Lahmina, Jivahakalpa, Himanshu tel.	*Hedychium spicatum* Sm.	Banhaldi	Rhizomes
125.	Shringyandi churn, Dashmularista, Kantakari Avaleha, Chyavanpras Nokufcough, Kasni.	*Pistacia khinjuk* Stocks	Kakarsingi	Leaves, petiole, and branches
126.	Som kalp, Swankalp.	*Ephedra gerardiana* Wall. ex Stapf	Tutgautha	Stem
127.	Sumenta, Mental, Menosan, Geriforte, Reosta, Liv-52 Mentat.	*Terminalia arjuna* (Roxb. ex DC.) Wight & Arn.	Arjuna	Bark
128.	Sumento	*Evolvulus alsinoides* L.	Sankhpuspi	Whole plant
129.	Surma (Yogi Pharmacy)	*Thalictrum foliolosum* DC.	Kirmuri	Roots

(*Continued*)

TABLE 22.3 (*Continued*)
Medicinal Plants of Uttarakhand Used in Various Ayurvedic Drugs

S.NO	Ayurvedic Formulation	Botanical Name	Local Name	Parts Used
130.	Tagardi kwath, Chandana lodhrasava, Sumaya (Valeriancaceae) Sarivadyasava, Chandrasekhar ras, Newrocardine, Nervoplex Mondo- valerian, Neocardial liquid, Sumenta.	*Valeriana jatamansi* Jones	Sumayan	Rhizomes
131.	Talisadi churn, Lavanbhaskar churn, Sudarshan churn, Kanksava	*Taxus baccata* L.	Thuner	Leaves and bark
132.	Tentex forte	*Strychnos nuxvomica* L.	Kuchla	Seed
133.	Tranquil	*Abrus precatorious* L.	Ratti	Seed and roots
134.	Triphala churn, Triphala kwath, Sanjivani vati, Kutajavaeleha, Herbotone Hepatogard	*Terminalia bellirica* (Gaertn.) Roxb.	Bahera	Fruits
135.	Vasakamadu, Vasavaleha, Vasakasva, Kasni, Kankasava Livomyn, M$_2$-Tone, Styplon, Diakof, Geriforte.	*Adhatoda zeylanica* Medic.	Vasaka	Leaves and roots
136.	Vidari churn, Laxamivilasras, Narsing churn, Ashwagandha churn, Saraswatista, Chyvanpras.	*Pueraria tuberosa* (Willd.) DC.	Siralu	Tuber
137.	Vigroll	*Abies pindrow* (Royle ex D. Don) Royle	Raga	Bark
138.	Vomiteb	*Cinnamomum zeylanicum* Blume	Dalchini	Bark
139.	Vomiteb, Mental, Himacospaze, Bonnisan, Abana	*Anethum sowa* Roxb.ex	Sowa	Leaves

Nowadays, the trade of medicinal plants is increasingly giving more importance to earning. Thus, local inhabitants exploit medicinal plants without considering traditional methods and mechanisms destroying the immense wealth of the Himalayas. This also involves the loss of oral tradition of passing information about traditional uses to subsequent generations.

22.6   THE PRESENT STATE OF KNOWLEDGE

Documentation of inventories of medicinal plants in the Himalayas, which includes information on the species, parts used, and distribution area, is one of the main contributions. Therefore, it is possible to find a comprehensive listing of medicinal plants in the country and their biological activity. Plants utilized in traditional medicines (Figure 22.3), formulation of traditionally used herbal drugs, and medicinal plants used in Ayurvedic formulation are listed in Tables 22.2 and 22.3 (Kala 2011, 2015; Negi et al., 2011; Pandey and Verma 2005; Prakash, 2014; Samant et al., 1998; Rawat and Vashistha, 2011).

22.7   CONCLUSION AND PROSPECTS

Deforestation and illicit collecting play a role in the decreased availability of several plant species in their native habitat. The knowledge flow has been disrupted by subsequent generations' lack of interest. The Indian Himalayan ethnic community's traditions and customs were impacted by socio-economic and cultural developments. The only people who still have information about medicinal herbs, traditional healthcare practices, and use as well as information about seeding, harvesting, extraction, and formulation are the elder people and, particularly, the Vaidyas.

Murraya koenigii L.   *Berberis aristata* DC   *Adhatoda vasica* L.

Calotropis procera Aiton   *Bergenia ciliata* (Haw.) Sternb.   *Rododendron arboretum* Sm.

Pyracantha crenulata (D.Don) M.Roem.   *Curcuma longa* L.   *Gloriosa superba* L.

FIGURE 22.3   Medicinal plants of Uttarakhand, India (pictures taken by Dr. Sarla Saklani).

22.8   COMPETING INTERESTS

The authors declare that they have no competing interests.

ACKNOWLEDGEMENTS

The authors are very thankful to all the authors whose work has been cited in this chapter. The authors are also grateful to the traditional healers and local medical practitioners (also known as *Vaidyas*) of Uttarakhand for providing the knowledge about the traditional healthcare system of Uttarakhand, India.

REFERENCES

Dhar, U., Rawal, R.S., and Upreti, J. (2000). Setting priorities for conservation of medicinal plants—a case study in the Indian Himalaya. *Biological Conservation, 95,* 57–65.

Dhawan, B.N. (1997). Biodiversity–a valuable resource for new molecules. In: *Himalayan Biodiversity: Action Plan.* Dhar U. (eds.), Gyanodaya Parkashan, Nainital, 111–114.

Gaur, R.D., Semwal, J.K., and Tiwari, J.K. (1983). A survey of high altitude medicinal plants of Garhwal Himalaya. *Bulletin of Medico-Ethno-Botanical Research, 4,* 102–116.

Hazlett, D.L. and Sawyer, N.W. (1998). Distribution of alkaloid-rich plant species in shortgrass steppe vegetation. *Conservation Biology, 12,* 1260–1268.

Kala, C.P. (2011). Medicinal plants used for dermatological disorders: a study of Uttarakhand state in India. *Australian Journal of Medical Herbalism, 23,* 132.

Kala, C.P. (2015). Medicinal and aromatic plants of tons watershed in Uttarakhand Himalaya. *Applied Ecology and Environmental Sciences, 3,* 16–21.

Kalakoti, B.S. and Pangety, Y.P.S. (1988). Ethnomedicine of Bhotiya tribe of Kumaun Himalaya, UA. *Bulletin of Medico-Ethno-Botanical Research, 9,* 11–20.

Mukherjee, P.K. (2001). Evaluation of Indian traditional medicine. *Drug Information Journal, 35*, 623–632.

Narayana, D.B.A., Katayar, C.K., and Brindavanam, N.B. (1998). Original system: search, research or research. *IDMA bulletin, 29*, 413–416.

Negi, V.S., Maikhuri, R.K., and Vashishtha, D.P. (2011). Traditional health care practices among the valley of Rawain, Uttarakashi, Uttarakhand, India. *Indian Journal of Traditional Knowledge, 10*,533–537.

Pandey, H.P. and Verma, B.K. (2005). Phytoremedial wreath: a traditional excellence of healing. *Indian Forester, 131*, 437–441.

Prakash, R. (2014). Traditional uses of medicinal plants in Uttarakhand Himalayan Region. *Scholars Academic Journal of Biosciences, 2*, 345–353.

Purohit, A.N. (1997). Medicinal plants-need for upgrading technology for trading the traditions. In: *Harvesting Herbs–2000*, Nautiyal, A.R. Nautiyal, M.C., and Purohit, A.N (eds.), Bishen Singh Mahendra Pal Singh, Dehradun, Uttarakhand, 46–76.

Rawat, R. and Vashistha, D.P. (2011). Common herbal plant in Uttarakhand, used in the popular medicinal preparation in ayurveda. *International Journal of Pharmacognosy and Phytochemical Research, 3*, 64–73.

Ros, F. (2011). *The Lost Secrets of Ayurvedic Acupuncture: An Ayurvedic Guide to Acupuncture.* (3rd edition), . Motilal Banarsidass Publication, Delhi, 223.

Saklani, D.P. (1998). *Ancient Communities of the Himalaya.* Indus Publishing, New Delhi, 196.

Saklani, S., Chandra, S., and Mishra, A.P. (2011). Evaluation of antioxidant activity, quantitative estimation of phenols, anthocynins and flavonoids of wild edible fruits of Garhwal Himalaya. *Journal of Pharmacy Research, 4*, 4083–4086.

Samal, P.K., Shah, A., Tiwari, S.C., and Agrawal, D.K. (2004). Indigenous healthcare practices and their linkages with bioresource conservation and socio-economic development in Central Himalayan region of India. *Indian Journal of Traditional Knowledge, 3*, 12–26.

Samant, S.S., Dhar, U., and Palni, L.M.S. (1998). *Medicinal Plants of Indian Himalaya: Diversity, Distribution, Potential Values.* Gyanodaya Prakashan, Nainital, Uttarakhand, 161.

Semwal, D.P., Saradhi, P.P., Kala, C.P., and Sajwan, B.S. (2010). Medicinal plants used by local Vaidyas in Ukhimath block, Uttarakhand. *Indian Journal of Traditional Knowledge, 9*, 480–485.

Sharma, J R, Mudgal, V., and Hajra, P K. (1997). Floristic diversity—review, scope and perspective. In: *Floristic Diversity and Conservation Strategies in India.* Mudgal, V.and Hajra, P.K (eds.), Botanical Survey of India, Ministry of Environment and Forests, Calcutta, 1–45.

Zohary, D. and Hopf, M. (2000). *Domestication of Plants in the Old World: The origin and Spread of Cultivated Plants in West Asia, Europe and the Nile Valley*, (3rd edition). Oxford University Press, New York, 328.

23 Physicochemical Characterization, Antioxidant, and Anti-Inflammatory Activities of Turmeric-Black Cumin Homegrown COVID-19 Herbal Mixture

Ayodeji Oluwabunmi Oriola, Gugulethu Mathews Miya and Adebola Omowunmi Oyedeji

CONTENTS

23.1 INTRODUCTION

The global index cases of the novel Severe Acute Respiratory Syndrome Coronavirus-2 (nSARS-COV-2) infection were reported on 13 December 2019 in Wuhan, China, alluding to the name COVID-19. It was declared a pandemic on 11 March 2020, spreading to more than 180 countries, including Nigeria

DOI: 10.1201/9781003206620-23

413

(Cai et al., 2020; Wang et al., 2020; WHO, 2020). Nigeria's COVID-19 index case was announced on 27 February 2020, causing the enforcement of a national lockdown on 30 March 2020 (NCDC, 2020).

Nigeria is a Sub-Sahara West African nation with a land mass of 923,768 km² and a population of about 213 million people—the most populated in Africa and the sixth most populated country in the world (Worldometer, 2021). The World Health Organization (WHO) categorised Nigeria among the top-13 high-risk African countries for COVID-19 because of the enormous human population and inadequate healthcare facilities (Amzat et al., 2020).

So far, Nigeria has not fared badly in terms of COVID-19 morbidity and mortality. As of 6 November 2021, about 213,000 cases have been reported, with a significant decline in daily reported cases to 116 people from a record high 2,464 new cases on 23 January 2021. As of 31 December 2021, a total 87,617 COVID-19 cases were reported in Nigeria with less than 2.4% death since the first (index) case on 27 February 2020. Within the stated period, a total 2,905 deaths were recorded, with a significant decline in mortality from about 22 new daily deaths in January 2021 to about three new daily deaths in December 2021 (NCDC, 2022). Though the capacity for testing in Nigeria cannot be compared with countries such as the USA, Brazil, and the UK, which have so far recorded higher and unprecedented deaths due to COVID-19, one thing is clear: There is a strong believe in folkloric medicine as a complementary and/or an alternative to orthodox medicine in Nigeria. Perhaps, this might be playing a significantly role by helping to reduce the burden of COVID-19 infection in such a densely populated country.

Nigeria has a rich flora diversity but inadequate modern healthcare facilities; hence, many resorted to nature rather than to the hospitals for health solutions. Interestingly, herbal medicine is not extant to the world, as it has been reported that about 80% of plant-derived drugs are linked to their original ethnomedicinal uses (Fabricant and Farnsworth, 2001). Currently, some immune-boosting herbs and spices are used in Nigeria as homegrown mono- and polyherbal remedies for the prevention and management of COVID-19 and other flu-related diseases. Among such are the rhizomes of turmeric and the seeds of black cumin often used in the form of an infusion (Orisakwe et al., 2020).

Turmeric, *Curcuma longa* L. (Zingiberaceae) is a rhizomatous, perennial herb that grows up to 5 ft. It has short stem, large oblong leaves, yellow flowers, and yellowish-to-brownish yellow pyriform rhizomes (Mukherjee et al., 2013). Black cumin or Black seeds, *Nigella sativa* L. (Ranunculaceae) is an annual herbaceous plant that grows up to 25 cm tall with finely divided linear leaves. The flowers are tender and are usually pale blue and white, while the fruits are in the form of inflated capsules comprising about five united follicles, each bearing blackish seeds (Farhan et al., 2021). Both turmeric and black seeds thrive in tropical and sub-tropical climates, and they are widely cultivated across the continents of Asia, South America, and Africa. They find many applications as culinary, food additives/preservatives, herbal medicines, and in cosmetics (Mukherjee et al., 2013). They are also known to be essential oil-bearing plants, used in the Asian ethnomedicine for the treatment of diabetes, jaundice, and other liver ailments (ElBahr et al., 2014) amidst others. In the Asian ethnomedicines, herbal infusions from these plants and the essential oils are used for the management of digestive, circulatory, respiratory, and reproductive ailments, while they are used as natural remedy for respiratory ailments, hypertension, diabetes, arthritis, wounds, cancer, fever, headache, memory loss, and dermatological-related ailments in the African ethnomedicines (Khan et al., 2011; ElBahr et al., 2014). Currently, the turmeric black seed herbal mix is claimed to be used ethnomedicinally for the management of COVID-19 in Nigeria (Table 23.1). Among such claims was the testament of the Executive Governor of Oyo State, in Southwest Nigeria, that he was healed of COVID-19 after about 1 week of treatment with the turmeric black cumin herbal mix (Punch Newspaper, April 7, 2020).

These extracts and essential oils have been reported to show biological potentials, such as antioxidant, immune-modulating, neuroprotective, cardioprotective, hepatoprotective, antiviral, anti-inflammatory, analgesic, anti-ulcer, anti-hyperglycaemic, antimicrobial, anti-cancer, antimalarial, and anti-fertility activities among others (Krup et al., 2013; Sambhav et al., 2014; Khan and Afzal, 2016; Ansari et al., 2020). Exposure of guinea pigs to infusions or aerosols of black cumin seeds protected the animals against citric acid aerosols-induced cough in a manner comparable with

TABLE 23.1

Acclaimed Use of Turmeric-Black Cumin Herbal Mixture by COVID-19 Survivors in Nigeria

Status of Claimant	Mode of Herbal Preparation/Use	References
Executive Governor of Oyo State	Infusion of the herbal mix or its oil with honey and vitamin C	Punch Newspaper, April 7, 2020
Journalist	Infusion or decoction of the herbal mix with alligator pepper, ginger, lemon, and bitter leaf	The Guardian Newspaper, August 13, 2020
A former Governor of Anambra State	Infusion of the herbal mix with ginger, lemon, steam inhalation	Vanguard Newspaper, April 24, 2020
A 22-year-old university graduate	Infusion of herbal mix with pineapple and ginger	The Guardian Newspaper, April 7, 2021
A Health professional and HR administrator	Infusion and decoction of the herbal mix	The Guardian Newspaper, April 7, 2021
A former President of the Federal Republic of Nigeria	A mixture of herbs comprising turmeric rhizomes and black cumin seeds	Sahara reporters, March 6, 2021

those of codeine aerosols (Boskabady et al., 2003; Boskabady et al., 2004). Furthermore, a co-administration of turmeric and black cumin has been reported to demonstrate significant clinical efficacy in metabolic syndrome, especially in reducing human cholesterol level, and their significant role in the management of diabetes had been emphasized (Amin et al., 2015). In recent times, the ethanol extract of black seed and the aqueous extract of turmeric have been reported to exhibit anti-viral activities individually against SARS-COV-2 infection (Uchejeso et al., 2021). An infusion of the herbal mixture is gaining more attention as an immune booster for the prevention and management of COVID-19 in Nigeria (Adeleye et al., 2019; Orisakwe et al., 2020).

Some bioactive compounds such as curcuminoids (curcumin, ar-curcumene, desmethoxycurcumin, monodemethoxycurcumin, bisdemethoxycurcumin, dihydrocurcumin and cyclocurcumin), sesqui-terpenoids (*ar*-tumerone, α- & β- turmerones, curlone, zingiberene, β-phellandrene, sabinene, and elemene), triterpenoids, alkaloids, and sterols (β-sitosterol and stigmasterol) have been identified in tur-meric rhizomes (Chattopadhyay et al., 2004; Sambhav et al., 2014). Also, antioxidant, anti-inflammatory compounds such as flavonol glycoside and dihydroflavonol glycoside have been reported in turmeric (Jiang et al., 2015). Essential oils such as thymol, thymoquinone, dithymoquinone, *p*-cymene; saponins (α-hederin); and alkaloids (nigellicimine, nigellicine, nigellidine, and nigellamine) have been reported in black cumin seeds with considerable biological activities, which include antioxidant, anti-inflamma-tory, and cytotoxic activities (Khan and Afzal, 2016; Yimer et al., 2019).

Currently, there is dearth of information on the chemical composition of the turmeric-black cumin herbal mix. It is expedient that the antioxidant and anti-inflammatory potentials of any locally made herbal remedies for COVID-19 be determined, since this novel infectious disease has been reported to be characterized by high level of inflammation, and oxidative damage to the lungs and other associated organs (Vollbracht and Kraft, 2022). It is on this backdrop that we investigated the turmeric black cumin homegrown COVID-19 herbal mixture. The physicochemical profiles of the EOs were obtained, while the *in vitro* antioxidant and anti-inflammatory activities were also determined, with a view to validate and potentiate its medicinal use.

23.2 MATERIAL AND METHODS

23.2.1 HERBAL RAW MATERIALS

The herbal raw materials (HRMs), turmeric rhizome powder, and black cumin seeds were purchased from the herbal store, SpicYem® Spices and Foods, Ile-Ife, Osun State 220005, Nigeria (Geographic latitude 7°33′00″N and longitude 4°33′00″E; Geomagnetic latitude 7°30″N and longitude 4°28″E), each registered with batch numbers RC: 875523 and RC: 875523, respectively. The HRM was obtained from *C. longa* L. (Zingiberaceae) and *N. sativa* L. (Ranunculaceae); both the plants confirmed with the http://www.theplantlist.org as kew-235249 and kew-2381679, respectively. The HRMs were combined in equal ratio (200 g each), homogenized, and kept in a Ziploc bag prior to extraction.

23.2.2 EXTRACTION

The homogenized HRMs (400 g) was hydro-distilled for its essential oils (EOs) on a Clevenger apparatus set at 70°C–100°C for 4 hours, according to Oyedeji et al. (2014). The extraction jar was allowed to cool, after which it was re-distilled to obtain six successive batches of EOs. The EOs were collected in 10 mL amber vials labelled I–VI, and they were stored in the cold room at −4°C.

23.2.3 PHYSICOCHEMICAL ANALYSES

The colour, odour, and percentage yield of the six EO batches were determined. Thin-layer chromatography (TLC) bioautography was carried out on silica gel 60 F254 GF plates (0.25 mm, Merck KGaA, Germany), with TLC plates developed in a pre-saturated tank of *n*-hexane–Ethyl acetate (9:1) in duplicate. The chromatograms were monitored at 254 and 365 nm wavelength under the UV-Vis spectrum. They were subsequently activated with 10% sulfuric acid and the replicate with 2,2-diphenyl-1-picrylhydrazyl (DPPH) radical (Sigma–Aldrich, St Louis, MO, USA), for a general detection of the phytoconstituents and their free radical scavenging potentials, respectively.

The EOs were analysed using Gas Chromatography Mass Spectrometry (GC–MS), according to the method of Miya et al. (2021). It involved the separation and characterisation of the EOs on Bruker 450 Gas Chromatograph connected to a 300 MS/MS mass spectrometer system (Germany). The GC–MS was operated in EI mode at 70 eV. The chromatogram comprised an HP-5 MS fused silica capillary system with 5% phenylmethylsiloxane as stationary phase. The capillary column parameter was 30 m length × 0.25 mm internal diameter × 0.25 μm film thickness. The column temperature was increased from 50°C to 240°C at a rate of 5°C/minutes, while the final temperature was maintained at 450°C for a duration of 66 minutes. The carrier gas was helium (1.0 mL/minutes flow rate), the scanning range was 35–450 amu, and the split ratio was 100:1. One microliter (1 μL) of the diluted oil (in hexane) was injected for analysis. The constituents were identified by comparison of their mass spectra data and retention indices with those available in our research group's GC–MS library as well as those in reported in the literature.

23.2.4 2,2-DIPHENYL-1-PICRYLHYDRAZYL (DPPH) SPECTROPHOTOMETRIC ASSAY

This was carried out according to a standard procedure that was previously described by Oriola et al. (2021). Here, 0.5 mL of 0.1 Mm DPPH radical in methanol was added to 0.5S mL of serial diluted test samples (EOs and *L*-ascorbic acid) at 3.125–100 μM concentration range in triplicate. The reaction mixture was incubated in the dark at 37°C for 30 minutes. The absorbance was measured at 515 nm on a 680-Bio-Rad Microplate Reader (Serial Number 14966, USA). The percentage inhibition of the radical was calculated thus:

$$\% \text{ DPPH inhibition} = \left(\frac{\text{ABSsample-ABScontrol}}{\text{ABScontrol}} \right) \times 100$$

where ABS sample is the absorbance of test sample, i.e., EOs or *L*-ascorbic acid, while ABS control is the absorbance of negative control (methanol).

The inhibitory concentration (IC_{50}) of each test sample was determined from the dose–response curve.

23.2.5 Nitric Oxide (NO) Inhibition Assay

The inhibitory effect of the essential oils against the NO radical was investigated using the method described by Jimoh et al. (2019). Here, 0.5 mL of test samples at varying concentrations (3.125–100 μM) were added to sodium nitroprusside (2 mL, 0.2 Mm) in triplicates. The reaction mixture was incubated at 25°C for 3 hours. Then, 0.5 mL of the mixture was mixed with Griess reagent (0.33% sulphanilamide dissolved in 20% glacial acetic acid and mixed with 1 mL of naphthylethylenediamine chloride (0.1%w/v)). The mixture of the complex and Griess reagent was then incubated at room temperature for 30 minutes. Thereafter, it was measured at an absorbance of 540 nm on a 680-Bio-Rad Microplate Reader (Serial Number 14966, USA). The percentage inhibition of NO radical was calculated as thus:

$$\% \text{ NO inhibition} = \left(\frac{\text{ABSsample-ABScontrol}}{\text{ABScontrol}} \right) \times 100$$

where ABS sample is the absorbance of test sample, i.e., EOs or L-ascorbic acid, while ABS control is the absorbance of negative control (sodium nitroprusside in methanol).

The IC_{50} of each test sample was also determined.

23.2.6 Hydrogen Peroxide Inhibition Assay

The ability of the essential oils to inhibit peroxyl ion radical was measured using standard colorimetric method, as described by Okeleye et al. (2015). The test samples (400 μL each) were serially diluted from 100 to 3.125 μM concentrations and mixed with 60 μL of hydrogen peroxide solution (4 mM) prepared in a 0.1 M phosphate buffer saline (pH 7.4) in triplicate inside a 96-well plate. The reaction mixture was incubated at room temperature (≈25°C) for 10 minutes. Thereafter, the absorbance was measured at 405 nm on a 680-Bio-Rad Microplate Reader (Serial Number 14966, USA). The percentage inhibition of the peroxyl radical was determined using the formula:

$$\% \text{ inhibition of peroxyl radical} = \left(\frac{\text{ABSsample-ABScontrol}}{\text{ABScontrol}} \right) \times 100$$

where ABS sample is the absorbance of test sample, i.e., EOs or L-ascorbic acid, while ABS control is the absorbance of negative control (1% DMSO).

23.2.7 Ferric-Reducing Antioxidant Power (FRAP) Assay

The procedure was based on the ability of the essential oils to reduce the greenish ferric ion (Fe^{3+}) 2,4,6-tri-(2-pyridyl)-1,3,5-triazine (TPTZ) to bluish ferrous ion (Fe^{2+}) at 593 nm absorbance measurement, as previously described by Benzie and Strain (1999). Thus, the ferric-reducing power of the essential oils was determined as ascorbic acid equivalent (AAE) from the calibration curve of the positive control (L-ascorbic acid) at concentrations 1000.00, 500.00, 250.00, 125.00, 62.50, and 31.25 μM in methanol.

23.2.8 In Vitro Anti-Inflammatory Test

This was performed using the egg albumin denaturation assay method by Chatterjee et al. (2012). Here, a reaction mixture comprising 0.2 mL of albumin content of fresh chicken egg, 2.8 mL of phosphate buffer saline (pH 6.4), and 2 mL each of the EOs at varying concentrations (6.25–100 μM) was prepared in triplicate. The mixture was incubated at 37°C for 15 minutes away from direct light and thereafter boiled at 70°C for 5 minutes in a thermostatic water bath. The resulting mixture was cooled, and the absorbance was measured at 655 nm on a 680-Bio-Rad Microplate Reader (Serial

Number 14966, USA). Diclofenac was similarly tested as the standard drug, and the percentage inhibition of the samples was calculated thus:

$$\% \text{ inhibition of protein denaturation} = \left(\frac{\text{ABSsample-ABScontrol}}{\text{ABScontrol}} \right) \times 100$$

where ABS sample is the absorbance of test sample, i.e., EOs or Diclofenac, while ABS control is the absorbance of negative control (vehicle), i.e., 1% dimethylsulfoxide (DMSO).

The inhibitory concentration (IC_{50}) of each test sample was determined from the dose–response curve.

23.2.9 STATISTICAL ANALYSIS AND WORKFLOW

Data obtained from the antioxidant and anti-inflammatory tests were expressed as mean ± standard error of mean (SEM) using Microsoft Excel version 365 (Microsoft Corporation, Washington DC, USA). The results were analysed using One-Way Analysis of Variance (ANOVA), followed by the Student–Newman–Keul's post-hoc test on a GraphPad Prism 5 (GraphPad Software Inc., San Diego, CA, USA). $P < 0.05$ was set as the level of significance. The experimental work done (workflow) is presented in Figure 23.1.

23.3   RESULTS

Six hydro-distilled batches of essential oils (I–VI) were obtained from the herbal mixture, as presented in Table 23.2. The EOs were generally woody in aroma. The aroma became roasty from batches III to V, and it was smoky in batch VI with increased colour intensity from yellow to dark brown upon re-distillation. A sequential increase in the EO yield from batches I to IV was observed, while there was also a reduction in yield from IV to VI. The Marc (herbal mixture) became foamy on the sixth hydro-distillation batch, which suggested an exhaustive extraction of the EOs, hence, the termination of the steam distillation process.

TLC bioautography revealed strong free radical scavenging potentials of the EOs with qualitative improvement in the DPPH radical scavenging property upon re-distillation from I to VI (Figure 23.2). Table 23.3 shows the chemical composition of the six EO batches. There were 14 major peaks on the GC chromatograms of the EOs representing 14 major constituents, based on the

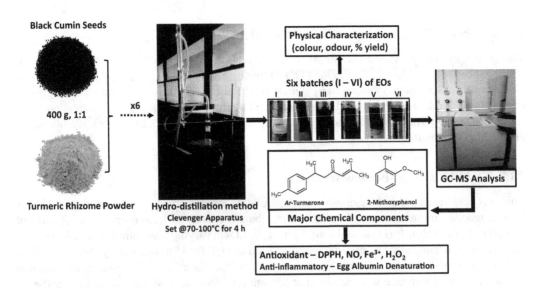

FIGURE 23.1   Flow chart for the study.

TABLE 23.2

Characteristics of Essential Oils Obtained from the Herbal Mixture (400 g)

EOs Batch	Colour	Sensory Characteristics	% Yield
I	Creamy	Earthy, woody	2.89
II	Golden yellow	Earthy, woody	2.95
III	Reddish	Woody, nutty, roasty	3.08
IV	Reddish brown	Woody, nutty, roasty	3.19
V	Reddish brown	Woody, nutty, roasty	3.01
VI	Dark brown	Woody, smoky	2.87

EOs: Essential oils; I–VI: Essential oil batches; % Yield expressed as w/w (g) × 100.

retention time and the peak area. They were categorized as 11, 6, 3, 2, 1, and 2 compounds (peaks) for batches I–VI, respectively (Figure 23.3). Reaction mechanisms for the elicitation of α-turmerone, p-hydroxy methylbenzene from ar-turmerone (Figure 23.4), and guaiacol from p-cymene (Figure 23.5) were proposed.

where EOs batches (I–VI); TLC solvent system: Hexane–EtOAc (75–25); antioxidant active (Ax), yellowish appearance against the purple 2,2-diphenyl-1-picrylhydrazyl (DPPH) free radical; and appearance of reddish-to-reddish brown oily substance suspected to be guaiacol (G)

$$KI_{exp} = 100 \left[\frac{(Rt - Rtz)}{(Rt(z+1) - Rtz)} + z \right]$$

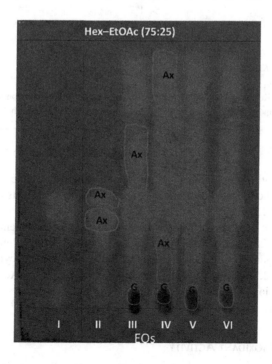

FIGURE 23.2 TLC-Bioautography of the essential oil batches.

[Essential oils (EOs) batches (I-VI); TLC solvent system- Hexane-EtOAc (75:25); antioxidant active (Ax)-yellowish appearance against the DPPH purple background; reddish brown oily substance suspected to be Guaiacol (G).]

TABLE 23.3

Chemical Profile of Essential Oils from Turmeric-Black Cumin Herbal Mixture

Peak	Constituent	Rt	KI exp	KI lit.	CAS No.	% Composition of EO Batches					
		Rt	KI_{exp}	$KI_{lit.}$	CAS No.	I	II	III	IV	V	VI
1	α-Phellandrene	6.94	1164	1168	99–83–2	-	1.54	-	-	-	-
2	p-Cymene	10.49	1354	1270	99–87–6	1.75	0.76	-	-		-
3	o-methoxyphenol (Guaiacol)	12.63	1485	1859	90–05–1	-	-	-	-	-	43.21
4	p-hydroxy methylbenzene	25.70	1965	-	-	-	27.46	-	-		
5	Ar-Curcumene	30.15	1941	1737	644–30–4	3.37	-	-	-	-	-
6	2-epi-α-Funebrene	30.71	1939	1419	65354–33–8	0.75	-	-	-	-	-
7	β-Sesquiphellandrene	31.86	1932	1771	20307–83–9	3.51	-	-	-	-	-
8	Ar-Turmerol	33.91	1922	1560	38142–57–3	0.57	-	-	-	-	-
9	1,2-Benzenediol, o-(2-methylbenzoyl)	34.82			#99434	1.25					
10	1,2-Benzenediol, o-(4-methylbenzoyl)	34.84			#99441	0.28					
11	Ar-Turmerone	37.22	1904	1664	532–65–0 #53122	53.69	56.39	70.27	98.66	99.10	55.88
12	α-Turmerone	37.46	1903	1654	180315–67–7	13.23	18.40	-	-	-	-
13	Curlone	38.62	1897	1650	87440–60–6	19.18	22.49	0.97	0.91	-	-
14	(E)-Atlantone	41.06	1884	1743	108645–54–1	2.33	0.17	-	-	-	-
	Total					99.91	99.75	98.70	99.57	99.10	99.09

$$KI_{exp} = 100\left[\frac{(Rt - Rtz)}{(Rt(z+1) - Rtz)} + z\right]$$

Where: KI_{exp} = Experimental Kovats index; Rt = Retention time of sample; Rtz = Retention time of lower n-alkane; and z = Number of carbons in the molecule; KI lit. = Kovats index reported in Literature (Babushok et al., 2011).

where KI_{exp} = experimental Kovats index; Rt = retention time of sample; Rtz = retention time of lower n-alkane; z = number of carbons in the molecule; and KI lit. = Kovats index reported in literature (Babushok et al., 2011).

where reactions steps: (1) thermal & enzymatic oxidation by p-cymene methyl monooxygenase (Pp-cymaA 1.14.15.25); (2) dehydrogenation by p-cumic alcohol dehydrogenase (Pp-cymB 1.1.1.M33); (3) dehydrogenation by cumic aldehyde dehydrogenase (Pp-cymC 1.2.1.29); (4 and 5) oxidation by p-cumate-2,3-dioxygenase (Pp-cmtAa Pp-cmtAb Pp-cmtAc Pp-cmtAd 1.14.12.25); (6) oxidative demethylation by demethylase; (7) decarboxylation by decarboxylase; and (8) methylation by methyl transferase (Defrank and Ribbons, 1977).

23.3.1 IN VITRO ANTIOXIDANT ACTIVITY

The result in Table 23.4 showed the level of antioxidant of the turmeric-black cumin essential oil batches I–VI, with respect to their ability to scavenge DPPH and hydrogen peroxide free radicals, the inhibitory activity against nitric oxide (NO) radical, and the ferric ion reducing ability, with reference to the positive control (L-ascorbic acid).

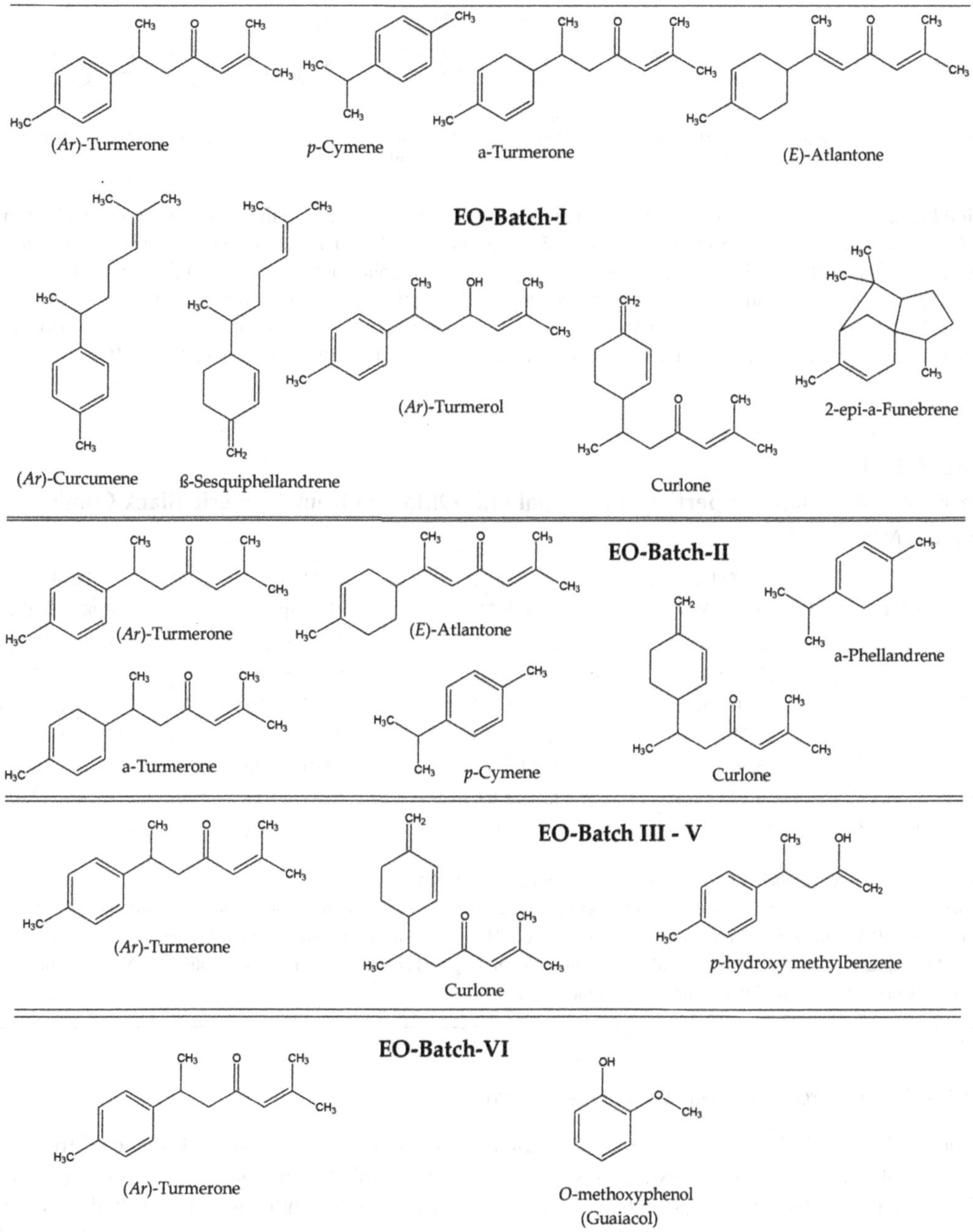

FIGURE 23.3 Chemical composition of essential oils from Turmeric-Black Cumin Combined Spice.

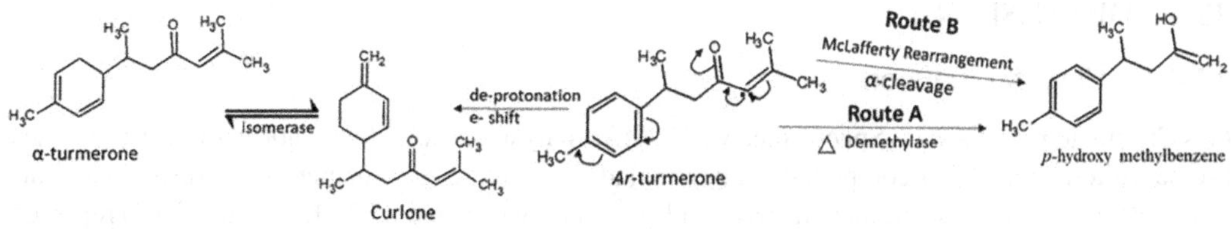

FIGURE 23.4 Plausible structural rearrangement for the elicitation of some major components of the essential oils.

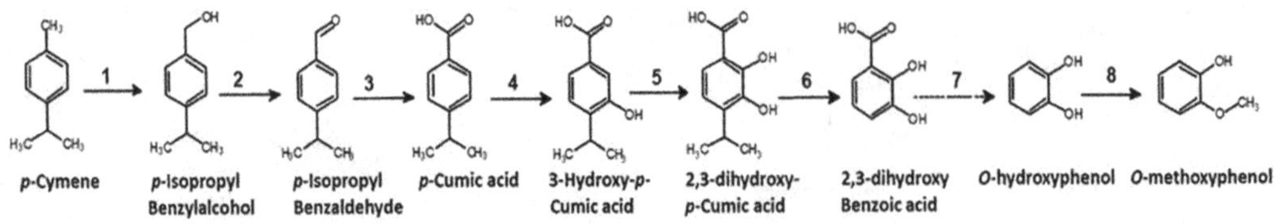

p-Cymene → 1 → p-Isopropyl Benzylalcohol → 2 → p-Isopropyl Benzaldehyde → 3 → p-Cumic acid → 4 → 3-Hydroxy-p-Cumic acid → 5 → 2,3-dihydroxy-p-Cumic acid → 6 → 2,3-dihydroxy Benzoic acid → 7 → O-hydroxyphenol → 8 → O-methoxyphenol

FIGURE 23.5 Plausible structural re-arrangement for the elicitation of Guaiacol (o-methoxyphenol) from p-Cymene. *Key Reactions Steps*: (1) thermal & enzymatic oxidation by p-cymene methyl monooxygenase (Pp-cymaA 1.14.15.25); (2) dehydrogenation by p-cumic alcohol dehydrogenase (Pp-cymB 1.1.1.M33); (3) dehydrogenation by cumic aldehyde dehydrogenase (Pp-cymC 1.2.1.29); (4 and 5) oxidation by p-cumate-2,3-dioxygenase (Pp-cmtAa Pp-cmtAb Pp-cmtAc Pp-cmtAd 1.14.12.25); (6) oxidative demethylation by demethylase; (7) decarboxylation by decarboxylase; (8) methylation by methyl transferase (Defrank and Ribbons, 1977).

TABLE 23.4
In Vitro Antioxidant Properties of Essential Oils Obtained from Turmeric-Black Cumin Herbal Mix

EO Batches	DPPH IC_{50} (µM)	NO IC_{50} (µM)	H_2O_2 IC_{50} (µM)	FRAP (mgAAE/g)
I	66.70 ± 0.83^f	45.06 ± 2.55^e	71.21 ± 2.84^f	258.19 ± 27.91^a
II	32.25 ± 0.68^d	49.09 ± 1.17^f	58.20 ± 3.07^e	493.64 ± 40.72^{bc}
III	47.10 ± 1.55^e	29.41 ± 1.23^d	30.61 ± 1.88^d	561.72 ± 33.67^c
IV	33.31 ± 1.44^d	19.42 ± 0.97^b	25.15 ± 1.37^c	455.11 ± 31.33^b
V	26.36 ± 0.95^c	21.51 ± 1.72^{bc}	29.01 ± 2.40^d	501.43 ± 26.08^{bc}
VI	16.10 ± 0.14^b	23.18 ± 0.63^c	13.17 ± 1.11^b	715.46 ± 21.65^d
L-ascorbic acid	10.23 ± 1.22^a	6.73 ± 0.39^a	8.09 ± 0.51^a	NA

Values with different alphabets in superscript are significant at $P < 0.05$.
where $n = 3$; data expressed as mean ± standard error of mean (SEM); concentration that inhibits the 2,2-diphenyl-1-picryl-hydrazyl (DPPH), nitric oxide (NO) or hydrogen peroxide (H_2O_2) radical by 50% (IC_{50}); ferric-reducing antioxidant power (FRAP) expressed as mg of ascorbic acid equivalent per gram (mgAAE/g) of test sample; not applicable (NA); Essential oil (EO) batches I–VI; and positive control (L-ascorbic acid).

23.3.2 *In Vitro* Anti-Inflammatory Activity

The result in Table 23.5 showed the anti-inflammatory activity of the essential oils obtained from the turmeric-black cumin herbal mixture. The bioactivities of the oils are presented in terms of their ability to inhibit the denaturation of protein under colorimetric reaction conditions. The result also showed concentration-dependent increase in the anti-inflammatory response of the EOs within 6.25–100 µg/mL concentration range.

23.4 DISCUSSION

23.4.1 Chemical Composition

Gas Chromatography–Mass Spectrometry (GC-MS) analysis showed that majority of the EO batches are sesquiterpenoid (C_{15}) compounds. They included aromatic (ar)-turmerone, α-turmerone, curlone, ar-curcumene, β-sequiphellandrene, and (E)-atlantone (Table 23.3). Liju et al. (2011) reported ar-turmerone, curlone, and ar-curcumene as the major EO components of turmeric rhizome, while p-cymene and β-sequiphellandrene were among the major EO components reported in black cumin

TABLE 23.5

Anti-Inflammatory Activity of Essential Oils Obtained from Turmeric Black Cumin Herbal Mix

EO Batches	Inhibition of Protein Denaturation (%)					$IC_{50} \pm SEM$ (µg/mL)
	100 µg/mL	50 µg/mL	25 µg/mL	12.5 µg/mL	6.25 µg/mL	
I	61.45 ± 16.35[ab]	13.97 ± 3.16[a]	5.66 ± 0.94[a]	2.91 ± 0.55[a]	0.53 ± 0.15[a]	103.63 ± 31.02[e]
II	71.26 ± 1.32[b]	59.33 ± 0.62[c]	49.87 ± 140[d]	34.04 ± 0.77[d]	18.65 ± 3.07[d]	44.21 ± 0.98[c]
III	97.26 ± 1.69[d]	86.56 ± 1.53[e]	73.56 ± 1.47[f]	27.06 ± 2.12[c]	10.26 ± 2.93[c]	31.22 ± 1.24[b]
IV	62.16 ± 1.09[a]	51.72 ± 1.96[b]	34.31 ± 0.69[b]	20.34 ± 1.24[b]	5.04 ± 1.53[b]	61.93 ± 1.77[d]
V	83.64 ± 2.04[c]	71.09 ± 1.77[d]	44.56 ± 2.32[c]	16.71 ± 5.45[b]	3.54 ± 2.26[b]	44.76 ± 2.51[c]
VI	97.88 ± 1.36[d]	87.89 ± 1.00[e]	75.15 ± 2.19[f]	37.75 ± 1.24[e]	20.34 ± 2.84[d]	25.36 ± 1.61[a]
Diclofenac	99.74 ± 0.15[e]	88.68 ± 4.14[e]	66.05 ± 1.77[e]	52.52 ± 4.12[f]	20.96 ± 5.30[d]	21.57 ± 2.66[a]

Where: $n = 3$; data expressed as mean ± standard error of mean (SEM); data with different alphabets in superscripts were considered significant at $P < 0.05$ down the columns; while those with same alphabets in superscripts were comparable ($P > 0.05$) when subjected to Student–Newman–Keul's post-hoc test; essential oil (EO) batches I–VI; Diclofenac (standard anti-inflammatory drug); and concentration that inhibits egg albumin (protein) denaturation by 50% (IC_{50}).

seed (Edris, 2009; Kiralan, 2014; Ashfaq et al., 2021). Thus, the result implies that turmeric rhizomes may be playing a major contributory role to the EO components of the herbal mixture.

Ar-turmerone and curlone were the major constituents identified in batches I–IV; batch V comprised majorly *ar*-turmerone (99.10%), while guaiacol (43.21%)—an *o*-methoxyphenol—was formed in batch VI in addition to *ar*-turmerone (55.88%) (Table 23.3). Guaiacol is a yellowish-to-brownish oily compound, often produced as an artefact of thermal oxidation and/or enzymatic degradation, reported in the industrial brewing of malt as a flavour during coffee and whisky productions, and in the pyrolysis of lignin (Dorfner et al., 2003; Gallegos, 2017). It is also found in the EOs of celery seeds and tobacco leaves (Burdock, 1995). *Ar*-turmerone (32.24%), zingiberene (27.70%), and β-sesquiphellandrene (13.14%) have been reported as the major component of turmeric oil grown in Korea (Hwang et al., 2016). *P*-cymene constituted the major (60.50%) EO component of the black cumin seed grown in Tunisia, according to Bourgou et al. (2010).

Water has been reported as an important condition for deprotonation (Hu et al., 2016). Therefore, since our herbal mixture stayed in water for long during the distillation process, deprotonation of *ar*-turmerone gave curlone (β-turmerone), which could attain stability by isomerization to form α-turmerone, thus, justifying the compositional increase of α-turmerone from 13.23% in batch I to 18.40% in batch II, as presented in Table 23.3. Furthermore, thermal degradation of *ar*-turmerone under the influence of demethylase enzyme might have brought about the elicitation of benzene, 1-methyl-4-(1-methyl-3-buten-3-ol)—another major compound identified in the batch III EO, classified as *p*-hydroxy methylbenzene. Furthermore, we proposed an alternative route by McLafferty structural rearrangement of *ar*-turmerone and consequently an α-cleavage of its propyl substituent to elicit the compound (Pavia et al., 2001) (Figure 23.4).

Guaiacol could be biosynthetically derived from *p*-cymene—a monoterpenoid (C_{10}) compound reported in black cumin seeds (Khan and Afzal, 2016). Thermal and enzymatic degradation of *p*-cymene to form dihydroxy-*p*-cumate and subsequently *o*-methoxyphenol were presented in Figure 23.5, based on the report of Defrank and Ribbons (1977). Here, thermal oxidation of *p*-cymene influenced by methyl monooxygenase enzyme (Pp-cymaA 1.14.15.25) could result in the production of *p*-isopropyl benzylalcohol, which, in turn, could undergo two-step dehydrogenation mediated by cumic alcohol, followed by cumic aldehyde dehydrogenase enzymes, to give *p*-cumate. An *ortho, meta*-dioxygenation of the latter by *p*-cumate-2,3-dioxygenases would result in 2,3-dihydroxy-*p*-cumate, which could, in turn, undergo oxidative demethylation and decarboxylation influenced by demethylase and decarboxylase enzymes, respectively, to form a hydroxyphenol. Finally, the single proton on the *ortho*-hydroxy functional group of hydroxyphenol could become substituted to form *o*-methoxyphenol under the influence of a methyltransferase enzyme (Figure 23.5). Perhaps, the structural rearrangement from *p*-cymene to *o*-methoxyphenol could only have been possible because of the repeated cooling and re-distillation that formed part of our experiments. Similarly, structural rearrangement from methyl eugenol to isoeugenol methyl ether, α-pinene to terpineol, and 1,8-cineol to terpineol upon cooling and re-distillation of three *Melaleuca* species for their EOs has been reported (Oyedeji et al., 2014).

23.4.2 Biological Activities

TLC-bioautographic profile of the EOs (Figure 23.1) indicated strong free radical scavenging property against the stable DPPH-free radical solution. Our result on the quantitative antioxidant activity (Table 23.4) showed concentration-dependent increase in antioxidant activity across the 6.25–100 µg/mL concentration range. Likewise, repeated HD appeared to have impacted significantly on the DPPH, NO, Fe^{3+}, and H_2O_2 radical scavenging properties of the EOs. The Batch VI oil was the most active with IC_{50} values of 16.10 ± 0.14 and 13.17 ± 1.11 µg/mL against DPPH and H_2O_2 radicals, respectively. It also showed considerable Fe^{3+}-reducing ability under acidic condition (pH 3.6). Thus, the FRAP activity was 715.46 ± 21.65 mg ascorbic acid equivalent per 1,000 mg of the sample, which translated into about 72% of the antioxidant power of *L*-ascorbic acid.

However, the fourth ($IC_{50} = 19.42 \pm 0.97$ µg/mL) and fifth ($IC_{50} = 21.51 \pm 1.72$ µg/mL) oil batches demonstrated comparable ($P > 0.05$) activities in terms of their ability to inhibit the nitric oxide radical. It is noteworthy that turmeric oil and its ethanol extract have been reported to exhibit IC_{50} values of 4.5 mg/mL and 27.2 ± 1.1 µg/mL, respectively, against the purple DPPH radical (Mallmann et al., 2017; Sabir et al., 2021). Also, black cumin oil had previously been reported to exhibit an *in vitro* DPPH radical scavenging activity of 460 µg/mL (Burits and Bucar, 2000), which is far less in activity than what has been obtained in this study.

The EOs of the individual plants have been reported to exhibit considerable antioxidant activities (Burits and Bucar, 2000; Kazemi, 2014; Eleazu et al., 2015; Tanvir et al., 2017; Awan et al., 2018; Akter et al., 2019); hence, establishing the capacity of the herbal mixture to scavenge free radicals for the prevention and management of diseases. *Ar*-turmerone, α-turmerone, and curlone identified in this study are known components in turmeric, regarded as natural antioxidants (Jayaprakasha et al., 2002; Liju et al., 2011). Turmerones have been implicated as immune modulatory, anti-inflammatory, antiviral, and more recently as an anti-SARS-COV-2 agents (Park et al., 2012; Paciello et al., 2020; Rattis et al., 2021). Curlone is an isomer of α-turmerone, and a natural antimicrobial and cytotoxic compound (Essien et al., 2015), while β-sequiphellandrene identified in the herbal mixture has been reported in turmeric as an antimicrobial and cytotoxic compound (Tyagi et al., 2015). *P*-Cymene and α-phellandrene elicited by black cumin seed in this study are known antioxidant, antinociceptive, anti-inflammatory, and antimicrobial compounds (Bourgou et al., 2010; Schuff et al., 2019). *p*-Cymene [1-methyl-4-(1-methylethyl)-benzene] is an alkyl-substituted naturally occurring aromatic hydrocarbon. It has been reported to increase the activity of antioxidant enzymes *in vivo*, thus, reducing oxidative stress (Quintans et al., 2013; de Oliveira et al., 2015). *p*-Cymene also showed anti-inflammatory activity, modulated cytokine production (tumour necrosis factor (TNF-α, interleukin-1β-IL-1β, interleukin-6-IL-6), *in vitro* (murine macrophage-like cell line RW 264.7), and *in vivo* (Female C57BL/6) by inhibiting nuclear factor-κB (NF-κB) and mitogen-activated protein kinase (MAPK) signalling pathways involved in synthesis of pro-inflammatory cytokines (Zhong et al., 2013).

Guaiacol (43.21%)—an artefact in the 6th EO batch in this study is an *ortho*-methoxyphenol proposed to be derived from *p*-cymene (Figure 23.4). It is noteworthy that guaiacol is an expectorant, a disinfectant, and an EC 1.1. 1.25 (shikimate dehydrogenase) inhibitor (Dionisio et al., 2018). This compound has been reported as an essential oil component of celery seeds and tobacco leaves (Burdock, 1995). It has also been reported to demonstrate symptomatic relief of coughs associated with colds, bronchial catarrh, influenza, and upper respiratory tract infections such as laryngitis and pharyngitis in mammals. It is used empirically as an expectorant to lessen the amount of mucous in the chronic stages of bronchitis and bronchiectasis. It is frequently prescribed in humans for its stimulant expectorant action as a constituent of a steam inhalant. It has also been employed as a local anaesthetic in dentistry, in a manner like thymol (Remington and Osol, 1975).

Our finding on the herbal mixture and the reported pharmacological properties of the individual plants may, in part, validate the ethnomedicinal use of the herbal mixture as an infusion for the treatment of SARS-CoV-2 infection and other flu-related ailments in Nigeria. However, in isolated cases of guaiacol self-medication, hypertension and general cardiovascular collapse have been reported (Clayton and Clayton, 1982). Therefore, there is the need for further preclinical and clinical studies to potentiate the herbal mixture in relations to its implicated local use as a remedy for SARS-CoV-2 infection.

This study marks the first report on the chemical composition, antioxidant, and anti-inflammatory activities of essential oils from combined turmeric-black cumin homegrown (Nigerian) COVID-19 herbal mixture.

23.5 CONCLUSION AND PROSPECTS

Exhaustive hydro-distillation of turmeric-black cumin herbal mixture afforded six EO batches. The EOs were generally woody in aroma with colour and yield variations. A total 14 chemical components were identified from the GC–MS profiling. Sesquiterpenoids such as *ar*-turmerone, α-turmerone, and curlone were identified as the major constituents of the EOs. The sixth EO batch elicited guaiacol in substantial amount. TLC bioautography of the EOs showed that they possess considerable free radical scavenging property. It suffices to mention that tumerones and curlone are natural antiviral agents, while guaiacol is a known expectorant; hence, it may justify the basis for the folkloric use of the herbal mixture as a remedy for COVID-19 and other flu-related ailments.

FUNDING

The authors are thankful to the Directorate of Research Development and Innovation, Walter Sisulu University (WSU), South Africa, for funding support and for awarding Postdoctoral Research Fellowship to the corresponding author.

ACKNOWLEDGEMENT

The corresponding author acknowledges the Drug Research and Production Unit, Faculty of Pharmacy, Obafemi Awolowo University, Ile-Ife, Nigeria, for study leave permission to WSU, South Africa.

DECLARATION OF COMPETING INTEREST

The authors declare that there is no conflict of interest.

REFERENCES

Adeleye, O.A., Femi-Oyewo, M.N., Bamiro, O.A., Bakre, L.G., Alabi, A., Ashidi, J.S., Balogun-Agbaje, O.A., Hassan, O.M., and Fakoya, G. (2019). Ethnomedicinal herbs in African traditional medicine with potential activity for the prevention, treatment, and management of coronavirus disease. *Future Journal of Pharmaceutical Sciences*, 7(1), 72. https://doi.org/10.1186/s43094-021-00223-5. Accessed on 13-10-2021.

Akter, J., Hossain, M.A., Takara, K., Zahorullslam, M., and Hou, D.X. (2019). Antioxidant activity of different varieties of turmeric (*Curcuma* spp): isolation of active compounds. *Comparative Biochemistry and Physiology Part C: Toxicology and Pharmacology*, 215, 9–17.

Amin, F., Islam, N., Anila, Nfn., and Gilani, A.H. (2015). Clinical efficacy of the co-administration of turmeric and black seeds (Kalongi) in metabolic syndrome—A double blind randomized controlled trial—TAK-MetS trial. *Complementary Therapies in Medicine*, *23*(2), 165–174. doi: 10.1016/J.CTIM.2015.01.008.

Amzat, J., Aminu, K., Kolo, V.I., Akinyele, A.A., Ogundairo, J.A., and Danjibo, M.C. (2020). Coronavirus outbreak in Nigeria: Burden and socio-medical response during the first 100 days. *International Journal of Infectious Diseases*, *98*, 218–224. https://doi.org/10.1016/j.ijid.2020.06.067.

Ansari, S., Jilani, S., Abbasi, H., Siraj, M.B., Hashimi, A., Ahmed, Y., Khatoon, R., and Rifas, A.M. (2020). *Curcuma longa*: a treasure of medicinal properties. *CellMed*, *10*(2), 9.1–9.7. doi: 10.5667/CELLMED.2020.0009.

Ashfaq, S., Khan, N.T., and Ali, G.M. (2021). *Nigella sativa* (Kalonji), its essential oils and their therapeutic potential. *Biomedical Journal of Scientific and Technical Research*, *33*(1), 25448–25454. doi: 10.26717/BJSTR.2021.33.005335.

Awan, M.A., Akhter, S., Husna, A.U., Ansari, M.S., Rakha, B.A., Azam, A., and Qadeer, S. (2018). Antioxidant activity of *Nigella sativa* seeds aqueous extract and its use for cryopreservation of buffalo spermatozoa. *Andrologia*,*50*(6), e13020. https://doi.org/10.1111/and13020.

Babushok, V.I., Linstrom, P.J., and Zenkevich, I.G. (2011). Retention indices for frequently reported compounds of plant essential oils. *Journal of Physical and Chemical Reference Data*, *40*(4), 043101. doi:10.1063/1.3653552.

Benzie, I.F.F. and Strain, J.J. (1999). Ferric reducing antioxidant power assay: direct measure of total antioxidant activity of biological fluids and modified version for simultaneous measurement of total antioxidant power and ascorbic acid concentration. *Methods in Enzymology*, *299*, 15–27. doi:10.1016/S0076-6879(99)99005-5.

Boskabady, M.H., Kiani, S., Jandaghi, P., Ziaei, T., and Zarei, A. (2003). Comparison of antitussive effect of *Nigella sativa* with codeine in guinea pig. *Iranian Journal of Medical Science*, *28*, 111–115.

Boskabady, M.H., Kiani, S., Jandaghi, P., Ziae, T. and Zarei, A. (2004). Antitussive effect of *Nigella sativa* in gunieapigs. *Pakistan Journal of Medical Science*. *20* (2004), 224-228

Bourgou, S., Pichette, A., Marzouk, B., and Legault, J. (2010). Bioactivities of black cumin essential oil and its main terpenes from Tunisia. *South African Journal of Botany*, *76*(2), 210–216. doi: 10.1016/j.sajb.2009.10.009.

Burdock, G.A. (1995). *Encyclopedia of Food and Color Additives*. CRC Press, Boca Raton, FL, 1244–1245, ISBN 978-0849394126.

Burits, M. and Bucar, F. (2000). Antioxidant activity of *Nigella sativa* essential oil. *Phytotherapy Research*, *14*(5), 323–328. doi: 10.1002/1099-1573(200008)14:5<323::aid-ptr621>3.0.co;2-q.

Cai, J., Sun, W., Huang, J., Gamber, M., Wu, J., and He, G. (2020). Indirect virus transmission in cluster of COVID-19 cases, Wenzhou, China. *Emerging Infectious Diseases*, *26*(6), 1343–1345.

Chatterjee, P., Chandra, S., Dey, P., and Bhattacharya, S. (2012). Evaluation of anti-inflammatory effects of green tea and black tea: a comparative *in vitro* study. *Journal of Advanced Pharmaceutical Technology and Research*, *3*, 136–138. doi:10.4103/2231-4040.97298.

Chattopadhyay, I., Biswas, K., Bandyopadhyay, U., and Banerjee, R.K. (2004). Turmeric and curcumin: Biological actions and medicinal applications. *Current Science*, *87*(1), 44–53. https://www.jstor.org/stable/24107978. Accessed on 07 November 2021.

Clayton, G.D. and Clayton, F.E. (1982). Patty's industrial hygiene and toxicology. *Toxicology*, vol. 2A, 2B, 2C, (3rd edition), John Wiley Sons, New York, 2603

De Oliveira, T.M., de Carvalho, R.B.F., da Costa, I.H.F., de Oliveira, G.A.L., de Souza, A.A., de Lima, S.G., and de Freitas, R.M. (2015). Evaluation of *p*-cymene, a natural antioxidant. *Pharmaceutical Biology*, *53*, 423–428.

DeFrank, J.J. and Ribbons, D.W. (1977). *P*-cymene pathway in *Pseudomonas putida*: initial reactions. *Journal of Bacteriology*, *129*(3), 1356–1364.

Dionisio, K.L., Phillips, K., Price, P.S. Grulke, C.M. Williams, A., Biryol, D., Hong, T., and Isaacs, K.K. (2018). Data descriptor: The chemical and products database, a resource for exposure-relevant data on chemicals in consumer products. *Scientific Data 2018*, 5, doi:10.1038/SDATA.2018.125.

Dorfner, R., Ferge, T., Kettrup, A., Zimmermann, R., and Yeretzian, C. (2003). Real-time monitoring of 4-vinylguaiacol, guaiacol, and phenol during coffee roasting by resonant laser ionization time-of-flight mass spectrometry. *Journal of Agricultural and Food Chemistry*, *51*(19), 5768–5773. doi:10.1021/jf0341767.

Edris, A.E. (2009). Evaluation of the volatile oils from different local cultivars of *Nigella sativa* L. grown in Egypt with emphasis on the effect of extraction method on thymoquinone. *Journal of Essential Oil-Bearing Plants*, *13*(2), 154–164. doi: 10.1080/0972060X.2010.10643805.

ElBahr, S., Taha, N., Korshom, M., Mandour, A., and Lebda, M. (2014). Influence of combined administration of turmeric and black seed on selected biochemical parameters of diabetic rats. *Alexandria Journal of Veterinary Sciences*, *41*(1), 19. doi: 10.5455/AJVS.154650.

Eleazu, C., Eleazu, K., Chukwuma, S., Adanma, I., and Igwe, A. (2015). Polyphenolic composition and antioxidant activities of 6 new turmeric (*Curcuma longa* L.) accessions. *Recent Patents on Food, Nutrition and Agriculture*, *7*(1), 22–7. doi: 10.2174/2212798407666150401104716.

Essien, E.E., Newby, J.S., Walker, T.M., Setzer, W.N., and Ekundayo, O. (2015). Chemotaxonomic characterization and *in vitro* antimicrobial and cytotoxic activities of the leaf essential oil of *Curcuma longa* grown in Southern Nigeria. *Medicines (Basel)*, *2*(4), 340–349.

Fabricant, D.S. and Farnsworth, N.R. (2001). The value of plants used in traditional medicine for drug discovery. *Environmental Health Perspectives*, *109*, 69–75. http://dx.doi.org/10.1289/ehp.01109s169.

Farhan, N., Salih, N., and Salimon, J. (2021). Physiochemical properties of Saudi *Nigella sativa* L. (Black cumin) seed oil. State and trends of oil crops production in China. *Oilseeds, Crops, fats and Lipids*, *28*, 11©N. https://doi.org/10.1051/ocl/2020075.

Gallegos, J. (2017). The best way to drink whiskey, according to science. The Washington posts. Guaiacol is what gives whiskey that smoky, spicy, peaty flavor. https://www.washingtonpost.com/news/speaking-of-science/wp/2017/08/17/the-best-way-to-drink-whiskey-according-to-science/. Accessed on 15 November 2021.

Hu, C., Shaughnessy, K.H., and Hartman, R.L. (2016). Influence of water on the deprotonation and the ionic mechanisms of a Heck alkynylation and its resultant E-factors. *Reaction Chemistry and Engineering*, *1*, 65–72. https://doi.org/10.1039/C5RE00034C.

Hwang, K.W., Son, D., Jo, H-W., Kim, C.H., Seong, K.C., and Moon, J-K. (2016). Levels of curcuminoid and essential oil compositions in turmerics (*Curcuma longa* L.) grown in Korea. *Applied Biological Chemistry*, *59*, 209–215. https://doi.org/10.1007/s13765-016-0156-9.

Jayaprakasha, G.K., Jena, B.S., Negi, P.S., and Sakariah, K.K. (2002). Evaluation of antioxidant activities and antimutagenicity of turmeric oil: A byproduct from curcumin production. Verlag der *Zeitschrift für Naturforschung C, Journal of Bioscience*, *57*, (9–10). Tübingen www.znaturforsch.com. Accessed on 12 October 2021.

Jiang, C-L, Tsai, S-F., and Lee, S-S. (2015). Flavonoids from *Curcuma longa* leaves and their NMR assignments. *Natural Product Communications*, *10*(1), 63–66. doi:10.1177/1934578X1501000117.

Jimoh, M.O., Afolayan, A.J., and Lewu, F.B. (2019). Antioxidant and phytochemical activities of *Amaranthus caudatus* L. harvested from different soils at various growth stages. *Scientific Reports*, *9*(1), 1–14. doi:10.1038/s41598-019-49276-w.

Kazemi, M. (2014). Phytochemical composition, antioxidant, anti-inflammatory and antimicrobial activity of *Nigella sativa* L. essential oil, *Journal of Essential Oil-Bearing Plants*, *17*(5), 1002–1011. doi: 10.1080/0972060X.2014.914857.

Kiralan, M. (2014). Changes in volatile compounds of black cumin (*Nigella sativa* L.) seed oil during thermal oxidation. *International Journal of Food Properties*, *17*(7), 1482–1489. doi: 10.1080/10942912.2012.723231.

Krup, V., Prakash, L.H., and Harini, A. (2013). Pharmacological activities of turmeric (*Curcuma longa* Linn): a review. *Journal of Homeopathy and Ayurvedic Medicine*, *2*, 133. doi: 10.4172/2167–1206.1000133.

Liju, V.B., Jeena, K., and Kuttan, R. (2011). An evaluation of antioxidant, anti-inflammatory, and antinociceptive activities of essential oil from *Curcuma longa* L. *Indian Journal of Pharmacology*, *43*, 526–531.

Khan, M.A. and Afzal, M. (2014). Chemical composition of *Nigella sativa* Linn: Part 2 recent advances. *Inflammopharmacology*, *24*, 67–79. doi: 10.1007/S10787-016-0262-7.

Khan, M.A., Chen, H.C., Tania, M., and Zhang, D.Z. (2011). Anticancer activities of *Nigella sativa* (black cumin). *African Journal of Traditional Complementary and Alternative Medicines*, *8*(S), 226–232. doi: 10.4314/ajtcam.v8i5S.10.

Mallmann, C.A., Brugnari, T., de Abreu Filho, B.A., Mikcha, J.M.G., and Machinski, M. (2017). *Curcuma longa* L. Essential oil composition, antioxidant effect, and effect on *Fusarium verticillioides* and fumonisin production. *Food Control*, *73*, 806–813. doi:10.1016/J.FOODCONT.2016.09.032.

Mukherjee, P.K., Nema, N.K., Pandit, S., and Mukherjee, K. (2013). Indian medicinal plants with hypoglycemic potential. *Bioactive Food as Dietary Interventions for Diabetes*, *2013*, 235–264. doi: 10.1016/B978-0-12-397153-1.00022-6.

Nigeria Centre for Disease Control, NCDC, (2020). Nigeria Centre for Disease Control: An Update of COVID-19 Outbreak in Nigeria. https://ncdc.gov.ng/diseases/sitreps/?cat=14&name=An%20update%20of%20COVID-19%20outbreak%20in%20Nigeria. Accessed on 07 November 2021.

Nigeria Centre for Disease Control, NCDC, (2022). Nigeria Centre for Disease Control: COVID-19 Nigeria, Progression. https://covid19.ncdc.gov.ng/progression/. Accessed on 13 August 2022.

Okeleye, B.I., Nongogo, V., Mkwetshana, N.T., and Ndip, R.N. (2015). Polyphenolic content and *in vitro* antioxidant evaluation of the stem bark extract of *Peltophorum africanum* Sond (Fabaceae). *African Journal of Traditional, Complementary and Alternative Medicines*, *12*, 1–8. doi:10.4314/ajtcam.v12i1.1.

Oriola, A.O., Aladesanmi, A.J., Idowu, T.O., Akinwumi, F.O., Obuotor, E.M., Idowu, T., and Oyedeji, A.O. (2021). Ursane-type triterpenes, phenolics and phenolic derivatives from *Globimetula braunii* Leaf. *Molecules*, *26*(21), 6528. doi:10.3390/molecules26216528.

Orisakwe, O.E., Orish, C.N., and Nwanaforo, E.O. (2020). Coronavirus disease (COVID-19) and Africa: acclaimed home remedies. *Scientific African*, *10*, 11. https://doi.org/10.1016/j.sciaf.2020.e00620.

Oyedeji, O.O., Oyedeji, A.O., and Shode, F.O. (2014). Compositional variations and antibacterial activities of the essential oils of three *Melaleuca* species from South Africa. *Journal of Essential Oil-Bearing Plants*, *17*(2), 1–12. ISSN Online: 0976–5026. http://dx.doi.org/10.1080/0972060x.2013.813221.

Paciello, F., Fetoni, A.R., Mezzogori, D., Rolesi, R., Di Pino, A., Paludetti, G., Grassi, C., and Troiani, D. (2020). The dual role of curcumin and ferulic acid in counteracting chemoresistance and cisplatin-induced ototoxicity. *Scientific Reports*, *10*(1), 1063. doi:10.1038/s41598-020-57965-0.

Park, S.Y., Kim, Y.H., Kim, Y., and Lee, S.J. (2012). Aromatic-turmerone's anti-inflammatory effects in microglial cells are mediated by protein kinase A and heme oxygenase-1 signaling. *Neurochemistry International*, *61*(5), 767–777. doi: 10.1016/j.neuint.2012.06.020.

Pavia, D.L., Lampman, G.M., and Kriz, G.S. (2001). Introduction to spectroscopy. In: *Mass Spectrometry: Some Fragmentation Patterns*, (3rd editon). Thomson Learning Inc., USA. ISBN 10987.

Punch Newspaper (2020) https://healthwise.punchng.com/i-fought-covid-19-with-vitamin-c-black-seed-oil-mixed-with-honey-gov-makinde-2/. Accessed on 13 November 2021.

Quintans-Junior, L.., Moreira, J.C.F., Pasquali, M.A.B., Rabie, S.M.S., Pires, A.S., Schröder, R., Rabelo, T.K., Santos, J.P.A., Lima, P.S.S., Cavalcanti, S.C.H., Araujo, A.A.S., Quintans, J.S.S., and Gelain, D.P. (2013). Antinociceptive activity and redox profile of the monoterpenes (+)-camphene, *p*-cymene, and geranyl acetate in experimental models. *ISRN Toxicology*, *2013*, 459530. doi: 10.1155/2013/459530.

Rattis, B.A.C., Ramos, S.G., and Celes, M.R.N. (2021). Curcumin as a potential treatment for COVID-19. *Frontiers in Pharmacology*, *12*, 675287. doi: 10.3389/fphar.2021.675287.

Remington, J.P., Osol, A., and Hoover, J.E. et al., (1975). *Remington's Pharmaceutical Sciences*, (15th edition). Mack Publishing Co., Easton, Pennsylvania, 1102.

Sabir, S., Zeb, A., Mahmood, M., Abbas, S., Ahmad, Z., Iqbal, N., and Kashmir, A. (2021). Phytochemical analysis and biological activities of ethanolic extract of *Curcuma longa* rhizome. *Brazilian Journal of Biology*, *81*, 737–740. doi:10.1590/1519-6984.230628.

Sahara Reporters (2021). How I Used Herbs to "cure" My COVID-19 Infection – Obasanjo. http://sahara-reporters.com/2021/03/06/how-i-used-herbs-cure-my-covid-19-infection-%E2%80%93-obasanjo. Accessed on 13 November 2021.

Sambhav, J., Rohit, R., Raj, U.A., and Garima, M. (2014). *Curcuma longa* in the management of inflammatory diseases: a review. *International Ayurvedic Medical Journal*, *2*(1), 33–40. http://www.iamj.in/posts/2014/images/upload/33_40.pdf. Accessed on 12 October 2021.

Schuff, S.W., Rasool, W., and Shafi, K. (2019). A comprehensive review of the antibacterial, antifungal, and antiviral potential of essential oils and their chemical constituents against drug-resistant microbial pathogens. *Microbial Pathogenesis*, *134*, 103580.

Tanvir, E.M., Hossen, M.S., Hossain, M.F., Afroz, R., Gan, S.H., Khalil, M.I., and Karim, N. (2017). Antioxidant properties of popular turmeric (*Curcuma longa*) varieties from Bangladesh. *Polyphenols and Food Quality*, *2017*. https://doi.org/10.1155/2017/8471785

The Guardian Newspaper (2020). Health: More Plants with Antiviral Properties Validated. https://guardian.ng/features/health/more-plants-with-antiviral-properties-validated/. Accessed on 09 November 2021.

The Guardian Newspaper (2021). Focus: Surviving COVID-19: Five Nigerians Tells Their Survival Stories. https://guardian.ng/features/surviving-covid-19-five-nigerians-tells-their-survival-stories/. Accessed on 09 November 2021.

Tyagi, A.K., Prasad, S., Yuan, W., Li, S., and Aggarwal, B.B. (2015). Identification of a novel compound (β-sesquiphellandrene) from turmeric (*Curcuma longa*) with anticancer potential: comparison with curcumin. *Investigational New Drugs*, *33*(6), 1175–1186. doi: 10.1007/s10637-015-0296-5.

Uchejeso, O.M., Okolo, R.C., Etukudoh, N., and Obiora, E.R. (2021). Use of turmeric against COVID-19 in Nsukka, the need for massive farming. *Direct Research Journal of Public Health and Environmental Technology*, *6*, 21–25. https://doi.org/10.26765/DRJPHET87125908.

Vanguard Newspaper (2020). Coronavirus Updates: COVID-19: People Treated with Herbal Cure Recovers Faster—Ezeife.https://www.vanguardngr.com/2020/04/covid-19-people-treated-with-herbal-cure-recovers-faster-%E2%80%95ezeife/. Accessed on 09 November 2021.

Vollbracht, C. and Kraft, K. (2022). Oxidative stress and hyper-inflammation as major drivers of severe COVID-19 and long COVID: implications for the benefit of high-dose intravenous Vitamin C. *Frontiers in Pharmacology, 13*, 899198. https://doi.org/10.3389%2Ffphar.2022.899198.

Wang, L., Li, J., Guo, S., Xie, N., Yao, L., Cao, Y., and Ji, J. (2020). Real-time estimation and prediction of mortality caused by COVID-19 with patient information-based algorithm. *Science of The Total Environment 727*, 138394. https://doi.org/10.1016/j.scitotenv.2020.138394.

World Health Organisation, WHO (2020). Origin of the SARS-COV-2 Virus. WHO-Convened Global Study of the Origins of SARS-CoV-2 (including annexes). https://www.who.int/emergencies/diseases/novel-coronavirus-2019/origins-of-the-virus. Accessed on 06 November 2021

Worldometer (2021). Nigeria Population Live. https://www.worldometers.info/world-population/nigeria-population/. Accessed on 07 November 2021. www.theplantlist.org. Search result for *Curcuma longa* L. http://www.theplantlist.org/tpl1.1/search?q=curcuma+longa, kew-235249 and Nigella sativa L. http://www.theplantlist.org/tpl1.1/search?q=Nigella+sativa, kew-2381679. Accessed 06 November 2021.

Yimer, E.M., Tuem, K.B., Karim, A., Ur-Rehman, N., and Anwar, F. (2019). *Nigella sativa* L. (Black Cumin): a promising natural remedy for wide range of illnesses. *Evidence Based Complementary and Alternative Medicines, 2019*, 1528635. doi:10.1155/2019/1528635

Zhong, W., Chi, G., Jiang, L., Soromou, L.W., Chen, N., Huo, M., Guo, W., Deng, X., and Feng, H. (2013). *p*-Cymene modulates *in vitro* and *in vivo* cytokine production by inhibiting MAPK and NF-κB activation. *Inflammation, 36*(3), 529–537. doi: 10.1007/s10753-012-9574-y.

24 Effect of Climate Change on Medicinal Plants and Traditional Medicine-Based Health Security in Nasarawa State, Nigeria

Oluwakemi Abosede Osunderu

CONTENTS

24.1 INTRODUCTION

Medicinal plants and their knowledge form a veritable stock of raw materials for production of not only medicines but also food. Worldwide, the estimated business volume of herbal medicine alone is over USD 60 billion. Ethnomedicine or Traditional Medicine is an integral part of any cultural habitat and its people. The use of remedies from plant sources, animal parts, minerals, or microorganisms for the relief of illnesses arising from disease conditions predates history, and about 70% of these traditional medicines are plant based. The World Health Organization (WHO) posits that medicinal plant refers to any part, tissue, or organ of a plant species containing substances usable for therapeutic purposes or which serve as templates for the synthesis of more useful drugs with minimal side effects (WHO, 2006). Medicinal plants and their products are derived from natural forests in form of tree barks, roots, leaves, flowers, fruits, and exudates. The bulk of the traditional medicine in Nigeria is prepared from plants, and less than 5% of medical preparations come from soil minerals, domestic, and wild animal sources. Medicinal plants are commonly traded in open markets (Osunderu, 2017). Trade in medicinal plants is a huge economic activity that supports an estimated population of over 124 million Nigerians on a regular basis (Soladoye et al., 2013). WHO has long been aware of the very vital role Traditional Complementary and Alternative Medicine

DOI: 10.1201/9781003206620-24

(TCAM) practitioners from member states play in health delivery systems, more importantly the rural areas of African Countries, and had encouraged member states in the use and exploitation of all available resources in tackling primary healthcare.

The WHO in its Alma-Ata declaration of 1978 gave due recognition to the role of traditional medicine (TM) and traditional medicine practitioners (TMPs) in achieving comprehensive and affordable healthcare delivery, previously tagged "Health for All in the 21st Century".

TM refers to health practices, approaches, knowledge, and beliefs incorporating plant-, animal-, and mineral-based medicines, spiritual therapies, manual techniques, and exercises, applied singularly or in combination to treat, diagnose, and prevent illness or maintain well-being (Onyiapat et al., 2017). TM plays a primary role in people's health, as they have been used as therapies for thousands of years, there is a wide range of therapies and practices, varying greatly from country to country and from region to region (Osunderu, 2017). The most well-known are the Ayurveda of India and traditional Chinese medicine, and these systems of medicines have now spread to other countries. TM practices include herbal medicine; it is also known as complementary/alternative medicine. TM has always maintained its popularity worldwide (WHO, 2022). For more than a decade, there has been an increasing use of TM (in form of complementary and alternative medicine (CAM)) in most developed and developing countries (Awodele et al., 2012).

Climate change is the significant increase in the earth's temperature over a long period of time. McLean and McMillan (2009) describe climate change as "a phenomenon where solar radiation that has reflected back off the surface of the earth remains trapped at atmospheric levels due to the build-up of carbon dioxide and other greenhouse gases rather than being emitted back into space". The effect of this is the warming of the global atmosphere. It is a long-standing phenomenon, as the mix up of the various gases that make up the earth's atmosphere has changed over long periods of time, so average global temperature has fluctuated. This is largely caused by human activities (anthropogenic factors) through the combustion of fossil fuels for industrial or domestic usages. Biomass burning has resulted to the production of greenhouse gases and aerosols that have affected the composition of the atmosphere Umezurike (2012). Consequently, the degradation of these natural resources (through deforestation and erosion) has impoverished the environment, leading to a poorer ecosystem function—less production of biodiversity (medicinal plants) and fewer pollinating insects for example. The land suffers; a lower level of carbon in the land means it is unable to support vegetation, because organic matter levels rich in carbon are reduced by erosion and the loss of trees. Organic matter is important, because it keeps the soil "healthy" and acts as a buffer against drought brought about by climate change.

24.2 CURRENT HERBAL MEDICINE DEVELOPMENT AND RESEARCH

The government of Nigeria desires to maintain and encourage the growth and development of traditional medical practice through coordination and control. The government agencies include NAFDAC, NIPRID, NGOs, and tertiary institutions. They work together to facilitate national and international collaborative research development and promotion of TMs. They research, collate, and document all TM practices and products to preserve the nation's positive indigenous knowledge on traditional medical, science and technology development, and promote safe, efficient traditional therapies and facilitate their integration into the national healthcare delivery system (WHO, 2022).

Traditional delivery, obstetrics, and gynaecology are services rendered by traditional birth attendants (TBAs) in Nigeria. They are widely practiced in the rural and urban areas of the country, and are highly patronised by over 75% of Nigerians because of the increasing costs of health care in our conventional health institutions, as emphasised by Essentials of Community Health. Their positive contributions to healthcare delivery system cannot be ignored. The services of TBAs are readily accessible, personalised, and relatively cheap. However, this practice is not perfect, and it has been improved upon (Osunderu, 2008).

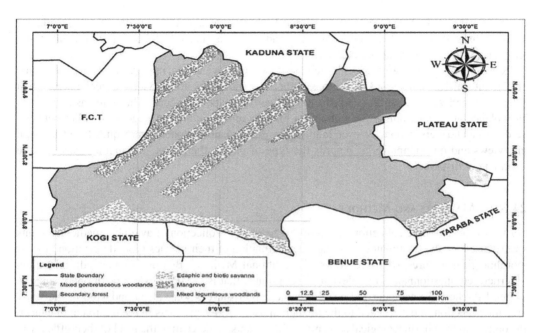

FIGURE 24.1 Vegetation map of Nasarawa State.

The type of method of diagnosis is related to the symptoms observed. (WHO, 2022). This is one of the deficiencies of TM practice, as the method of diagnosis is not adequate in most cases, although they still enjoy increasing popularity (Duru et al., 2019).

Forest products are the primary source of medicines used by TBAs in Nigeria. Several medicinal plants of global importance originated in Africa. The number of TBAs far outnumber that of their allopathic counter parts. The National Demographic and Health Survey report (NHS, 2021) indicate that only 37% of births take place in conventional health centres or hospitals.

24.3 METHODOLOGY

24.3.1 AREA OF STUDY NASARAWA (NORTH CENTRAL)

Nasarawa state is a major contributor to the agricultural and forestry subsector of the Nigerian economy. The vegetation map of Nasarawa State is shown in Fig 24.1. Nasarawa state is located at Coordinates: 8°32′N 8°18′E in the North Central region of Nigeria, bordered to the east by the states of Taraba and Plateau, to the north by Kaduna State, to the south by the states of Kogi and Benue, and to the west by the Federal Capital Territory. Named for the historic Nasarawa Emirate, the state was formed from the west of Plateau State on 1 October 1996 (GDL, 2021) The state has 13 local government areas, and its capital is Lafia—located in the east of the state, while a key economic centre of the state is the Karu Urban Area—suburbs of Abuja—along the western border with the FCT.

Of the 36 states of Nigeria, Nasarawa is the fifteenth largest in area and the second least populous with an estimated population of about 2.5 million as of 2016 (NBS, 2022). Geographically, the state is mostly within the tropical Guinean forest–Savanna mosaic ecoregion. Important geographic features include the river Benue forming much of Nasarawa State's southern borders and the state's far northeast containing a small part of the Jos Plateau.

Nasarawa State is inhabited by various ethnic groups, including the Koro and Yeskwa in the far northwest; the Kofyar in the far northeast; the Eggon, Gwandara, Mada, Ninzo, and Nungu in the north; the Alago, Goemai, and Megili in the east; Eloyi in the south; the Tiv in the southeast; the Idoma in southwest; and the Gade and Gbagyi in the west while the Hausa and Fulani live throughout the state. Nasarawa is also religiously diverse, as about 60% of the state's population are Muslim, with around 30% being Christian and the remaining 10% following traditional ethnic religions (Sa'adatu and Abubakar, 2012).

Economically, Nasarawa State is largely based around agriculture, mainly of sesame, soybeans, groundnut, millet, maize, and yam crops. Other key industries are services, especially in urban areas, and the livestock herding and ranching of cattle, goats, and sheep. The state has been beset by violence at various points throughout its history, most notably the ongoing conflict between herders and farmers primarily over land rights. Several practitioners were questioned through interviews and questionnaire, as regards the type of plants, their uses, and methods used in the treatment of diseases.

24.3.2 MATERIALS AND METHODS

For the purpose of data collection in this study, field trips, collection of available medicinal plants species used for the treatment of diseases, determination of their species type, description of their morphological features, oral interviews of Traditional Medicine Boards officials, administration of structured questionnaires on relevant target groups, that is, TMPs, CAM Practitioners, and the General Public (GP) were carried out. Ethnomedicinal surveys were also conducted in the study area for collection of data related to the medicinal use of forest products in the treatment of cancer, in addition to the pharmacological screening of the plants, to determine the level of their efficacy in the treatment of cancer, and to validate the claims of the TMPs. Pharmacological screening of the plants could not be completed due to lack of funds.

To identify the locations with high concentration of TMPs in the study area, primary data were obtained through oral interviews of the officials of the Department of Hospital Management of the State's Ministry of Health. Multistage sampling technique was employed. By personal interview through administration of structured questionnaires to respondents who are TMPs, CAM Practitioners, and the GP in the study area. Information obtained through administration of questionnaires was based on sampling frame that was obtained from officials at relevant Government Ministries and Management of Boards of Traditional Complementary and Alternative Medicine Practitioners in the state

24.3.3 SAMPLING TECHNIQUE

One hundred TMPs within the study areas were selected using multistage sampling technique. The second stage involved purposive selection of three Senatorial Districts [Karu (Northwest Senatorial District) and Doma (North South Senatorial District) of Nasarawa State], given the prominence of TM practice and utilisation of TM products according to the Nasarawa State board of TMPs. The third stage, that is, the selection of local government areas was based on the prominence given to Traditional Medicine Practice and utilisation of TM products according to the states board of TMPs. The fourth stage employed random selection of ten wards in each of the three selected local government areas in each of the Senatorial District using the list of wards available at the offices of the Independent National Electoral Committee in the state. This was used as the sampling frame. Thirty (30) wards were then sampled in the state. The fifth stage involved random selection of TCAM Practitioners and participants from the public in each of the 30 wards making a sample size of 100.

Primary data collected included, among others, the socioeconomic characteristics such as age, gender, state of origin, occupation, marital status, religion, ethnicity, language understood, method of diagnosis, availability of medicinal plants, educational qualification, year of experience, and types of medicinal plants used. The data also covered healing activities as well as constraints to practitioners. Data that were collected were used to determine the available plants, popularity, and acceptance of TM in Nigeria.

24.3.4 ANALYTICAL TOOLS/TECHNIQUES

Data collected for this research work were analysed using descriptive statistics; frequencies and percentages as well as means were adopted to describe the socioeconomic characteristics of the TMPs, orthodox, and the GP in the study area. The data also described the pattern of practice, and identified constraints and the interrelationship between the different study groups within the study areas.

24.5 RESULTS

24.5.1 ANALYSIS OF THE DATA FROM NASARAWA STATE

The raw questionnaire data are shown in Appendix 24.1.

Thirty species of medicinal plants were identified from the information supplied by the TMPs. Table 24.1 shows the distribution of the species in relation to the source, status, parts of the plant used, and availability within the study area.

The life forms of these plants (Table 24.1) show that the trees constituted the highest number (66%), followed by shrubs (20%), herbs (11%), and rhizomes (3%) In all, the family Leguminosae was dominant with four species. This was followed by Annonaceae, Anacardiaceae, Euphorbiaceae, and Caesalpinioideae (three species each). The existence of other plant families in Table 24.1 demonstrates the rich forest diversity in North Central Nigeria. This also shows the dynamism in ecosystem maintenance. A number of these are also used for economic purposes and are consumed as food in one way or the other (Mallik and Panigrahi, 1998). Some of these include: *Anacardium occidentalis, Mangifera indica, Musa sapientum, Citrus medica, Vernonia amygdalina*, etc.

Majority of the TMPs source their medicinal plants from free areas and rarely cultivate them (Appendix 24.1). Data also shows that some of the plants are already scarce, and species

---

TABLE 24.1

Availability of Medicinal Plants Used for the Treatment of Diseases in Nasarawa State, Nigeria

S/No	Species	Family	Floral Type	Source	Status of Availability	Parts Used
1.	*Anona senegalensis* Pers	Annonaceae	Shrub	Free areas, forest	Abundant	Fruit and leaves
2.	*Xylopia aethiopica* (Dunal) A. Rich	Annonaceae	Tree	Free areas,	Abundant	Leaves and branches
3.	*Khaya ivorensis* A. Chev.	Meliaceae	Tree	Free areas	Rare	Stem, branches, and bark
4.	*Citrus medica* Linn.	Rutaceae	Shrub	Free areas, forest	Abundant	Leaves
5.	*Spondias mombin* Linn.	Anacardiaceae	Fruit Tree	Farmland, free areas, forest	Scarce	Fruits and bark
6.	*Enantia chlorantha* Oliv.	Annonaceae	Tree	Free areas, forest	Scarce	Bark
7.	*Alstonia boonei* De Wild	Apocynaceae	Tree	Free areas, forest	Scarce	Leaves, bark, and root
8.	*Ricino Dendron* heudelotii (Baill) Heckel	Euphorbiaceae	Tree	Free areas, forest	Scarce	Leaves, bark
9.	*Morinda lucida* Benth.	Rubiaceae	Tree	Free areas, forest	Abundant	Leaves

(Continued)

TABLE 24.1 (*Continued*)
Availability of Medicinal Plants Used for the Treatment of Diseases in Nasarawa State, Nigeria

S/No	Species	Family	Floral Type	Source	Status of Availability	Parts Used
23.	*Zanthoxylum zanthoxyloides*	Rutaceae	Herb	Forest\wild, cultivate	Abundant	Branches, stem, and bark
24.	*Agerantum conyzoides*	Compositae	Shrub	Wild	Abundant	Root
25.	*Allium* sativum Linn	Liliaceae	Rhizome	Forest\wild, cultivate	Abundant	leaves, Branches, stem, and root
26.	*Helianthus annuus*	Asteraceae	Shrub	Forest/wild, cultivate	Abundant	Leaves
27.	*Securine gavirosa*	Euphorbiaceae	Shrub	Forest\wild, cultivate	Abundant	Leaves and stem
28.	*Vitellaria*	Sapotaceae	Tree	Forest\wild, cultivate	Abundant	Leaves, stems and root
29.	paradoxa	Poaceae	Shrub	Forest\wild, cultivate	Abundant	Fruit
30.	*Saccharum*	Piperaceae	Shrub	Forest/wild, cultivate	Abundant	Leaves, stems and roots
31.	*officinarum*			Forest\wild, cultivate		
	Piper *guineensis*			Forest\wild,		Leaves, stems, roots and fruits
	Garcinia koli	Guttiferae	Tree	cultivate	Abundant	Fruits and leaves
10.	*Vitex doniana* Sweet	Verbenaceae	Tree	Free areas, forest	Abundant	Fruit and leaves
11.	*Magnifera indica* Linn.	Anacardiaceae	Fruit tree	Free areas, forest, plantation	Abundant	Leaves, fruits, bark, branches, and stem
12.	*Piliostigma thinningi* Milne Redhead	Leguminosae Sub: Mimosoidae	Shrub	Free areas, forest	Abundant	Leaves
13.	*Azadirachta indica* A. Juss	Meliaceae	Tree	Free areas, plantation	Abundant	Leaves, branches, and stem
14.	*Ficussur Forssk.*	Moraceae	Tree	Free areas, forest	Abundant	Fruit and bark
15.	*Margaritaria discoidea* (Baill.) Webster	Euphorbiaceae	Tree	Free areas, forest, dry outliers	Scarce	Leaves, branches, stem, bark, and roots
16.	*Erythrophleum suaveolens* (Gull. and Perr.)	LeguminosaeSub: Caesalpinioideae	Tree	Forest	Rare	Leaves, branches, stem, bark, and root
17.	*Cordia millenii* Bak.	Bignoniaceae	Tree	Free areas, forest	Scarce	Leaves, branches, and stem
18.	*Vernonia amygdalina* (Schreb) Del.	Asteraceae	Tree	Free areas, forest	Abundant	Leaves, branches, bark, and root
19.	*Daniellia oliveri* Rolfe	Leguminosae Sub: Caesalpinioideae	Tree	Savannah forest, re-growth	Abundant	Branches, stem, bark, and root
20.	*Zingiber officinale* Rossae.	Zingiberaceae	Herb	Free areas, forest	Abundant	Rhizome
21.	*Sida acuta*	Malraceae	Herb	Forest\wild, cultivate	Abundant	Leaves, branches, stem, and root
22.	*Mirabilis nyctaginea*	Nyctaginaceae	Herb	Forest\wild, cultivate	Abundant	Leaves, branches, stem, and root

regeneration is by growing in the wild. According to the reports by Gbile et al. (1981) and Amaede et al. (2016), the Nigerian ecosystems are at greater risk of extinction, if urgent attention is not given to the cultivation of medicinal plants. The results of the present study also show that about 90% of the TMPs use the whole plant for treatment, that is, they make use of the fruits, stems, barks, and leaves at the same time. Furthermore, this study revealed that forest products used for the treatment of cancer are multipurpose; they are used as firewood, medicine, foods, chewing sticks, and animal feeds (*Agerantum conyzoides*). This corroborates the works of Soladoye et al. (2013).

The results identified seven species belonging to seven different families: Rutaceae, Asteraceae, Anarcardiaceae, Annonaceae, Meliaceae, Guttiferaceae, and Leguminaceae topping the TMPs priority list. The list of plants used by the TMPs are in consonant with earlier reports of Soladoye et al. (2014), who have reported extensive use of plants in Southwest Nigeria. The supply of medicinal plants and other forest products have contributed immensely to the traditional healthcare and income to stakeholders in the traditional healthcare sector. Availability of medicinal plants, particularly of choice species in medical preparations, has been a determining factor to the prices of traditional healthcare delivery.

Forest products exploitation from the study area was mainly of wild stocks, where the species occur naturally in the forest. This has resulted into special investment problems with medicinal plants, because most medicinally important species that might seem worth developing as new crops are undomesticated. These often have a limited yield, mature irregularly, are very variable, are not suited to current planting and harvesting machinery, or present many other problems such as:

- Seeds are often difficult to obtain;
- expertise and knowledge are often in very short supply (and often are regarded as trade secrets);
- plant breeding and silviculture have become necessary;
- management techniques will have to be developed; and
- markets will have to be located or developed.

The application of ecosystem services and goods in the well-being of Nigerians is extensive, although traditional agriculture in Nigeria is having difficulties, and there is a great need for initiative that involves new crops and new products. Medicinal plants are mostly obtained in rural settings—both from wild plants and from plants cultivated on small farms; therefore, there is a need for governmental initiatives that encourage rural development in a sustainable fashion. Private sector interest alone is often insufficient to initiate or expand projects such as the medicinal plant initiative suggested, and government funding is needed as a catalyst.

24.5.2 LIMITATIONS OF THE RESEARCH PROJECT

- The major limitation of the research project was lack of adequate fundings.
- Limited cooperation from practitioners due to suspicion and the previous unfulfilled promises.
- Dearth of information on TM practices in Nigeria.

24.6 CONCLUSION AND PROSPECTS

From the study carried out in Nasarawa, North Central Nigeria, majority of Nigerians depend directly on medicinal plants. TM is patronised because it is easy to gain access to healers, and treatments are cheap (Sofowora, 1993; NBS, 2020). The bulk of the TM in Nigeria is prepared from plants, and less than 5% of medical preparations come from soil minerals, domestic and wild animal sources. Medicinal plants are commonly traded in open markets. These plants and their products are derived from natural forests in form of tree barks, roots, leaves, flowers, fruits, and exudates. Trade in medicinal plants is a huge economic activity that supports an estimated population of over 124 million Nigerians on a regular basis (Tade, 2020). Also, there is a strong compelling need to preserve the natural biodiversity for continuity's sake and taking cognizance of the fact that the very vast flora in Africa remains a source for future drug discovery and new ethnopharmaceuticals, especially contraceptives.

Forest products exploitation from the study areas was mainly wild stocks where the species occur naturally in the forest. Therefore, there is a need for government to ensure the uniformity and regulation of herbal medicine practices to achieve the millennium development goals on health. The following recommendations are suggested:

- For the government to achieve the millennium development goals on health, attention must be given to the role played by TCAM Practitioners in the treatment of disease.
- Government should put policies in place to encourage youths to learn the practice by setting up ethnoforestry institutions. They should also provide incentives such as IPR (Intellectual Propriety Rights) protection for the older TMPs, which will encourage them to share information.
- There should be the standardisation of morphological/biochemical methods to identify the medicinal plants and uniform method to analyse the effect of cultivated species on drug quality.
- The research department should be encouraged to establish a herbal research unit for drug production research, teaching, and training.
- TMPs should be allowed to offer consultation in our general hospitals alongside the orthodox doctors so that the patient has a wide choice of healthcare to choose from; this is of practice in other countries such as Ghana.
- The time has come for economic and financial policymakers, the international financial community, and/or international domestic investors to start large-scale cultivation of medicinal plants, especially those with proven efficacy and safety. Most of these plants grow well even on deteriorating soil.
- There is need for government to recognise and formulate policies that will maximise the medicinal properties of forest products.

REFERENCES

Awodele, O., Agbaje, E.O., Abiola, O.O., and Awodele, D. (2012). Doctors' attitudes towards the use of herbal medicine in Lagos, Nigeria. *Journal of Herbal Medicine*, 2(1), 16–22.

Duru, C., Nduka, I., and Obikeze, O. (2019). Complementary and alternative medicine use for treatment of acute illnesses in children living in Yenagoa, Nigeria. *Journal of Complementary and Alternative Medical Research*, 8(4), 1–9.

Gbile, Z.O., Ola-Adams, B.A., and Soladoye, M.O. (1981). Endangered species of the Nigerian flora. *Nigerian Journal of Forest*, 8, 14–20.

Sa'adatu, H.L and Abubakar, W.S.I. (2012). *Muslims of Nasarawa State: A Survey*, NRN Background Paper No. 7. University of Oxford, London, UK.

McLean I. and McMillan A. (2009) *Oxford Dictionary of Politics*. Oxford University Press, India.

Onyiapat, J. L., Okafor, C., Okoronkwo, I., Anarado, A., Chukwukelu, E., Nwaneri, A., and Okpala, P. (2017). Complementary and alternative medicine use results from a descriptive study of pregnant women in Udi local government area of Enugu state, Nigeria. *BMC complementary and alternative medicine*, 17(1), 189.

Osunderu, A.O. (2008). *Sustainable Production of Traditional Medicine in Africa; Appropriate Technologies for Environmental Protection in the Developing World.* In: Yanful, E.K. (ed.), Springer, Dordrecht, 45–51.

Osunderu, O. (2017). Medicinal and Economic Values of Forest Products in the Treatment of Cancer in Southwest Nigeria. *Natural Products Chemistry and Research, 15*(6), 1000288.ISSN:2329-6836.

Sofowora, A. (1982). *Medical Plants and Traditional Medicine in Africa.* John Wiley & Sons. UK.

Soladoye, M.O., Ikotun, T., Chukwuma, C., Ariwaodo, J., Ibhanesebor, G.A., Agbo-Adediran, O.A., and Owolabi, S.M. (2013). Our plants, our heritage: preliminary survey of some medicinal plant species of South-western University Nigeria Campus, Ogun State, Nigeria. *Annals of Biological Research, 4*(12), 27–34.

Soladoye, M.O., Chukwuma, E.C., Sulaiman, O.M., and Feyisola, R.T. (2014). Ethnobotanical survey of plants used in the traditional treatment of female infertility in Southwestern Nigeria. *Ethnobotany Research and Applications, 12*, 81–90.

Tade, O. (2020). What's triggered new conflict between farmers and herders in Nigeria. *The Conversation.* Accessed 18 January 2022. https://theconversation.com/whats-triggered-new-conflict-between-farmers-and-herders-in-nigeria–145055.

Umezurike, C. (2012). The Implications of climate change for democratic governance in Nigeria. *The Social Sciences, 7*(3), 412–423.

World Health Organization. (2002). Traditional Medicines – Growing Needs and Potential. WHO policy respective on Medicine, Geneva (WHO/EDM/2002. 4)

World Health Organization (2006). WHO Report on Traditional Medicine. my documents/WHO traditional medicine.htm 15/06/2006.

World Health Organization. (2019). WHO Global Report on Traditional and Complementary Medicine. https://www.who.traditionalmedicine/2019. Accessed 8 July 2022.

APPENDIX 24.1

AGE

Variable	Frequency	Percentage
Below 20 years	0	0
20–29 years	4	7
30–39 years	4	7
40–49 years	28	48
50–59 years	10	17
60 years and above	12	21
Total	58	100

SEX

Variable	Frequency	Percentage
Male	54	93
Female	4	7
Total	58	100

RELIGION

Variable	Frequency	Percentage
Christianity	20	34
Muslim	38	66
Others	0	0
Total	58	100

MARITAL STATUS

Variable	Frequency	Percentage
Single	0	0
Married	38	65
Divorced	8	14
Widowed	12	21
Total	58	100

EDUCATIONAL LEVEL

Variable	Frequency	Percentage
Primary	18	31
Secondary	26	45
Post-Secondary	14	24
Total	58	100

USE OF FOREST PRODUCTS

Variable	Frequency	Percentage
Yes	54	93
No	4	7
Total	58	100

DURATION OF PRACTICE

Variable	Frequency	Percentage
Less than 2 years	6	10
2–5 years	4	7
6–10 years	12	21
11 years and above	36	62
Total	58	100

HOW DID YOU BEGIN PRACTICE?

Variable	Frequency	Percentage
Inheritance	40	69
Apprenticeship	4	7
Hobby	10	17
Others	4	7
Total	58	100

CAN YOUR PRACTICE TREAT/CURE ALL DISEASES?

Variable	Frequency	Percentage
Yes	28	48
No	26	45
I don't know	4	7
Total	58	100

AREA OF SPECIALITY

Variable	Frequency	Percentage
Fertility problems	8	14
Mental cases	4	7
General	10	17
Others	36	62
Total	58	100

WHERE DO YOU HOLD YOUR CONSULTATION?

Variable	Frequency	Percentage
Clinic	14	24
Home	36	62
Others	8	14
Total	58	100

HOW MANY PATIENTS DO YOU SEE IN A WEEK?

Variable	Frequency	Percentage
Less than 10	26	45
10–20	16	28
20–50	6	10
>50–above	10	17
TOTAL	58	100

DO YOU KEEP RECORD OF PATIENTS?

Variable	Frequency	Percentage
Yes	34	59
No	24	41
Total	58	100

WHERE DO YOU SOURCE YOUR HERBS?

Variable	Frequency	Percentage
Forest	24	41
Garden	6	10
Market	16	28
Others	12	21
Total	58	100

HOW FREQUENTLY DO YOU HARVEST YOUR HERBS?

Variable	Frequency	Percentage
Monthly	20	34
Weekly	26	45
Daily	12	21
Total	58	100

MAJOR WAY(S) OF HARVESTING

Variable	Frequency	Percentage
Plucking	22	38
Uprooting	18	31
Cutting	18	31
Total	58	100

AVAILABILITY OF FOREST PRODUCTS

Variable	Frequency	Percentage
Yes	22	38
No	32	55
Indifferent	4	7
Total	58	100

WHAT FOREST PRODUCTS DO YOU USE?

Variable	Frequency	Percentage
Water	4	7
Animal parts	8	14
Minerals	4	7
Plants	42	72

HOW LONG HAVE YOU BEEN PRACTICING?

Variable	Frequency	Percentage
Over 20 Years	22	38
15 Years	6	10
10 Years	12	21
5 Years	8	14
Indifferent	10	17
Total	58	100

Index

Printed in the United States
by Baker & Taylor Publisher Services